THE MEASURE OF THE UNIVERSE
A HISTORY OF MODERN COSMOLOGY

BY
J. D. NORTH

DOVER PUBLICATIONS, INC.
New York

Copyright © 1965 by Oxford University Press.
Preface to the Dover Edition copyright © 1990 by J. D. North.
All rights reserved under Pan American and International Copyright Conventions.

Published in Canada by General Publishing Company, Ltd., 30 Lesmill Road, Don Mills, Toronto, Ontario.

Published in the United Kingdom by Constable and Company, Ltd., 3 The Lanchesters, 162–164 Fulham Palace Road, London W6 9ER.

This Dover edition, first published in 1990, is a corrected, unabridged republication of the work originally published by Oxford University Press in 1965 and reprinted with corrections in 1967. This edition adds a new Preface.

Manufactured in the United States of America
Dover Publications, Inc., 31 East 2nd Street, Mineola, N.Y. 11501

Library of Congress Cataloging-in-Publication Data

North, John David.
   The measure of the universe : a history of modern cosmology / J. D. North.
      p.   cm.
   "A corrected unabridged republication of the work originally published by Oxford University Press in 1965 and reprinted with corrections in 1967"—T.p. verso.
   Includes bibliographical references and index.
   ISBN 0-486-66517-8 (pbk.)
   1. Cosmogony. 2. Cosmology. I. Title.
QB981.N77   1990
523.1—dc20   90-39942
                                                                                   CIP

PREFACE TO THE DOVER EDITION

When I began this work in the late 1950s, scientific cosmology was still predominantly geometrical, or rather geometrico-gravitational, in character. This is not to say that it was entirely without astrophysical clothing, only that this was diaphanous in comparison with the heavy layers it now carries. The book was finally published at about the time of the first of the great empirical discoveries of the 1960s. There were quasars, for example, recognized through their unusual radio emissions, by T. A. Matthews, C. Hazard and others, and then found by M. Schmidt to have extraordinarily large red-shifts. Although my history was more or less restricted to the first half of the century, I was able to refer to those particular discoveries (see pp. 248-9), but I was less fortunate with the microwave background radiation, found by A. A. Penzias and R. W. Wilson at Bell Laboratories in 1965, while they were working on ostensibly unrelated problems. Their discovery had immediate repercussions on cosmology—in particular on the steady-state debate.

Such topics as these may have a prehistory of sorts within my period, but by the nature of things I could hardly touch upon it. Quasars were actually recorded photographically in the last century, but were not then distinguished from ordinary stars. Certain interstellar absorption lines observed from 1940 onwards only later found their explanation in the microwave background radiation; George Gamow and his associates (cf. pp. 256-8 below) had over the same period been arguing that if the universe had an explosive origin vestigial radiation should persist; G. Lemaître had made a similar point at a still earlier date; and R. H. Dicke and his colleagues quite independently hypothesized the background radiation shortly before it was detected, and were thus ready to offer at once a cosmic interpretation of it. Writing today, one could produce many comparable examples. My admiration for G. Lemaître should be obvious enough from the pages that follow, but like most writers thirty years ago I did not properly appreciate his very considerable physical intuition. A letter he wrote to *Nature* in 1939, offering an argument for the idea of the Primeval Atom from the principles of quantum theory in the compass of three or four deceptively easy sentences, now has a prophetic look about it. If it proves any one thing of historical value, however, it is that the quality of prophecies is easier to judge the longer one waits.

Such antecedents of the new wave of cosmology as these might now be thought right for inclusion in a history covering the first half of the century, if only as instances of dormant ideas. They were ideas waiting in a limbo that falls between received opinion on the one hand and lost causes on the other. I was not alone in paying little attention to them, and if cosmologists are not prophets, neither are historians.

There are of course many passages in this volume that I should write differently today, even within my original terms of reference. Many of the topics discussed in the final chapters, for instance, have an ancient and medieval ancestry deserving more attention than is usually paid to it in current discussions. I also left much unsaid about the structure of the intellectual society that created modern cosmology—but it has to be remembered that when this book first appeared, scientific cosmology was in many quarters not even acknowledged as having an intellectual existence in its own right, let alone a social existence. To appreciate how important are the social threads in the fabric of some of the ideas here discussed, we need only consider the vital role of W. de Sitter as diplomatic intermediary between Germany and the English-speaking world, during and after the First World War. Just as to de Sitter's ambassadorial function, so I should have given more attention to A. S. Eddington's genius for imparting a missionary zeal to his satellites. And so I could go on; but heaven forbid that this be seen as a plea for turning intellectual history into nothing more than a branch of sociology. That particular bandwagon is in no great need of horses.

Eddington once said that familiarity with the history of a subject is an obstacle to creative research in that subject. Cosmologists may consider themselves warned. I never thought of myself as writing primarily for them, although their responses to the original edition were the ones I most valued. History and philosophy have lives of their own, and this work has inevitably been tacitly supplemented on points of detail by more recent studies—although I think it is true to say that since its original publication, no monograph has appeared from quite the same perspective. In a sense, it can never be so: consider the change in the level of seriousness with which the steady-state theory was treated, for instance, after R. Penrose's studies of singularities in the regions of space-time known as black holes, and after his and S. W. Hawking's time-reversed proofs that Friedmann-like expanding universes within general relativity must— given the matter we observe—have begun from a singularity. The fact that Hawking now believes this to be an illusion, stemming from cosmologists' disregard of quantum effects, should serve to remind us of yet another quantum shift, in cosmology itself, during the 1970s. There are evident singularities in intellectual history, too. One cannot gainsay those ardent despisers of Whig history, those who profess to find later historical developments irrelevant to what went before, any more than one can deny the accuracy of a description of a cheerful army that is on the point of marching unwittingly over the edge of a cliff. Had I foreseen the events of the late 1960s and after, I know that for good or ill I should have written my book somewhat differently. This confession will no doubt be greeted with a shriek of horror, but chiefly by those who consider periodization more important than human achievement. *Plus ça change* .... A generous Soviet review of the original edition ended with only one reproach: I had overlooked that great cosmologist of our time, Lenin. If that review confirms me in any one belief, it is that one cannot produce an opium suited to all people for all time. Here, then, are some of the reasons for making what follows a straightforward reprinting of my original study, with a very few trivial emendations.

J. D. N.

Groningen
May 1989.

PREFACE TO THE FIRST EDITION

The main concern of this book is with the development of cosmology during the first half of this century. It is divided into two parts. The first is a fairly straightforward historical and theoretical narrative. The second is the sort of thing Lucretius might have had in mind when he spoke of 'the outer view of the inner law': it is a discussion of conceptual problems which are to be found at various points in the first part of the book.

Modern theories of cosmology were almost all prompted by theories of gravitation, and such peripheral subjects as relativistic thermodynamics and theories of a unified field are here barely touched upon. The methods of radio astronomy are to some extent ignored, falling largely outside our period. It might appear inconsistent that Part I begins with three chapters dealing briefly with some of the historical antecedents of modern gravitational cosmology. On the other hand, without such an excursion into earlier centuries, some of the curious convolutions of later history would appear meaningless. Speaking generally, it is the *language* of cosmology to which most attention is paid. Not that I am unmindful of the overwhelming importance of the methods of the observatory, but that they are simply not the prime concern of this book.

It is surprising that no historical study of the sort has been attempted before; for the subject abounds in controversies of a kind relevant to the whole of physical science. Perhaps there is a general feeling that the scientist's intellectual arrogance has carried him too far in his pretending to study the universe. ('It's a great deal bigger than I am', as Carlyle remarked, adding that 'People ought to be modester'.) Perhaps it is rather that scientific historians are reluctant to comment upon matters which are almost as controversial today as thirty years ago. Or perhaps it is that the body of cosmological writing is so very extensive, and its boundaries so ill-defined. At all events, if an apology for the book is needed, it must simply be that the 'great dust heap called history' will not be much the worse for a new kind of dust.

For assistance and criticism I owe a great deal to Professor G. Temple, Dr. A. C. Crombie, Mr. H. R. Harré, Dr. G. J. Whitrow, and, not least, my wife. I must thank not only them, but also those who, at the Clarendon Press and elsewhere, are taking so much care in producing this volume. Lastly I must thank the Nuffield Foundation, whose award

PREFACE TO THE FIRST EDITION

of a Fellowship allowed me the time needed to complete this work, which was written for the most part during my spare moments of the past five years.

Oxford
October 1964.

J. D. N.

AUTHOR'S NOTE TO THE CORRECTED REPRINT

I should like to take this opportunity of thanking Professor G. C. McVittie of the University of Illinois Observatory for help he has given with the corrections incorporated in this reprint.

Oxford
April 1967.

J. D. N.

CONTENTS

PREFACE TO THE DOVER EDITION — i
PREFACE TO THE FIRST EDITION — v
ABBREVIATIONS — xi
NOTATION — xix
INTRODUCTION — xxiii

Part I. ORIGINS OF MODERN COSMOLOGY

CHAPTER 1. NINETEENTH-CENTURY ASTRONOMY: THE NEBULAE — 3
1. The apparent structure of the Milky Way and the nebulae — 3
2. The Island Universe theory established — 10

CHAPTER 2. COSMOLOGICAL DIFFICULTIES WITH THE NEWTONIAN THEORY OF GRAVITATION — 16
1. Seeliger's and Neumann's objections to a Newtonian infinite — 17
2. Charlier's hierarchic hypothesis — 18

CHAPTER 3. FIELD THEORIES AND THE WISH TO REPHRASE NEWTON'S THEORY OF GRAVITATION — 24
1. Gravitation and the field concept — 25
2. Gravitation and the hydrodynamical analogues — 32
3. The Le Sage-Thomson explanation of gravitation — 38
4. Reasons for wishing to revise the Newtonian theory. Reality and the field theories — 41
5. The attempts to amend Newton's law — 43
6. Gravitation and Lorentz invariance — 49

CHAPTER 4. THE ORIGINS AND CONCEPTUAL BASIS OF EINSTEIN'S THEORY OF GRAVITATION — 52
1. Gravitation and light. The Principle of Equivalence — 52
2. Relativity and covariance — 56
3. The Riemannian framework and the calculus of tensors — 58
4. Einstein and the concept of gravitational force — 63
5. Einstein's field equations — 65
6. Observational tests — 67

CHAPTER 5. EARLY RELATIVISTIC COSMOLOGY — 70
1. Cosmological considerations — 70
2. A closed universe. W. K. Clifford and others — 72
3. The Einstein world — 81
4. The cosmological constant — 83
5. The de Sitter world and 'Mach's Principle' — 87
6. The de Sitter effect and Weyl's Principle — 92
7. Material content — 104
8. The geometry of the de Sitter world — 106

CONTENTS

CHAPTER 6. THE EXPANDING UNIVERSE: THE FORMAL ELEMENT — 110
1. The non-static line-element — 111
2. The work of Friedmann, Lemaître, and Robertson — 113
3. Eddington and the Lemaître model — 122
4. The stability of the Einstein world — 125
5. Types of relativistic models — 129
6. The relativistic models: further points of comparison. The expanding universe — 135

CHAPTER 7. THE EXPANDING UNIVERSE: THE ASTRONOMER'S CONTRIBUTION — 142
1. The K-term and Hubble's Law — 142
2. Minor revision of the recession factor — 145

CHAPTER 8. KINEMATIC RELATIVITY AND THE REVIVAL OF NEWTONIAN COSMOLOGY — 149
1. The conceptual basis of Milne's theories — 149
2. Milne's Cosmological Principle — 156
3. The simple kinematic model. Milne on 'gravitation' — 158
4. The development of Kinematic Relativity by Milne and Whitrow — 165
5. The relatively stationary substratum — 168
6. Milne on gravitation and galactic evolution — 170
7. Kinematic and General Relativity — 173
8. The neo-Newtonian cosmology of Milne and McCrea — 176
9. Sources of criticism — 180

CHAPTER 9. THE THEORIES OF GRAVITATION OF BIRKHOFF AND WHITEHEAD — 186
1. Birkhoff's theory of gravitation — 186
2. Whitehead's theory of gravitation and its extension by J. L. Synge — 190
3. The Rayner-Whitehead cosmology — 194

CHAPTER 10. CONTINUAL CREATION AND THE STEADY-STATE THEORIES OF BONDI, GOLD, AND HOYLE — 198
1. Continual creation: some early arguments — 198
2. Energy conservation. The Bondi-Gold theory — 208
3. Hoyle's steady-state theory and McCrea's interpretation of it — 212
4. Creation as a physical process — 217

CHAPTER 11. THE RECEDING GALAXIES: APPEARANCE AND REALITY, THEORY AND OBSERVATION — 223
1. The time-scale difficulty — 223
2. The reality of the expansion. Alternative explanations of the red-shifts — 229
3. The decision: empirical considerations — 234

CONTENTS

4. The theoretical (δ,m) relation in relativistic cosmology, and the Stebbins-Whitford effect	242
5. The theoretical (N,m) relation	245
6. The theoretical (N,δ) relation	250
7. Other criteria	251
8. Cosmology and the formation of galaxies	255

Part II. PHILOSOPHICAL ISSUES

CHAPTER 12. 'FACT' AND THE 'UNIVERSE' — 265
1. Basic objects — 265
2. Fact — 268
3. Objectivity and the uniqueness of the universe — 271

CHAPTER 13. THE ELEMENT OF CONVENTION IN COSMOLOGY AND SCIENCE — 276
1. 'Conventionalism': an ambiguous term. Poincaré and his critics — 276
2. The Duhem-Quine thesis — 285

CHAPTER 14. GENERALITY, SIMPLICITY, AND THE COSMOLOGICAL PRINCIPLES — 290
1. Simplicity — 291
2. Generality — 293
3. Deduction and extrapolation — 296
4. Homogeneity and isotropy. The cosmological principles — 300
5. Cosmological theories and models — 311

CHAPTER 15. CONCEPTUAL PROBLEMS: (i) DISTANCE AND COORDINATES — 319
1. Distance, coordinates, and a possibly circular argument — 320
2. Coordinates. Whitehead's challenge of circularity — 324
3. Concepts of distance. Operationalism — 330
4. Which distance-concept is fundamental? — 337

CHAPTER 16. CONCEPTUAL PROBLEMS: (ii) ABSOLUTE AND RELATIVE — 349
1. Newton and Leibniz on 'space' — 350
2. Absolute and universal time — 355
3. Cosmic time and modern cosmology — 357
4. The exaggerated philosophical involvement of theories of natural cosmology — 361

CHAPTER 17. CONCEPTUAL PROBLEMS: (iii) INFINITY AND THE ACTUAL — 371
1. The concept of a potential infinity and its inadequacy — 372
2. Cantor's transfinite cardinals — 377
3. Infinities in cosmology — 379

CONTENTS

CHAPTER 18. CONCEPTUAL PROBLEMS: (iv) CREATION AND THE AGE OF THE UNIVERSE 384
   1. The age of the universe 386
   2. The First Event 389
   3. Creation 399
   4. Cause and creation 402

CONCLUSION 407

APPENDIX 408

SELECT BIBLIOGRAPHY 425

INDEX 428

ABBREVIATIONS

A. PERIODICALS

A.J.	*Astronomical Journal*, Dudley Obs., Albany, 1849–.
Am. J. Math.	*American Journal of Mathematics*, Baltimore, 1878–.
Ann. Astrophys.	*Annales d'astrophysique*, Paris, 1938–.
Ann. Inst. Poincaré	*Annales de l'Institut Henri Poincaré*, Université de Paris, 1930–.
Ann. Math., Princeton	*Annals of Mathematics Studies*, Princeton University, New Jersey, 1940–.
Ann. Obs. Paris	*Annales de l'Observatoire de Paris: mémoires publiées par U. J. LeVerrier*, Paris, 1855–.
Ann. Phys. Lpz.	*Annalen der Physik*, Leipzig, 1799–.
Ann. Phys. Paris	*Annales de physique*, Paris, 1914–.
Ann. Soc. Sci. Brux.	*Annales de la Société Scientifique de Bruxelles*, Brussels, 1877–.
Annali di Mat.	*Annali di Matematica pura ed applicata*, Rome, 1858–.
Ap. J.	*Astrophysical Journal*, University of Chicago, Chicago and New York, 1895–.
Arch. Néerl.	*Archives Néerlandaises des Sciences Exactes et Naturelles*, Harlem, 1866–.
Archiv. Math. Phys.	*Archiv der Mathematik und Physik*, Leipzig, 1841–.
Archives des Sciences	*Bibliothèque Universelle: Archives des Sciences Physiques at Naturelles*, Geneva, 1846–.
Arkiv. Mat. Astr. Fys.	*Arkiv für Matematik, Astronomi och Fysik ut givet av. K. Svenska Vetenskademien*, Stockholm, 1903–.
Astr. Nachr.	*Astronomische Nachrichten*, Kiel, 1823–.

ABBREVIATIONS

Aus. J. Sci.	*Australian Journal of Scientific Research*, A. Physical Sciences, Melbourne, 1948–52. *Australian Journal of Physics*, 1953–.
B.A.N.	*Bulletin of the Astronomical Institute of the Netherlands*, 1922–.
B.A. Rp.	*Report of the Meeting of the British Association for the Advancement of Science*, London, 1831–.
B.J.P.S.	*British Journal for the Philosophy of Science*, Edinburgh, 1950–.
Berlin Mém.	*Mémoires de l'Académie Royale des Sciences de Berlin*, 1770–1804.
Berlin Monats.	*Monatsberichte der K. Preussischen Akademie der Wissenschaften zu Berlin*, 1856–81.
Berlin Sitz.	*Sitzungberichte der K. Preussischen Akademie der Wissenschaften zu Berlin*, 1882–1938, 1948– as *Sitzungberichte der Deutschen Akademie der Wissenschaften zu Berlin*.
Bode Jb.	*Astronomisches Jahrbuch nebst einer Sammlung der neusten in die astronomischen Wissenschaften einschlagenden Abhandlungen, Beobachtungen, und Nachrichten*, Bode, Berlin, 1776–1829.
Bol. Soc. Math. Mex.	*Boletín de la Sociedad matemática mexicana*, Mexico.
Bordeaux Mém.	*Mémoires de la Société des Sciences Physiques et Naturelles de Bordeaux*, 1855–.
Bull. Acad. Brux.	*Bulletins de l'Académie Royale des Sciences etc. de Belgique*, Brussels, 1834–.
Bull. Am. Math. Soc.	*Bulletin of the American Mathematical Society*, New York, 1894–.
Bull. Calc. Math. Soc.	*Bulletin of the Calcutta Mathematical Society*, Calcutta, 1909–.

Bull. Sci. Math.	Bulletin des Sciences mathématiques, Paris, 1885–.
C.R.	Compte Rendu hebdomadaire des scéances de l'Académie des Sciences, Paris, 1835–.
Camb. and Dubl. Math. J.	The Cambridge and Dublin Mathematical Journal, Cambridge, 1846–54.
Canad. J. Math.	Canadian Journal of Mathematics, Toronto, 1949–.
Christiania Skrifter	Skrifter udgivne af Videnskabsselskabet i Christiania. Mathematik-Naturvidenskabelig Kl., 1894–. (A continuation of Forhandlinger i Videnskabs-Selskabet i Christiania 1859–93).
Crelle's J.	Journal für die reine und angewandte Mathematik, Crelle, Berlin, 1826–.
Czech. J. Phys.	Czechoslovak Journal of Physics, Prague, 1950–.
Gött. Nachr.	Nachrichten von der Gesellschaft der Wissenschaften in Göttingen, 1845–1938 and Nachrichten Akademie der Wissenschaften in Göttingen, 1945–.
Hamburg Abh. Math.	Abhandlungen aus dem Mathematischen Seminar der Universität Hamburg, 1922–.
Harvard Bull.	Harvard College Astronomical Observatory Bulletin, Cambridge, Mass., 1898–.
Harvard Circular	Harvard College Astronomical Observatory Circular, Cambridge, Mass., 1895–.
Helv. Phys. Acta	Helvetica Physica Acta, Basle, 1928–.
Isis	Isis. Revue consacrée a l'Histoire de la Science, Brussels, 1913–.
J. Am. Chem. Soc.	Journal of the American Chemical Society, New York, 1879–.
J. Franklin Inst.	Journal of the Franklin Institute of the State of Pennsylvania, Philadelphia, 1828–.

ABBREVIATIONS

J.L.M.S.	*Journal of the London Mathematical Society*, London, 1926–.
J. Math. Phys.	*Journal of Mathematics and Physics*, Boston, 1921–.
J. Phys.	*Journal de Physique théorique et appliquée*, Paris, 1872–.
J. Phys., U.S.S.R.	*Academy of Sciences of the U.S.S.R.: Journal of Physics*, Moscow, 1939–.
J. Univ. Bombay	*Journal of the University of Bombay*, 1932–.
J. Wash. Acad. Sci.	*Journal of the Washington Academy of Sciences*, Washington, 1911–.
Jahrb. Rad. Elek.	*Jahrbuch der Radioaktivität und Elektronik*, Leipzig, 1904–24.
Leipzig, *Abh. Math. Phys.*	*Abhandlungen der Mathematische-Physischen Classe der K. Sächsischen Gesellschaft der Wissenschaften*, Leipzig, 1852–.
Lund Ann.	*Annals of the Observatory of Lund*, 1926–.
M.N.	*Monthly Notices of the Royal Astronomical Society*, London, 1827–.
Math. Ann.	*Mathematische Annalen*, Leipzig, 1869–.
Math. Gaz.	*The Mathematical Gazette*, London, 1894–.
Math. Nat. Ber. Ungarn.	*Mathematische und Naturwissenschaftliche Berichte aus Ungarn*, Berlin, 1887–.
Math. ZS.	*Mathematische Zeitschrift*, 1918–.
Mém. Acad. Brux.	*Mémoires de l'Académie Royale des Sciences des Lettres, et des Beaux-Arts de Belgique*, Brussels, 1820–.
Mem. Accad. Bologna	*Memorie della Accademia della Scienze dell' Instituto di Bologna*, 1850–.
Münch. Sitz.	*Sitzungberichte der K. Bayerischen Akademie der Wissenschaften zu München*, Munich, 1871–.

ABBREVIATIONS

N.Z.I.T.	Transactions and Proceedings of the New Zealand Institute, Wellington, 1868–.
Nat.	Nature, London, 1869–.
Naturwiss.	Naturwissenschaften, Berlin, 1913–.
Nova acta Leopoldina	Miscellanea Curiosa. Verhandlungen der K. Leopoldinische-Carolinischen Akademie der Naturforscher, 1757–1928; Nova acta, 1932–.
Nuovo Cim.	Nuovo Cimento, Pisa, 1855–.
Obs.	Observatory, London, 1878–.
Op. Phys-Math. Fenn.	Finska Vetenskaps-societeter: Commentationes Physico-mathematicae, Helsinki, 1927–.
P.A.S.P.	Publications of the Astronomical Society of the Pacific, San Fransisco, 1889–.
P.L.M.S.	Proceedings of the London Mathematical Society, 1865–.
P.N.A.S.	Proceedings of the National Academy of Sciences, Washington, 1915–.
P.R.I.	Proceedings of the Royal Institution of Great Britain, London, 1929–.
P.R.I.A.	Proceedings of the Royal Irish Academy, A. Mathematics, Astronomy and Physics, 1902–.
P.R.I. Not.	Notices of the Proceedings of the Royal Institution, London, 1858–1928.
P.R.S. (A).	Proceedings of the Royal Society, Series A., 1905–.
P.R.S. Edin.	Proceedings of the Royal Society of Edinburgh, 1845–.
Palermo Rend.	Rendiconti del Circolo Matematico di Palermo, 1887–.
Phil Mag.	London, Edinburgh and Dublin Philosophical Magazine and Journal of Science, London, 1832–44, Philosophical Magazine, 1845–.

ABBREVIATIONS

Phil. Rev.	The *Philosophical Revue*, Boston, 1892–.
Phil. Trans.	*Philosophical Transactions of the Royal Society*, Series A, 1887–.
Phys. Rev.	*Physical Review*, Cornell University, Ithaca, 1893–.
Phys. Rev. Suppl.	*Physical Review Supplement*, Minneapolis, 1929–.
Phys. ZS.	*Physikalische Zeitschrift*, Leipzig, 1899–.
Pop. Astr.	*Popular Astronomy*, Minnesota, 1894–.
Proc. Acad. Amst.	*Proceedings of the Royal Academy of Sciences Amsterdam*, 1898–.
Proc. Am. Phil. Soc.	*Proceedings of the American Philosophical Society*, Philadelphia, 1840–.
Proc. Camb. Phil. Soc.	*Proceedings of the Cambridge Philosophical Society*, 1843–.
Proc. Phil. Soc. Glasgow	*Proceedings of the Philosophical Society of Glasgow*, 1841–.
Proc. Phys. Soc.	*Proceedings of the Physical Society of London*, 1874–.
Proc. Phys-Math. Soc. Japan	*Proceedings of the Physico-Mathematical Society of Japan*, Tokyo, 1919–44.
Publ. Obs. Lyon	*Publications de l'Observatoire de Lyon*, (1st series), 1932–.
Q.J.M.	*Quarterly Journal of Mathematics*, Oxford, 1930–.
Rend. Lincei.	*Atti della R. Accademia Nazionale dei Lincei, Rendiconti*, Rome, 1884–.
Rep. Prog. Phys.	*Reports on Progress in Physics*, London, 1934–.
Repertorium Math.	*Repertorium der literarischen Arbeiten aus dem Gebeite der reinen und angewandten Mathematik*, Leipzig, 1877–1879.
Rev. Mét. Morale.	*Revue de Métaphysique et Morale*, Paris, 1892–.

ABBREVIATIONS

Rev. Mod. Phys.	*Reviews of Modern Physics*, Minneapolis, 1930–.
Rev. Philos.	*Revue philosophique de la France et de l'Etranger*, Paris, 1876–.
Ricerca Scientifica	*La Ricerca Scientifica ed il Progresso Tecnico nell'Economia Nazionale*, Rome, 1931–.
Rivista Sci.-Ind.	*Rivista Scientifico-Industriale delle principali scoperte ed invenzione fatte nelle scienze e nelle industrie*, Florence, 1869–.
Sci. Am.	*The Scientific American*, New York, 1845–.
Sci. Mon.	*Scientific Monthly*, New York, 1915–.
Sci. Prog.	*Scientific Progress in the Twentieth Century*, London, 1906–.
South Afr. J. Sci.	*South African Journal of Science*, 1909–.
Stock. Acad. Hand.	*Kongliga Svenska Vetenskaps-Akademiens Handlingar*, Stockholm, 1739–.
Trans. Am. Math. Soc.	*Transactions of the American Mathematical Society*, New York, 1900–.
Trans. Camb. Phil. Soc.	*Transactions of the Cambridge Philosophical Society*, 1822–.
Trans. R.I.A.	*Transactions of the Royal Irish Academy*, Dublin, 1787–.
Trans. R.S. Can.	*Transactions of the Royal Society of Canada*.
Wien Sitz.	*Sitzungsberichte der Mathematisch—Naturwissenschaftlichen Classe der K. Academie der Wissenschaften*, Vienna, 1848–.
ZS. Ap.	*Zeitschrift für Astrophysik*, Berlin, 1930–.
ZS. Math. Phys.	*Zeitschrift für Mathematik und Physik*, Leipzig, 1856–1917.
ZS. Naturforsch.	*Zeitschrift für Naturforschung*, Wiesbaden, 1946–.
ZS. Phys.	*Zeitschrift für Physik*, 1920–.

B. BOOKS

G.R.C.	G. C. McVittie, *General Relativity and Cosmology* (Chapman and Hall, 1956).
H.T. Ae. E.	E. Whittaker, *A History of the Theories of Aether and Electricity*, 2nd ed. (Nelson, 1951), two vols.
Jubilee R.T.	*The Jubilee of Relativity Theory* (Basle, 1956).
K.R.	E. A. Milne, *Kinematic Relativity* (C. P. Oxford, 1948).
M.T.R.	A. S. Eddington, *The Mathematical Theory of Relativity*, 2nd ed. (Cambridge, 1924).
P.R.	A. Einstein et al., *The Principle of Relativity* (Methuen, 1922), reprinted by Dover Books.
R.T.C.	R. C. Tolman, *Relativity, Thermodynamics, and Cosmology* (C.P. Oxford, 1934).
S.T.M.	H. Weyl, *Space-Time-Matter*, 4th ed., trans. H. L. Brose (Methuen, 1922), reprinted in Dover Books.
Vistas.	A. Beer, (ed.), *Vistas in Astronomy* (Pergamon), vol. i, 1955; vol. ii, 1956; vol. iii, 1960; vol. iv, 1961.
W.S	E. A. Milne, *Relativity, Gravitation and World-Structure* (C.P. Oxford, 1935).

NOTATION

(Special notation is explained in the text. Notation which is in common use and which may be unexplained in the text, is given here.)

(i) *Scalar quantities*

c	Velocity of light ($= 2 \cdot 99793 \times 10^{10}$ cm sec^{-1})
D	Luminosity-distance
e	Electronic charge ($= 4 \cdot 8025 \times 10^{-10}$ abs. e.s.u.) ($= 1 \cdot 6020 \times 10^{-28}$ abs. e.m.u.)
E	Energy
γ	Newton's constant of gravitation ($= 6 \cdot 670 \times 10^{-8}$ dyne cm^2 gm^{-2})
h_1, H	The Hubble factor ($h_1 = R'_0/R_0$, where appropriate; H is usually used of the Steady-State model).
h_2	The acceleration factor ($= R''_0/R_0$)
h	Planck's constant ($= 6 \cdot 624 \times 10^{-27}$ erg sec)
k	k is used to indicate the sign of the 'spatial' curvature, i.e. of the 3-space orthogonal to the world lines of the fundamental particles (in relativistic cosmology). In Newtonian cosmology the interpretation is different.
l	Proper distance
m or m_e	Electronic mass ($= 9 \cdot 1078 \times 10^{-28}$ g)
m_n	Neutron mass ($= 1 \cdot 67469 \times 10^{-24}$ g)
m_p	Proton mass ($= 1 \cdot 6725 \times 10^{-24}$ g)
p	Pressure
r	A radial coordinate. For r_e and r_s see p. 88, note 49
$R = R(t)$	The expansion factor of a cosmological model
R', R'', etc.	Differential coefficients of R with respect to t
R_e	The Einstein radius
t	A time coordinate
W	Work
x, y, z	Spatial coordinates
x^1, x^2, x^3, x^4	Coordinates of a four-space

NOTATION

α	$\begin{cases}\text{Right ascension}\\ \text{Number of sources per unit volume}\end{cases}$
γ	Newton's constant of gravitation ($= 6{\cdot}670 \times 10^{-8}$ dyne cm^2 g^{-2})
δ	$\begin{cases}\text{Fractional spectral displacement}\\ \text{Declination}\end{cases}$
ε	Base of natural logarithms ($= 2{\cdot}71828$)
θ, ϕ, χ	Polar coordinates
κ	$8\pi\gamma/c^2$ ($= 1{\cdot}864 \times 10^{-27}$ cm g^{-1})
λ	$\begin{cases}\text{Wavelength}\\ \text{Cosmological constant}\end{cases}$
ν	Frequency
ξ	Distance by apparent size
ρ	Density
τ	$\begin{cases}\text{Period}\\ \text{A time coordinate}\end{cases}$
ϕ	A scalar potential

Note. A zero superscript or subscript frequently indicates a proper quantity, as measured by a local observer. The suffix 'e' occasionally indicates the value corresponding to the Einstein model (but see p. 88, note 49).

(ii) *Tensor quantities*

(Indices: Latin ($i, j, k, \ldots m, n$, etc.) take the values 1 to 4.

Greek ($\alpha, \beta, \ldots \lambda, \mu, \nu, \sigma$, etc.) take the values 1 to 3)

ds	Invariant interval
$\delta^{i_1 \ldots i_m}_{j_1 \ldots j_m}$	Kronecker delta (rarely generalized as here)
$\left.\begin{array}{l}\varepsilon_{r_1 \ldots r_m}\\ \varepsilon^{r_1 \ldots r_m}\end{array}\right\}$	Permutation symbols
F^{mn}	A field tensor
g_{mn}, g^{mn}	Fundamental metrical tensor. The normalized minor of g_{mn}
g	Determinant $|g_{mn}|$
R^m_{npq}	Riemann-Christoffel tensor
R_{mn}	Contracted Riemann-Christoffel tensor

R	Invariant obtained from Riemann-Christoffel tensor (not to be confused with R used for $R(t)$)
T^{mn} and T_{mn}	Energy-momentum tensor

The summation convention is usually assumed. Covariant differentiation is indicated by a comma (e.g. $T_{mn,s}$)

$[ij, k]$	Christoffel symbol of the first kind
$\begin{Bmatrix} l \\ ij \end{Bmatrix}$	Christoffel symbol of the second kind

In keeping with the usage of certain authors, energy units are occasionally chosen in such a way that the symbol (c) for the velocity of light does not enter the equations of general relativity.

INTRODUCTION

THIS work is meant as a history of modern scientific cosmology with almost the entire emphasis on the first half of the present century. As a history it relies not only on those records which are available, but on that set of principles, tacit or otherwise, by which they are interpreted. As a history it is the outcome of a series of questions which, in a sense, define its very subject matter. The past, as elicited by the historian, is not something which was simply 'there' and now awaits description. It is a product of the minds of both author and reader, and hence of the circumstances under which it was written and read. Different individuals may, of course, ask the same questions and answer them differently, and to say that each defines his own subject matter is not to deny the possibility of bad histories. It is rather to claim that one can 'cover the entire field' only in so far as the field is loosely determined by the questions which are asked. *A fortiori* it follows that there is nothing illogical either in singling out aspects of a historical situation which are now forgotten, or in ignoring those with which men of the past were most preoccupied. It is necessary to make this last point because almost all modern natural cosmology is the fortuitous outcome of scientific studies which at the outset were not in any strict sense cosmological.

The establishment of priorities and the detection of influence are separate matters, and it is not easy to strike a balance between them. In each we are presumably concerned with illuminating not only the past, but also the present. This should be even more obvious in the study of formal systems than elsewhere; the connexion between scientific beliefs currently held and those held in the past being usually clearer. Yet even the curiosities of history have their value—and recent cosmology abounds in them. Even so, it is hoped that these have been kept in their proper place.

The historical questions which we have asked, and which define the field of discussion, concern issues which either arose or were revived in the first half of this century. They concern, for example, the concepts Age of the Universe, Creation, Distance, the assignment and significance of coordinates, spatial and temporal measurements, and so forth. In Part I, which is a simple narrative with exegesis its main concern, these issues are kept in the background: they are dealt with more explicitly in Part II. Have they influenced the course of modern cosmology? To some it appears that influences should be the prime concern of a history, and that such influences are, to men of sufficiently sharpened perception, clear and unambiguous. 'The main outlines of the actual historical development . . . are pretty clear', writes one historian of a related subject. To take an example, it might be held that the history of recent cosmology turns ultimately on the fortunes of the 'cosmological constant'. It might be explained how Einstein adopted it to avoid difficulties at infinity; how others found a non-static model to be possible without it; how Einstein therefore regarded it as superfluous to any simple theory, whilst others retained it on account of the generality which it allowed . . . and so forth. It is not difficult to find such threads running through the historical fabric: the difficulty is in singling out those which seem more important than the rest. But (to continue the metaphor) the pattern in the fabric is something which the historian imposes. To do so he must act on some set of principles. In what follows, the first broad principle to be accepted will be that there should be something of philosophical interest.

Such interest will no doubt be found wherever there has been a wholesale introduction of ordinary language into cosmological theory. In this way, in fact, two broad classes of question arise, one traditional, the other ephemeral. Problems in the first class are often self-perpetuating, for although they may be framed in common enough language yet the language associated with their 'solution' is often highly contrived. The tendency is for this contrived and artificial element to be transferred, after a time, to ordinary language—thus adding mystery to mystery. At least a part of the problems of Mind, of Perception, and of Substance seems to be of this sort. Although these might once have fallen within the domain of cosmology, they are not the concern of this book.

Problems such as those of Creation, of the Infinite, and of the nature of Time and Space—all of which find a place both in philosophical and cosmological writings—might be placed in the first category, but they seem to fall more naturally into the second. Problems like these have an ephemeral appearance to the extent that they are, as often as not, associated with a particular scientific theory. When the associated theory disappears, the perplexities engendered by it tend to die away. On the other hand, their death is, all too often, an indecently protracted affair, and many are the tentative physical hypotheses which have the misfortune to be embalmed as philosophical principles. No problem of this kind can be fully appreciated if separated from the circumstances which gave rise to it, although any confusion to which it leads is likely to be the outcome of an admixture of everyday language. Such problems have the habit of disappearing in the solvent of a new theory only to be precipitated later. The context of these problems, which is essential if they are to be examined, is in a state of perpetual flux and is best studied by methods which are in some sense historical.

It is no longer so fashionable as it was to censure past ages on moral grounds; and yet a history of any formal discipline will be the poorer without an analogous form of criticism. We may, however, take warning from what are conceived to be the mistakes of past historians: accepted moral principles change, whether we like it or not. So, too, do the principles by which we condemn scientific theories. Historical criticism does not consist of a monotonous disproof of all theories but one in the light of the principles by which that one theory was established. It should not, primarily, comprise an examination of the inner consistency of those theories which are examined. There should be a wider concern with *types* of scientific validity and invalidity. As far as possible our conclusions should apply indifferently to as many scientific assertions as possible.

The conceptual apparatus necessary for any analysis is bound to be artificial. Whatever its comprehensiveness, or its symmetry, its depth, and its clarity, yet its choice must inevitably be something of an arbitrary matter, with personal satisfaction the ultimate court of appeal. But conceptual apparatus and historical perspective apart, it remains true that (in Aristotle's words) 'he who is familiar with various conflicting theories, like one who has listened to both sides in a law suit, is best qualified to judge between them'. Much of our work will be concerned with setting out the points of conflict of the various theories, and few references will be made to the characters of their authors, great as is the temptation.

In what sense is this book meant to be philosophical? One might expect it to reflect back a little on philosophical analysis to the extent that the statements of cosmology, fragments of the theories into which they are ordered and, more especially, many of the terms which are their components have a propensity for wandering in and out of neighbouring realms of thought. (If only for this reason a historical treatment again seems a natural one.) On the other hand, one also looks to philosophy both to give an account of the linguistic and theoretical forms which have hitherto been conceived, and to propose alternative ones. This work is neither meant to be an account of past scientific methodology as such, nor a proposal of new meta-linguistic forms. The nearest approach we shall make to philosophy, in either of these senses, will be in advocating the acceptance of a form of conventionalism. This can scarcely be called a philosophical innovation, for the idea has been an object of philosophical disapproval for half a century and more. In science and philosophy, just as in history, how we should order our beliefs is an evaluative question. As such it is a matter for debate, but it is one which cannot be decided solely by reference either to the subject matter which is being ordered, or to the way in which others choose to do so. If this were more generally recognized the volume of controversial writing in cosmology would be greatly reduced. A bald statement of the differences between one kind of dogma and another is of far less interest than a comparative study of the reasons which authorized them in the first instance, and of the stages at which the argument assumes this evaluative character. These reasons are not ineffable, for the very concept of reason seems to require the possibility of our referring to some *expressible* standard of rationality. Thus although the procedure which we follow may be loosely knit, it is certainly not meant as an imitation of current linguistic methods whereby (one supposes) it is secretly hoped that, by describing at great length the use of words, insights into the supposedly pre-existing conceptual apparatus of language will be exhibited.

The purpose of any conceptual system is to help us to reach what must serve as an understanding. Thus the philosopher can scarcely lay claim to an understanding of any feature of a language unless he can classify that feature in some way; or at least unless he knows how and under what circumstances it should be applied. The fact that it can be classified in several different ways, without one way being inherently more correct than another, only serves to show how misguided is the hope of attaining to a complete understanding in terms of a single analysis. Once again, however, this is not a widely held view, and the mistaken hope that there might be some set of propositional patterns capable of dispelling all the intractabilities of the problems of perception and knowledge still, one suspects, survives. A similar situation is in danger of arising in regard to the description of the whole of scientific method. Owing to the great influence of a handful of recent books, there is the danger that a new scholasticism may appear in writings on this subject. It is hoped that we have managed to avoid the new dogmatism, but if not we can only plead that the new tradition is, in so many respects, superior to what went before. It is certainly not final.

The notion that philosophical perplexity is the outcome of a shallow insight is fortunately losing favour—or at least the enigmatic 'insight into reality' is being replaced by 'insight into the use of a particular sector of language'. (It is one of the virtues of contemporary philosophy that, in emphasizing the multitude of uses to which language can be put, it has provided a long-needed corrective to those attitudes which the Vienna Circle disseminated so ruthlessly and which gave the impression that mathematics and formal logic were the sole arbiters of linguistic propriety. Theirs was a persuasive thesis on account of the aura of authority and the general unfamiliarity of mathematical and scientific works, which contrived to make both immune from a great deal of criticism.) The fact that many uses of language can be distinguished, and that no single kind of use is fundamental to the rest, need not, however, detract from the important therapeutic value of attempts to codify any language in a system of rules, even if the system be peculiar to this language. Now any writer who wishes to perform such a service for cosmology will inevitably encounter the question: Is cosmology not an ordinary science? With one or two minor qualifications, the answer given here is that cosmology *is* an ordinary science. Cosmologists differ amongst themselves, of course, as regards their meta-scientific attitudes and prejudices, but it is doubtful whether the differences are more marked than, say, amongst other physicists. The rules of the scientific game, by and large, are the rules of cosmology. Are the concepts of cosmology common to the rest of science? Clearly not, for the concepts and the methods of handling them in some way mark out a subject which, at least for the time being, tends to be separated from the rest of physics.

Even if there were such a thing as a Philosophical Method, it might be asked what this could possibly have to do with the stern formalities of scientific cosmology. Philosophy begins, if it does not end, with a description of the commonplace of language—with its anachronisms, its inconsistencies, its exceptions. As it is frequently pointed out, albeit apologetically, the philosopher has little choice when dealing with an unformalized language capable of evolution and retrogression. This description is undoubtedly valid, but one of the purposes of this book is to make it clear that this could equally well be said of the transient theories of natural science. The best ways of elucidating such concepts as Space and Creation are likely to have much in common with those for elucidating the concepts Good and True. But at the other extreme, the view that it is impossible to deal with the fluidities of language—even of scientific language—overlooks the principal agent of all intellectual advancement, namely the elaboration of rules and requirements of a general sort to which arguments must conform. There must be such rules, although it is not denied that they may change with time and author. It is precisely because they, like scientific theories themselves, do so change that a form of conventionalism (for want of a better word) provides an invaluable qualification for the principles of historical and philosophical interpretation.

PART I
ORIGINS OF MODERN COSMOLOGY

'I have a bit of FIAT in my soul,
And can myself create my little world'.

T. L. BEDDOES, Death's Jest Book

CHAPTER 1

NINETEENTH-CENTURY ASTRONOMY: THE NEBULAE

IT HAS been said that the Greeks found the universe a mystery and left it a polis. If cosmologists can be said to agree on any single point it is that they hope to do likewise. They do not, of course, *qua* cosmologists, speak of form and symmetry without first specifying the individuals exhibiting these properties. Before discussing the more important sources of modern cosmology, a brief indication will be given of the different points of view as to what it was proper to consider as the elements of the cosmologist's world—the stars, the various kinds of nebula, the globular clusters, and so on—and the ways in which they were, at the beginning of this century, arranged into a tentative conceptual scheme.

1. THE APPARENT STRUCTURE OF THE MILKY WAY AND THE NEBULAE

Not until the middle of the eighteenth century did astronomers consider the possibility that the stars might be arranged into a system of one or more 'Island Universes'. The very remarkable works of Thomas Wright,[1] Immanuel Kant,[2] Jean Lambert,[3] and an even earlier work by E. Swedenborg,[4] are by now too well-known to require more than mention. None of the speculation in which these authors indulged was at first seriously entertained by any astronomer of note, although it is said to have provided the French court with a topic of conversation. Not until William Herschel was any of the 'Island Universe' hypotheses given any observational support.

The outcome of Herschel's studies of 'the structure of the heavens' appears in retrospect to be the more remarkable as they rested upon hypotheses which were essentially unsound.[5] The greatest obstacle was

---

[1] *An Original Theory, or New Hypothesis of the Universe, founded upon the Laws of Nature, etc.* (London, 1750).
[2] *General Natural History and Theory of the Heavens* (1755) to be found as *Kant's Cosmogony*, trans. W. Hastie (Glasgow, 1900). Kant tells us, by the way, that it was the cosmological problem which led him to his theory of knowledge.
[3] *Kosmologischen Briefe* (Augsburg, 1761). Cf. p. 19 *infra*.
[4] *Principia Rerum Naturalium* (Dresden, 1734). Trans. by J. R. Rendell and I. Tansley as *The Principia* (2 vols.) (London, 1912). See ii, 160 and Appendix A.
[5] Herschel's work will be found in *The Scientific Papers of Sir William Herschel*, ed. J. L. E. Dreyer (London, 1912).

in regard to distances. Hoping to effect measurements of annual parallaxes through the relative displacements of 'optical' double-stars, he set out to catalogue all the double-stars he could find. At the same time he made measurements on the relative brightness of stars, assuming negligible variations between their intrinsic luminosities. His hypotheses ultimately reduced to the assumption that distance is directly proportional to magnitude. Information obtained by John Michell in 1767, which could have shown that he was interpreting the traditional stellar magnitudes wrongly, Herschel chose to ignore. Not until a few years before his death in 1822 did he have any accurate way of directly measuring brightness.[6]

This mistake of Herschel's was trivial, however, by comparison with his assumption that stars vary little amongst themselves. Once again, Michell had shown that this was almost certainly wrong, for the stars of the Pleiades are by no means equal in brightness and yet it would be too much of a coincidence to suppose that their grouping is a chance optical effect. Herschel again obstinately ignored this sort of argument, but when in 1802 he found that many of his double-stars were true binaries and yet contained stars of disparate brightness, he grudgingly acknowledged the point. But he still often fell back upon his uniformity hypothesis.

Other hypotheses of crucial importance to his famous chart of the Milky Way were: (i) that the stars are uniformly distributed in space, and (ii) that his telescopes were capable of penetrating to the outermost regions of this assemblage of stars. His 'star-gages', that is counts of the number of stars visible on a given area in a given direction, thus gave him an indication of the extent of the Milky Way in that direction. Later, his great 48-inch (f/10) reflector revealed no limit to the distribution of stars. It revealed, too, more information as to the nature of the nebulae.

Herschel had for long been intrigued by the nebulae. By now the catalogue compiled by Charles Messier was generally available, and Herschel was able to use his great telescopes to put to the test Messier's hypothesis that many of his nebulae were not composed of stars. The planetary nebulae were particularly puzzling. In 1790 Herschel found one in which a central star was to be seen, by comparison with which the rest looked like a gaseous atmosphere. He was now much less confident that all 'milky nebulosities' would ultimately be resolved as star-systems. His confidence was also shaken in the validity of the hypotheses

[6] He used a twin telescope, one half of which was of variable aperture.

NINETEENTH-CENTURY ASTRONOMY: THE NEBULAE 5

used to determine the shape of the Milky Way. If this extended indefinitely in certain directions, the 'Island Universe' theory to which he had previously inclined must be abandoned. The list of nebulae which he one day expected to be resolved into stars grew shorter.[7] This was partly, perhaps, because he was now eager to accommodate as many systems as possible within a new theory of nebular evolution (1811 and 1814). His distance criterion could offer no assistance in deciding whether an individual nebula was a system of stars. And here the matter rested at his death; contemporary astronomers were sceptical of his theories, if not of his observations; he himself had lost faith in the simple Wright-Kant theory and in the lenticular form which he had previously ascribed to the Milky Way; and only the authors of popular works were inclined to prefer the more systematic views of his earlier years.

The younger Herschel is notable for the meticulous care with which he revised his father's catalogues, continuing the work on star gauges into the southern hemisphere. Having made counts of 70 000 stars, he confirmed his father's conclusion that the Milky Way represented the 'fundamental plane of the stellar system'. His observations of the two Magellanic Clouds did not shake his belief. These systems he described as *sui generis*, without analogues in our hemisphere, and very rich in nebulae and star clusters. They looked like fragments detached from the Milky Way. Perhaps this flaw in the symmetry of the universe was merely an indication that the Milky Way was breaking up. This, as his father had said, affords a proof that it cannot last for ever and that its past duration cannot be infinite.[8]

The first person to measure the parallactic displacement of a star was F. W. Bessel, who in 1838 gave the annual parallax of the star 61 Cygni as $0''\cdot31$ (modern value: $0''\cdot30$).[9] This advance was not, however, the key to a better description of the form of the universe. The star in question was one of the closest to the earth. F. G. W. Struve, using the star Vega, which is at little more than twice the distance of 61 Cygni,

---

[7] Of Messier's 29 'nebulae without stars' which Herschel resolved, 18 are globular clusters, 6 are galactic clusters and 5 are galaxies. Of the 9 which Herschel expected to resolve, 2 are globular clusters, 4 are galaxies, 1 is the Crab nebula and 2 are planetary nebulae. (These figures were kindly provided by Dr. M. A. Hoskin.)

[8] For the Herschels see M. A. Hoskin, *William Herschel and the Construction of the Heavens* (Oldbourne Press, 1964); C. A. Lubbock, *The Herschel Chronicle* (Cambridge, 1933).

[9] As early as 1812 he had noticed that this star had a very large proper motion which suggested that it was near to us. Its faintness caused many to reject Herschel's distance-magnitude relation.

was even so in error by a factor of two.[10] At the end of the century there were fewer than sixty reliable measurements of stellar parallax available, all concerned with relatively near stars—certainly not sufficient to provide anything like a satisfactory model of our stellar system in three dimensions.

Struve devoted a great deal of his time to investigating the structure of the Milky Way. Most of his work was concerned with justifying a misguided supposition that the Milky Way extended to infinity in its central plane, the distant regions being obscured by the progressive absorption of light by interstellar matter. This view, if not originating in a misreading of William Herschel, at any rate bears a strong resemblance to the ideas in Herschel's later writing.

A remarkable and little-known paper, 'On the distribution of nebulae in space', which was based upon a study of Sir John Herschel's *General Catalogue of Nebulae and Clusters of Stars*, appeared in 1867.[11] Its American author, Cleveland Abbe, suggested that (i) the (ordinary) clusters are 'members of the Via Lactea and are nearer to us than the average of its faint stars'. He regarded these clusters as being situated at the same distance as the planetary nebulae. (ii) 'The nebulae resolved and unresolved lie in general without the Via Lactea, which is essentially stellar'. (iii) 'The visible universe is composed of systems, of which the Via Lactea, the two Nubeculae [i.e. the Magellanic clouds], and the Nebulae, are the individuals, and which are themselves composed of stars (either simple, multiple, or in clusters) and of gaseous bodies of both regular and irregular outlines'. Two years later R. A. Proctor, despite Abbe's conclusions, used the same evidence to draw exactly the opposite conclusions.[12] Having constructed charts of the nebular distribution, which indicate clearly the tendency of all but gaseous nebulae to avoid the plane of the Milky Way, he concluded that since it is unlikely that such an arrangement should be accidental, all must be part of the same system. As he pointed out, William Herschel had taken the same view. Abbe's astonishingly perspicacious view soon lost favour, and well into the present century many astronomers believed the Milky Way to comprise the whole universe. Others admitted that the matter could not be decided until the nature of the nebulae was more fully understood.

[10] For further details see Agnes Clerke's *A Popular History of Astronomy during the Nineteenth Century* (Black, London, 1902) 4th ed. More than seventy-five years after its first publication, this remains the only comprehensive and well-documented work on its subject. The first half of the century is covered equally carefully by R. Grant, *History of Physical Astronomy* (Baldwin, London, 1852).
[11] *M.N.*, xxvii (1867) 257.   [12] *M.N.*, xxix (1869) 337.

William Herschel had, as we have seen, resolved many of the nebulae which we should term 'globular clusters' into stars, being eventually convinced that not all nebulous masses could be resolved in this way. In 1845, after ten years at the task, William Parsons, third Earl of Rosse, completed work on a 72" speculum. (This was to be the world's largest telescope for the next seventy years.) After three months of use the instrument had resolved many more nebulae into stars and had also revealed, for the first time, a nebula with a spiral structure. Once again it had become an open question whether all nebulae, spirals included, could not be resolved into stars, given an instrument of sufficient power.

Lord Rosse's results could not be easily confirmed, and for some time there was a widespread disbelief in the Irishman's truthfulness. There is little enthusiasm to be found in the brief remark announcing Rosse's discovery in the *Monthly Notices*: 'Lord Rosse had stated to the Astronomer Royal that the nebula H.131 exhibited a well-marked spiral structure and that 2241 had a central hollow.'[13] He published sketches,[14] and W. Lassell in Malta confirmed them after 1861.[15] A section of the astronomical world remained generally incredulous until photographs were eventually obtained. One of Rosse's many well-executed drawings, that of M.99 (a spiral in the constellation of Coma), was used to illustrate—by analogy—a new theory of the structure of the Milky Way.

The vortical structure of the spiral nebulae prompted speculation on the subject of stellar evolution. By many the swirling masses were assumed to be evidence of the truth of Laplace's nebular hypothesis.[16] Isaac Roberts, for example, an amateur astronomer who discovered the spiral structure of the great nebula in Andromeda by photographing it (exposure three hours), gave the published photograph the subtitle: 'A pictorial representation of the past of our system' (i.e. the solar system).[17] To Rosse the spiral structure seemed to indicate 'the presence of dynamical laws which we may fancy to be almost within our grasp'. The dynamical laws in question were, however, to be overlooked in the excitement aroused at the end of the next decade by the birth of a new study—astrophysics.

Auguste Comte was by no means alone in his belief when he offered

[13] *M.N.*, x (1850) 21.   [14] *Phil. Trans.*, cxl (1850) 499.
[15] *Memoirs of the R.A.S.*, vol. xxxvi.
[16] Writers of the time do not appear to have referred back to the vortical-galaxy theory of Swedenborg. (See n. 4, p. 3 *supra*.)
[17] *M.N.*, xlix (1889) 65.

as a perfect example of unattainable knowledge that of the chemical composition of the stars.[18] As far as most astronomers were concerned, one could learn more about the nebulae only by building telescopes of larger aperture. (In the words of G. P. Bond of Harvard, writing in 1857, it was 'simply a question of finding one or two hundred thousand dollars'.)[19] In 1859, however, the situation changed when G. R. Kirchhoff and R. Bunsen gave the physical interpretation of emission and absorption spectra and, in particular, of the spectral lines so carefully charted by J. Fraunhofer forty-five years previously.

These discoveries were immediately extensively applied by astronomers, notably to the Sun's spectrum. In 1860, G. B. Donati's attempt to obtain the spectrum of a star failed.[20] Nevertheless, within two years Huggins, Father Secchi, and L. M. Rutherfurd were independently successful, although not for a further twelve years were spectra found which were clear enough to show the anticipated Fraunhofer lines.[21]

Sir William Huggins was the first to apply the methods of spectrum analysis to the nebulae (this he did in 1864). Speaking long afterwards of his work, he confidently announced that 'the riddle of the nebulae was solved'. The answer, he added, 'which had come to us in the light itself, read not an aggregation of stars, but a luminous gas'.[22] Over the next few years he modified his views, finding that only about a third of the observed brighter nebulae gave spectra with a few emission lines. These, which included planetary, diffuse, and annular nebulae, he called 'green nebulae'—in contrast to the 'white nebulae' of continuous spectra, the most famous of which is the nebula in Andromeda, M.31.

Were the 'white nebulae' aggregations of stars? Evidence was conflicting. Photography, an innovation almost as important to the astronomer as spectrum analysis, had been in astronomical use for half a century when in 1899 J. Scheiner used it to obtain a spectrogram of M.31, suggesting that it appeared to display the spectrum of a 'cluster of sun-like stars'.[23] Two years previously, however, Sir William Huggins and his wife saw 'bright and dark bands intermixed'.[24] Long

[18] *Cours de philosophie positive* (Paris, 1835) vol. ii. 19th leçon. (Book IV, Ch. I of Harriet Martineau's translation (1853). Cf. the last sentence of Book II of the latter: '... the field of positive philosophy lies wholly within the limits of our solar system, the study of the universe being inaccessible in any positive sense'.)
[19] *P.A.S.P.*, ii (1890). [20] See Huggins, *Phil. Trans.*, cliv (1864) 414.
[21] Huggins announced the discovery of stellar velocities in the line of sight in *M.N.*, xxxii (1872) 359 and xxxiii (1873) 238. In *M.N.*, xxxiv (1874) 201, he stated that he was unable to detect corresponding velocities for any nebula.
[22] *Nineteenth Century Review*, June, 1897.
[23] *Ap. J.*, ix (1899) 149. See also Scheiner, *Populäre Astropysik* (Leipzig, 1908).
[24] Huggins, *Atlas of Stellar Spectra* (1899) p. 125.

after the turn of the century many thought it eminently rash to draw Scheiner's conclusions, which is not altogether surprising when it is borne in mind that even so conspicuous a 'white nebula' as M.31 was not resolved into stars until 1924.

Whilst these astrophysical issues were being debated, work on the spatial structure of the Milky Way continued—for might this not comprise the totality of things? Sir John Herschel had long previously pointed out that much of the Milky Way appeared to be in the form of streams of stars of, perhaps, circular cross-section. These streams were, he proposed, nearer to us than the main body of the Milky Way. Giovanni Celoria of Milan developed these ideas into a much simpler theory when in 1879 he proposed a finite universe, whose depths had already been sounded, in the form of two more or less concentric rings. Hugo von Seeliger disposed of the simpler forms of ring theory between 1884 and 1898, when (using the well-known *Durchmusterung* compiled by F. W. A. Argelander and E. Schönfeld, together with the Cape photographic survey) he showed them to be inconsistent with the apparent distribution of stars in depth and position. This was not the last word on the subject, for in 1900 the Dutch astronomer C. Easton reintroduced Celoria's ideas,[25] varying the position of the Sun and the centres of the rings and their inclinations in his model. Easton regarded this model merely as a first approximation to a more accurate version and spent the next decade in producing a photographic map of the Milky Way. In 1913 he published his findings, concluding that the universe was probably a spiral closely resembling M.51 and M.101.[26] It is worth adding that, although Easton likened the Milky Way to the spiral nebulae, he did not suppose the latter to be independent. They were, in his words, 'small eddies in the convolution of the great one'.

One can see how, as the nineteenth century wore on, there was a gradual return to William Herschel's ideas on the general form of the universe. His insight appears, in retrospect, the more remarkable as he relied almost entirely on observations of the densities of stellar distribution, whereas towards the end of the nineteenth century a great deal of information as to the *motions* of the stars was available. In 1895 H. Kobold found that stellar motions were not, as was generally believed, random, but that they showed a preference for the plane of the Milky Way. W. H. S. Monck in Dublin had, three years previously, observed that stars of a certain spectral type (Secchi's second spectral

---

[25] 'A new theory of the Milky Way', *Ap. J.*, xii (1900).
[26] *Ap. J.*, xxxvii (1913) 105.

type) were generally found to have greater proper motions than certain others (Secchi's first type). In 1904 J. C. Kapteyn found the solution, namely, that the majority of the easily visible stars moved in two distinct streams towards different parts of the sky, the motions of Sun and Earth being allowed for.[27] Between 1906 and 1922 he continued William Herschel's star counts and came, to the surprise of his contemporaries, to conclusions very similar to the better-known ideas of Herschel regarding the stellar distribution. The Sun, moreover, Kapteyn placed near the centre of the galaxy (650 pcs distant). This was in sharp conflict with the ideas of Harlow Shapley, who in 1918 identified the centre of the Milky Way with the centre of the system of globular clusters and estimated that this lies 15 000 pcs from the Sun (in the direction of Sagittarius). Kapteyn's system appears, needless to say, to have been the more favourably received.

2. THE ISLAND UNIVERSE THEORY ESTABLISHED

The last phase of the argument as to the nature of the 'white nebulae' was entered at about this time. Despite strong support for the Island Universe theory by such men as A. S. Eddington and F. W. Very, there was still considerable opposition to it in the early 1920s. This opposition was founded upon three important arguments. The first argument, and perhaps the strongest of the three, purported to show that the dimensions of the nebulae were insignificant, by comparison with the dimensions of the Milky Way. Their distances would of themselves, of course, be immaterial to any argument relating to the status of the nebulae so long as these were symmetrically disposed about us. (One would not otherwise tend to regard them as subordinate.) The first argument, then, began with the internal motions of many of the spirals. Evidence for very great velocities within at least one spiral was presented by Adriaan van Maanen in 1916.[28] Van Maanen had derived, from an examination of photographs taken at intervals of some years, an annual rotation for M.101 of the order of $0''{\cdot}02$. His results were later more or less confirmed, and results of the same order were found for several other spirals.[29] These angular rotations, in conjunction with the relative linear velocities of the parts, spectroscopically measured, yielded (using a method rather like the familiar

[27] Kapteyn's great discovery is well described in an obituary notice by A. van Maanen, *Ap. J.*, lvi (1922) 145.

[28] *Ap. J.*, xliv (1916) 210 = *Mt Wilson Cont.*, no. 118.

[29] See K. Lundmark, *Ap. J.*, lvii (1923) 264 = *Mt Wilson Cont.*, no. 260; *Ap. J.*, lxiii (1926) 67 = *Mt Wilson Cont.*, no. 308. Both were concerned with M.33. There was agreement within a factor of 2.

NINETEENTH-CENTURY ASTRONOMY: THE NEBULAE

method of statistical parallaxes)[30] a distance of 6520 light years (i.e. a parallax of 0″·0005) for M.33.[31] The parallaxes of the larger spirals were all, it seemed at the time, between a few thousandths and a few ten-thousandths of a second of arc. Other evidence supported this. Sir James Jeans, for example, developed in the early 1920s a mathematical theory of the separation of condensations in the arms of the spiral nebulae.[32] Together with the apparent angular separations, this was sufficient to allow Jeans to deduce parallaxes of 0″·0006, 0″·0011, and 0″·0065 for M.31, M.101, and M.51 respectively. Parallaxes such as these had an important bearing upon the general attitude towards the spirals. It was thought that at such distances they must have diameters of, at the most, a few hundred light years. They were thus thought to be 'not at all comparable with the Milky Way system' (van Maanen)—a system which was, according to contemporary authorities, between 20 000 and 300 000 light years in diameter.

It is seldom now remembered that although these arguments were ignored for many years, yet they were not refuted until 1935.[33]

The second argument against the galactic nature of the 'white nebulae' was levelled against those who maintained that since the spectrum of each of the 'white nebulae' was crossed by Fraunhofer lines, these nebulae were therefore 'aggregations of sun-like stars'. If this was so, might they not be of similar status to the Milky Way, even though smaller? In 1912 V. M. Slipher found nebulae having spectra which had the appearance of stellar spectra only because light of stellar origin had been reflected from the otherwise diffuse material of the nebulae. For a time, therefore, the argument from spectrum alone was discredited.

The third argument was that, according to work by Shapley, the dimensions of the Galaxy—that is, the system based on the Milky Way—were sufficiently large to accommodate the spirals, many of which were firmly believed to have fairly large parallaxes. This point of view was supported by another of Shapley's beliefs (to which we shall

[30] See, for example, R. J. Trumpler and H. F. Weaver, *Statistical Astronomy* (Univ. of California Press, Berkeley, 1953) ch. 3.

[31] Using the same plates, Lundmark put the distance of M.33 at something between 40 000 and 160 000 light years (op. cit.). He was, however, aware of the inferences from Cepheid measurements leading to a figure of 300 000 pcs. The argument ultimately fell when, having repeated van Maanen's observations with a greater time interval and having found quite contrary results, Hubble showed the earlier findings to have been influenced by some kind of systematic error.

[32] See *The Nebular Hypothesis and Modern Cosmology* (C.P., Oxford) 1923 (Halley Lecture, 1922).

[33] See Hubble, *Ap. J.*, lxxxi (1935) 334 = *Mt Wilson Cont.*, no. 514.

return in Chapter 7), namely, that there is a dynamical relationship between this, our own galactic system, and the radial velocities of the spirals. These in any case seemed to avoid low galactic latitudes.[34]

Here, then, were three broad arguments for the view that the nebulae were smaller than, different in composition from, and in position and motion in some way subordinate to the Milky Way. A knowledge of the distances of the spirals was obviously of critical importance to at least two aspects of the debate, and the evidence was unfortunately ambiguous. In 1885 a 'new star' was discovered near the nucleus of M.31. Attaining the seventh magnitude, it accounted for one-tenth of the entire luminosity of the nebula. The enormous luminosity which this implies was not, of course, obvious to astronomers of the time, who were uncertain as to the distance and nature of M.31 and the 'white nebulae' generally. In 1917, however, G. W. Ritchey discovered a fainter 'nova' in a spiral nebula (N.G.C. 6946).[35] An immediate examination by G. W. Ritchey and H. D. Curtis of plates of the Andromeda and other nebulae showed that novae were by no means infrequent occurrences.[36] Curtis at the Lick Observatory and Shapley at the Solar Observatory, Pasadena, were the first to exploit this discovery as a criterion of the distances of the spirals. An early confusion sprang from the fact that two distinct kinds of nova had been observed, that of 1885 being several thousand times brighter than all others, being now distinguished as a 'supernova'. In a three-cornered discussion Curtis, Shapley, and Lundmark[37] considered which criterion of distance was appropriate, although at first the distinction between the two classes of object was not made. Shapley was initially convinced that each nova was to be explained as the penetration of nebulosity by a star.[38] He concluded that the spirals could not be of a size comparable with the size of the Galaxy; for, in this case, the absolute magnitudes of the novae they contained would be much larger, 'at the distances computed', than those of novae in the Galaxy as determined by van

[34] See the brief historical sketch of more recent work on galactic obscuration, &c., in *Vistas*, iii, p. 1574 (article by C. D. Shane).
[35] *Harvard Bull.*, no. 641 (July, 1917).
[36] See *P.A.S.P.*, xxix (1917) 180, 210.
[37] Curtis, *J. Wash. Acad. Sci.*, ix (1919) 217; Shapley, *P.A.S.P.*, xxxi (1919) 261. This paper provides an excellent summary of the evidence then obtained for and against the Island Universe hypothesis. See also Shapley and Curtis, *Bulletin of the National Research Council*, no. 11 (1921); Lundmark, *Stock. Acad. Hand.*, Bd. lx (1920) no. 8.
[38] *P.A.S.P.*, xxix (1917) 213; cf. Curtis's paper at p. 206, and *P.A.S.P.*, xxx (1919) 43: 'The most generally accepted explanation of galactic novae is the penetration of nebulosity by a star with considerable velocity. Inverting this, the novae in spirals may be considered as the engulfing of a star by rapidly moving nebulosity'. (op. cit., p. 53.)

Maanen.[39] Once the distinction between 'dwarf' and 'giant' novae was clearly made, Lundmark and Curtis sought out the fainter non-Galactic novae as being comparable with those ordinarily observed in the Galaxy (i.e. excluding the supernovae of 1054 and 1572: F. Zwicky estimates that the average galaxy experiences two or three such outbursts per millennium). This choice was soon to be generally accepted as correct. For the time being, however, although Shapley believed the spirals to be small by comparison with the Galaxy and to be of the nature of satellites to it, all authors agreed that on these grounds the spirals appeared to be at least fifty or more times as remote as the unambiguously 'Galactic' nebulae.

To some it seemed that the 'riddle of the nebulae' had been solved. Shapley drew up a rather simple plan for the Galaxy, 'a single, enormous, all comprehending unit'. Near the Galactic plane lay the open clusters, diffuse and planetary nebulae, the naked-eye stars, and so on. The globular clusters, by virtue of their systematically avoiding this plane, could not, it was said, be independent objects. The spirals too were subordinated to this single scheme. Their high velocities, which to some suggested their independence, were to Shapley and others comparable with the velocities of objects 'in our own system': but 'high' meant only of the order of a few hundred km/sec.

Notwithstanding the arguments of van Maanen, Jeans, and Shapley, a far more satisfying means of ascertaining the distances of the nearer spirals than any yet proposed became available in the early 1920s. The existence of variable stars in a spiral nebula was first recognized by Duncan in 1922.[40] Within a year, a year of considerable activity at Mount Wilson, the first Cepheid-type variable was detected in M.31.[41] Almost as soon, Shapley, who exploited this criterion of distance above all others, was satisfied that the distances of extragalactic nebulae were to be measured in millions, rather than thousands, of light years.[42] Especial attention was paid to those Cepheids in the nearby spirals M.31 and M.33 and to the irregular nebula N.G.C. 6822. The period-luminosity curves of these distant Cepheids were perfectly typical, and

---

[39] *P.A.S.P.*, xxxi (1919) 234. The average distance of the brighter spirals was given as 20 000 light years.

[40] *P.A.S.P.*, xxxiv (1922) 290. In 1904 and the years immediately following, Miss Leavitt had carried out her well-known work on 992 variable stars in the two Magellanic clouds. (24-inch Bruce telescope at Arequipa—see *Harvard Annals*, lx (1908) no. 4 and *Harvard Circular*, clxxiii (1912).)

[41] *Annual Reports of Mt Wilson Observatory* (1923–4).

[42] He took the distance of N.G.C. 6822 as about 10^6 light years, *Harvard Bull.*, no. 796 (Dec. 1923). Cf. Hubble, *Ap. J.*, lxii (1925) 409 = *Mt Wilson Cont.*, no. 304.

if the calibration of the curves represented the greatest uncertainty, at least the distance of one or other of the Magellanic clouds would, for many purposes, suffice as a unit of distance. One of the first achievements of the Cepheid-method was to provide a rough confirmation of the (by now familiar) nova-method of estimating the distances of the spirals.[43] The two methods thus lent each other support.

Now for some years past there had been a discussion of the possibility of a systematic recession of the nebulae from the Sun. This will be treated more fully in Chapter 7, and here we are concerned only with the *status* of the nebulae—a problem bequeathed by the nineteenth century to the twentieth. As we shall see, however, it was only when the Cepheid- and nova-methods of measuring distances became available that the investigation of a connexion between recession and distance could be profitably resumed. Hubble cautiously proposed that even N.G.C. 6822, whose distance he had quoted as 230 000 pcs, was beyond the limits of our Galaxy: perhaps it was the nearest independent system. As to the fainter spirals there was no doubt that they were well beyond the stellar limits of the Milky Way. That they were stellar systems in their own right was morally certain, as was the fact that they were of the same order of size as our own. (Admittedly, Shapley's value for the diameter of the Galaxy fitted uneasily into Hubble's scheme, and the sceptic was provided, as recently as 1930, with an excuse for refusing to liken the spirals to the Milky Way.)

Thus after a hundred and fifty years of dispute, the realization that the vast majority of the nebulae actually observed lie far outside the Galaxy came only slowly. This realization was a necessary, but far from sufficient, condition of modern cosmology. The new cosmology was therefore, *qua* astronomical science, in its infancy little more than thirty years ago. Conceptually, on the other hand, it was very well prepared. And far more important to its early progress than descriptive astronomy were two other developments. The first of these was the discovery that Newton's theory of gravitation, when applied to problems on a cosmic scale, led to inconsistencies and otherwise unacceptable conclusions. The second, and most important of all, was Einstein's

[43] This confirmation did not settle the argument as to the relative sizes of Galaxy and galaxy, an argument which was resolved only when obscuration within the plane of the Milky Way was taken into account—thus showing the diameter of our own system to have been overestimated—and when the faint globular clusters were found, surrounding the external galaxies, and included in a reckoning of their diameter (as globular clusters had been included in our own).

General Theory of Relativity. This, however, came in the wake of an influential group of writers who wished to rephrase Newton's theory of gravitation in terms of the theory of the field. Einstein took much from the past, but he also gave much to the future. His theory of gravitation, in making use of Riemannian geometry and the calculus of tensors, gave to cosmology methods which have been widely used by writers in different, albeit derivative, traditions. The earlier influences will be discussed in the next three chapters.

The rest of this book is mainly concerned with the leading concepts of cosmology. Although little attention is given to the part played by the observing astronomer, the aim is not to belittle his contribution: without it, after all, cosmology would not be a science. Without going further, however, we can detect a quiet but definite change in the procedure followed by the cosmologist. Nowadays he tends to begin with a theory raised on a very slender empirical foundation. In the past the astronomer often merely described what he saw in the hope that theoretical reasons for the observed state of things would be obvious. He naïvely thought that it was possible to have an open mind. He might make some slight progress with a minimum of prior theoretical reasoning. He might even have believed himself free from theoretical involvement, but if so he was shortly to be disillusioned.

CHAPTER 2

COSMOLOGICAL DIFFICULTIES WITH THE NEWTONIAN THEORY OF GRAVITATION

WE HAVE seen that in the course of the last century astronomers were discussing the nature and internal structure of individual nebulae rather than the wider cosmological problem. Probably the greatest single obstacle to progress in the wider context was an inability to appreciate what it would mean to decide between a finite and an infinite universe. How could one possibly recognize an end to the system of nebulae? Do space and the nebulae end together? How, on the other hand, could one recognize that there was no end to the system? Telescopes alone were unable to resolve these difficulties. The non-Euclidean geometries offered a way out but, although a few would admit that they might be legitimately employed, fewer still were aware of a means whereby they could be introduced into astronomy. With the first really satisfactory solution of this problem came a renewal of interest in cosmology. The recovery was sudden and unexpected. Rarely does a subject present itself, unsought, in a completely unfamiliar light, but this was how scientific cosmology re-emerged in the first quarter of the present century. Simple observations on the over-all distribution of matter were not enough, even when combined with a theory of gravitation which no one seriously doubted: nor was it even sufficient to be aware (as, for example, Bernhard Riemann and W. K. Clifford were fully aware) that the world could be significantly treated as finite and yet without bounds. Only when such geometrical conceptions were combined with a *related* theory of optics and gravitation did the astronomer's cosmology become deserving of its title. The revival was completed only with Einstein's General Theory. The movement can be said to have begun, however, in 1895-6 when two authors separately raised the difficulties to be found in a theory of an infinite Newtonian universe.[1] In the present chapter we shall therefore outline the reasons given for the belief that Newton's theory of gravitation was cosmologically unacceptable. We shall also outline other objections to an

---

[1] H. Seeliger, *Astr. Nachr.*, cxxxvii (1895) 129 and *Münch. Sitz.* (1896) 373; C. Neumann, *Allgemeine Untersuchungen über das Newtonsche Prinzip der Fernwirkungen* (Leipzig, 1896). Neumann had actually discussed the problem many years previously (Leipzig *Abh. Math. Phys.*, xxvi (1874) 97).

infinitely extended universe, together with some of the reactions to these criticisms.

1. Seeliger's and Neumann's Objections to a Newtonian Infinite

In 1895 Seeliger began by protesting that as the volume (V) of a Newtonian distribution of matter of finite density tends to infinity, the gravitational potential at any point[2] can be assigned no definite value; added to which the expression for the gravitational force also becomes indefinite. Carl Neumann, faced by the same difficulties, proposed that Poisson's equation should be adjusted so as to permit a uniform and static distribution of matter throughout space. For the gravitational potential they took expressions of the usual Newtonian form, multiplied by an additional factor $\varepsilon^{-\alpha r}$, where α is a quantity sufficiently small to make the modification insignificant, except for large distances. This way of modifying the potential function was thought to be preferable to the somewhat more obvious way of taking it to be of the form $\phi(r) = r^{p-2}$ (where $0 \leqslant p < 1$), which, for almost any useful value of p, would have invalidated the highly successful classical theory of perturbations.[3]

This form of the potential function had, in effect, been taken by Asaph Hall as capable of accounting for the anomaly in the motion of Mercury. The idea was not new, for not only had Newton himself used a similar expression on at least one occasion,[4] but George Green had also used it in his 'Mathematical Investigations concerning the Laws of the Equilibrium of Fluids, &c.,' in 1832.[5] On the other hand, neither were Seeliger and Neumann first with the exponential law: Laplace had taken this very law fifty years before.[6] In all the earlier cases, however, the concern was in only a narrow sense cosmological.[7]

The effect of the exponential modifying factor is to introduce a cosmical repulsion capable, at large distances, of exceeding the usual gravitational forces. As will be seen in due course, the introduction of

---

[2] $\phi = \int (\rho/r) dV$.   [3] Pointed out by Seeliger, *Münch. Sitz.*, xxxvi (1906) 595.
[4] *Principia* (1686 edition) p. 139.
[5] *Trans. Camb. Phil. Soc.* (1835), see his *Collected Mathematical Papers* (1871).
[6] *Mécanique Céleste* (1846 edition), vol. v, bk. xvi, ch. 4. p. 481.
[7] Neumann takes both Green's and his own form of the potential-function as special cases of the *Exponentialgesetz*

$$\phi(r) = (A\varepsilon^{-\alpha r}/r) + (B\varepsilon^{-\beta r}/r) + (C\varepsilon^{-\gamma r}/r).$$

It should be noted that both Neumann and Green were dealing with both gravitational and electrical subjects, and that no treatment of the latter had ever enjoyed the same measure of general confidence as had Newton's theory of gravitation.

what was to be known as the 'cosmological term' into the later gravitational field equations of Einstein is reminiscent of Neumann's modification of Poisson's equation. Both, needless to say, have been attacked on the grounds of their *ad hoc* character, just as have those other modifications of the Newtonian law which will be discussed in the next chapter. (The phrase '*ad hoc*' has a variety of polemical uses, but is here best interpreted as 'without empirical consequences apart from those it was devised to explain'. In a sense, the general antipathy to *ad hoc* theories militates against the multiplication of theories with an explanatory range as great as, but no greater than, the range of those theories already in existence. For if two theories have what might roughly be called 'the same empirical basis', that which came second is bound to fall under this definition of '*ad hoc*'.)

2. CHARLIER'S HIERARCHIC HYPOTHESIS

Another objection to an infinitely extended universe had been raised, perhaps independently, by the astronomers J.-P. de L. de Cheseaux in 1744[8] and H. W. M. Olbers in 1823.[9] They calculated that if the universe contained infinitely many bright stars the night sky would shine with a brightness corresponding to their average surface brightness. De Cheseaux began by assuming that the Sun is a typical first-magnitude star. He was thus able to calculate the distance of the remaining stars of the first magnitude (24×10^4 a.u.) and their apparent diameter ($0''\cdot 008$). Stars a quarter as bright would (assuming them to be of the same intrinsic luminosity) be at twice the distance. Their apparent diameters would be halved, but as there would be four times as many of them (on the hypothesis of their uniform distribution) the area of

[8] *Traité de la comète qui a paru en décembre* 1743, etc. (Lausanne and Geneva, 1744). Appendix.

[9] Bode Jb.,(1826)110. The paper is dated 7 March 1823. It is well known through Bondi's writing. It is probably not widely known that substantially the same paper is to be found in both the *Bibliothèque Universelle* for February 1826, and in the *Edinburgh New Philosophical Journal* for April–October 1826, p. 141. De Cheseaux's paper is mentioned by P. G. Tait in an *Encyclopaedia Britannica* article (1881 edition). I owe this reference to an article by Otto Struve (*Sky and Telescope*, March 1963). F. G. W. Struve appears to have been the first to recognize (in 1847) that Olbers' argument had already been given by de Cheseaux.

The similarities in the arguments offered by the two men are very striking. Olbers puts the first-magnitude stars at 35×10^4 a.u., and assigns to them a diameter of $0''\cdot 005$. Struve (*Études d'Astronomie Stellaire*, 1847) points out that Olbers' library included a copy of de Cheseaux's book.

Actually Edmond Halley had hinted at the paradox more than a century before Olbers, but his solution is confused. He refers to this as 'an Argument I have heard urged, that if the number of Fixt Stars were more than finite, the whole superficies of their apparent Sphere would be luminous, for that those shining bodies would be more in number than there are Seconds of a Degree in the area of the whole Spherical Surface, which I think cannot be denied'. (*Phil. Trans.*, xxxi (1720) 23.)

sky they occupied would be the same as that of the first-magnitude stars. Continuing this argument it was clear that the sky should be filled completely with radiance as bright as that from the Sun, and that without proceeding to an infinite layer of stars. De Cheseaux concluded, in fact, that it was necessary to consider stars within a distance of 76×10^{10} astronomical units.

In order to account for the darkness of the night sky, de Cheseaux assumed the existence of an interstellar fluid of unknown composition, with a transparency 33×10^{16} times as great as that of water. Olbers adopted the same course, making the arbitrary assumption ('but I do not imagine it to be far from the truth') that space is transparent only to the extent that, 'of 800 rays emitted by Sirius, 799 attain to the distance at which we are placed from that star'. He argued that, in this case, stars more than 3×10^4 times as distant as this star would make no contribution to the brightness of the night sky.[10] As it turned out, no satisfactory evidence for the absorption hypothesis was to be given for well over half a century. That this explanation is itself insufficient, was pointed out by Herman Bondi.[10] The gas must be heated in the process of absorption until it reaches a temperature at which it radiates as much as it receives; in this way it would sustain the average density of radiation.

Neither de Cheseaux nor Olbers ever questioned the other assumptions, namely that the stars are distributed more or less homogeneously in space, that space is Euclidean, and that the stars are subject to no systematic movement in time. The truth of the first assumption was questioned by Thomas Wright, by Kant, and in an especially interesting way by Jean Lambert.[11] Lambert loosely speculated that the Sun was one of a large system of stars revolving round a centre, that there were many such systems, 'and several systems taken together have a centre in common; and where shall we stop?' Groups of these compound systems he supposed to have their own centres. Finally he believed there to be 'a universal centre for the entire Universe around which all things revolve'.[12] It is worth adding that although Lambert regarded the nebulae as comprising groups of stars, he did not regard them as external universes.

Lambert's hierarchic hypothesis could scarcely be counted as more than a curiosity of astronomical literature until it was revived and

[10] *Cosmology*, 2nd ed. (Cambridge, 1960), p. 21.
[11] *Kosmologischen Briefe* (Augsburg, 1761).
[12] More curiously, these centres were 'not void but occupied by opaque bodies'. He conjectured that 'perhaps the pale light seen in Orion is our centre'.

treated more thoroughly by C. V. I. Charlier in 1908 and 1922.[13] Although by now Lambert's assumption of Euclidean space had been questioned, and that of a static universe was shortly to be so, yet Charlier's methods found several supporters, largely because they were thought to be simple and at the same time effective in removing the objections raised by de Cheseaux and Olbers, Neumann, and Seeliger.[14]

Charlier took for simplicity *spherical* systems, of radii R_0, R_1, R_2 . . . and so on for successively greater orders. If the two sorts of objection are to be avoided, two sets of algebraic inequalities involving the radii are entailed. These two sets of inequalities turn out to be one and the same (namely, $(R_i/R_{i-1}) > \sqrt{N_i}$, where N_i is the number of $(i-1)^{\text{th}}$ order galaxies in a system of order i).

Charlier worked out the scheme in considerable detail. One feature which may be mentioned was his prediction of a high collision rate between the nebulae of a single metagalaxy (1 per 1000 years). This high rate, he suggested, explained the high incidence of spiral nebulae which he supposed to be created in such a process.[15]

In the same year as Charlier published these results, his general argument was supported by Frederick Selety.[16] Seeliger's objection is invalid, according to Selety, if a system can be described in such a way that the acceleration at each point tends to a finite limit. If the mean density at a point decreases more rapidly than the reciprocal of its distance from any other point, he said, this condition is fulfilled. He went on to show (using the statistical method first applied by Darwin to meteorite swarms)[17] that a system in equilibrium can be described in which the mean velocity is small, but where the potentials and mass are infinite. His 'hierarchical molecular world' need have no centre (in this respect the model may be contrasted with Borel's); its mean density would tend to zero, whilst the local density would be everywhere finite.

A polemical discussion with Einstein followed, which hinged upon

[13] *Arkiv. Mat. Astr. Fys.* (1922), Bd. 16, no. 22, pp. 1–34. Preliminary discussion, ibid., (1908), Bd. 4, no. 24. In point of time E. E. Fournier d'Albe (*Two New Worlds*, London, 1907) should be given priority, but scientifically the work of this self-styled 'Newton of the soul' is worthless.

[14] For a short account of Charlier's work see L. Silberstein, *Theory of Relativity* (1924), Appendix H.

[15] V. G. Fessenkov has discussed ways of avoiding Olbers' paradox, given a hierarchic universe with diffusion. (*Comptes Rendus Doklady de l'Academie des Sciences, U.S.S.R.*, xv (1937) 123—in French.)

[16] *C.R.*, clxxvii (1923) 104, 250; *Ann. Phys.* Lpz., lxviii (1922) 281, lxxiii (1924) 291. Cf. A. Costa, *C.R.*, clxxv (1922) 1190 and E. Borel, *C.R.*, clxxiv (1922) 977 on the same subject.

[17] *P.R.S.* (A), xlv (1888) 12.

the issue as to whether the so-called 'Mach's Principle' was essential to any satisfactory theory.[18] We shall return to this issue later: for the time being we remark only that Selety gave grounds for subscribing to it, but that they are not always clear. He emphasized, however, that his model allowed for *irregularities* in density, and that it utilized a simple form of geometry, that is to say, Euclidean.

Charlier's static and Selety's quasi-static solutions of these difficulties were achieved only at the expense of introducing complex types of homogeneity for which no independent evidence had been adduced. Also, their arguments can hardly be said to have *explained* why the nebulae are to some extent clustered: for them, this clustering was merely a sufficient condition for an infinite distribution of matter to remain in a static Newtonian equilibrium. This is not, however, the most telling argument against them. It has been shown by E. Finlay-Freundlich that in an *expanding* universe with hierarchic structure and with a velocity of expansion proportional to the distance, new singularities arise, the density tending to infinite values.[19] He points out that a slight 'absorption of gravitation in space', which at first sight seems to offer an escape from infinite gravitational potentials, is powerless to rule out such singularities. They can be removed by introducing a repulsive force into the Newtonian law of gravitation, although there are, as we have seen, good reasons for not taking this course by choice.

Against the background of changing opinion as to the status of the nebulae it is easy to see that such a view as Charlier's was unlikely to be widely entertained before, say, the publication of Hubble's galactic observations of 1923–5 and their subsequent confirmation by E. Öpik and K. Lundmark. In his statistical investigation of 400 extra-galactic nebulae, Hubble took what was no more than a working hypothesis, namely, that the spacing of the nebulae was approximately uniform. He was soon led to the idea that our Galaxy belongs to a 'Local Group', thought at this time to have half a dozen members.[20] Clusters of nebulae were, of course, known; but the belief was still widespread that there was a dynamical connexion between the Galaxy and the general system

---

[18] Einstein, *Ann. Phys.* Lpz., lxix (1922) 436; Selety, *Ann. Phys.* Lpz., lxxii (1923) 58 and lxxiv (1924) 291. Einstein was at a disadvantage in the sense that (like most of his contemporaries) he did not believe that the Galaxy could be likened to the white nebulae.

[19] See the article 'Cosmology' in *The International Encyclopedia of Unified Science* (Chicago, 1955) vol. i, pt. ii, p. 505, in which Finlay-Freundlich gives a simple account of an alternative theory of hierarchical structure.

[20] Early observation on the Local Group: Hubble, *Ap. J.*, lxii (1925) 409; lxiii (1926) 236; lxix (1929) 103; lxxvi (1932) 44. (= *Mt. Wilson Conts.* nos. 304, 310, 376, 452 respectively.)

of galaxies then observed.[21] Such clusters could hardly be construed as providing conclusive evidence for Charlier's wider hypothesis, and Lundmark, who regarded the latter favourably, was bound to admit that there was no evidence for clustering of the third order.[22] Methods of observation improved, and some slight evidence was later forthcoming. Opinion has since divided, however, and (although there are one or two astronomers who would disagree) the evidence which once seemed as though it might favour the Charlier-Selety hypothesis, seems to do so no longer.[23]

The Newtonian treatment of the infinite, as given by Neumann and Seeliger, led to unacceptable conclusions. That there remained the possibility of deriving a non-hierarchic 'Newtonian' cosmology was claimed by E. A. Milne, who proposed in 1934 a Newtonian analogue of a particular relativistic model of the 'expanding universe'.[24] In the same year Milne and W. H. McCrea put forward the view that Newtonian analogues could be found for a whole range of relativistic models.[25] Over twenty years later they were generalized still further by O. Heckmann and E. Schücking[26] who, dropping Milne's assumption of isotropy, gave not only a Newtonian model analogous to Gödel's relativistic model,[27] but a model for which there was no analogue. These developments will be dealt with in later chapters. How it came about that the efforts of these authors met with success—limited as it was—where Seeliger and Neumann failed to make any headway, is to be explained in terms of the abandonment in the late 1920s of the idea that the universe must necessarily be static. The delay in appreciating the possibility of a Newtonian solution was perhaps an outcome of the relative success of the Einsteinian method and the fact that it proved to be so mathematically inviting. As we have already indicated, the most important conceptual advances in modern cosmology came with Einstein's General Theory, and this was well under way by the time Charlier and his followers were publishing their amended versions of

[21] Cf. Gustav Strömberg, *Ap. J.*, lxi (1925) 353 (= *Mt. Wilson Cont.*, no. 292). There was, he said, no correlation between distances and radial velocities. 'The only definite correlation is one between radial velocity and position in the sky, indicating a solar motion of 300 or 400 km/sec in about the same direction for both classes of objects' (viz. 'globular clusters' and 'non-galactic nebulae').

[22] *Meddel. Uppsala*, no. 30 (1927).

[23] For further details see Appendix, Note I.   [24] *Q.J.M.*, v (1934) 64.

[25] Viz. those with zero pressure. See McCrea and Milne, *Q.J.M.*, v (1934) 73. Cf. *W.S.*, 299–321.

[26] *ZS. Ap.*, xxxviii (1955) 95 and xl (1956) 81.

[27] *Rev. Mod. Phys.*, xxi (1949) 447. Cf. p. 363 *infra*.

Newtonian cosmology. We mentioned these writers at the outset not merely because they came first in point of time—their ideas were, after all, much longer in maturing than Einstein's—but because they also showed signs of an unusual intellectual daring. It is easy to speak of the infinite, as every theologian knows, but it is difficult to speak of it meaningfully.

CHAPTER 3

FIELD THEORIES AND THE WISH TO REPHRASE NEWTON'S THEORY OF GRAVITATION

IN THE first chapter we examined nineteenth-century nebular astronomy—a *sine qua non* of relativistic cosmology, but virtually without influence upon its early development. We next considered some early promptings of a truly cosmological kind, again with only slight influence on Einstein's thought. In what follows we shall consider the concept of the gravitational field and the extent to which he incorporated it in his General Theory. It is often said that one of the characteristics of this, which distinguishes it from much of subsequent physical theory, is Einstein's reversion to the intuitively appealing idea of *contiguity*, in explaining gravitation. In this he is said to have owed much to the development during the nineteenth century of the mechanics of continuous media. The matter is worth examining, for although contact *forces* are not an element of the General Theory of Relativity, yet the concept of the *field* associated with such forces might well have been responsible for the more elaborate concept of the so-called 'metrical field'. Although Einstein made a break with force-theories, we shall find that he can yet be historically regarded as a product of the movement to translate Newton's theory into such concepts as are found in Maxwell's theory of the electromagnetic field. This general movement, and the reactions within it, will be considered first.

It is interesting to speculate on the fate of Einstein's theory had it been no more than a philosophically acceptable picture of the Newtonian theory of gravitation. Such was the opposition to it in point of fact, that had it not been observationally confirmed in respects in which the older theory was in difficulties, it would in all likelihood have been rejected out of hand. By the end of the century any theory of gravitation which was to oust Newton's must be not only 'philosophically' sound: it must explain away the 'anomalies' in the motions of the planets. The first part of the present chapter will deal with some of the chief points in the development of the concept of the gravitational field, and especially the Maxwellian analogue of Newton's theory. In the second part, further attempts to discover the true nature of gravitation will be discussed (and especially the 'hydrodynamical analogies' which, one suspects, it was hoped might one day turn out to be more than analogies). In the last part of the chapter we shall consider reasons for believing that Newton's theory was not only philosophically unsound, but also untenable on purely scientific grounds.

1. Gravitation and the Field Concept

Action at a distance had often been a source of embarrassment to Newton and his followers. 'Hypotheses non fingo' was, after all, Newton's apology for offering no hypotheses as to the mode of operation of gravitation.[1] Newton had more than once denied that he supposed gravitation to be an essential property of matter,[2] although Roger Cotes, in his preface to the second edition of the *Principia*, speaks as though it were. If it is not an occult quality then, so Leibniz maintained, 'the attraction of bodies, properly so-called, is a miraculous thing'.[3] According to Leibniz, such attraction at a distance was unacceptable, being 'inexplicable, unintelligible, precarious, groundless and unexampled'. Like so many of his contemporaries, Leibniz was much happier explaining the motion of the planets in terms of vortices in a Cartesian aether, and had such a theory been worked out in detail with the success of Newton's, there can be little doubt as to which would have prevailed. Even Samuel Clarke, Newton's protagonist, felt compelled to admit that action at a distance is impossible. Leibniz, he wrote, would earn the thanks of the learned world if he could find a *mechanism* for gravitation. How strong was this general feeling, on the Continent in particular, can be seen from the fact that mathematicians there persisted in trying to use the vortex theory as a means to explaining away the inverse square law of gravitation. Thus in 1730 John Bernoulli was awarded a prize by the Académie des Sciences, for a derivation of Kepler's third law from a vortex hypothesis. As late as 1782 G. L. Le Sage postulated a corpuscular aether closely resembling Descartes's earliest version. With one or two additional hypotheses of a fairly innocuous but imprecise nature, he derived Newton's inverse square law of gravitation.[4] The vague mode of derivation could hardly serve to recommend it, and mathematicians of the latter half of the

[1] *Principia*, 2nd ed., General Scholium.
[2] *Opticks*, preface to second edition, 1717; letter to Bentley 1693, &c.
[3] Third letter to Clarke, 1716.
[4] *Berlin Mém.* (1784) 404. Le Sage is supposed to have spent sixty-three years working on this one problem. Newton had put forward a similar hypothesis, but with more caution (cf. *Opticks*, Query 21, &c.). Euler developed a theory of the same kind (*Opera Omnia*, 3rd ed., iii, pp. 4, 149) only to be reproved by Daniell Bernoulli for adhering to Cartesian principles. (Letter of 4 February 1744.)

eighteenth century were indeed too preoccupied with the development of Newton's dynamics to worry about its philosophical pedigree. When eventually new ground was broken in the form of a precise dynamics of a continuum, philosophical arguments as to the impossibility of forces at a distance were scarcely ever introduced.

Despite its immensely successful application, especially to celestial mechanics, Newtonian dynamics was hardly cast in a form suitable for the treatment of problems on elastic solids. In 1821 C. L. M. H. Navier treated solids as aggregations of minute particles whose interactions were effective only at short range.[5] Their particulate nature was not obvious from the form of Navier's final results, which were obtained by an averaging procedure. The forces between Navier's particles were 'Newtonian' in the sense that they did not depend on the mediation of any material between the particles.

Shortly after Navier's results were published, A. L. Cauchy, one of the great French analysts, considered the lack of evidence for a corpuscular view of matter and was led to discard it—and with it the idea of forces between non-contiguous material. Instead he looked upon matter as a real continuum, his forces being stresses across surfaces and spreading, not instantaneously, but with a finite velocity.

We must now turn aside for a moment from the concept of a mechanical continuum to that of a medium suited to a theory of optics. Both A. J. Fresnel[6] and Thomas Young[7] had suggested that light consists of a transverse vibration in a luminiferous medium which is capable of resisting distortion. Both applied the wave hypothesis, in various forms, with great success to most of the recognized problems of optics. Young, in his first paper, even pointed out that experiment might conceivably show this luminiferous aether to be identical with the 'undeniable' aether of electricity.

Working from the differential equations for the motion of an elastic solid, Cauchy supposed the luminiferous aether to be of the same general nature, and set himself the purely analytical task of deducing Fresnel's results. There are many ways of making these deductions, each involving specific boundary conditions and elastic properties of the medium. Cauchy paid little attention to these physical considerations, and his various solutions (c. 1830) were little more than mathematical exercises. P. G. Tait, many years afterwards, wrote to Sir William

---

[5] *Mém. Acad. Brux.*, vii (1827) 375.
[6] *Annales de Chimie* (2), i (1816) 239. See Fresnel's *Oevres* for later papers.
[7] *Phil. Trans.*, xc (1800) 106, &c. See Young's *Works* for later articles.

Hamilton: 'I should very much like to know your opinion of Cauchy's investigations in the Undulatory Theory—for I have found it possible by apparently legitimate uses of his methods to prove almost *anything*.'[8] George Green, pursuing a dynamical investigation of the conditions which could reasonably be supposed to exist at an interface during the reflexion of light, seems to have put Cauchy—and others—back on the empirical path, and over the following twenty to thirty years 'elastic-solid' theories of the aether multiplied.[9]

We may mention here that Riemann tells us that whilst reading Brewster's *Life of Newton*, he discovered the letter to Bentley in which Newton speaks unfavourably of the notion of action at a distance. Not surprisingly, therefore, Riemann attempted to give a mathematical formulation of the properties of an aether capable of transmitting optical, magnetic and electrical effects.[10] He did not get very far. Like Cauchy he failed to take into account the known properties of dielectric and magnetic media.

These lines of research were not alone responsible for what have since been termed 'field' theories of electromagnetism. For reasons of their obvious nature, Faraday preferred contact forces to Coulombian action at a distance, and his well-known use of the idea of lines of electric and magnetic forces conforms to this preference. Using the idea of tensions within, and mutual repulsions between, these lines of force, he was able to give a rough qualitative account of Coulomb's law.[11] William Thomson lent mathematical support to Faraday's general point of view and suggested a model for the propagation of electric forces, likening this both to the flux of heat from sources distributed in the same manner as the supposed electric particles, and to the propagation of displacements through an elastic solid.[12] Here were the first hints that there were at least two ways of arriving at the known formulae of static electricity.

[8] C. G. Knott, *The Life and Scientific Work of P. G. Tait* (Cambridge University Press, 1911) p. 122.

[9] Important names in this subject are those of F. Neumann, J. MacCullagh, G. Green, and Lord Kelvin. See *H.T. Ae. E.*, vol. i.

[10] Note of 1853 and paper of 1858 to Göttingen Academy. Both published posthumously in *Werke*, 2e Aufl., pp. 526 and 288.

[11] Faraday, it may be noted, attached great importance to the idea that *all* actions might be transmitted through the medium of contiguous particles, although in his 'Thoughts on Ray Vibrations' (*Phil. Mag.*, xxviii (1846) 345) he 'endeavours to dismiss the aether, but not the vibrations.' In this very informal lecture he spoke of a universe of atoms, each with a set of infinitely extended rays along which the vibrations could pass. These ideas are interesting enough, but they were not amongst those to be taken up by his followers.

[12] *Camb. and Dubl. Math. J.*, ii (1847) 61.

Faraday and Thomson having prepared the ground, J. Clerk Maxwell, who had spent several years working with analogies between almost all parts of physics, gave a mathematical treatment of the propagation of electrical and magnetic effects which closely resembled Cauchy's treatment of waves in elastic media.[13] 'His great object', wrote P. G. Tait, 'was to overturn the idea of action at a distance.'[14] Once Maxwell's genius had led him to postulate the existence of 'displacement currents', he saw that his revised system of equations had solutions which could be interpreted as representing progressive waves with a finite velocity of propagation—equal, in fact, to the ratio of the unit of charge measured electrostatically to its value measured magnetically. Rudolph Kohlrausch and W. Weber at about the same time (1856) found this ratio to be very nearly the same as the accepted value for the velocity of light.[15] The electromagnetic theory of light was thus inaugurated.

Maxwell's theory was inaugurated, but it was not readily accepted. It rested, to be sure, upon a conception of the aether which by the middle of the century was acceptable enough. Faraday had lent his authority to a view of the aether as something in a state of varying stress, these stresses being represented by lines of force. Contemporaries objected, however, to Maxwell's counting as true electric currents the rates of change in the electric displacement. As a mere piece of terminology, 'displacement currents' might have been accepted, but Maxwell claimed that displacement currents might be present even in the absence of ponderable matter, and that in all cases they produced magnetic fields in the same way as the conventional electric current. (In this way he removed an obvious lack of symmetry between the electrical and magnetic quantities and, at the same time, made the current vector circuital.) Neither Kelvin nor Helmholtz would at first subscribe to the displacement current 'paradox' and there were many who remained unconverted until 1887, when Hertz first detected non-visible electro-magnetic waves.

The concept of the field flourished, however, within the new scheme and has persisted, in one form or another, down to the present day.

---

[13] Maxwell's important work began in about 1855. A polished version of his theory of the electromagnetic field was given in 1864. See *Phil. Trans.*, clv (1865) 459.

[14] Weber's law of action at a distance between charged particles involved their velocities. Maxwell wrote: 'The mechanical difficulties, however, which are involved in the assumption . . . are such as to prevent me from considering this theory as an ultimate one. . . .' Maxwell endeavours 'to explain the action between distant bodies without assuming the existence of forces capable of acting directly at sensible distances'. (*Phil. Trans.*, clv (1865) 460.)

[15] Kirchhoff seems to have been the first to point out the fact.

There is unfortunately no standard meaning for the phrase 'field theory'. In a weak sense it is one which involves the continuous distribution in space of the values of some physical magnitude. At times one suspects that it is used of any theory the fundamental laws of which may be expressed as partial differential equations with spatial and temporal coordinates as independent variables. A more recent and extreme view is that what may be called 'corpuscular' phenomena must be derivable from such laws in any 'true field theory'. In the first and weakest sense we have the concept of a field as introduced into the classical theory of gravitation, the 'physical magnitude' being force. Maxwell's laws fall in with the requirement that it should be possible to embody them in partial differential equations. But so, likewise, can Poisson's (Newtonian) law. No doubt a stipulation would be generally added according to which no 'true field theory' can incorporate the idea of the instantaneous propagation of actions. (Poisson's equation may be applied to continua, but any change in the density of matter in one region is propagated instantaneously to distant regions.) With the growth of Maxwell's theory, something else was added. The idea grew up that the field has at least as much reality as the electric charges and magnetic poles with which it interacts. It may possess energy, both kinetic and potential, regardless of the presence of charge and pole. Did this not prove the reality of the field? Even today there are those who would admit this sort of argument. But the concept of an aether, which might seem to have been a necessary concomitant of it, has recently, together with the concept of absolute space, been characterized as superfluous and otherwise logically unsound. Now the *logical* problems associated with absolute space scarcely ever presented themselves to the man of the nineteenth century because his aether, if not exactly tangible, was to him a commonsensical requirement. It was simply regarded as the medium necessary for the propagation of all kinds of action, if the common-sense requirement of contiguity was to be satisfied. However illogical, it contributed much to the unofficial trappings which surround any theory and which make for its imaginative development. If Maxwell's theory was slow to find supporters, it was perhaps because his aether was a little too aetherial. As we saw, the reluctance to accept it was partly due to some of the unpalatable consequences of his theory of electric displacement currents, and this was a critical concept in the development of the general notion of the field. If the contiguity doctrine were to be retained, and if displacement current were to exist in the absence of ponderable matter, it was more important than ever that there should be some *immaterial* seat of electric actions. This could no doubt have been conceived as a medium with an atomic nature, but the greater part of nineteenth-century physics pointed to an all-pervading *continuum*, all regions of which are of significance to the remainder.

What Maxwell had achieved was to bring two groups of phenomena, electrical and magnetic, within a single scheme. To bring unity to the entire scientific scheme was an old ambition which now seemed to be within an ace of being realized.[16] Maxwell himself actually worked out the correspondence between his electromagnetic theory and Cauchy's dynamics of elastic media, his equivalences supplying a prototype for similar ones in later field theories.[17] Of these we are only concerned with the field theories of gravitation.

It was not long before attempts were being made to design an aether suitable for the representation of (Newtonian) gravitation. Coulomb's law of electrostatics can be regarded as a consequence of the differential laws

$$\varepsilon^{rst} E_{t,s} = 0,$$

and
$$e\, E^r_{,r} = 4\pi\sigma. \qquad (1)^{18}$$

The action of the entire surrounding medium can then be regarded as responsible for the force on a charge, in keeping with the ideas we have been discussing. Now Coulomb's law is a close analogue of Newton's law of gravitation. Might not gravitational effects be likewise ascribed to the action of the neighbouring medium, with differential expressions corresponding to those of electrostatics?

Maxwell, in the first satisfactory draft of his ideas on the electromagnetic field, had turned his attention to this problem, but only to meet with insuperable difficulties.[19] He began by remarking upon the fact that gravitating bodies attract each other whilst like electrically charged bodies (or like magnetic poles) repel. He argued by analogy with the magnetic case. If E were the intrinsic energy of a given field,

---

[16] Cf. the way in which Kepler had likened gravitation to magnetism, three centuries earlier.

[17] Thomson had already compared electric force to the displacement in an elastic solid—hence the term 'electrical displacement'.

[18] The condition must be added that the field E vanishes at an infinite distance from the charge. The first law is the Maxwell–Faraday law of induction, in the absence of magnetic fields. The second is 'Gauss's Theorem', where e is the dielectric constant and σ is the charge density.

[19] *Phil. Trans.*, clv (1865) 492.

R the resultant gravitational force per unit mass, and dV an element of volume, then

$$E = \text{const.} - \Sigma(1/8\pi)R^2 dV. \qquad (2)$$

From this result it is clear that, as the intrinsic energy of the field is 'essentially positive', at those places where there is no resultant gravitational force the intrinsic gravitational energy density per unit volume must be greater than $(1/8\pi) \cdot \bar{R}^2$, where $\bar{R}$ is the 'greatest possible value of the gravitating force in any part of the universe'. These consequences were unacceptable: the intrinsic energy of space was, it seemed, to be supposed very great indeed, dense bodies merely serving to diminish it. 'As I am unable to understand in what way a medium can possess such properties,' wrote Maxwell, 'I cannot go further in this direction in searching for the cause of gravitation.'

Over forty years later, Max Abraham looked into this question of a possible analogy with Maxwell's theory of electromagnetism, only to find further difficulties.[20] Taking the hypothesis that a vibrating material particle emits transverse gravitational waves with the velocity of light, it followed that once the gravitational field reached a region hitherto undisturbed it must there have the effect of diminishing the energy density. The energy must be supposed to stream towards the vibrating particle, the equilibrium of which is thus unstable.

The fact that, despite these obstacles, Abraham and others persisted in trying to use Maxwell's theory as the prototype of a truly satisfactory theory of gravitation, speaks very clearly of the almost ritual manner in which the concept of action at a distance was dismissed. It speaks also for the widespread misunderstanding of the role of a physical theory. In these terms one can perhaps better understand Einstein's strong feelings on the subject of the field concept. This all represents a great triumph for the kind of physical explanation Maxwell had offered. Only a few years before Maxwell's best work was completed, G. B. Airy could write: 'I contemplate [gravitation] as a relation between . . . two particles, and not as a relation between one particle and the space in which the other finds itself for the moment.'[21] These sentiments were shared by most of Airy's profession. Like most astronomers, he was so accustomed to this method of 'contemplating' gravitation, that the fact of his being able to do so was almost tantamount to a proof that gravity really did act as he supposed. Thirty years after the work

---

[20] *Rend. Lincei*, xx (1911) 678; *Phys. ZS.*, xiii (1912) 1, 4, 176, 310–11, 793; *Archiv. Math. Phys.*, xx (1913) 193.

[21] Bence Jones, *Life and Letters of Faraday*, vol. ii, p. 348 (London, 1870).

of Maxwell, those whose attitudes coincided with Airy's were in a minority.

The attempts to formulate gravitational theory by analogy with Maxwell's theory of electromagnetism came to nothing, and in 1880 P. G. Tait could say that 'the mechanism of gravitation is still to us, as it was to Newton, an absolute mystery'.[22] In this he was tacitly dismissing three other types of 'explanation' which were being investigated at the time, namely, those based on the revived ideas of first Boscovich and then Le Sage, and those based upon an analogy with hydrodynamics. We shall take the hydrodynamical analogies first.

2. GRAVITATION AND THE HYDRODYNAMICAL ANALOGUES

In 1869 Kirchhoff remarked that, under certain restrictions, the apparent forces which are created between two rigid and infinitely thin rings, when they move in a frictionless, infinite, and incompressible fluid, are equal to the forces with which the rings would act on each other were certain constant electrical currents to circulate in them.[23]

In a series of papers Thomson pursued this line of thought.[24] In the course of enquiring into the behaviour of two solids, one of them oscillating, and both of them surrounded by an incompressible fluid, he found that under certain special conditions there would be a force, as of attraction, between the solids. He does not appear to have followed up the possibilities of this curious theorem and yet, strangely enough, the Swede C. A. Bjerknes was, quite independently, being led to similar conclusions.

Bjerknes, using Hamilton's Principle, in 1870 investigated the movement of a solid of revolution in a fluid and found that, when there are no external forces, the components of its velocity can be given by elliptic functions of the time. He also considered the movement of rigid bodies of general shape and of arbitrary mass distribution. In 1871 he presented to the Royal Scientific Society of Christiania a memoir on the simultaneous displacement of spherical bodies in an infinite and incompressible fluid. Over the next three years he made a thorough study of this and its associated problems. After generalizing an earlier solution (by Schering) of the equations of motion of an

[22] In a lecture at Glasgow. (*Life and Scientific Work of P. G. Tait*, p. 296.)
[23] Equal, in fact, to the circulation through the aperture. See *Berlin Monats.* (1869) 881 and *Crelle's J.*, lxxi (1869) 263.
[24] *Papers in Electricity and Magnetism* (1870–2), sections 573, 733, 751. Cf. his letter to Guthrie, *Phil. Mag.*, xli (1871) 427.

ellipsoid in an infinite and incompressible fluid, he considered cases involving translations, rotations and changes in the shape of the ellipsoid. By 1874 he realized that in the case of pulsating solids there was the possibility of an analogy with the inverse square law of gravitation. A year later he came across the work of Kirchhoff and Thomson through reading a new textbook of mechanics, written by Kirchhoff in 1874. Again independently, Kirchhoff had given what was, in effect, the substance of Bjerknes' memoir of 1871.[25]

Bjerknes continued his work for the next decade. In 1877 he presented many of his new findings.[26] He pointed out that the vibrations of a spherical body are capable of making a second spherical body oscillate when both are immersed in an incompressible fluid. If the spheres vibrate with the same period then forces of the second and higher powers of the ratio of their radii to their mean separation will be obtained: for these forces the principle of equality of action and reaction holds good. He found that if the spheres were to pulsate in phase, then forces, as it were of interaction between the spheres, would be produced inversely proportional to the separation of their centres. If the phases of the pulsations were to differ by an odd multiple of π, the forces would be unchanged in magnitude but would be forces of apparent repulsion. For a phase difference equal to an odd multiple of $\pi/2$ there would be no interaction.[27]

At the Paris exhibition of 1881 these effects were convincingly demonstrated.[28]

A. H. Leahy extended several of Bjerknes' results, considering the case of spheres pulsating in an elastic medium in which a disturbance is created of wavelength greater than four times their separation.[29] The inverse square law was still found to hold, but in the case of like phases the law became a law of repulsion. Conversely, it became a law of attraction when the pulsations were exactly out of phase.

Leahy drew attention to some of the difficulties which an explanation of gravitation, based upon these ideas, would encounter. On the assumption that motion or stress was transmitted through an intervening

---

[25] Noticing what Bjerknes did not, namely that the series for velocity potential was convergent. Bjerknes' work is to be found in *Christiania Skrifter*, vols. for 1863, 1871, and 1875.

[26] *Repertorium Math.*, i (1877) 268.

[27] W. M. Hicks developed a simple method of determining rigorously the actions on the two spheres by a 'method of mass-images'. *Proc. Camb. Phil. Soc.*, iii (1879) 276 and iv (1880) 29.

[28] See the *Journal of Telegraph Engineers* for 1882.

[29] *Trans. Camb. Phil. Soc.*, xiv (1884–5) 45 and 188.

medium, then either this was the same aether as that which served for the propagation of light—and was then elastic—or it was not. If it were elastic, then there would be repulsion between bodies at distances greater than a quarter of the wavelength of the disturbance, and there was at that time no evidence of repulsion between gravitating masses. If it were, as Bjerknes had said, incompressible, then all pulsations would be instantaneously diffused through space. This possibility Leahy all but dismissed, arguing that a fluid aether 'can hardly be supposed to be absolutely incompressible', and falling thereby into the error of following an analogy too closely.

In the second of his articles Leahy developed a theory of the mutual action of oscillatory twists in tubular surfaces within an electric medium, using this as a model for magnetic and electrical action. The fluid gravitational aether was by this time generally forgotten. If it had never been received with much enthusiasm that was probably because it gave rise to more problems than it solved. A. Korn tried to revive it when, in 1898, he modified Bjerknes' theory of spheres pulsating in an incompressible fluid, just as Hicks had done twenty years before. He made the amplitude of the pulsations of the spheres proportional to their masses, and by this means the occurrences of the masses in Newton's law of gravitation was 'explained'.

Anyone offering an explanation of the Newtonian theory based on Bjerknes' principles is committed to the view that all atoms pulsate in phase. For this view there was, of course, no independent evidence. Hicks had hoped that he might explain gravitation in terms of an interaction between the 'vortex atoms', the theory of which Thomson developed. This was thought to explain (amongst other things) the conservation of matter, its spectroscopic properties and its elasticity.[30] Since one consequence of the Hicks-Korn theory was that the time of pulsation of all particles is the same, it was therefore hardly capable of explaining the observed spectroscopic properties of the vortex atoms (or indeed of any other kind of atom) so long as these were identified with gravitating particles.

Vortex atoms were first conceived several years before the principal attempts to explain gravitation in hydrodynamical terms. The idea, most clearly elaborated by Thomson, was first distinctly stated by Sir Humphry Davy and J. P. Joule. The idea owed much to W. J. M. Rankine, who aimed to reduce elasticity and thermal expansion to 'mechanical principles' by means of 'an hypothesis called that of

[30] See, for example, *Phil. Mag.*, xxxiv (1867) 15; *P.R.S. Ed.*, vi (1869) 94.

Molecular Vortices'.[31] The hypothesis was 'that each atom of matter consists of a nucleus or central point, enveloped by an elastic atmosphere which is retained in its position by attractive forces, and the elasticity due to heat arises from the centrifugal force of those atmospheres, revolving or oscillating about their nuclei'. Quantity of heat was the *vis viva* of the molecular revolutions or oscillations'.

Only a few years before discovering a paper by Helmholtz on the theory of vortex rings, Thomson had firmly rejected the atomic hypothesis. Regarding the world as constituted of atomic vortices in a perfect and homogeneous fluid was, however, a short step from believing it to be a continuum. It was a step Thomson took after having been greatly impressed by some experiments carried out by P. G. Tait on smoke rings.[32] The rings had many of those qualities which an atom was supposed to possess, above all that of permanence. Thomson now sought out Helmholtz's work. To generate or destroy Helmholtz's *Wirbelbewegung* in a perfect fluid could only be, as Thomson put it, 'an act of creative power'. As Hicks later added, 'the permanence of a vortex filament with its infinite flexibility, its fundamental simplicity, with its potential capacity for complexity, struck the scientific imagination as the thing which was wanted'.[33] In addition to its permanence the vortex atom could explain the spectroscopic properties of matter in so far as it could be shown to have definite fundamental modes of vibration. It had to meet with two objections. As Maxwell saw, it could not account for the variation in density between different substances. Vortex rings have, moreover, the property that as their velocity decreases their energy increases. This state of affairs is clearly incompatible with that envisaged by the authors of the kinetic theory of gases, a theory which few were prepared to sacrifice.

But Thomson never managed to prove the stability of the vortex atoms,[34] and despairing of ever using them to explain gravitation, he resuscitated Le Sage's theory. The Thomson-Le Sage theory will shortly be outlined.

The vortex atoms required the existence of some sort of fluid aether. If this were to be identified with the electromagnetic aether and to carry transverse vibrations, then some sort of quasi-elasticity must be

---

[31] *P.R.S. Ed.*, ii (1850) 275. On an even more speculative note, one could again refer to Swedenborg's writings, in particular the *Minor Principia*.

[32] Tait was demonstrating some theorems of Helmholtz, late in 1866.

[33] *B.A. Rp.* (1895) 595.

[34] The problem is noteworthy as having led mathematicians to some of the topological properties of knots.

inherent in it. The conception of a vortex atom had given considerable impetus to the general theory of vortex motion, and there eventually emerged the conception of the aether as a 'vortex sponge'—a perfect fluid with an overall elasticity which it owed to its turbulent condition. This analogy, as developed by Thomson, G. F. Fitzgerald, and Hicks, for example, must be regarded as one of the most mathematically ingenious models ever provided for a physical theory. The phenomenon of gravitation, however, still resisted incorporation in such a scheme.[35]

There is one further class of hydrodynamical models of gravitational effects which deserves mention. Thomson's vortex atoms had the characteristic that they always consisted of the same elements as the aether. A different point of view was discussed by Karl Pearson, who proposed that an atom be regarded as a point source of aether or, as he called it, an 'ether squirt'.[36] Thomson, following Kirchhoff's lead, had in 1870 remarked that a bar magnet may be compared to a straight tube immersed in a perfect fluid, the fluid passing through the tube outside which its motion was directed along the lines of magnetic force.[37] The analogy was not a very good one, for the like ends of such 'magnets' attract, whilst the unlike ends repel. Their laws of interaction were, on the other hand, identical in most respects to those of magnets.

Pearson seized upon the effect as providing a means of explaining gravitational phenomena. Thomson had shown that two sources or sinks of incompressible liquid will attract each other in accordance with the inverse square law, but he no doubt shrank from the idea (necessary to an analogy with gravitation) of a source of fluid seemingly isolated from any supply. Pearson was immune from such fears. 'We shall be ignorant,' he claimed, '. . . just as long as we try to find realities corresponding to geometrical ideals.' His 'ideals' had at least simplicity to recommend them. For example, by making the rate of supply of aether at a point proportional to the mass of the atom at that point, he was able to explain the mass terms in Newton's law.

Pearson would probably have objected to this last statement. For him, the all-important question was *how* bodies move. Rather than providing a mechanistic basis for the Newtonian theory he probably

[35] No one appears to have extended Bjerknes' results to a vortex sponge medium. Had this been done, perhaps some of Leahy's objections to the vortex atoms might have been met. M. J. M. Hill's discovery of spherical vortices likewise appears to have remained unexploited (*Phil. Trans.*, 1894), perhaps because the physicists of the new generation were unsympathetic towards the quest for an 'explanation' of the true nature of gravitation.

[36] *Am. J. Math.*, xiii (1891) 309; *P.L.M.S.* (1), xx (1888) 38, 297.

[37] loc. cit.

thought himself to be supplying a superior theory, superior because of its capacity for unifying physics. By such theoretical devices as the introduction of a periodic variation in the rate of 'squirting', he could infer 'many of the properties of electromagnetism, light, cohesion, and chemical action'. To most of his contemporaries his scheme was a sheer romance. Assuming the constancy of the quantity of matter in the universe, Pearson's theory was held to imply the existence of *aether sinks*, which corresponded to negative matter. (It seems that this is a consequence of the conservation of aether, rather than of the conservation of matter; but, as we have seen, the mass density was related to the rate of supply of aether.) The fact that Carnelley had postulated an element of negative atomic weight could hardly be considered to favour the theory, since it would have disappeared from our neighbourhood long ago, there being between source and sink a mutual repulsion. Sir William Schuster light-heartedly emphasized the alternatives.[38] Matter might be an 'endless stream constantly renewing itself and pushing forward the boundaries of our universe', or, as he prophetically indicated, worlds may have formed from 'anti-matter' so far indistinguishable from our own. 'Some day,' he wrote, 'we may detect a mutual repulsion between different groups of stars and obtain a sound footing for what is only a random flight of the imagination.'[39] But the main question for most people was: What happens to the aether after entering at the anti-matter point and before emerging from the source atoms? Even Pearson had to acknowledge the possibility, albeit remote, of his 'conceptual atoms becoming perceptual ones'. Where did the aether go? Pearson hinted that it might well pass into a fourth dimension,[40] an idea which he owed to W. K. Clifford. For Schuster this was no more than 'a Holiday Dream'. For the majority of those who troubled to think about it, it was even less.

Paul Tannery was another who failed to make a lasting impression on his public—despite, or perhaps because of, his generally enlightened attitude towards the purpose of scientific thought. He began, in no spirit of optimism, by seeking to substitute for attractions and repulsions acting at a distance between material molecules the action upon the molecules of an intervening fluid medium.[41] By making many wild approximations for one or two simple cases, he found a number of possible laws, one of which was—not surprisingly—the inverse square

[38] *Nat.*, lviii (1898) 367, 618. [39] See also A. Föppl, *Berlin Sitz.*, (1897) 93.
[40] K. Pearson, *The Grammar of Science*, Everyman edition, p. 229.
[41] *J. Phys.*, vi (1877) 242. Also *Bordeaux Mém.*, ii (1878) 95.

law of attraction. This was all very much in the current fashion, but when he had connected gravitation with '*les radiations astrales*', he did not hesitate to draw the conclusion that the law of gravitation must change with time to the point when these radiations cease at the limit of the world's entropy. Sixty years later the logical possibility of laws which vary in time was to be seriously debated.

3. The Le Sage-Thomson Explanation of Gravitation

Not all attempts to provide a fundamental explanation of the phenomenon of gravitation were made in terms of the new theories of continua. Of the important attempts to construct a different sort of mechanical explanation of the Newtonian theory of gravitation that of Le Sage was amongst the first,[42] although he makes acknowledgements to both Nicolas Fatio[43] and one Redeker.[44] Le Sage's theory does not appear to have made any impression on his contemporaries, but it achieved some notoriety when Thomson revived it almost a century after its publication.[45]

Le Sage supposed that all ponderable matter was composed of indivisible particles which were of the nature of empty cages of various regular shapes. Amongst these, and indeed throughout all space, certain corpuscles were moving at great speeds. The only matter in the particles was in the bars themselves, and these were so thin that any given corpuscle intercepted one of them on an average only once in several hundred thousand years. The '*corpuscules ultramondains*' were inelastic, but rather than adhere to each other on collision they conveniently slipped apart (although their average velocities were thereby reduced). Any two particles of ordinary matter would clearly screen each other from corpuscular bombardment, and Le Sage showed that this would give rise to forces of attraction proportional to the inverse square of the separation.[46]

Le Sage's greatest difficulty was that of the diminution of the mean corpuscular velocity (and hence of the gravitational 'constant') with time, owing to the inelastic impacts.[47] This objection he was reduced to

---

[42] Given in a paper entitled 'Lucrèce Newtonien' presented to the Royal Berlin Academy in 1782. Published in *Berlin Mém.* (1784).
[43] Letter to Leibniz, 30 March 1694. [44] Latin dissertation, 1736.
[45] A. Picart gave an explanation of gravitation which, although without acknowledgement, is essentially that of Le Sage. *C.R.*, lxxxiii (1876) 577.
[46] See Books i and ii of Le Sage's posthumous *Traité de Physique Mécanique* ed. Prévost (Genève et Paris, 1818).
[47] This must be the earliest occasion on which the possibility was considered.

dismissing as unimportant '*pourvu que cet obstacle ne contribue pas à faire finir le monde plus promptement qu'il n'auroit pas fini sans lui*'.[48]

Now the problems of the high temperatures which would inevitably be generated at collision between corpuscles and matter, and of the diminution in the forces of gravitation with time, were solved by Thomson in a most ingenious way.[49] He supposed that despite an all-round diminution in velocity of translation, the particles must carry away the same dynamical energy as they brought, the loss in kinetic energy of translation being made up by the vibrational and rotational energy of the atoms.[50]

Many other writers showed an interest in these questions[51] but in 1898 C. C. Farr presented a more telling objection to the principles used.[52] The theory can only predict forces proportional to the individual masses of two gravitating particles if the matter of which they are composed has a sufficiently open structure. W. Nernst's experimental findings on intermolecular spaces in liquids showed that Le Sage's corpuscles would be incapable of penetrating far into liquids. Not only should Newton's law be abandoned in hydrodynamics, but it should also be dependent on temperature and even vary with the physical state of the gravitating material.

Before finally passing into oblivion, Le Sage's theory was transformed almost beyond recognition by H. A. Lorentz.[53] Whereas in the mid-century a mechanical basis for electromagnetic actions was sought, Lorentz now attempted, conversely, to explain gravitation in terms of electromagnetism. It was known from Maxwell's theory that electromagnetic waves were capable of exerting pressures against bodies in their path. If such waves were to replace Le Sage's corpuscles, Farr's objections might be overcome by supposing the waves to be of very high frequency and hence highly penetrating. Lorentz realized, however, that electromagnetic energy must persistently disappear on this theory and therefore he discarded it in favour of an idea previously developed by O. F. Mossotti,[54] W. Weber, and F. Zöllner.[55] All gravitating matter was supposed to be an aggregation of molecules, these comprising groups of negatively charged corpuscles. If the force of attraction between corpuscles of opposite charge is somewhat greater than that of

[48] op. cit. [49] *P.R.S. Ed.*, vii (1872) 577.
[50] He owed the idea to Clausius's hypotheses on the energy of gas molecules.
[51] e.g., S. J. Preston's thesis on the subject (Ph.D. Munich, 1895).
[52] *N.Z.I.T.*, xxx (1898) 118. [53] *Proc. Acad. Amst.*, ii (1900) 559.
[54] *Sur les forces qui régissent la constitution intérieure des corps* (Turin, 1836).
[55] *Erklärung der universellen Gravitation* (Leipzig, 1882).

repulsion between corpuscles of like charge, then the gravitational attraction between particles of neutral matter could be accounted for on the supposition that the fractional discrepancy between these electrical forces is of the order of 10^{-35}.[56]

Although these hypotheses were proposed very tentatively, and were entirely without independent support, they found several followers in their day. R. Gans at first favoured the principles involved, on the basis of which he found that gravitational equations of the Maxwellian type may be obtained.[57] The energy flux of the gravitational field can then be expressed as a vector corresponding to the Poynting vector, but having the opposite sense. Gans found that any acceleration which a neutral particle possessed would, in consequence, be *increased* by the gravitational radiation, and that its equilibrium would thus be unstable.[58] The theory never recovered from this setback: as it happened it was soon to be abandoned with the advent of Einstein's General Theory.

The idea that all scientific problems could be reduced to problems concerned with some variety of continuum by no means held a monopoly. There was, for example, a vogue for the doctrines of R. G. Boscovich, who explained all physical effects in terms of action at a distance between point atoms.[59] J. G. McVigar was perhaps responsible for reviving Boscovich's principles:[60] at any rate Thomson in 1899 cautiously observed that they suggested several problems pertinent to 'the real molecular structure of matter'.[61] At small distances Boscovich had supposed there to be a repulsion between atoms which was infinite at infinitely small distances and which changed to an attraction at measurable distances. On these grounds he was able to explain cohesion between particles of matter, the mutual pressure between bodies in contact, and even 'chemical affinity'. Thomson investigated at great length the equilibrium conditions of various groupings of Boscovichian atoms, and even made use of the implied possibility of the mutual penetration of two pieces of matter. Despite all this, he was never at home with 'the most fantastic of paradoxes, *contact does not exist*'. Like most of his generation, he preferred the 'not unnatural dogma . . . *matter cannot act where it is not*'.

---

[56] It is easy to see that the constant in the Newtonian law would differ as between the cases of attraction between charged matter on the one hand and neutral matter on the other.

[57] *Phys. ZS.* (1905) 803.   [58] *H. Weber, Festschrift* (1912), p. 75.

[59] *Theoria Philosophiae Naturalis* (Vienna, 1758 and Venice, 1763).

[60] *Proc. Phil. Soc. Glasgow*, iv (1860) 52. (Written in 1857.) Cf. D. Cipoletti, *Rivista Sci.-Ind.*, iv (1872) 139, 174, 252, 344.

[61] *B.A. Rp.* (1889) 494.

4. Reasons for Wishing to Revise the Newtonian Theory. Reality and the Field Theories

Once again Tait might have affirmed that the mechanism of gravitation was as much of a mystery as it had been to Newton. Of the many attempts to rephrase Newton's theory of gravitation, how many were meant to represent the 'true mechanism' of gravitation? There is no doubt that this was an almost universal ambition, but to leave the matter there would be to misrepresent the case. These writers were as conscious as any of our contemporaries of the value of creating analogies and models in assisting theoretical advance. They were looking for more than a transcription of one theory—perhaps one which was unsatisfactory because of its introduction of action at a distance—into an intuitively more appealing one. They were in fact usually aiming at a *unification* of many scientific modes of expression.[62] Throughout the century it seemed that they might well achieve both objects at one and the same time. The theory of vortex atoms, for example, at first seemed capable of explaining much of what was known of light, electricity, magnetism, heat, and gravitation, not to mention the fundamental structure of all matter.

But neither a unification of the several parts of physics nor the search for analogies (either for their own sake or for the sake of their suggestiveness) provided the main inspiration for Maxwell, Thomson, Bjerknes, and the rest. They were, in most cases, seeking the true cause of gravitation and usually confessed as much. By the end of the century the general view was almost surely that the most conspicuous clue to the mystery was the discovery of the electromagnetic aether. The physical properties of this may have been something of a problem, but of its *reality* the nineteenth century was never in doubt. The sceptical were merely asked what became of the light from the Sun during the eight minutes of its journey. 'When they consider that, they observe how necessary the ether is.'[63] More explicitly, the reality of the *field* was imagined to be a consequence of the principles that energy is conserved, that it is propagated with finite velocity, that it is itself real and that the field in question is its seat of action. Such 'self-evident principles' as these died hard, and when the notion of an aether was at last generally thought to have been discredited (in the 1920s), several

---

[62] The full title of Boscovich's book is worth quoting in this connexion: *Theoria Philosophiae Naturalis Redacta ad Unicam Legem Virium in Natura Existentium*.

[63] G. F. Fitzgerald, *B.A. Rp.* (1888) 557. The argument is frequently found.

of the older physicists refused to proscribe an old friend: for was it not obvious 'how necessary the ether is'?

We recall that the gravitational aether was often called upon in the first place merely to avoid the supposed absurdity of action at a distance. A gradual and subtle transfer of meaning now led some to say that it was the gravitational field, first cousin of the aether, which held the real responsibility for saving us from this 'absurdity'. The whole question was, in addition, confused with the causality idea. This mistaken confusion has a long history which it is not possible to discuss here. Suffice it to say that Hume was probably above all responsible for emphasizing that the causal principle requires spatial contiguity of the 'links' in the causal chain. This is not essential to the main causal concepts, namely of productive determination and lawfulness, and the concept of action at a distance can no doubt be forced into the straitjacket of an analysis in terms of cause and effect. Hume was *not*, however, of the opinion that science is the search for efficient causes, any more than was Berkeley. Like Locke, the third member of the Holy Trinity of British empiricism, both seem to have thought that if science is to pursue a programme of conceptual reduction, there is not much to choose between action at a distance on the one hand, and elasticity, cohesion, contact pressure and impact on the other. In both respects, these philosophers were neither typical nor influential. Probably no exclusively philosophical writer of this period influenced gravitational theory in an important way. This includes Kant, who argued for the plausibility of action at a distance, notwithstanding his requirement of continuity: phenomena might be continuous, and yet be accounted for by a theory admitting discontinuity.

Just as the physicist of the first two decades of this century began to regard the field quantities appearing in Maxwell's equations as being almost as substantial as they were elementary, so, later on, the so-called gravitational field of the General Theory of Relativity was to be looked upon as a real entity. The field, according to Einstein and Infeld, is not merely a mathematical device for describing the motion of bodies but is 'the ultimate reality'. Perhaps the reason for this attitude is to be found in the dislike of a dualism of matter and field.[64] 'Field', by the

---

[64] Einstein, in the early formulation of his General Theory, seems to have been scarcely conscious of the issue. There is no qualification added to his remark that 'we make a distinction between "gravitational field" and "matter" in this way, that we denote everything but the gravitational field as "matter"' (*Ann. Phys.* Lpz., xlix (1926) 769). But he found the dualism of field and particle 'disturbing to any systematic mind' when, later, he was faced with the question of whether or not particles should be allowed to introduce singularities into his continuum.

beginning of this century, had come to mean something with a twofold relation to matter: not only was it created and modified by matter, but it acted upon matter. For long it had seemed desirable to eliminate this dualism. Boscovich in a sense avoided it with his point-atoms, each penetrable and yet surrounded by an infinite sphere of influence. Faraday, whose reputation was considerably greater than that of Boscovich, at one time held very similar views.[65] Sir Joseph Larmor, prompted by the need to make the aether not only a medium through which energy is transmitted across space, but also 'a receptacle of very high energy densities', conceived electrons as 'nuclei of strain' in the medium.[66] He tentatively proposed that electrons might have *vacuous* nuclei and hinted that 'by orbital motion of electrons in the atom' he was able to derive a 'representation of an atom as a fluid vortex'. Whether the field was to be looked upon as derived from matter, or whether material objects were to be regarded as point singularities in the field, was a question they never asked; but as time went on and people became aware of the field-particle dualism, the question of primacy arose. Here, as elsewhere, the theory of electromagnetism was a source of great persuasion. And amongst other things, it persuaded natural philosophers that the mechanism of gravitation was an even greater mystery than the Newtonians, without this dualism, had realized. 'Mechanism' is perhaps the wrong word in the new context. For three centuries, a mechanistic natural philosophy had prevailed. Action at a distance, as found in Newton's theory, had for long been looked upon as one of the principal flaws in the most successful elaboration of this philosophy. It is somewhat ironical that the notions of aether and field, which were largely developed as a means of removing this flaw, should at last threaten to displace the very mechanistic outlook which it was originally hoped they would reinforce.

5. THE ATTEMPTS TO AMEND NEWTON'S LAW

Although we shall fail to discern more than a shadow of the older field-concepts in the General Theory of Relativity, the subjects discussed so far in this chapter are not without significance for modern cosmology—which almost invariably begins with questions of gravitation. Einstein's theory did not occur spontaneously: there was a continuity in the history of the displacement of Newton's theory, and those theories which are expressly concerned with the gravitational

---

[65] *Phil. Mag.*, xxviii (1846) 345.
[66] *Phil. Trans.*, clxxxv (1894) 719 ff. See especially p. 810.

field play their part in it. But quite apart from a philosophical uneasiness in the presence of, for example, the idea of action at a distance, it was believed that other flaws had been found in Newton's theory. There were two sorts of reason for thinking that this was so. The first was that Newton's law appeared to be unsuited to an explanation of certain 'anomalies' of planetary motions. This resulted in many attempts to make small amendments to it. The work on the 'anomalies' is important to cosmology, being contingent upon the empirical basis of any satisfactory theory of gravitation, and thus of gravitational cosmology. The second, 'cosmological', reason for dissatisfaction with Newton's theory has aleady been discussed in Chapter 2. The first reason will now be considered at some length.

Anyone who hoped to cast a new theory of gravitation in the form of Maxwell's electromagnetic field equations was committed to forces which could not be supposed to act simultaneously over finite distances. Certainly, according to R. A. Proctor, most astronomers thought of gravitation as acting instantaneously. The question of whether or not it did so was nevertheless generally regarded as meaningful.[67] There was at least one person who, even at the outset of the nineteenth century, had considered the possibility that gravitation is propagated with a finite velocity. Laplace took the hypothesis that gravitation is produced by the impulse of a fluid directed towards the centre of the attracting body.[68] He proceeded to derive an expression for the secular acceleration of the Moon's mean motion arising from this impulse, using an expression which he had first obtained as giving the effect of the impulse of the Sun's light upon the planets.[69] His expression for the new effect can be written as

$$\frac{3}{2}\frac{a}{a'}\,(n/N)^3\,N^2 t^2\,(\overline{V}/c)\,(c/V). \tag{3}$$

(V is the velocity of the fluid; a and a' are the mean distances of the Earth from Moon and Sun respectively; $\overline{V}/c$ is the ordinary constant of aberration; t is the time in Julian years; Nt is the Earth's mean sidereal motion about the Sun and nt is the corresponding expression for the Moon's motion about the Earth.)

[67] J. Waterton, *Phil. Mag.*, xxxiv (1867) 55, seriously considers the legitimacy of the question before he decides that 'the power acts with a velocity that is practically infinite', and that therefore the acceleration of a body is virtually independent of its velocity.

[68] *Mécanique céleste*, Bk. x, ch. vii, paras. 20–2.

[69] The secular acceleration of the mean motion of the Moon had been discovered by Halley in 1693, using Ptolemaic and Arab eclipse records. (*Phil. Trans.*, xvii (1693) 913.)

NEWTON'S THEORY OF GRAVITATION

Since the observed value of the Moon's secular acceleration could be quoted by Laplace as 31″·424757 per Julian century, a lower limit could be stated for (V/c), the only term the value of which was not known independently. The velocity of the gravitational fluid could not, it seemed, be less than 7×10^6 times as great as the velocity of light. It seemed, indeed, that it must be considerably greater than this value, for most of the observed secular equation could be accounted for on far surer grounds than these. Laplace's conclusion was, for his contemporaries, the last word on the matter. 'Mathematicians may therefore suppose, as they have done hitherto, that the velocity of the gravitating fluid is infinite.'

An interesting feature of Laplace's hypothesis is that it can be looked upon as involving a modification of Newton's law of gravitation. This is brought about by introducing a retarding force tangential to the orbit of the attracted body, proportional both to its velocity and to the Newtonian gravitational force. It would be easy to attach too great an importance to these ideas. There is nothing to suggest that Laplace thought himself to be challenging the Newtonian law. Even the original hypothesis he treated lightly, introducing it as one of several, and hoping that one or more of them would throw some light upon what, as we saw, he thought to be a secular acceleration of the Moon's motion. It is more than likely that the hypothesis was suggested to him by his investigation of the effect on the planetary motion of the pressure of the Sun's light. There is no evidence that he ever regarded it as more than a possibility, and after he had come nearer to the solution of the problem of the Moon's motion by other means, the hypothesis was effectively forgotten.[70]

It might be added, in parenthesis, that it is not difficult to unearth 'laws of gravitation' apparently rivalling Newton's in the century after his death. (We discount the work of the Cartesians who—according to Voltaire at least—like Mairan and the Cassini, shared with the poles the honour of having been flattened by Maupertuis.) Maupertuis, for example, often took an inverse nth power law and chose to specialize his results at the end of his working. Henry Cavendish later did the same thing for the electrical law of force. There was no challenge to Newton here, nor was any implied by A. C. Clairaut who, in his *Théorie*

---

[70] The acceleration in the motion turned out to be a periodic, and not a secular quantity. Indeed, Laplace had earlier proved an extremely important theorem of celestial mechanics, namely, that the mean motion of any member of a system of uniform and perfectly spherical planets cannot experience a secular acceleration as the result of mutual attractions.

de la figure de la terre (1743), found the same oblateness for the Earth whatever the law of attraction. Indeed, Newton himself provided a precedent; for, in proposition 80 of Book xii of the *Principia*, he considered a law of attraction which is any function of the distance. Cavendish found that only if the law of electrical attraction contains alone an inverse square power-term in the distance can a particle inside a spherical shell be in equilibrium.[71] Laplace found the same result for the case of a gravitating shell. Relaxing the requirement of equilibrium, he found that the following law makes the resultant attraction the same as that were the mass of the shell concentrated at its centre:

$$F = Ar + B/r^2. \qquad (4)[72]$$

No doubt there were other examples than these, but it seems likely that few of them were meant as more than a mathematical exercise. Certainly there was no effective challenge to Newton before our own century, although by the middle of the last century the theories of electromagnetism had already begun to show how some of Newton's ideas might be undermined. They too provided inspiration for another group of challenges in the last years of the nineteenth century.

The first of these came in 1870 when G. Holzmuller asked whether Newton's law might not be modified in much the same way as that in which Weber, in 1846, had modified Coulomb's law for electrical charges.[73] Holzmuller proposed, for reasons which may seem more cogent than Laplace's, that the expression for the gravitational force between two particles should contain a term in the radial component of their relative velocity. If γ is the gravitational constant and r the distance of separation of masses m_1 and m_2, then he suggested that the gravitational force between them should be taken in the form

$$F = (\gamma m_1 m_2/r^2)\{1 + (2r/h^2)\,r'' - (1/h^2)\,r'^2\}, \qquad (5)$$

where h was the velocity of propagation of gravitation. This term he took to be a constant of the same order as the velocity of light.

F. Tisserand, two years later, took the same law of force, applying it to the problem of the motion of the planets.[74] Realizing that the new

---

[71] *Phil. Trans.*, lxi (1772) 584.
[72] *Mécanique céleste*, Bk. ii, ch. xii.
[73] *ZS. Math. Phys.* (1870) 69. (Weber proposed as his law of force
$$F = (e_1 e_2/r^2)\{1 + (r/c^2)r'' - (1/2c^2)\,r'^2\}$$
(with obvious notation). The differential equation of motion of a charge moving under such a central force was integrated by Seegers in 1864, using elliptic integrals.)
[74] *C.R.*, lxxv (1872) 760; cf. J. Bertrand, *Leçons sur la théorie mathématique de l'électricité* (Paris, 1890) p. 183.

law could be integrated rigorously using elliptic functions, he preferred the simpler expedient of writing it in the form

$$F = (\gamma m_1 m_2/r^2) + F', \tag{6}$$

regarding F' as a perturbing force. Tisserand made two alternative hypotheses as to the value of h, namely, that it was equal to the velocity of light and that it was equal to a constant determined by the experiments of Weber in conjunction with his electrical law. He was led to expect that in either case the perturbations of the elements would be null or imperceptibly periodic with the exception of that in the longitude of perihelion, which contained a secular part. For the case of Mercury, such a secular motion had been observed (38″ per century over and above that which could be explained on the grounds of ordinary perturbations by the remaining planets, according to LeVerrier's calculation),[75] but Tisserand's theory could not account for more than about a third of the unexplained part of observed motion.

Weber's was not the only electrodynamic law to serve as a guide for a new theory of gravitation. Riemann had proposed a law differing from Weber's in that the relative velocity of the two charges replaced the radial component of this velocity. Maurice Lévy in 1890 found that, by combining this law with Weber's in an *ad hoc* manner, the observed anomaly could be obtained by calculation.[76] Needless to say, this approach to the problem was recognized for what it was worth, not least by its author. In the same year Joseph Bertrand[77] and Tisserand[78] took yet another law, ascribed to Gauss, namely,

$$F = (\gamma m_1 m_2/r^2) \left\{ 1 + \frac{1}{h^2}(2u^2 - 3r'^2) \right\} \tag{7}$$

(u being the relative velocity of the two masses).
They now obtained, as an approximate figure for the unexplained part of the perihelion advance, two-thirds of the observed motion—the same result as is obtained from Riemann's law.

For Laplace and Tisserand the introduction of a finite velocity of propagation of gravitation was a means to an end. In the first case it was supported by a quasi-hydrodynamical analogy, in the second by an electrodynamical one. One feels that neither would have introduced the idea of such a finite velocity had it not promised to 'save the

---

[75] *C.R.*, xlix (1859) 379. *Cf.* p. 69 *infra* for later values. LeVerrier predicted the existence of a hitherto unobserved planet ('Vulcan'), thus initiating half a century of futile searching.
[76] *C.R.*, cx (1890) 545.   [77] op. cit.   [78] ibid., p. 313.

(astronomical) appearances'. In the case of Bertrand, Tisserand, and Lévy, there was no independent reason for assuming the velocity to be equal to that of light. After Einstein had accounted for an 'anomaly' which was, more or less, that observed, there was a renewal of enthusiasm for these quasi-electrical laws of gravitation. The aim was not so much to save the appearances as to detract from Einstein's achievement.[79]

The history of the nineteenth-century attempts to solve the problem of the anomalous motion of Mercury is long and involved, but one further broad class of solutions will be mentioned. The perturbing effect of the planets on Mercury was first calculated by U. J. J. LeVerrier—joint discoverer of Neptune—in about 1850.[80] As we have seen, there was a discrepancy between his calculation and the observed perihelion advance. Before long it was suggested that perturbation by an unobserved inter-Mercurial planet might explain the anomaly: even if such a planet could not have escaped observation, a cloud of particles or a ring of asteroids might equally well have accounted for the phenomenon. Seeliger put forward a hypothesis in 1906 that there are two ellipsoidal distributions of interplanetary matter (namely those giving rise to the zodiacal light), and that they are responsible for perturbing Mercury's orbit. One of these was said to contain the Earth's orbit, the other to lie between the orbits of the Earth and Mercury.[81] Seeliger later affirmed that the inner cloud was the only important one, as far as its effect was concerned,[82] and one or two articles were written in his support. The majority of astronomers were uneasy, however, with an hypothesis which could be tested only in terms of the phenomenon which it was devised to explain, and nothing has been heard of it since the early 1920s. Even today, there appears to be no theory capable of accounting for the anomaly in the motion of Venus. It should be added, however, that the observational problem is unusually difficult in the case of Venus. Having an orbit of very small eccentricity—that is, nearly circular—its perihelion is very difficult to determine with sufficient accuracy. It is hoped that a more critical set of observations will soon be available from a study of the orbit of Mars,

[79] See, for example, Gaston Bertrand, *C.R.*, clxxiii (1921) 440; L. Lecornu, *C.R.*, clxxiv (1922) 341; clxxvi (1923) 205, 795; L. Decombe, *C.R.*, clxxv (1922) 1194; Lense, *Wien Sitz.*, cxxvi (1917) 1038. The last paper is of interest in that it replaces the distance-term r in the Newtonian law by the 'distance' $R\sin^{-1}(r/R)$, where R is the radius of space, supposed spherical or elliptical. Lense's predictions were, however, quite unacceptable.

[80] See *Ann. Obs. Paris*, v (1859) 104. S. Newcomb and Doolittle confirmed the calculation in outline. (Newcomb's paper is now known to have contained an error.)

[81] *Münch. Sitz.*, xxxvi (1906) 595; *Viert. Astr. Ges.*, xli (1906) 234.

[82] *Astr. Nachr.*, cci (1915) 273; cf. E. Weichert, *Ann. Phys. Lpz.*, lxiii (1920) 301.

whose perihelion advance, although much smaller, is much better defined.

6. Gravitation and Lorentz Invariance

At the beginning of this century Henri Poincaré examined the same question, but this time from a completely different point of view.[83] H. A. Lorentz had said, in effect, that the definition of any sort of force should be made so that the force is affected by what is now known as a Lorentz transformation of coordinates in the same way as are electromagnetic forces. Poincaré, in subscribing to this, sought the modification which he would have to make to Newton's law of gravitation. Like Laplace, he took a finite velocity for the propagation of gravitation, but unlike Laplace he did not make this his starting-point. Taking the components of any gravitational force in the direction of the axes as functions of the relative position vector and of the components of the velocities of two attracting particles, he found it possible to determine these functions in such a way that Lorentz's condition is satisfied and Newton's law recovered, when the squares of the ratios of the velocities to that of light are neglected.

Poincaré's theory of gravitation, or rather his method of treating gravitational problems, can hardly be called a field theory, nor does it involve direct analogy with any other physical theory.[84] Had the general sympathy not been so emphatically in favour of a field theory of gravitation, Poincaré's memoirs might well have been a turning point in the history of the subject. Even as things were, Hermann Minkowski[85] followed Poincaré's lead in not regarding gravitational action as a field action in any stronger sense than that it involved the existence of a force-component at each point of space. Unlike Poincaré, he worked by analogy with a law of force for electrons in motion. This led him to the differential equations

$$\frac{d^2x}{d\tau^2} + \frac{h^2Mx}{r^3} = 0, \&\text{c.}, \qquad (8)$$

for the motion of a planetary mass (i.e. one which is negligible in comparison with the central body). The form of this equation is exactly that of the equation describing Keplerian motion in planeto-centric time. There is, however, a periodic but imperceptible deviation from

[83] *C.R.*, cxl (1905) 1504; *Palermo Rend.*, xxi (1906) 129.
[84] He occasionally uses the phrase '*l'onde gravifique*' but the idea is by no means necessary to his work.
[85] *Gött. Nachr.* (1908) 53. See in particular pp. 110–11.

Keplerian motion when heliocentric time is used. The secular terms in the perturbation of the elements of the motion being found to be multiplied by the planetary mass, are negligible in Minkowski's theory —and this was probably considered a point in its favour.

The gravitational forces of the theory were four-vectors, constructed out of the components of ordinary force: they are now known as 'Minkowski forces'.[86] The same type of four-vector was used by Lorentz when he proposed another law of gravitation.[87] This law, when expressed in terms of ordinary Newtonian forces, has no term containing the planetary velocity and it was for this reason that Lorentz favoured it.[88] When the planetary mass is ignored, Lorentz's law leads to differential equations of planetary motion whose solution (found approximately by the method of quasi-perturbation) suggests a secular motion of the perihelia of planets, even for a planet of infinitesimal mass. The anomaly in the motion of the perihelion of Mercury was still, however, far from being completely explained. Admittedly, if contentions such as Seeliger's were correct, and the motion was in part the outcome of a perturbation by such matter as gives rise to the zodiacal light, then it seemed that here might be the explanation of the residual motion. Astronomers realized, however, that to argue in this way, so long as there were uncertainties of the order of those introduced by Seeliger, meant that in observations on the advance of the perihelion they had no longer a critical test of *any* theory.

Theories of gravitation had, it seemed, strained observational techniques to the very limit. This might not have seemed so serious had the choice been between two or three alternative theories; but, as Poincaré realized, the expressions given by himself, Minkowski, and Lorentz for the gravitational force could be multipled by any power of a certain expression and the restricted principle of relativity would still be satisfied.[89] This meant that any multiple of the values already calculated for the perihelion advance could be upheld as compatible with the requirements of Lorentz's invariance. Clearly some more restrictive condition should be sought. The condition ultimately to be

---

[86] Defined by the equation $F^m = c^2(d/ds)(m \cdot dx^m/ds)$, with s for Minkowski's proper-time τ.

[87] *Phys. ZS.*, xi (1910) 1234. See especially p. 1239.

[88] The law had already been given—for different reasons—by H. Wacker (Inaugural dissertation, Tübingen, 1909).

[89] *Palermo Rend.*, xxi (1906) 129. The expression is $\dfrac{1 - \xi_1\xi_2 - \eta_1\eta_2 - \zeta_1\zeta_2}{\sqrt{\{(1 - \phi_1^2)(1 - \phi_2^2)\}}}$ where $\phi = \xi^2 + \eta^2 + \zeta^2$, (ξ, η, ζ) being the components of the velocity of the planet with respect to the central mass.

imposed was that of covariance with respect to a *general* point transformation. When this was eventually introduced by Einstein all previous theories of gravitation appeared trivial by comparison, although Einstein's debt to them was not insignificant. In the next chapter we shall discuss the immediate sources, the conceptual innovation and some of the consequences of Einstein's new theories of gravitation.[90]

[90] Attempts to alter Newton's law of gravitation did not, of course, cease when the worth of Einstein's theory became appreciated—rather the opposite. The aim was usually to offer a *simpler* explanation than Einstein's of the things his theory predicted. Of the exceptions to this rule, a paper by Arrigo Finzi is fairly typical. (*M.N.*, cxxvii (1963) 21.) By changing Newton's law to $F = \gamma m_1 m_2 / \sqrt{\rho r^3}$, where $r \gg \rho$, where r is the separation of the masses, and where ρ is a 'characteristic length which we take to be half a kiloparsec', he aims to explain a problem which has long worried astrophysicists. The several observed clusters of galaxies have an appearance of stability, and yet the velocities of their members are larger than conventional theories will allow as compatible with stability. (See F. Zwicky, *Helv. Phys. Acta*, vi (1933) 110 and S. Smith, *Ap. J.*, lxxxiii (1936) 23.) There is much unsatisfactory in Finzi's theory, but it has one notable characteristic—it begins with an otherwise unsolved problem and is thus not *ad hoc* in a sense explained earlier.

CHAPTER 4

THE ORIGINS AND CONCEPTUAL BASIS OF EINSTEIN'S THEORY OF GRAVITATION

1. Gravitation and Light. The Principle of Equivalence

The advent of the restricted theories of relativity in the early twentieth century promised new methods of approaching the problems of gravitation. Three such methods have already been discussed. The ways in which Poincaré, Minkowski, and Lorentz modified Newton's account of gravitational attraction do not appear, in retrospect, to be wildly out of sympathy with the older tradition. The hypothesis that the inertia of a body depends upon its energy content was firmly incorporated in the Special Theory of Relativity by Einstein in 1905.[1] The equality of gravitational and inertial mass had, furthermore, been looked upon as verified from the time of the well-known experiments of R. von Eötvös.[2] What might be said to mark the dividing line between the theories of this early period and those which followed is that hypothesis proposed by Max Planck, according to which, since all forms of energy have inertia, they must also have gravitational properties.[3] Henceforth no theory of gravitation could afford to ignore the rest of physical science.

At the end of 1907 Einstein discussed the influence of gravitation on the propagation of light, and at the same time he first introduced the Principle of Equivalence. Here are two of the four ideas most important to the growth of the General Theory of Relativity—the others being the introduction of Riemannian space-time and the Principle of Covariance. In the final presentations of his theory the four ideas are interwoven to a greater extent than was the case

---

[1] Poincaré (*Arch. Néerl.* (2), v (1900) 252) seems to have been the first to suggest clearly the identification of a distribution of electromagnetic energy with one of mass, but this was itself only a short step from J. J. Thomson's theory that to electromagnetic energy-momentum there corresponds a mechanical momentum. Poincaré seems to have attached little importance to his result. As Whittaker points out, both Thomson (1881) and F. Hasenöhrl (1904) had spoken of, respectively, a charged spherical condenser and a box filled with radiation *as though* their masses were increased by constant fractions of the corresponding energies. (*H.T.* Ae. E., ii, 51–2.) Despite this challenge to the priority of Einstein's work, it is notable that later references to the relation '$E = mc^2$' were generally made to Einstein's paper of 1905 (*Ann. Phys.* Lpz., xviii (1905) 639).

[2] *Math. Nat. Ber. Ungarn*, viii (1891) 65.

[3] *Berlin Sitz.* (1907) 542 and *Ann. Phys.* Lpz., xxvi (1908) 1.

historically. We begin, therefore, by considering the Principle of Equivalence and its historical importance in the transition from Newton's theory to Einstein's.

Suppose that an observer in a closed chamber observes that a free object within the chamber tends to move to one side of it. The observer might choose to explain the phenomenon in terms of a field of force, or he might suppose the chamber to be in motion. What was to be known as Einstein's Principle of Equivalence asserts that the two explanations are to be regarded as one and the same. An early statement (1911) of this principle is found in *The Principle of Relativity* (p. 100).[4] K and K' are systems of coordinates, K being stationary with respect to some material system, and K' moving with a freely falling object. Einstein calls the systems 'physically exactly equivalent'; that is, he assumes 'that we may just as well regard the system K as being in a space free from gravitational fields, if we then regard K as uniformly accelerated. . . . This assumption of exact physical equivalence makes it impossible for us to speak of the absolute acceleration of the system of reference.' As we shall see, this appears to have been one of the routes by which Einstein was led to the Principle of Covariance. Its present purpose was of a practical nature, however, and for the time being covariance was in the background.

Having now asserted, in effect, that there should be agreement between the lawlike statements of a uniformly accelerated observer in the absence of a gravitational field and those of a stationary observer in such a field, from a simple calculation of the results to be expected by the former observer, he deduced the way in which any periodic effect, such as light, would be seemingly influenced by the gravitational field in which the latter observer was situated. In 1907 and 1911 Einstein found that light emitted with frequency ν_1 from a place of gravitational potential Φ, arriving at a place of relative gravitational potential zero, will appear to the receiver to be of frequency ν_2, where

$$\nu_1 = \nu_2 \left(1 + \Phi/c^2\right). \tag{1}$$

Displacements towards the red of the solar spectral lines had, in fact, already been noted by L. F. Jewell in 1897,[5] and by C. Fabry and H. Boisson in 1909,[6] but they had been explained in part as a pressure effect in the absorbing layers of the Sun, and in part as a Doppler

[4] The paper in question is a translation of *Ann. Phys.* Lpz., xxxv (1911) 898.
[5] *J. Phys.*, vi (1897) 84. One way to the classical result will be clear if we bear in mind that the potential energy of the photon at the Sun is $\Phi \times$ photon mass $= \Phi h \nu_1/c^2$.
[6] *C.R.*, cxlviii (1909) 688.

effect in the descending currents of these layers. It was soon recognized that a comparison of the predicted fractional displacement (($\Delta \nu/\nu \simeq 2 \times 10^{-6}$) with the recorded displacements involved an intricate problem in interpretation, not to mention immense practical difficulties—which have only recently been overcome.

We shall later have to question not the historical role of Einstein's Principle of Equivalence, but its validity within the General Theory of Relativity, in the development of which it played such an important part.

In the paper of 1911 Einstein extended his argument and showed that the velocity of light could no longer be taken as constant, but that it must depend on the gravitational potential. From this, together with Huyghens' Principle, he concluded that light rays propagated across a gravitational field should be deflected. Although he recommended the problem to astronomers, the matter was more or less ignored for the next seven or eight years.

By this time the problem of adapting a theory of gravitation to the results arising out of the Special Theory of Relativity was being investigated by several writers. Max Abraham, for example, began by considering what to many seemed the most promising starting point, namely the relation between the mass of a body and its total energy content ($E = mc^2$). Abraham remarked that the energy of a body depends to some extent upon the gravitational potential at the point where it is situated and that therefore one or both of the remaining terms in the equation (namely m and c) must do likewise. Einstein's hypothesis that the velocity of light depended on the gravitational potential was in fact Abraham's choice.[7] The theory of gravitation which followed stemmed thus from Abraham's 'Postulates':[8]

(i) The surfaces $c =$ const. coincide with the equipotential surfaces of the gravitational field, the negative gradient of c giving the direction of the force of gravitation.
(ii) An observer belonging to the material system he observes, should the system be moved to a region where c has a different value, will be unable to perceive the fact.
(iii) The forces which act on two bodies at the same place in the gravitational field are proportional to their energies.

---

[7] *Rend. Lincei*, xx (1911) 678; *Phys. ZS.*, xiii (1912) 1, 311, cf. also pp. 4, 176, 310, 793; *Archiv. Math. Phys.*, xx (1913) 193 (Lecture of October, 1912). Einstein's result (1911) is in *P.R.*, p. 107.

[8] The 1912 lecture, pp. 195 ff.

Abraham, with Einstein, concluded from (i) that there should be an appreciable deviation of any light ray grazing the Sun's surface. Taking next the unit of length as independent of c he was led to suppose that the period of electromagnetic oscillations from an antenna of given length is inversely proportional to c. By postulate (ii), neither change in this nor a change in c itself, can be known to a local observer. (This is reminiscent of Einstein's relativity-postulate.) Although the observed value of c will always be the same, the gradient of c—the gravitational force—will of course be locally determinate. An observer not belonging to a system is, however, quite able to discover the influence of the gravitational potentials on the periods: he can compare the Fraunhofer lines in the Sun with the corresponding lines from a terrestrial source. Here too the expected fractional shift in frequency agreed with the expression Einstein had already published.

Postulate (iii) is really a statement of the equality of inertial and gravitational mass, bearing in mind the relation between mass and energy with which Abraham began.

Deriving the law of force between a point-mass and a sphere (at rest), Abraham found that it differed from the familiar law by a term in the inverse cube of the separation. This small deviation he looked upon as arising from the existence of the energy in the external gravitational field. As we have seen, some such deviation was needed if the anomalies of planetary theory were to be explained. What was perhaps more surprising was that the Special Theory of Relativity, which was Abraham's starting point, ultimately 'falls into the dust': invariance of the theory with regard to Lorentz transformations was preserved only *in vacuo*, it being necessary to relinquish this even in infinitely small regions when matter is present.

Shortly after the first of Abraham's publications on these lines, Einstein put forward a similar theory containing the hypothesis that the gravitational field depends—this time in a less simple way than Abraham suggested—upon the variations of c with both time and place.[9] The merits of the rival schemes were argued at some length by their authors,[10] but the whole issue was undoubtedly overshadowed by the prospect of a unitary theory of physics such as that being currently presented by Gustav Mie.[11] This contrasted with almost all previous gravitational theories, in so far as it introduced field variables of a

---

[9] *Ann. Phys.* Lpz., xxxviii (1912) 355, 443.
[10] ibid., p. 1056 (Abr.); 1059 (Einst.); xxxix (1912) 444 (Abr.); 704 (Einst.).
[11] *Ann. Phys.* Lpz., xxxvii (1912) 511; xxxix (1912) 1; xl (1913) 1.

purely electromagnetic nature—twenty in number. The Maxwell-Lorentz theory was recognized to be invalid when applied to the interior of the electron, whereas Mie's theory seemed to offer an opportunity of deriving the 'matter' of the electron from the composite field. The system was very impressive, not least because it could be condensed into the form of Hamilton's Principle.[12]

The conceptual upheaval prompted by Mie's work has lasted, in one form or another, to the present day, although physics has so far avoided a satisfactory unified treatment. But ideas equally unforseen, and immeasurably more successful, were shortly to be announced by Einstein. Between 1912 and 1914 he worked with the Swiss mathematician Marcel Grossmann. The quadratic differential form was by this time a firmly established part of physics, following the work of Minkowski[13] and H. Bateman,[14] who not only referred their theories to a four-dimensional manifold, but who had begun to perceive the great power of the methods of the absolute differential calculus.[15] Einstein collated these ideas and adapted them to the subject of gravitation, introducing for the first time the ordinary and null geodesics of a *Riemannian* space-time, as representing the paths of free particles and light rays respectively.[16] The remarkable exclusion of the concept of gravitational force was achieved, but at the expense of the simple scalar potential function—which was now replaced by the ten 'gravitational potentials', the functions g_{mn}. The additional complexity seemed unwarranted to many people at the time, but Einstein's ideas were here to stay. This, the second stage in the development of the General Theory of Relativity, will be discussed in the following sections.

2. Relativity and Covariance

As is well known, according to the 'Special Principle of Relativity', it is required of physical laws that if they hold good when referred to one system of coordinates they should also hold good when referred to any other system moving in uniform translation with respect to it. The postulate is also true of Newtonian dynamics, and what distinguishes the Special Theory of Relativity from Newton's 'relativity' is

---

[12] Weyl's work was greatly influenced by Mie's theory and Max Born contributed to the theory as recently as 1934. See *P.R.S.* (A), cxliii (1934) 410.
[13] *Gött. Nachr.* (1908) 53.   [14] *P.L.M.S.* (2), viii (1910) 223.
[15] *H.T. Ae. E.*, ii, 64, 154.
[16] *Phys. ZS.*, xiv (1913) 1249; *Archives des Sciences*, xxxvii (1914) 5; *Berlin Sitz.* (1914) 1030. With Grossmann, *ZS. Math. Phys.*, lxii (1913) 225; lxiii (1914) 215.

not so much this, but the postulate of the constancy of the velocity of light. From the two postulates one can deduce the Lorentz transformation, the principle of relative simultaneity, the laws for the behaviour of moving clocks, and so on. Now Einstein was greatly influenced by what he called an 'epistemological defect' in both the Special Theory of Relativity and classical mechanics, a defect indicated by Mach. If two spheres at constant distance are conceived as beginning to rotate (relatively) about the line joining them, and one remains spherical whilst the other becomes ellipsoidal, to what can we assign the cause? The Newtonian would, of course, have referred to a privileged space in which the spherical object is at rest. This 'factitious cause' was disliked by Einstein and Mach alike. They might conceivably have looked upon it as some sort of 'theoretical construct', but instead they sought as 'reason' what Einstein called an 'observable fact of experience'.[17] His simple epistemology at this stage of his career was wholeheartedly Machian. 'The law of causality', he wrote, 'has not the significance of a statement as to the world of experience, except when *observable facts* ultimately appear as causes and effects.'[18] What seems to be the reason for his wanting to extend the simple relativity postulate can be summarized in the following rough argument: If 'privileged spaces' are invoked to explain such a situation as outlined by Mach, then we are merely providing factitious causes for observable facts; but such causes are ruled out by our epistemology; therefore privileged spaces are unacceptable. 'The laws of physics', he concluded, 'must be of such a nature that they apply to systems of reference in any kind of motion.'

In Part II the force of his argument for this principle will be questioned, but of its real value there can be no doubt. Whatever its origins, it is a regulative principle of great importance. It provides us with a criterion for the significance of our scientific laws: it is, in a sense we shall use later, a meta-law.

All coordinate systems were, according to Einstein's precepts, to be on an equal footing. What is sometimes known as the Principle of Covariance he wrote as follows:[19]

> The general laws of nature are to be expressed by equations which hold good for all systems of coordinates, that is, are covariant with respect to any substitutions whatever (generally covariant).

This is clearly a sufficient condition of the Principle of General Relativity (which simply deals with the relative motion of three-dimensional

[17] *P.R.*, p. 113.   [18] loc. cit.   [19] op. cit., p. 117.

systems of coordinates—a subclass of the general transformation in four dimensions). Most of the observations of importance to the confirmation of physical laws concern the intersection of world-lines: they concern, that is to say, coincidences in space and time. The Principle of Covariance could be said to assert that the form of any law is not to be affected by deformation of world-lines. The topological aspect of physical laws is being emphasized: we are to reject any that does not preserve the order of the intersections of world-lines.[20]

As many writers have pointed out, the task of finding laws which conform to the requirements of the Principle of Covariance is not quite so hopeless as it might seem. It is, indeed, embarassingly easy, and the Principle of Covariance can hardly be regarded as a means of justifying the rejection of hypotheses: for laws which are not covariant can generally be made so. In 1901 C. G. Ricci and Tullio Levi-Civita demonstrated a theorem[21] which can be interpreted for our purposes as follows: from almost any law, covariant or not, it is possible to derive a law which is covariant and from the point of view of its interpretation to observation, is indistinguishable from it. Despite this result, the heuristic function of the Principle is obvious, for it obliges us to express our laws in covariant form before other considerations are introduced. On the other hand it is clearly untrue to claim that a law which can be readily expressed in covariant form is more likely to be true than one whose covariant form is complex and awkward. Such a law is, however, more likely to be accepted, and perhaps that is what is really important.

3. The Riemannian Framework and the Calculus of Tensors

Einstein needed a means of formulating generally covariant laws and he found it in the 'absolute differential calculus' (calculus of tensors) of Ricci and Levi-Civita. Riemann's analytical development of non-Euclidean geometry was essential to this. As these related concepts provide the very fabric of relativity and its derivative cosmology, we shall give an indication first of their origins and then of the way in which Einstein exploited them.

Had Euclid ever regarded the 'Parallel Postulate', the fifth postulate of his first book of *Elements*, as self-evident, he would have designated

---

[20] The preoccupation with tensors (and tensor-densities and pseudo tensor-densities) which followed Einstein's eventual success, has led some writers to make the Principle of Covariance require of a law that it be expressible in *tensor* form. In fact, of course, this is not at all necessary—as is evidenced by, for example, the later use of spinors.

[21] *Math. Ann.*, liv (1901) 125.

EINSTEIN'S THEORY OF GRAVITATION

it a 'common notion'.[22] Many of his successors, nevertheless, hesitantly regarded it as self-evident, whilst others attempted to give what Euclid, by implication, had claimed was impossible; namely, a derivation of the proposition from the remainder of the geometry. Those who tried to give a demonstration of the proposition, especially during the seventeenth and eighteenth centuries, achieved not this, but something of even greater importance.[23] They found that the postulate could be replaced by one of several others with no resultant change in the derived geometrical theorems. This, at first, seemed merely to confirm the majority in their belief that the truths of Euclid were *a priori* truths, a belief to which Kant lent the weight of his reputation.

Gauss was one of the first to realize that if the Parallel Postulate were to be replaced by an axiom which was *not* equivalent to it, and which was not incompatible with the rest, then the resulting geometry might yet be a consistent deductive system, important in its own right.[24] Although Gauss appears to have derived one or two seemingly paradoxical theorems of a non-Euclidean character, the first comprehensive accounts of a geometry of this kind were given by N. I. Lobachevsky and J. Bolyai, independently of Gauss and of each other, in the early 1820s. Their work was either obscured by the barrier of language—Lobachevsky wrote in Russian—or its publication delayed by the failure of most mathematicians to appreciate, and hence support, the innovations of the two authors.[25] Only when R. Balzer, writing in 1867, drew attention to it, remarking that the subject had the approval of no less a figure than Gauss, did the widespread study of this new geometry begin.[26]

[22] Cf. Heath's edition (Cambridge, 1908). This is to assume that Euclid's distinction is much the same as Aristotle's. Proclus gives two other ways of distinguishing axioms from postulates and the three are not equivalent.

[23] Wallis, 1693, Ludlam, 1785, Gauss, 1799.

[24] See, for example, some of the quotations from his letters at p. 75 *infra*. For the history of Gauss's thoughts on the matter, see Stäckel and Engel, *Die Theorie der Parallellinien* (Leipzig, 1895). Gauss seems to have been the first with the phrase 'non-Euclidean geometry' (or 'anti-Euclidean geometry').

[25] In 1826 Lobachevsky delivered a lecture (MS now lost) at the new university of Kazan, dealing with what Klein later termed a *hyperbolic* geometry. The substance of the lecture he gave in a memoir 'On the Principles of Geometry', *Kazan Bulletin* (1829–30). His first work in German was *Geometrische Untersuchungen zur Theorie der Parallellinien* (1840) translated into English by Halsted in 1891. In an appendix to the work *Testamen* (1832) written by his father, J. Bolyai presented his work on non-Euclidean geometry (once again hyperbolic) the main ideas of which he had worked out more than nine years previously.

[26] Balzer brought about the publication of French translations in 1866–7. For what are probably the best histories on the subject see R. Bonola, *La geometria non-euclidea: esposizione storico-critica del suo sviluppo* (Bologna, 1906) trs. H. S. Carslaw (Chicago, 1912), and the article 'Geometry' by Russell and Whitehead in the *Encyclopaedia Britannica* (11th ed.).

The discoveries made by Bolyai and Lobachevsky were followed, at a rather leisurely pace, by a very thorough revision of geometrical procedures. For cosmology it is their analytical aspects which are of greatest importance. The non-Euclidean geometries were, until the work on their consistency was undertaken, very largely neglected. Before 1868 no one had shown even Euclidean geometry to be self-consistent: failing to find to the contrary, mathematicians had simply assumed it to be so. In 1868 E. Beltrami showed hyperbolic geometry to be *as* consistent as its Euclidean forerunner by the device of coordinating the two geometries, making the one, that is to say, a model for the other.[27] He showed that Lobachevsky's plane geometry holds in Euclidean space on certain surfaces of constant negative curvature (the pseudospheres) and that these could be conformally represented on a plane (geodesics becoming straight lines, for example). This proof is probably the oldest proof of relative consistency in any formal study. It was the fortuitous outcome of Beltrami's search for such a representation as he gave.[28]

In calling attention to the non-Euclidean geometries Beltrami was concerned with relating them to an analytical scheme, formulated by Riemann in the 1850s. Riemann's thoughts on the geometry now known by his name seem to begin with his work on conformal mapping in the theory of complex functions.[29] In 1854 he made use of an earlier conclusion to the effect that a complex function could be specified entirely if its local behaviour were known: he now supposed that geometrical spaces could be similarly specified.[30] To show this he introduced the quadratic differential form; but clear and well-directed as the dissertation was, the associated transformation theory of quadratic forms was not given until an essay of 1861, and even then it was introduced incidentally rather than as a subject of importance in its own right.[31]

[27] *Saggio di interpretazione della geometria non-euclidea* (1868).

[28] It was later shown by D. Hilbert (*Trans. Am. Math. Soc.*, ii (1901) 86) that the plane geometry of Bolyai and Lobachevsky cannot be *wholly* represented on any analytic surface free from singularities in a three-dimensional Euclidean space. H. Liebmann proved a similar result for plane elliptical geometry. This can only be represented on a *closed* surface. The only closed surface of positive constant curvature is the sphere—on which geodesics have two points in common, whilst in elliptic space they have only one. (*Gött. Nachr.* (1899) 44.)

[29] Inaugural dissertation, Göttingen, 1851 = *Riem. Ges. Werke*, iii (1892) pp. 3–43.

[30] *Über die Hypothesen, welche der Geometrie zu Grunde liegen*. This was Riemann's *Habilitationschrift*, read before the Philosophical faculty at Göttingen. *Werke*, ii (1892) pp. 272–87. For later analytical notes see *Werke*, i (1876) pp. 384–91; English trans. by W. K. Clifford in *Nat.*, viii (1873).

[31] The theory was more fully developed by Beltrami, Christoffel, Lipshitz and Ricci. Riemann's memoir is '*Commentatio mathematica, qua respondere tentatur quaestioni ab Illma Academia Parisiensi propositiae "Trouver . . . &c"*.' (1861), *Werke*, ii (1892) p. 391.

Riemann's work is important for its use of the idea of a manifold, the elaboration of which, if not entirely above reproach, could be readily generalized by others. (The idea was probably not original, for it might be said to have been anticipated by H. G. Grassmann in his controversial *Ausdehnungslehre*.)[32] Starting with a system of points whose coordinates were already assigned, and making the assumption that 'figures are freely movable', Riemann was led to an expression for the length (ds) between the points $P(x, y, z)$ and $Q(x + dx, y + dy, z + dz)$—namely[33]

$$ds = \frac{\sqrt{(dx^2 + dy^2 + dz^2)}}{1 + (\alpha/4)(x^2 + y^2 + z^2)}. \qquad (2)$$

This expression was derived largely by analogy with Gauss's formula for the intrinsic curvature of a surface. The constant α was called by Riemann the 'curvature' (*das Krümmungsmass*) of space. It is worth noticing that whereas Bolyai and Lobachevsky dealt only with space of *negative* curvature, Riemann considered only space of zero or *positive* curvature. It was left to Beltrami to connect these two fields of study.

Riemann's paper can be looked upon as inaugurating the study of the differential geometry of spaces of more than three dimensions. Such spaces were fully accepted later in the century, chiefly because they were useful in interpreting the theory of algebraic and differential forms in more than three variables. Gauss's paper on the geometry of curved surfaces in Euclidean space used a notation which did not lend itself to further generalization. Once Riemann's work was published, it attracted the attention of many mathematicians, in particular Christoffel, Lipschitz and Beltrami and his associates.[34] It is not possible to make more than the briefest mention of the development of this subject, but we can single out two aspects of it which were subsequently of the greatest importance to Einstein. One was the concept of invariance, the other was a progressive simplification of methods of calculation. The latter began with Beltrami's introduction of his 'differential parameters', in 1868. The trend continued when Christoffel and Lipschitz introduced the idea of covariant differentiation in 1869 and when Christoffel introduced symbols for his two linear combinations of the first derivatives of the coefficients of the line-elements. The years which followed saw a generalization of most of the important

---

[32] First edition, 1847.   [33] *Werke*, ii (1892) p. 282.
[34] The political activities of one or two geometers at the time of the *Risorgimento* led to a mild academic dictatorship, during which time the study of geometry in Italy flourished as never before.

results of classical differential geometry, but a second movement towards the simplification of the underlying calculus came with the work of Ricci and Levi-Civita. In 1884 Ricci introduced a new invariant symbolism, originally created to deal with the transformation theory of partial differential equations. This at the same time lent itself to the study of quadratic differential forms. The resulting Ricci-Calculus (or 'absolute differential calculus', now known as the 'tensor calculus') was responsible for uniting the large number of invariant symbolisms then in vogue. It was very extensively developed by the two authors over the following twenty years. Ricci saw clearly that properties of configurations in Riemannian geometry are properties of covariant or contravariant tensors. In the 1890s and later, Riemannian geometry developed rapidly. This period saw the investigation of much which was to be of vital importance to Einstein later on: the contracted Riemann-tensor (i.e. the 'Ricci-tensor'), the generalization of the theorem of Stokes, the Bianchi identity, and so on. This was the foundation upon which Einstein built.

Running through most of this work is the concept of invariance, but this has a much longer history. The invariance of cross-ratios under projection, for example, was known to the Greeks. As a fairly natural consequence of the invariance of cross-ratio, involution, polarity, and so forth under the projective group, there was developed, early in the nineteenth century, a theory of algebraic invariance which continued to occupy some of the best mathematical minds of the later century. (The subject is usually traced to Lagrange, Gauss, and Boole, but any list of important names must include those of S. Aronhold, Arthur Cayley, J. J. Sylvester and Alfred Clebsch.) With the work of Riemann began the investigation of a new sort of invariance: instead of the simple affine and projective transformations with algebraic forms we now find mathematicians looking for the invariants of general analytic transformations with quadratic differential forms. (The theory of covariants of course grew alongside that of invariants: perhaps the first explicit covariant could be taken as the Jacobian (1841).) For the importance of the notion of invariance in the history of geometry one need only cite Felix Klein's use of it in his *Erlanger Programm*. What geometrical relations are invariant under such and such a group of transformations? What group of transformations will keep such and such a geometrical relation unchanged? These were the sorts of question that Klein suggested the geometer should be answering. 'Geometrical objects' thus assumed a secondary position: if two sets of 'objects' are transformed by the same group in the same way then they must be essentially the same objects. Geometrical *structure* came to the fore, and this preoccupation with structure we find nowhere more marked than in Einstein's General Theory of Relativity.

The person responsible for introducing Einstein to these mathematical ideas was Marcel Grossmann, with whom he worked between 1912 and 1914. These ideas had been related to physical theory by several authors in the nineteenth century and the first decade of the twentieth, Minkowski and Bateman being historically the most important names. In 1908 Minkowski introduced into the context of the Special Theory of Relativity the metric now known by his name:[35]

$$ds^2 = dt^2 - (1/c^2)(dx^2 + dy^2 + dz^2). \qquad (3)$$

The null geodesics of Minkowski space-time represent the paths followed by light. Now in dealing with gravitational fields it was realized that it is impossible to find coordinate axes such that the above metric could be used over the entire field: in such a field two sets of axes with uniform relative motion cannot in general be both inertial. Bateman saw that, nevertheless, coordinates (t, x, y, z) could be taken so that (3) held locally; and that if at each position the local differentials (dt &c.) were expressed in terms of differentials of any coordinate set (x^r) covering the entire field, then (3) becomes expressible in the general Riemannian form[36]

$$ds^2 = g_{mn} dx^m dx^n. \qquad (4)$$

The coefficients g_{mn} are characteristic of the requisite field. It was, in fact, Bateman's aim in this paper of 1909 to determine a set of electromagnetic equations unaltered under a general transformation. The notion of the requirement of *general* covariance was clearly very much in the air some years before Einstein's memoirs of 1913–16.

4. Einstein and the Concept of Gravitational Force

Einstein and Grossmann now built upon this foundation. In a series of papers[37] they drew two analogies of very great importance. Just as the motion (rectilinear) of a free particle in Minkowski space-time follows a geodesic line, so they proposed to take the path of a material

---

[35] *P.R.*, p. 75.   [36] *P.L.M.S.* (2), viii (1910) 223.
[37] *ZS. Math. Phys.*, lxii (1913) 225; lxiii (1914) 215; *Vierteljahr. Nat. Ges. Zürich*, lviii (1913) 284; *Archives des Sciences* (4), xxxvii (1914) 5; *Phys. ZS.*, xiv (15 December, 1913) 1249; *Berlin Sitz.* (1914) 1030.

particle in a gravitational field as a geodesic in the space-time of form (4). Just as light travels along a null geodesic of Minkowski space-time, so, following Bateman's lead, the null geodesics of (4) were taken as representing light paths.

We can now see how radical Einstein's new theory was proving itself. The coefficients of the metric tensor (g_{mn}) now have a dual purpose: they specify both the 'gravitational field' (and hence are often called the 'gravitational potentials') and the scale of time- and distance-intervals. Gravity is no longer a force but a change in the metrical structure of space-time, which is no longer, like Euclidean space-time, either homogeneous or isotropic.

We can now also see the extent to which Einstein incorporated in his work those ideas which we found to be central to previous theories of the field. It is obvious that the extent to which he did so is slight. In Chapter 3 it was explained how, at their inception, these 'field theories' were usually bound up with the metaphor of pressures between contiguous pieces of matter. A satisfactory account of the General Theory of Relativity has neither need nor place for such a metaphor. It is possible to retain the concept of force, but only at the expense of simplicity. If defined as 'rate of change of momentum', for example, acceleration and force turn out to be no longer in the same direction. Failing a unified theory, electrical, magnetic, and nuclear fields of force must, on the other hand, be added to Einstein's theory in an alien manner. The way in which the terminology of 'gravitational field' is applied, moreover, differs substantially between the relativistic and the older cases. Einstein's field, for example, is not unique; for a coordinate transformation between relatively accelerated axes would be interpreted as a transformation to a new gravitational field. The propriety of this way of putting the matter is occasionally questioned. Is a coordinate transformation not a subjective process, and is a change to a new gravitational field not an objective change? How is it possible to alter something in nature simply by changing coordinate systems?

The answer is simply that a gravitational field is not 'something in nature' in any straightforward sense, even in Newtonian mechanics—where the concept can be entirely excluded, granted the necessary circumlocutions. Of Einsteinian mechanics this is true *a fortiori*, for gravitational forces simply are not known to it. Of course one can speak in a hybrid language. In speaking of 'gravitational potentials' we saw that Einstein kept some of the older terminology, and in his discussion of the Principle of Equivalence he does, after all, mention 'force'.

(This principle is of historical importance, although it need be given no place in a non-historical exposition of the General Theory of Relativity. Its greatest value seems to lie in its convincing hardened workers in the older tradition of the conceptual simplicity of the new theory.) In this hybrid language one might say that the gravitational field we experience is no more than an accident of our motion and, as such, it could be called 'subjective'. If a paraphrase of 'gravitational field' is called for, we may give an implicit definition: to specify a gravitational field is to specify the aggregate of all local inertial frames. The problem of representing this aggregate was solved with Einstein's introduction of what is often called the Geodesic Principle. 'Force' has gone, and to reintroduce it one may perhaps take some such approach as Eddington indicated: the field of force may be taken as the discrepancy between the 'natural geometry' and the geometry which is 'arbitrarily assumed to hold'.[38] But a field can hardly be called 'objective' when to specify it we rely upon 'arbitrary' assumptions.

5. Einstein's Field Equations

Not until early in 1915 could Einstein lay claim to a complete theory of gravitation; not, that is to say, until he was in possession of equations corresponding to Poisson's equation. These he presented late in the same year,[39] making full use of the methods of Ricci and Levi-Civita and also making liberal use of what he generously designated 'Mach's Principle'. Einstein again proceeded by analogy with his Special Theory, making the supposition that the components of the metric tensor must, in the case of a region devoid of matter, have constant values. The necessary and sufficient condition for this to be true is that the Riemann-Christoffel tensor (R^p_{qrs}—denoted variously by Einstein, e.g. by $B^\alpha_{\beta\gamma\delta}$ and (ik, lm)) should vanish.[40] It might thus have seemed reasonable to take

$$R^p_{qrs} = 0 \qquad (5)$$

as the equations of the matter-free field, but Einstein saw that this would entail the possibility of transforming away the gravitational field in the neighbourhood of a material point. As this was clearly out of the question he took instead the vanishing of the contracted

---

[38] *M.T.R.*, pp. 37–9. He is presumably referring to Euclidean geometry.
[39] *Berlin Sitz.* (1915) 778, 799, 831, 844. The '*Formale Grundlage der allgemeinen Relativitätstheorie*' was given the previous year (ibid. p. 1030). For a self-contained account see *Ann. Phys.* Lpz., xlix (1916) 769 (*P.R.*, 111).
[40] Lipschitz, *Crelle's J.*, lxx (1869) 71.

Riemann-Christoffel tensor[41] as the equations of this field. Apart from this tensor there is no tensor of second rank, formed from g_{mn} and its first and second derivatives, which is linear in these derivatives. (These conditions are satisfied by the left-hand side of Poisson's equation.) He concluded that there was a 'minimum of arbitrariness' in his choice of equations.

The field equations for space without matter correspond to Laplace's equation ($\nabla^2 \phi = 0$) of the classical theory. Einstein next sought those which correspond to Poisson's equation ($\nabla^2 \phi = 4\pi\gamma \cdot \rho$). From the Special Theory it was known that inertial mass could be regarded as a form of energy, which could be represented by a symmetrical tensor (the energy-momentum tensor T^{mn}) of second rank. The equivalence of mass and energy meant, conversely, that an electromagnetic field, for example, will create a gravitational field. Einstein therefore first examined the possibility that R_{mn} was proportional to T_{mn}. This would have conflicted with the principle of conservation of energy and momentum which, within the Special Theory, requires the divergence of T_{mn} to vanish; whereas the divergence of R_{mn} is not in general zero. He gave instead equations which may be written[42]

$$R_{mn} = -\kappa(T_{mn} - g_{mn}T/2). \tag{6}$$

The equations can also be written

$$R_{mn} - g_{mn}R/2 = -\kappa T_{mn}. \tag{7}$$

In either case they amount to ten equations for the ten unknown 'potentials' g_{mn}. That these ten equations cannot be independent was first shown by Hilbert.[43] If the field equations are to be covariant under any transformation, it must be assumed that the quantities on the left-hand side satisfy four identities. There will be four arbitrary functions contained in the solutions of the equations for g_{mn}.[44]

[41] Namely the 'Ricci tensor', generally denoted by R_{mn}. Einstein uses 'G_{mn}', reserving 'R_{mn}' for that part which remains when the condition $\sqrt{(-g)} = 1$ is imposed.

[42] He actually wrote, on the left of the equation

$$\frac{\partial}{\partial x_\alpha} \Gamma^\alpha_{\mu\nu} + \Gamma^\alpha_{\mu\beta} \cdot \Gamma^\beta_{\gamma\alpha},$$

where $-\Gamma^\alpha_{\mu\nu}$ is the Christoffel symbol of the second kind. He had to add the condition $\sqrt{(-g)} = 1$ to account for the vanishing of the remaining part of the Ricci tensor. $T(= T^m_m = g^{mn}T_{mn})$ was known as Laue's scalar.

[43] Gött. Nachr. (1915) 395.

[44] We shall later see that coordinates are often chosen such that $g_{\mu 4} = -\delta_{\mu 4}$ throughout the space-time.

We may now briefly revert to the 'Principle of Equivalence'. If this is to mean that the effects of a gravitational field are indistinguishable from those of an observer's acceleration, Einstein's theory clearly denies it. The 'existence of a gravitational field', on this theory, is linked with the non-vanishing of the Riemann-Christoffel tensor. But the values of this can be seen to have nothing whatever to do with the space-time path of an observer (i.e. with his acceleration) unless it is denied that non-inertial motion is possible (i.e. unless forces of every kind are excluded).

Having presented field equations which are, in general, non-linear partial differential equations in the functions g_{mn}, Einstein did not let the matter rest. In his outstanding contributions of 1915 he showed that Newton's theory followed as a first approximation, if the g_{mn} were assigned values differing from $\pm \delta_{mn}$ by quantities which could be ignored in higher powers than the first.[45] These small differences were taken to vanish at 'spatial infinity', with a suitable choice of coordinates.

The whole question of *weak* gravitational fields was considered in more detail in 1916 when, from a set of linear differential equations approximating to the general equations above, Einstein found expressions for the differences between the g_{mn} and the corresponding constant values in the Special Theory of Relativity, $\bar{g}_{mn}$. In doing so he founded a difficult, but now extensively studied, sector of the General Theory. This is concerned with the application of approximation methods to problems of intermediate scale analogous to many of the problems of classical mechanics. All too often effects are predicted which are below the threshold of detection (for example, in the influence of the rotation of a central body on the motion of its satellites). These methods concern us here only when they bear upon the question of the validity of the theory generally.

6. Observational Tests

We have already mentioned one observational test of Einstein's theory, namely, the gravitational red-shift. By 1915 the astronomer E. Freundlich had made a preliminary examination of the spectra of several massive stars, but could offer no categorical assurance that the

---

[45] Actually there is one respect in which Newton's theory cannot be said to follow from Einstein's. The former makes gravitational forces attractions, whereas the latter, as usually presented, is indifferent as between attraction and repulsion.

predicted effect was to be found. In his 1915 papers Einstein gave two further ways in which his theory might be put to the test. He predicted, first, the deflexion of light by a massive body. The calculated amount, by way of example, was $1''{\cdot}7$ for the case of light passing the Sun at grazing incidence.[46] The likelihood of such an effect had previously been maintained by Newton and Laplace, but their thesis had never been widely entertained. J. Soldner in 1801 had actually calculated the extent of the deflexion.[47] Had he not made a slip in his working, he would have obtained the value $0''{\cdot}84$, and there is thus a clear conflict between the classical and the later Einsteinian accounts. A long series of observations beginning with those at the British Sobral and Principe eclipse expeditions (May 1919) undoubtedly rule out the former. (Had they been available at the turn of the century it would have been quite obvious that a more drastic revision of Newton's law of attraction was needed than those which were then being proposed.) Although the 1919 observations were, at the time, thought to provide excellent evidence for the truth of Einstein's theory, yet later observations made at the eclipses of 1922, 1929, 1936, 1947, and 1952 were less convincing. Opinions on the best way of both reducing and of weighting the observations are very discordant, but S. A. Mitchell[48] gives good reasons for accepting the figure of $1''{\cdot}79 \pm 0''{\cdot}06$, a figure supported by van Biesbroek's findings at Khartoum in 1952.[49]

Einstein's third proposed test of his theory related to the predicted advance of the perihelia of the planets, notably of Mercury. Simple Newtonian theory was capable of explaining a perihelion displacement of $5557''{\cdot}18 \pm 0''{\cdot}85$ per century for this planet, this originating chiefly as the result of perturbations by the remaining planets.[50] The discrepancy between this figure and the observed value of $5599''{\cdot}74 \pm 0''{\cdot}41$

[46] *Berlin Sitz.* (1915) 834. A more carefully calculated value is $1''{\cdot}751$. This was not the first occasion on which Einstein had worked at the problem. In 1908 and 1911, starting from the Principle of Equivalence, he worked out the deviation of light passing through a *Newtonian* (solar) gravitational field, obtaining the value of $0''{\cdot}87$ (*Jahrb. Rad. Elek.*, iv (1908) 411 and *Ann. Phys.* Lpz., xxxv (1911) 898). Freundlich sought evidence of such an effect in examining old photographic plates, but without success (*Astr. Nachr.*, cxciii (1913) 369). A Cordoba Observatory expedition of 1912, which aimed to detect any deviation, was unsuccessful owing to bad weather. The war seems to have put an end to this early activity, although A. F. and F. A. Lindemann suggested the possibility of day observations (*M.N.*, lxxvii (1916) 140).

[47] *Astronomische Jahrbuch* (Berlin, 1804) p. 161. See *Ann. Phys.* Lpz., lxv (1921) 593 for a reprint of parts of the text with a foreword by P. Lénard.

[48] For bibliography (eclipse observations) see *Vistas*, iii, p. 58; *G.R.C.*, p. 191. See also the table analysing the observations, p. 93.

[49] *Eclipses of the Sun* (New York, 1951), ch. xviii.

[50] See G. M. Clemence, *Rev. Mod. Phys.*, xix (1947) 361; *Proc. Am. Phil. Soc.*, xciii (1949) 532.

per century has already been remarked upon. The anomaly (42″·56 ± 0″·94 per century—LeVerrier from the first had as close a value as 38″ per century and later quoted a figure of 45″ per century) had for a long time been well outside the expected errors of observation, and what was perhaps generally regarded as the most unambiguous mark of Einstein's success was his calculation of a value of 43″ per century.[50] Although for the other planets, and especially Venus, there remained discordances between the astronomical data and theoretical predictions,[51] yet the new explanations contrasted sharply with the *a posteriori* appeals to the data of observation which were so essential to many of the 'explanations' already discussed in Chapter 3.

Shortly after these results were published they were obtained in a far more elegant manner by K. Schwarzschild, who solved Einstein's field equations for the case of a space-time occupied only by a single spherically symmetrical particle of constant mass.[52]

We have now indicated the conceptual basis of Einstein's General Theory, together with some of the local requirements of that theory. In the following chapter we leave the history of these lines of investigation and turn to the cosmological issues in which Einstein suddenly found himself to be involved.

[50] *Berlin Sitz.* (1915) 831–39.   [51] See *M.T.R.*, pp. 89–90.
[52] *Berlin Sitz.* (1916) 189. See Appendix, Note II.

CHAPTER 5

EARLY RELATIVISTIC COSMOLOGY

1. Cosmological Considerations

In the last chapter we saw that the field equations

$$R_{mn} - \frac{1}{2} R g_{mn} = -\kappa T_{mn} \qquad (1)$$

were designed to lead in the first approximation to the Newtonian law, when weak fields are being considered. The same is to be expected, however, of any set of equations of the kind in which the left-hand side is replaced by a tensor with vanishing divergence. Any such set of equations would be consonant with the existing astronomical data, although less simple than the set written above. In 1917 Einstein was led to abandon the simple alternative in favour of a new law:

$$R_{mn} - \frac{1}{2} R g_{mn} - \lambda g_{mn} = -\kappa T_{mn}. \qquad (2)$$

His reasons for doing so were presented in an important paper—entitled '*Kosmologische Betrachtungen zur allgemeinen Relativitätstheorie*'[1]— which initiated at least one controversy lasting to the present time.

Einstein began by remarking that Poisson's equation, taken together with Newton's laws of motion, is not sufficient to determine completely motion in a gravitational field: these must be supplemented by 'boundary conditions at spatial infinity, if it is essential that we treat the universe as of infinite spatial extent'.[2] We have seen that, in another problem, both he and Schwarzschild had found similar boundary conditions to be necessary, both taking the 'gravitational potentials' g_{mn} to be constant 'at spatial infinity'. Einstein's paper discusses the validity of such a step when larger portions of the universe are considered than a solar system away from external influence. The Newtonian theory did not seem to offer much help, for the general belief in a constant value for the gravitational potential at infinite distance entailed several odd conclusions. Some of these Einstein outlines: he makes the dubious assumption, however, that there must be a place about which the gravitational field possesses spherical symmetry. He

[1] *Berlin Sitz.* (1917) 142. [2] ibid., p. 143.

EARLY RELATIVISTIC COSMOLOGY 71

also consistently speaks of 'infinity' as though it were a 'place'. For example, radiation and even complete celestial bodies (with finite kinetic energy) may 'pass radially outwards, becoming ineffectual and lost in the infinite'.

As a preparation for the relativistic arguments to come, Einstein next showed a way in which Poisson's equation could be slightly modified in order to escape from the difficulties which he supposed the Newtonian argument to encounter. If Poisson's equation were to be replaced by

$$\nabla^2 \phi - \lambda \phi = 4\pi\gamma\rho \quad (\lambda \text{ a 'universal constant'}), \tag{3}$$

then a solution would be

$$\phi = -4\pi\gamma\rho_0, \tag{4}$$

where ρ_0 Einstein interpreted as the mean density of matter distributed uniformly and 'infinitely extended'. Adding to this model inhomogeneities in the form of masses of much greater density than the mean, it was easy to see that (as in their neighbourhood $\lambda\phi$ is to be small by comparison with $4\pi\kappa\rho$) the field will there approximate to one with the Newtonian values. Such a theory, said Einstein, would not run counter to statistical mechanics, for it would explain how matter can be in equilibrium without the introduction of negative internal pressures for which there was no obvious interpretation. Other means of avoiding these pressures—and λ—were later to be found, that is to say when the static metric was dispensed with.

Einstein then turned to the vexed question of the boundary conditions needed for the differential field equations of his General Theory, conditions without which the g_{mn} would not be completely determined by the overall distribution of mass and energy. He argued from two principles, first, that 'there can be no inertia *relatively to space* but only an inertia of masses *relatively to one another*'; and secondly, that it is possible to choose a system of coordinates so that the gravitational field at *all* points is spatially isotropic. The second principle, not vital but 'introduced for the sake of clarity', meant that the metric could be written in the form

$$ds^2 = -A\{(dx^1)^2 + (dx^2)^2 + (dx^3)^2\} + B(dx^4)^2, \tag{5}$$

whilst the first principle he interpreted as meaning that the expression $mA/\sqrt{B}$ (which corresponds to the rest mass in the expression for the components of momentum $\{(mA/\sqrt{B})\,dx^1/dx^4\}$ &c.) should vanish 'at infinity'. Were this so, it would follow that the potential energy

($m\sqrt{B}$) 'becomes infinitely great at infinity',[3] and thus 'a point mass can never leave the system'. The same is true for radiation.

Satisfactory as this seemed, the possibility of boundary conditions of the kind it implied had to be rejected 'for the system of fixed stars'. The grounds given were that for 'an appropriate choice of the system of coordinates, the stellar velocities are very small by comparison with the velocity of light';[4] that therefore

$$ds^2 \simeq g_{44}\,(dx^4)^2, \qquad (6)$$

and hence that all other components of the gravitational energy tensor $T^{mn}\{=\rho(dx^m/ds)\,(dx^n/ds)\}$ are small in comparison with T^{44}. The supposition of small stellar velocities could not, however, be reconciled with the chosen boundary conditions, for 'wherever there are fixed stars the gravitational potential (in our case $\sqrt{B}$) can never be much greater than here on Earth'.

It is now possible to see what was Einstein's most palpable obstacle. It was, once again, his confidence in a more or less static world. This obstacle was, in a sense, fortunate; for it led him to look beyond the current vague conceptions of 'the edge of the world' to the idea of a universe *closed with respect to its spatial dimensions*. He went on first, however, to reject two further possibilities, namely, that either the g_{mn} assume, at infinity, the Lorentz-Minkowski values, or that one must—as de Sitter had already contended—refrain from asserting boundary conditions of general validity. The first Einstein deemed unsatisfactory, as it contravened both the principle of relativity (it required a special choice of coordinate system) and the 'principle of the relativity of inertia' (the single mass point would possess inertia, and inertia would therefore be *influenced* but not completely *governed* by matter).

2. A Closed Universe. W. K. Clifford and Others

Einstein did not solve the boundary value problem but rather dissolved it, by regarding the universe as a continuum, closed with respect to its spatial dimensions He was not by any means the first to consider the possibility that space may in point of fact be curved. In this section we shall show the extent to which he was anticipated in

---

[3] In the *static* case, that is. The significance of this stipulation only fully emerged some years later.

[4] The 'most important fact to be drawn from experience as to the distribution of matter'.

this respect. Consider, first, W. K. Clifford, who was prompted to publish some very tentative and incomplete remarks which have occasionally been generously interpreted as an anticipation of some of Einstein's ideas. Part of an abstract by Clifford will show this more clearly:[5]

> Riemann . . . says although the axioms of solid geometry are true within the limits of experiment for finite portions of our space, yet we have no reason to conclude that they are true for very small portions[6] . . . I hold in fact
>
> (i) That small portions of space *are* in fact of nature analogous to little hills on a surface which is on the average flat; namely, that the ordinary laws of geometry are not valid in them.
>
> (ii) That this property of being curved or distorted is continually being passed on from one portion of space to another after the manner of a wave.
>
> (iii) That this variation of the curvature of space is what really happens in that phenomenon which we call the *motion of matter*, whether ponderable or etherial.
>
> (iv) That in the physical world nothing else takes place but this variation, subject (possibly) to the law of continuity.[7]

Had Clifford not died in 1879 at the age of thirty-four, something might well have come of these lines of thought. He was responsible for introducing Riemann's geometry into this country through his translation of Riemann's paper of 1854[8] and was much less of a wild speculator than many of the philosophers who took his ideas as their starting point.

Few astronomers took the possibility of curved space seriously: notable exceptions were K. Schwarzschild and the American Simon Newcomb.[9] But it would be wrong to suppose that the methods of non-Euclidean geometry were not linked with arguments from theoretical physics before the time of Einstein's General Theory of Relativity. In the half century ending in 1915 one can find at least eighty papers devoted to non-Euclidean statics, dynamics and kinematics, not to mention over a score on the subject of gravitational attraction and

---

[5] *Proc. Camb. Phil. Soc.*, ii (1876) 157. (Read 21 February 1870.) Reprinted in his *Mathematical Papers* (London, 1882) p. 21.

[6] We notice that Clifford supposes that the deviation from Euclidean space is on the small, rather than on the large scale.

[7] Similar views were expressed in his *Common Sense of the Exact Sciences*.

[8] *Nat.*, viii (1873) 14, 36–7.

[9] Newcomb's first paper is *Crelle's J.*, lxxxiii (1877) 293. Newcomb used Klein's distinction between spherical and elliptical space. Cf. p. 108 *infra*. Schwarzschild's work is described below. A paper by M. P. Barbarin (*Proc.-verbaux des séances de la soc. des sci. phys et nat, Bordeaux* (1898–9) p. 71) may also be referred to, although its author appears to have been curiously out of touch with astronomical practice.

potential, Laplace's equation, and so on. Even the so-called 'geometrization of gravitation' had been brought about before the turn of the century; for Darboux had investigated the shapes of trajectories by reducing the problem to that of determining geodesics corresponding to an appropriate metric. This was not, of course, a metric in spacetime, and indeed Darboux should perhaps be excluded from a list of those who thought of non-Euclidean geometry as something more than a useful analytical device. But hydrodynamics, elasticity and Maxwellian electrodynamics were all discussed at length in non-Euclidean terms, and one can even find a discussion of non-Euclidean billiards—by a Frenchman of course. Predictions made by the authors of these papers usually differed from the so-called classical predictions by quantities beyond the threshold of measurement. Much of the time this scarcely seemed to matter, for as often as not these were purely intellectual exercises beloved of those who preferred to think of themselves as applied mathematicians rather than theoretical physicists.

As against these somewhat remote lines of thought it will be of some relevance if we consider a simpler problem. This is one well known as having occupied the thoughts of at least one of the founders of non-Euclidean geometry, namely Lobachevsky. It is the problem of determining from astronomical observations not the radius of curvature of space but merely a lower limit to it. There are two or three interesting arguments to be found which were used to this end, and it is perhaps worth bringing them into the light if only because a recent work has done them less than justice, as well as seriously misrepresenting one of them.

We might begin with a few remarks on the score of priority, by way of supplementing section 3 of the last chapter. Already in the eighteenth century Girolamo Saccheri and J. H. Lambert were on the verge of discovering the two versions of non-Euclidean geometry, now called hyperbolic and elliptic. But the influence of western philosophy was too strong, and neither was able to cast off the feeling that here was something of a heresy. As is well known, even Gauss kept his thoughts to himself and a few friends, and to this day most of the credit for the discovery of non-Euclidean geometry goes to the Russian Lobachevsky and the Hungarian Bolyai. But for all this, Lambert and Saccheri saw two alternatives to Euclid whereas Gauss, Lobachevsky, and Bolyai recognized only one—hyperbolic geometry. In 1786 Lambert made the observation that on what might be called the hyperbolic hypothesis there is an absolute unit of length which could obviate the need for a standard of length in the Archives in Paris. It never occurred to him, needless to say, that this standard set by the universe might be changing with respect to that already placed in the Archives.[10]

Again one finds Gauss in 1816 writing to C. L. Gerling in the same vein on the subject of the hyperbolic hypothesis:

> It seems something of a paradox to say that a constant length could be given *a priori*, as it were, but in this I see nothing inconsistent. Indeed it would be desirable that Euclidean geometry were not valid, for then we should possess a general *a priori* standard of measure.[11]

Eight years later, in a letter to F. A. Taurinus, he writes:

> Should non-Euclidean geometry be true, and this constant bear some relation to magnitudes which come within the domain of terrestrial or celestial measurement, it could be determined *a posteriori*.[12]

In the interval between these letters, F. W. Schweikart, a professor of law at Marburg, had asked Gerling, his colleague, to obtain Gauss's opinion on a memorandum in which he too distinguishes between Euclidean and hyperbolic geometry—what he calls *astralische Grössenlehre*. Schweikart also introduces a constant which is easily shown to be related to those of Gauss and Lambert in a simple way.

All were aware that experiments were needed if the value of the constant were to be found, although before Lobachevsky none but Gauss himself appears to have made any effort whatsoever to obtain the necessary measurements. Gauss is said to have made a careful triangulation between three mountain tops,[13] finding that within experimental limits space is Euclidean; that is to say, the angle sum of his mountain-top triangle turned out to be 180°. The conclusion could only be that the radius of the universe is very great by comparison with the dimensions of the Earth. In fact a careful reading of Schweikart's memorandum can tell us as much. Clearly celestial measurements are needed. One cannot, of course, take three stars as a triangulation points, but the position is not so hopeless as it might seem, as Lobachevsky was the first to explain clearly.

Lobachevsky often referred to his geometry as *imaginary geometry*, which might seem to suggest that he was reluctant to suppose that it

---

[10] He believed that the other hypothesis (that of elliptic geometry) might be rejected on the grounds that it led to the seemingly false conclusion that a single straight line could enclose a space. As we have seen, not until Riemann, in the 1850s, were the consequences of this alternative to be worked out fully.

[11] Gauss. *Werke* vol. viii, p. 169.   [12] op. cit., p. 187.

[13] These were the Brocken, the Hoher Hagen and the Inselberg. The sides of the triangle were approximately 69, 85, and 107 km. [Since this was written, A. Miller has shown that Gauss's real purpose was only to improve geodesy. See *Isis*, lxiii (1972) 345-8.]

could possibly be applied to reality. This was never his intention. In 1825 we find him asserting that the truth or falsity of Euclid is not contained in the data themselves, but must be rested on experiment or on astronomical observations, 'as in the case of the other laws of nature'. There is nothing of this sort in Bolyai, who was writing at very much the same time. For Bolyai, geometry takes on a more stern and formal appearance: it is as close to a purely deductive system as any to be found at this period of history. At all events he did not commit much to writing which could be included under the heading 'philosophy of space'.

At last, in 1829, Lobachevsky published the first result of real practical significance. In his memoir *On the Principles of Geometry*[14] he demonstrates in effect that in a hyperbolic world the minimum value for any directly measured parallax is

$$\text{arc tan } (a/R),$$

where a is the radius of the Earth's orbit (that is, the so-called astronomical unit). It follows that R, measured in astronomical units, must be greater than the cotangent of the smallest measured parallax. Now Lobachevsky had only imperfect data to hand: as we mentioned before, the first successful parallax measurement was of a nearby star, 61 Cygni, and was not made until nine years later. The figure quoted by Lobachevsky was for the parallax of Sirius, and was three or four times too large,[15] but at least it allowed him to state with certainty that if his was the true geometry, the curvature constant must be greater than $1 \cdot 66 \times 10^5$ a.u. It was, of course, morally certain that most of the millions of stars fainter than Sirius, our brightest star, would be much more distant, that is, they would have much smaller parallaxes. Thus Lobachevsky's figure for R was just about as conservative as it could be. His argument was not only more precise but more profound than the commonplace contention that the sufficiency of Euclidean geometry for our practical needs is self-evident. Lobachevsky gives us a *numerical measure* of the degree of accuracy which attends the use of Euclidean geometry in conjunction with classical physics. As he wrote in 1840: 'there is no means other than astronomical observation for judging of the exactness which attaches to the calculations of ordinary geometry'.

---

[14] See F. Engel's *N. I. Lobatschefskij: Zwei geometrische Abhandlungen*, &c. (Leipzig, 1899) pp. 22–4 and 248–52. The memoir, in translation, covers pp. 1–66 of the volume. It first appeared in the *Kazan Bulletin*.

[15] The figure which he used (1″·24) was taken from the Comte d'Assa-Montdardier's *Mémoire sur la détermination de la parallaxe des étoiles* (Paris, 1828).

For thirty years there was something of a lull in the study of non-Euclidean geometry, a lull which was hardly disturbed even by Riemann's paper in which spherical non-Euclidean geometry was investigated analytically. Important obstacles were no doubt the Kantian doctrine of space—against which even Gauss campaigned only in private correspondence—and the barrier of Lobachevsky's language. The early 1860s at last saw the publication of much of the relevant correspondence from Gauss and translations into French and German of many of Lobachevsky's writings.[16] As explained in the last chapter, Gauss's reputation lent weight to the writings of the little-known mathematicians from Russia and Hungary. At all events, mathematicians began to take non-Euclidean geometry seriously by about 1870. The English-speaking world owes most to Clifford and Cayley for its introduction. Wherever it took root it slowly began to undermine prevailing attitudes towards the nature of geometry in particular and mathematics in general. Thus, for example, the part played by non-Euclidean geometry in leading Poincaré to his conventionalism is well-known.[17] But more than twenty years before, C. S. Peirce had arrived at much the same opinions for very similar reasons. Peirce, however, discussing the problem again in 1891, expressed his confidence in the belief that 'our grandchildren will surely know whether the three angles of a triangle are greater or less than 180°—that they are *exactly* that amount is what nobody ever can be justified in concluding'. Quite clearly he has changed from his former conventionalist position.[18] As for his optimism, as things turned out it was hardly justified.

The conventionalist thesis is mentioned briefly here simply because the practical problem cannot be entirely divorced from it. One must come to a decision over the question: Can an astronomer retain a given form of geometry, come what may, if he is always prepared to modify the so-called physical section of his theory in the face of unpalatable evidence? Without digressing further we return to astronomical issues,

[16] Actually he himself had published his *Pangeometrie* not only in Russian but also in French, in 1855, but the work had not obtained a wide circulation. The same might be said for a short summary written in German in 1840.

[17] Cf. Chapter 13, *infra*.

[18] On the same occasion, Peirce held that the parallax of the furthest star lay between *minus* 0″·05 and plus 0″·15. (See *The Monist*, i (1890–1) 173–4.) Negative parallaxes are to be expected in spherical space, but not in elliptical space. Peirce knew of this, but inclined to attribute these negative parallaxes to errors of observation. 'It would be strange indeed', he wrote, 'if we were able to see, as it were, more than half-way round space, without being able to see stars with larger negative parallaxes.' At the same time, he asks whether the publication of other negative parallaxes might not have been suppressed.

merely remarking that none of the three remaining writers to be considered was conscious of this question.

For many years Simon Newcomb interested himself in non-Euclidean geometry. In 1898 he wrote:

> If our space is elliptical, then, for every point in it—the position of our sun, for example,—there would be, in every direction, an opposite or polar point whose locus is a surface at the greatest possible distance from us. A star in this point would seem to have no parallax. Measures of stellar parallax, photometric determinations and other considerations show conclusively that if there is any such surface it lies far beyond the bounds of our stellar system.

In writing this he was merely restating an idea which he had first discussed twenty years before. There is no question of his having introduced empirical data except in a negatively existential sense. But almost simultaneously the astronomer Schwarzschild showed how even this sort of data could be of value. His memoir represents a high-water mark of practical non-Euclidean geometry as linked with classical astronomical physics.[19] Schwarzschild's arguments are three in number. For the first, which although unacknowledged is essentially Lobachevsky's, he supplies new astronomical data to derive the result

$$R > 4 \times 10^6 \text{ a.u.,}$$

should space be hyperbolic.[20] He concluded that the curvature of hyperbolic space would be so trifling as to be without influence on measurements made within the planetary system.

Like Newcomb he realized that physical arguments were likely to be more important than metrical. Thus he showed, by way of example, how too small a figure for R would imply an absurdly high density for nearby stars. Unfortunately it was not until the mid-1920s that the status of our own Galaxy in relation to that of the nebulae was resolved in anything like a satisfactory way. But before then de Sitter was to adapt Schwarzschild's argument, each starting from the premiss that space (volume $\pi^2 R^3$) must at least be large enough to contain our own Galaxy.[21] If the mass of this is known (M), and also the average density

---

[19] *Viert. Astr. Ges.*, xxxv (1900) 337.

[20] Perhaps it is worth adding that if we were to use Lobachevsky's argument with the best modern data, we could go little further than Schwarzschild. The techniques used in measuring parallaxes *directly* have improved little since his day.

[21] He used the argument to discredit Einstein's 'cylindrical world', the metric of which was essentially the same as that of the elliptic space which Schwarzschild visualized. De Sitter's paper makes it clear that he at least had very strong doubts about the plausibility of a one-Galaxy universe. See *Proc. Acad. Amst.*, xx (1917) 235.

throughout space, then Schwarzschild's assumption is equivalent to

$$\rho V > M, \tag{8}$$

that is to say

$$R > (M/\rho\pi^2)^{1/3}. \tag{9}$$

(For reasons which will be appreciated later, de Sitter was obliged to give a different expression; for Einstein's field equations gave him a relation Between M and ρ which contradicts (9).)

Schwarzschild did not take the argument so far as this. Had he done so, any figure given for the density must of necessity have been worthless in the absence of knowledge as to the status of the nebulae. Indeed it is this very failing which invalidates his second main argument.

This next argument is of great interest in that it involves a hypothesis which would today be called a 'Cosmological Principle'. Here it is: *Give to the 10^8 stars with a parallax less than $0''\cdot 1$ a million times as much space as the 100 stars with a parallax greater than this figure.* Of course this particular form of the hypothesis breaks down once it is realized that the nearest hundred stars are in a region of unusually high density, the rest of space being populated by discrete groups of stars (viz. galaxies) which are widely separated. His mistake was in applying the principle to stars, rather than to galaxies. However, he goes on to conclude 'from a simple calculation' that

$$R = 16 \times 10^7 \text{ a.u.}$$

if space is elliptic.[22] This was thought to be satisfactory in that it was not incompatible with an earlier cosmological claim by Seeliger that there are 4×10^7 visible stars in a space of diameter 10^8 a.u.

Schwarzschild's last argument was again physical, and again he lacked sufficient data to enable him to make full use of it. Like Newcomb he pointed out that the impossibility of seeing the rear of the Sun by looking in a diametrically opposite direction might, granted an elliptical space, be explained on the assumption that interstellar absorption is sufficient to obliterate this image.[23] Estimating that an absorption of 40 stellar magnitudes is sufficient to obscure the image, in the knowledge

---

[22] In fact the calculation, which he omits, is more difficult than any other in the paper. It is necessary to make two or three approximations, and Schwarzschild's answer is not exact.

[23] Schwarzschild and de Sitter, both fully aware of the distinction between spherical and elliptical space, chose to phrase the argument in terms of elliptical space, although it is perfectly easily adapted to the alternative case. One would then speak of seeing (or not seeing) the Sun's antipodal image, rather than the Sun itself.

that the longest possible light path in an elliptic space is of length $\pi R/2$, he could have obtained another lower limit for R had he been in possession of any reasonable information on interstellar absorption. Nearly twenty years later de Sitter used this argument, with Shapley's figure for absorption of $0^{m}\cdot 01$ per kpc, to give

$$R > 2\cdot 5 \times 10^{11} \text{ a.u.}$$

As de Sitter pointed out, however, the argument breaks down completely if the age of the Sun is less than the time for the light to travel round the universe.[24]

Before returning to Einstein it should be added that although Schwarzschild's account might seem to have been far ahead of the time, yet it must be remembered that he was simply dealing, as Newcomb before him, with a geometrical possibility. This was suggested to him neither by any independent physical consideration nor by the thought that it was likely to lead to a physics differing from the existing physics (except metrically). The purely geometrical possibility was almost certainly in the minds of several of Einstein's contemporaries. One who, like Einstein, saw that it might be used to smooth out some of the 'illogicalities' in the classical physics was Paul Ehrenfest. In a conversation with de Sitter in 1916 he proposed the idea of making the four-dimensional world spherical in order to avoid the necessity of assigning boundary conditions.[25] This was not developed at the time.

The seemingly paradoxical 'curvature of space-time', and the analogies between spherical surfaces and hyperspherical surfaces, caught the popular imagination during the 1920s and early 1930s. Judging by what was written in this connexion it seems that the word 'curvature' was an unfortunate choice. As a mathematical term the word was selected long before it was used by Gauss, but it was he who defined the curvature of a surface in terms of properties *intrinsic* to the surface.[26] Riemann took Gauss's procedure and extended it to the case when the (two-dimensional) surface in question was constituted of a pencil of geodesics—determined for any two fundamental directions—in an

---

[24] Cf. L. Silberstein's 'cosmic day'. Actually, unknown to de Sitter, W. B. Frankland had already considered the same problem. (*Math. Gaz.* (1913) 136.) He drew the conclusion that 'the antipodal image of the Sun [in spherical space] would only be a rather inferior star, even if space were perfectly transparent'. (ibid, p. 139.) For de Sitter's argument see *Proc. Acad. Amst.*, xx (1917) 234.

[25] See *Proc. Acad. Amst.*, xix (1917) 1219, footnote.

[26] Memoir of 1827, reproduced in *Ostwalds Klassiker der exacten Wissenschaft*, V (Leipzig, 1889).

n-dimensional manifold.[27] Riemann's memoir was not published until 1868, but then it immediately attracted the attention of L. Kronecker and Beltrami. Beltrami dealt very thoroughly with the various curvature properties of an n-dimensional manifold,[28] using freely the ideas of 'Riemannian curvature' and the curvature invariant (scalar curvature). Kronecker first investigated the question of the curvature of a hypersurface of a space (in his case Euclidean) of n-dimensions.[29] The mathematical concept of curvature has been enlarged in many other ways, but these are scarcely relevant to an understanding of the Einstein and de Sitter theories, apart from a theorem by F. Schur which will be introduced in another connexion.

Writers who have had the task of explaining the mathematicians' use of the phrases 'curved space' and 'deviation from the Euclidean metric' have explained them, more often than not, in *extrinsic* terms. Often, if not invariably, the chosen metaphor has involved properties which are related to an embedding space. The mistake was already of long standing when de Sitter cautiously wrote that the term mathematically 'interpretable as a curvature of space . . . only serves to satisfy a philosophical need felt by many . . . it has no real physical meaning'.[30] De Sitter's scepticism may be excused, bearing in mind the wide currency of several errors which all pivot upon the idea that if a space of three dimensions is to be 'curved' there must be a fourth dimension into which it can curve. Those chiefly responsible for disseminating the idea amongst the philosophers were Helmholtz[31] and C. H. Hinton.[32] Their mistake was in failing to recognize that the newly-discovered representations of (two-dimensional) non-Euclidean geometries upon curved surfaces in a three-dimensional Euclidean space were representations and no more, and that even these 'curved surfaces' need not be regarded as immersed in a Euclidean space of higher dimension.

3. The Einstein World

We mentioned Ehrenfest's suggestion that boundary-condition problems might be avoided if the four-dimensional world were supposed spherical. This too was Einstein's starting point. Content with the idea

[27] *Gesammelte Mathematische Werke*, 2nd ed. (1892) pp. 272 ff.
[28] *Annali di Mat.* (2), ii (1868) 232 and *Math. Ann.*, i (1869) 575.
[29] *Berlin Monats.* (1869) 159, 688.
[30] *Proc. Acad. Amst.*, xix (1917) p. 1224.
[31] *Mind*, i (1876), iii (1878) in English.
[32] *Scientific Romances*, i (London, 1886); *The Fourth Dimension* (London, 1904).

of a static and spatially finite continuum, he made the following special assumptions:

(i) There is a system of reference relative to which matter may be looked upon as being permanently at rest. Since
$$T^{mn} = \rho \cdot (dx^m/ds) \cdot (dx^n/ds)$$
it follows that $T^{mn} = 0$ unless $m = n = 4$, and $T^{44} = \rho$.

(ii) The density ρ is independent of locality.

(iii) g_{44} is independent of locality (in order that a material particle can remain at rest in a static field) and is independent of the time-coordinate (so that all magnitudes are independent of time). He therefore took $g_{44} = 1$.

(iv) $g_{14} = g_{24} = g_{34} = 0$.

(v) Assumption (iii) was taken as implying that the 'potentials' $g^{\mu\nu}$ can be written
$$g^{\mu\nu} = -\{\delta^{\mu\nu} + x^\mu x^\nu/(R^2 - r^2)\} \tag{10}$$
where $r^2 = (x^1)^2 + (x^2)^2 + (x^3)^2$.

These sets of values for g^{mn} and T^{mn} were, in fact, incompatible with the field equations of gravitation; and yet by modifying these equations in a simple way already explained—a way which preserves their general covariance—the hypothesis of a spatially finite and static continuum could be reconciled with Einstein's General Theory. The field equations were now of the form
$$R_{mn} - \lambda g_{mn} = -\kappa \left(T_{mn} - \frac{1}{2}g_{mn}T\right), \tag{11}$$
with λ, later known as the cosmological constant, a universal constant of unknown value. This constant was known by the success of the Einstein-Schwarzschild theory of planetary motion to be very small.[33] It was now explained how the equations could be derived from Hamilton's Principle, and hence that they were associated with a set of conservation laws. It was further explained how λ was related to the mean density (ρ), the volume $(2\pi^2 R^3)$, radius (R) and mass (M) of a distribution of matter in equilibrium ('the universe') in spherical space. 'Since all points of our continuum are on an equal footing, it is sufficient to carry through the calculation for *one* point', for example, for one of the two points (spherical space) with all coordinates zero. The g_{mn} in (11)

---

[33] For the case of empty space the new field equations reduce to $R_{mn} = \lambda g_{mn}$, and this must be indistinguishable from $R_{mn} = 0$ on the solar scale.

then take Minkowskian values, and calculating the values of R_{mn} on this basis, equations (11) leave just two relations:

$$-(2/R^2) + \lambda = -\frac{\kappa}{2}\rho \text{ and } -\lambda = -\frac{\kappa\rho}{2}.$$

We can thus write:

$$\lambda = \frac{\kappa\rho}{2} = (1/R^2) \tag{12}$$

$$M = \rho 2\pi^2 R^3. \tag{13}[34]$$

A knowledge of only one of the quantities M, R, ρ, and λ obviously entails a knowledge of the other three. Weyl objected to this sweeping correlation: that there should be a relation between λ and M, he said, 'obviously makes great demands on our credulity'.[35] Such a model as that described was later known as an 'Einstein universe'. It was also distinguished as 'cylindrical', taking the time coordinate into account.

Einstein's extension of the field equations is analogous to the extension of Poisson's equation with which he opened his paper: which came first to him is not clear. For the most part, his followers were content to give a loose interpretation of the Einsteinian theory in Newtonian terms, and it was not difficult to see in the λg_{mn}-term a small repulsion from the origin directly proportional to the distance and acting over and above the ordinary gravitational attraction between masses. There would, loosely speaking, be a distance at which such a repulsion would be in equilibrium with the gravitational attraction. As Einstein admitted, the extension of the field equations was 'not justified by our actual knowledge of gravitation' but was merely 'logically consistent'.

4. THE COSMOLOGICAL CONSTANT

The 'principle of the relativity of inertia' was acknowledged to have been prompted by a reading of Mach's *Die Mechanik in ihrer Entwicklung*.[36] (Mach held it to be no accident that the inertial frames of Newtonian mechanics turn out to be those with respect to which distant

---

[34] G. J. Whitrow has pointed out the interesting fact that the ratio (M/R) obtained from (12) and (13)—viz. $4\pi^2/\kappa$—is numerically nearly the same as the ratio of the mass and radius of a homogeneous sphere which has condensed from infinite diffusion and whose intrinsic energy (relativistic) is equal to the negative gravitational potential energy (classical). (Gravitational and inertial masses identified.) Cf. *Nat.*, clviii (1946) 165. Whitrow's ratio is $40\pi/3\kappa$, which is $10/3\pi$ times Einstein's.

[35] *S.T.M.*, p. 279.

[36] 1893. Translated as *The Science of Mechanics* (Chicago, 1902). Cf. pp. 229–38, and especially p. 232.

objects, statistically speaking, have no rotation. In his view, inertial forces were actually caused by the distant masses.) It was not long, however, before Einstein gave an alternative mode of presentation, making λ appear as 'a constant of integration rather than as a universal constant peculiar to the fundamental law'.[37] This is contained in a paper concerned with the role of the gravitational field in the structure of subatomic particles. In the interior of the particles there is a negative pressure $(R_0 - R)$, where R is the scalar of curvature and R_0 is the value of R outside the particles. The negative pressure was said to be responsible for the equilibrium of the electrical charge on the particle (a problem which had puzzled physicists for some years). The cosmological constant, λ, was identified with $R_0/4$, which is essentially a constant of integration. This theory of the electron, which rivalled Mie's theory, was generally believed unsatisfactory to the extent that *any* spherically symmetrical distribution of electrical charge appeared to be capable of remaining in equilibrium.

The name 'cosmological constant' is misleading in the sense that the λg_{mn}-term might just as easily have occurred in an argument (as in the 1919 paper) not concerned with happenings on a cosmical scale. It is now customary to introduce it as a constant of integration in a rather general argument. When setting up the field equations the energy tensor is equated to a tensor depending on the g_{mn} and having zero vectorial divergence. Contracting twice the Bianchi identity the original equation

$$R^{\alpha}_{\beta\gamma\delta,\varepsilon} + R^{\alpha}_{\beta\delta\varepsilon,\gamma} + R^{\alpha}_{\beta\varepsilon\gamma,\delta} = 0 \tag{14}$$

reduces to

$$\left\{ R^{\alpha}_{\gamma} - \frac{1}{2}(\delta^{\alpha}_{\gamma})(R^{\eta}_{\eta} - 2\lambda) \right\}_{,\alpha} = 0. \tag{15}$$

The tensor in curly brackets (the 'Einstein tensor') is of a very general form and has the required property, where λ is an arbitrary constant, as of integration. It is then equated (with raised indices) to a certain multiple of the energy tensor and Einstein's modified field equations are obtained. This treatment, while formally correct, gives very little idea of the way in which this particular problem was originally solved: it also gives, *a fortiori*, little insight into the procedural methods of the subject as a whole. It does, however, show clearly how the inclusion of a non-zero λ increases the generality of the field equations.

Einstein's relation between λ and M had the consequence, which

[37] *Berlin Sitz.* (1919). See paras. 2 and 3.

seemed strange to Eddington, that 'the creation of a new stellar system in a distant part of the world would have to propagate to us, not merely a gravitational field, but a modification of the law of gravitation itself'.[38] The same sort of thing would happen in the case of the destruction of matter, an example of which, Eddington suggested, might be found in the coalescing of positive and negative electrons.[39] But his argument is directed against the whole conception of the 'Einstein universe' and not against the λg_{mn}-term in the revised field equations. We have elsewhere referred to Eddington's very strange reasons for believing that 'a fundamental necessity of physical space . . . established the existence of λ . . .' Since R (or $1/\sqrt{\lambda}$) has the dimensions of length, it must, he said, be the natural unit of length. Later he offered arguments for the 'existence' of the cosmological (or 'cosmical') constant from the microphysical realm.[40]

Eddington, on another occasion, looked for a meaning for the λg_{mn}-term as it occurred in the new field equations. This term, he said, corresponds to the 'absolute energy in a standard zero condition'; for he took T_{mn} to be the 'reckoned energy' and the remaining terms the 'actual absolute energy'.[41] When, as here, he states that λ fixes a zero 'from which energy, momentum and stress are measured', although he seems to believe that he is arguing for the need for the λ-term, he does not explain why the 'energy in the zero condition' should not in fact be zero. The essence of the argument is to be found later: 'λ cannot be zero because the zero condition must correspond to a possible rearrangement of the matter of the universe.' It is impossible to empty the whole universe and it would be 'absurd to define our reckoning of energy by reference to a fictitious process which conflicts with the most important property of matter, namely its conservation'.

It is difficult to know what to make of the assertion that one cannot empty the whole universe. Do we refer to a physical impossibility, or to some sort of conceptual impossibility. Are we merely denying meaning to a sequence of words? One is reminded of the seventeenth-century discussions on the impossibility of a vacuum. We may feel that it would be useful to have a 'self-consistent definition of the zero of energy reckoning', without its being necessary to visualize a state of affairs directly corresponding to such a definition. The terms of a theory need not be provided with physical interpretations in this fragmentary way.

[38] M.T.R., p. 166. In our notation, $M = (4\pi^2/\kappa\sqrt{\lambda})$. Cf. Weyl's remarks.
[39] op. cit., p. 167.
[40] P.R.S. (A), cxxxiii (1931) 605. See Appendix, Note III.
[41] Sci. Prog., xxxiv (1939) 225.

Georges Lemaître has advanced views not unlike these more recently.[42] He considers it desirable that the zero-level of energy should be arbitrary and that the theory should provide some possibility of adjustment. 'To suppose', he wrote, 'that $\lambda = 0$, would mean that the conventional level from which physicists are used to count energy is more fundamental than any other they could have chosen just as well.' This seems to be a reasonable request for the retention of a useful parameter, quite apart from the interpretation with which it is provided. There have been many interpretations of the unrelated 'λ'.[43] The term can be examined in the light of its relation to the energy tensor in the field equations, as it was examined by Eddington and Lemaître. Alternatively, it can be given a more familiar meaning in the light of its survival, under certain conditions, as the field equations degenerate to their Newtonian equivalent. Then, it has been said, the λg_{mn}-term points to a material substratum with old-fashioned ways. Again, it can be looked upon with a view to its metrical and astronomical function, once a line-element has in some way been related to the field equations in which the term originates. This way of discussing its interpretations seems to be least fraught with the sort of dangers which accompany all attempts to give a piecemeal analysis of meaning.[44]

Despite all the arguments for the retention of the cosmological term, there was, as will be seen more fully later, a large body of opinion in favour of its rejection. De Sitter, who retained it, nevertheless referred to it as a term which 'detracts from the symmetry and elegance of Einstein's original theory, one of whose chief attractions was that it explained so much without introducing any new hypothesis or empirical constant'.[45] By 1919 even Einstein had taken up the cry, holding that the introduction of λ was 'gravely detrimental to the formal beauty of the theory'.[46] He held this view to the end of his life and always looked with favour upon the Friedmann metric as avoiding this 'ad hoc addition' to the field equations. He finally discarded the term in 1931, and in doing so deliberately restricted the generality of his theory.

In the light of the principles which Einstein avowedly used in selecting his *previous* gravitational field equations it is hard to see how

---

[42] Article in *Albert Einstein, Philosopher-Scientist* (ed. P. A. Schilpp) 1949.

[43] Cf. V. V. Narlikar, *Nat.*, clxxvii (1956) 1138 and Schlüter, *A.J.*, lx (1955).

[44] Weyl developed some interesting but uninfluential views on λ. See Appendix, Note IV.

[45] *Proc. Acad. Amst.*, xix (1917) 1225. His reference to no 'new empirical constant' suggests that he thought of Einstein's theory as a generalization of Newton's rather than as a separate entity.

[46] *P.R.*, p. 193 (= *Berlin Sitz.* (1919)).

the omission of the λg_{mn}-term can be regarded as anything but an oversight. At all events, in assigning the value *zero* to the constant λ, loss of generality has seemed, to the majority, too high a price to pay for brevity and elegance. Einstein's removal of the term seems, in retrospect, to have been a far more arbitrary action than his inclusion of it. If, on the other hand, the original reasons for including the term are withdrawn owing to a change in hypothesis elsewhere, then the complexion of the issue changes. We recall that Einstein sought to preserve the tensorial nature of his field law in order to preserve its general covariance. He also maintained that energy is conserved. Several recent authors have rejected the second hypothesis in the establishment of their own theories. In Chapter 10 we shall turn to this issue and, once again, to the function of the cosmological constant.

5. The de Sitter World and 'Mach's Principle'

Einstein, in introducing λ into his field equations, clearly believed that they had no possible solution for empty space. Such a solution would have shown the equations to be incompatible with the 'principle of the relativity of inertia' which, as already explained, requires that inertia should be fully and exclusively determined by matter and that it should, consequently, be impossible to determine the g_{mn} (and hence the inertial behaviour of a test particle) in the complete absence of matter. Einstein was very soon shown to be mistaken in supposing there to be no solution for this case, for early in 1917 de Sitter found such a solution.[47] The importance of de Sitter's work was recognized from the very first: witness the large and immediate following it found in astronomical and, more especially, mathematical journals. Its importance for the problem of inertia was at first one of its main sources of interest, although circumstances alter cases, and five or six years later astronomers began to find a new significance in de Sitter's theory.

De Sitter, like Einstein and most of his contemporaries, believed that any large-scale astronomical problem might be referred to as an 'unchanging background of space'. In Einstein's words, 'that term

[47] The relevant papers by de Sitter will be referred to by the following abbreviations:

Proc. Acad. Amst., xix (1917) 1217 = deS. (i).
     xx (1917) 229 = deS. (ii).
     xx (1917) 1309 = deS. (iii).
M.N., lxxviii (1917) 3   = deS. (iv).

It will shortly be seen that Minkowski's space-time metric is a solution of the field equations which, just as much as de Sitter's, violates Einstein's principle. That this fact was not commented upon at the time is hard to explain—perhaps cosmologists concentrated on de Sitter's metric as being not only scandalous but interesting.

[λg_{mn}] is necessary only for the purpose of making possible a quasi-static distribution of matter, as required by the fact of the small velocities of the stars'.[48] He does not appear to have been disturbed by the thought that his static 'universe' provided a standard of rest: by virtue of its homogeneity it allows of a single and universal set of inertial frames. Even de Sitter, with his extensive knowledge of current astronomy, had no hesitation in holding that the field should be regarded as *static* and *isotropic*, both conditions being 'closely followed in Nature'.

De Sitter used the conditions of isotropy and the static state, together with the further condition that the spatial section should be of *constant curvature* (of radius R), to show that the field equations allowed of three solutions for the coefficients of the general line-element:

$$ds^2 = -a dr^2 - b(d\psi^2 + \sin^2\psi d\theta^2) + f dt^2. \tag{16}$$

The resulting line-elements were, respectively,

(A) $ds^2 = -dr^2 - R^2 \sin^2(r/R)(d\psi^2 + \sin^2\psi d\theta^2) + c^2 dt^2$
with $\kappa\rho = 2\lambda$ and $\lambda = 1/R^2$ and zero pressure

(B) $ds^2 = -dr^2 - R^2 \sin^2(r/R)(d\psi^2 + \sin^2\psi d\theta^2) + \cos^2(r/R) c^2 dt^2$
with $\rho_0 = 0$ and $\lambda = 3/R^2$ and zero pressure

(C) $ds^2 = -dr^2 - r^2(d\psi^2 + \sin^2\psi d\theta^2) + c^2 dt^2$
with $\rho_0 = 0$ and $\lambda = 0$ and zero pressure.

(17)

The first of these solutions was the one already given by Einstein, whilst the last is the metric of Minkowski space-time. The second was de Sitter's original contribution to the subject, a solution which he often referred to as 'the general solution for the case of a static and isotropic gravitational field in the absence of matter'.[49] To substantiate

---

[48] *Berlin Sitz.* (1917) 152. (Last sentence.)

[49] *Notation:* Although there are many variants in the coordinate notations used by different authors, there is no difficulty in sorting out the different conventions as to 'angular' coordinates. The use of the 'radial' coordinate (r) is very confusing, however, and we shall use the following convention: the 'r' of de Sitter's work will henceforth be denoted by 'r_s'. This has generally been interpreted as appropriate to rod measurements. Most writers of the 1920s use de Sitter's 'r', but Eddington uses the same symbol (here distinguished as 'r_e') in a sense corresponding to 'parallax distance'. The relation between r_e and r_s is $(r_e/R) = \sin(r_s/R)$.

The de Sitter metric can now be written

$$ds^2 = -dr_e^2/(1-r_e^2/R^2) - r_e^2(d\psi^2 + \sin^2\psi d\theta^2) + c^2(1-r_e^2/R^2)dt^2.$$

Robertson and others made this the more commonly accepted form. Many of the formulae given below have been altered slightly to make them conform to one of these two conventions.

his claim, he generalized a proof published by Levi-Civita in May of the same year.[50] Levi-Civita, discarding the condition of isotropy and allowing that the coefficient f in the general metric may be a function of r, ψ, and θ, gave a general solution for the differential equation in f. This equation had been formed, however, from Einstein's earlier field equations, and de Sitter now extended the proof to include the λg_{mn}-term.[51] Finally, introducing the condition for isotropy, he showed that the general solution for f (under these conditions) agreed with his own, namely

$$f = \text{const} \times \cos^2(r/R). \tag{18}$$

The question next arose of the need for the principle of the relativity of inertia. De Sitter asked himself: 'If no matter exists apart from the test body, has this inertia?' The system (B) ostensibly contradicts the neo-Machian answer to this question (for the density is zero throughout the de Sitter world), and a consistent Mach might well have chosen to follow Einstein in postulating some sort of 'world-matter' as an essential feature of any satisfactory model.[52] As regards the 'world-matter', de Sitter was exaggerating when he wrote that it 'serves no other purpose than to enable us to suppose it not to exist', for it 'existed' in order to elicit a solution to the field equations which had many satisfactory features. He believed that he could not reconcile his own solution with Mach's ideas for the simple reason that, although devoid of material, it predicted inertial properties for test objects. He did, however, offer a 'mathematical postulate' in place of what he called the 'material postulate of the relativity of inertia'. This 'makes no mention of matter, but only requires the g_{mn} to be zero at infinity', although it 'does not appear to admit of a simple physical interpretation'.[53] It is not clear why de Sitter felt conformity to this principle to be so desirable a feature of his own account.

The three solutions were always regarded as *idealizations* of the 'true state of affairs' (a phrase which Einstein frequently used). The high degree of idealization worried de Sitter, who noted that only if the g_{mn} were to deviate from the values indicated in the expressions for the line-elements could they be said to exhibit 'gravitational' effects.[54] It was not that he had failed to absorb the concept of a *forceless* gravitation, but that he was looking for *differential* gravitational effects. That

---

[50] *Rend. Lincei*, xxvi (1917) 519.
[51] Levi-Civita was aware of the λ-term. See p. 105 *infra*.
[52] Actually Mach—who had died in 1916—opposed even Einstein's Special Theory.
[53] deS. (iv), pp. 6, 18.   [54] deS. (ii).

the model contained matter producing 'inertia but not gravitation' (de Sitter) does not appear to have been the cause of much concern on Einstein's part. The Machian issue remained, however, and what no one doubted was that the principle of the relativity of inertia had been undermined.

It will shortly be seen that other forms of the de Sitter metric were discovered, indicating the same unfortunate conclusion. As an example which is easy to follow, consider the metric

$$ds^2 = c^2 dt^2 - \exp(2t/R) \cdot (dx^2 + dy^2 + dz^2), \qquad (19)$$

where R is constant. Since at least one component of the Riemann-Christoffel tensor is non-zero and constant,[55] the tensor is not null and the space-time with the given metric is 'curved'. Calculating the values of the components of the Ricci tensor from the coefficients of the metric, and deducing the corresponding components of the energy tensor from Einstein's modified field equations, it emerges that all these components are identically zero. There is, moreover, no singularity (none, that is to say, which cannot be removed by a coordinate transformation) at which the expressions for the components of the energy tensor are mathematically undefined and which might have been interpreted as some material event. The space-time, therefore, represents no distribution of matter and yet it possesses a quite definite inertial structure, in the Einsteinian sense.

More recently, A. H. Taub[56] again demonstrated the inadequacy of Einstein's efforts at implementing what he had called 'Mach's Principle'. Whereas it was formerly thought that the null character of de Sitter's energy tensor was the direct result of introducing the cosmological term,[57] Taub drew attention to conditions under which the *unmodified* field equations yield curved space in the absence of matter. That the space-time is always curved can be shown by calculating two components of the Riemann-Christoffel tensor from the (non-static) metric in question,[58] and showing that there is no value of t for which both of

---

[55] For example, $R^{\alpha}_{44} = \lambda/3$, where $\alpha = 1, 2, 3$.

[56] *Ann. Math.* Princeton, liii (1951) 472.

[57] De Sitter himself sometimes spoke as though it were. (If $\lambda = 0$, de Sitter's metric coincides with that of Minkowski space-time, but this has zero curvature and aroused no cosmological interest.)

[58] Taub, op. cit., p. 481.

$$ds^2 = \gamma dt^2 - \gamma_{11}(dx^1)^2 - (\gamma_{11} \sin^2 x^1 + \gamma_{33} \cos^2 x^1)(dx^2)^2 \\ - 2\gamma_{23} \cos x^1 dx^2 dx^3 - \gamma_{33}(dx^3)^2,$$

where $\gamma = \gamma_{11}^2 \gamma_{33}$ and $\gamma_{11} = \{k \cosh(kt + \alpha)\}/\{4 \cosh^2(kt + \beta/2)\}$ and $\gamma_{33} = k/\cosh(kt + \alpha)$ and α, β, and k are constants. See Appendix, Note V.

these components (functions of t) are simultaneously zero. The components of the Ricci tensor are, nevertheless, identically zero, and thus if there is no λg_{mn}-term in the field equations the energy tensor must also be identically zero. As there are, in addition, no singularities, there is nothing which corresponds conventionally to the 'gravitational field'. There is no longer any excuse (if there was ever an excuse) for the parlance 'the curvature of space-time is the cause of gravitation'; for whether or not the cosmological constant is included in the field equations—de Sitter included it, Taub did not—there are curved spaces corresponding to null-energy tensors. Conversely, however, the field equations in either form imply that a non-zero energy tensor must be associated with a non-zero Ricci tensor; and consequently with a non-zero Riemann-Christoffel tensor; and hence with a space-time of definite curvature.

It was not our purpose in this section to give an exegesis of those of Mach's writings which so influenced Einstein, nor is there much to be gained by asking whether he distorted them unduly. There is, nevertheless, an interesting logical difference between the earlier and later ideas. Rotation with respect to distant material may well cause inertial forces as Mach suggested. But it does not follow from this that the non-existence of distant masses is incompatible with such forces. In the language of cause and effect, we can say that one effect may have many causes. Why, then, was the discovery of the de Sitter model such a blow to Einstein's hopes? It was that he needed more than a mere cause-effect relation: he required an *equation* functionally relating the inertial properties of space-time to the material distribution. (The logical difference here is roughly that between material implication and equivalence.) The inertial structure of space-time was, he said, to be 'exhaustively conditioned and determined' by the distribution of material throughout the universe. And in this case the prediction of inertial properties in a universe without material is incompatible with Einstein's version of Mach's Principle.

One can still ask, however, why Einstein ever thought his basic gravitational theory to have been compromised by de Sitter's solution. What is incompatible with the Mach-Einstein principle is, after all, a cosmological model. Such models require for their development at least two important principles which are quite extraneous to the General Theory of Relativity. (These are what will later be referred to as a 'cosmological principle' and 'Weyl's Principle'.) Why did Einstein, de

Sitter, and those of their contemporaries who wished to preserve the Mach-Einstein principle not choose to place the blame on one of these? This was probably because the part played by such principles was as yet unclear; and also because the principles—so far as they were clearly perceived—had a very secure look about them. But another reason for thinking the General Theory to have been compromised, and not a very creditable one, seems to have been that the theory went so far as to *permit* the Mach-Einstein principle to be contradicted. This is rather like saying that projective geometry is unsound because it is indifferent to the parallel postulate.

6. The de Sitter Effect and Weyl's Principle

The supposition that the universe is, on the whole, static did not survive into the 1930s. Had it done so it would have had to face conflict with the accepted theories of thermodynamics; for the known emission of radiant energy in any region is far in excess of the known absorption of energy within the same region.

On the other hand, the 'de Sitter effect', the lowering of the frequency of light from distant sources 'in the course of its journey', was appreciated from the first.[59] It was explained as a 'slowing down of time at great distances from the observer', manifesting itself as an apparent slowing down of atomic vibrations, and hence as a spectral shift. The effect had its paradoxical side, for it was seen that *every* observer in a de Sitter world would have a 'horizon' at which any finite value of ds would correspond to an infinite value of dt. Nature should there appear to be at rest; although this would not be the view of someone on the horizon, who would, of course, have a different horizon of his own. Many objections were raised in this connexion. Einstein, for example, held that a solution of the field equations was not physically admissible if it admitted discontinuities at finite distances.[60] In particular he claimed that the determinant g must, for all points at finite distances, be non-zero. De Sitter, whose solution (B) did not fulfil these conditions, argued that they amounted to a 'philosophical postulate'.[61] He observed that his solution conformed to a corresponding 'physical postulate' in which the words 'for all points at finite distances' were replaced by the words 'for all physically accessible points'. The discontinuity in the form of a horizon (at $r_s = \pi R/2$), whilst it was 'at a finite distance in space', was nevertheless 'physically inaccessible'.[62] Generally speaking,

[59] See deS. (i)–(iv).   [60] *Berlin Sitz.* (1918) 270.
[61] deS. (ii) 230. By 'philosophical' he means something like 'not empirically verifiable'.
[62] Cf. deS. (iv) 17–18.

fewer paradoxes would have been acknowledged had the distinction between proper time and coordinate time been carefully drawn.

For ten years or so the de Sitter effect was discussed; and yet at the end of this time there were so many apparent discrepancies between the results obtained, that subsequent texts have preferred to derive a result only in the relatively simpler case of a de Sitter metric with cosmic time where Weyl's Principle is adopted.[63] De Sitter himself, as we shall see, used a hypothesis which was of value only in so far as it led to a simple result: no other reason could be adduced for holding to it. In order to draw together the many loose ends we shall therefore first give a fairly general account of the theory of the de Sitter effect.

We begin by determining the equations of light paths in the de Sitter model, and we shall show that there is a slight formal advantage to be had in using r_e rather than r_s. We take the line-element in the form (B):

$$ds^2 = \cos^2(r/R) \cdot dt^2 - dr^2 - R^2 \sin^2(r/R) \cdot (d\psi^2 + \sin^2\psi \cdot d\theta^2).$$

Here, as in most of this section, r is to be identified with r_s. We shall suppose that by symmetry it is clear that if the particle be at any time moving in the plane $\psi = \pi/2$ it will remain so (this can easily be proved). The first integrals of the differential equations of the geodesics are without much difficulty found to be:

$$\frac{dr}{ds} = \pm \sqrt{\{\alpha^2 \sec^2(r/R) - 1 - \beta^2 \operatorname{cosec}^2(r/R)\}}, \tag{20}$$

$$\frac{d\theta}{ds} = \beta \operatorname{cosec}^2(r/R), \tag{21}$$

and $$\frac{dt}{ds} = \alpha \sec^2(r/R), \tag{22}$$

where α and β are constants of integration, peculiar to a single particle. Since we want a positive increase in t to correspond to a positive increase in s, (dt/ds), and hence by the last equation α, must be positive. As the geodesics tend to null geodesics, the last two equations show that α and β then tend to infinity. Imposing this condition we have as the differential equation of a light ray:

$$\frac{dr}{d\theta} = \pm \frac{1}{R^2} \sin 2(r/R) \cdot \sqrt{[(\alpha^2/\beta^2) \cdot \{\tan^2(r/R) - 1\}]}. \tag{23}$$

[63] Weyl's Principle is discussed on pp. 100 ff. *infra*.

This has the solution

$$\operatorname{cosec}(r/R) = a\cos\theta + b\sin\theta, \qquad (24)$$

where $\qquad (\alpha^2/\beta^2) = (a^2 + b^2 - 1)$

(i.e. we have introduced only one constant of integration). Now it is obvious that (24) would represent a straight line if the left-hand side were inversely proportional to a measure of radial distance. This is precisely the formal advantage which obtains if we use r_e instead of r_s (for which we here use the symbol r). Since light rays are rectilinear in the (r_e, θ, ψ)-space, triangulation methods can be used to determine the radial coordinate. This is what writers meant when they said, in effect, that r_e represents parallax distance.[64]

Equation (22) tells us the relation between the time interval proper to an observer at a distant point and the corresponding coordinate interval (assuming that the point is in the plane $\psi = \pi/2$). From the expression for the line-element it is clear that the coordinate-time for a photon to reach an observer at the origin from a source at distance r is

$$\int_0^r \sec(r/R)\,dr \qquad (25)$$

$$= f(r), \text{ say.}$$

(This being equal to $R\log(1+\eta)/(1-\eta)$, where $\eta = \tan(r/2R)$, it is infinite when the upper limit equals $\pi R/2$—hence the 'horizon'.) Using a suffix 1 for emission and 2 for reception, this is equal to $(t_2 - t_1)$. The time intervals characteristic of a periodic phenomenon such as a light wave will then be related by the equation

$$\Delta t_2 - \Delta t_1 = f'(r)\cdot\frac{dr}{dt}\cdot\Delta t_1, \qquad (26)$$

where (dr/dt) is the coordinate velocity of the particle at the time of emission. We also have (by (22))

$$\Delta s_1 = \Delta t_1/\{\alpha\sec^2(r/R)\} \qquad (27)$$

and $\qquad \Delta s_2 = \Delta t_2, \qquad (28)$

---

[64] It is occasionally held that light paths are rectilinear in terms of the coordinates later introduced by Robertson and Lemaître, viz. $(\bar{r}, \theta, \psi, t)$. This is in fact only true if radial distance is defined as $(\bar{r}/R)\exp(t/R)$.

wherefore there is a spectral shift given (exactly) by

$$1 + \delta = \Delta s_2/\Delta s_1 = \alpha(1 + vf') \sec^2(r/R)$$
$$= \alpha\{1 + v \sec(r/R)\} \sec^2(r/R), \qquad (29)$$

using (26)–(28) and writing (dr/dt) as v. (In this equation α has nothing to do with the light-path, but is the parameter fixing the orbit of the source.) There will clearly be a displacement tending to infinity as r tends to $\pi R/2$. What does not appear to have been noticed before is that a particle with a finite velocity of approach (however small this may be) placed at a greater coordinate distance than $R \cos^{-1}|v|$, will exhibit a violet shift and that the shift *increases indefinitely* as the particle nears the horizon, however small its velocity.[64*] In other words, unless we make the highly artificial hypothesis that all particles have no coordinate velocity (and this is not to be true statistically but precisely) then there should be, on the de Sitter model, some very large violet shifts amongst nebular spectra. This is not found to be the case. It is not simply a question of red-shifts predominating over the violet: apart from a handful of local nebulae there have been found no displacements to the violet at all. As it happens, the orbits of particles as calculated from (20)–(22) are such that particles will spend more time receding from an observer than otherwise, as several authors discovered. This, however, cannot save the de Sitter model; nor can any plausible transformation of the form $r' = f(r)$, $t' = g(t)$. (No such transformation is acceptable unless the functions are monotonic increasing. Under such transformations v cannot become negative, and whatever finite absolute value it assumes, a value of the new radial coordinate can always be found such that the source will exhibit a violet shift to some observer.) One of the first to emphasize the probable astronomical significance of the de Sitter effect was Eddington, writing in his *M.T.R.* (1922). Eddington, like all astronomers, spoke of 'radial velocity' when he was fully aware that de Sitter's theory led to a twofold explanation of the spectral shifts. He gave an expression for the de Sitter effect corresponding to (29),[65] but this differs from (29), not merely superficially. Eddington has one fewer power of $\sec (r_s/R)$ on the right of his equation 70.4. The source of the difference is that instead of equation (22)—and (27)—he works from the expression for the line-element, writing (dt/ds) as the equivalent of $\sec (r_s/R)$. But the proper time interval for an observer moving with the source cannot be related

---

[64*] [P. D. Noerdlinger points out to me that my definition of v makes it a quotient of a physical distance and—as normally interpreted—a coordinate-time. Taking instead a differential of clock-time, as measured by a local stationary observer, there will be large red-shifts that cannot be compensated by any small velocity.]

[65] *M.T.R.*, p. 164. The notation differs somewhat from ours.

to the coordinate time interval unless we know the velocity components of the source. Eddington cannot have been oblivious of this, for at the vital stage in the calculations he adds the words 'neglecting the square of the velocity of the atom'.[66]

As we have seen, the first term on the right of equation (29) depends on the position of the source at the time of emission: in the second term, position is more or less unimportant by comparison with coordinate-velocity. The formula may be looked upon as showing how to calculate the coordinate-velocity from an observed spectral shift when the radial coordinate is known, and clearly there was no justification for a straight acceptance of the astronomers' 'radial velocity' as ordinarily calculated. It should now be obvious that the 'de Sitter effect' of (29) cannot, by itself, be interpreted as indicating a general nebular recession. (We notice that even a shift to the red may indicate a 'velocity' of approach.) Some such 'real recession' was, however, soon held to be a concomitant of de Sitter's theory. Eddington was probably the first to maintain that 'a number of particles initially at rest (in the de Sitter world) will tend to scatter', and to realize the astronomical significance of this behaviour.[67] A straightforward calculation of the equations to the geodesics of the space-times (A) and (B) (such as led to equations (20)–(22)) showed that the radial acceleration of a particle is zero in the former case, but is given by

$$\frac{d^2r}{ds^2} = \frac{\lambda r}{3} \tag{30}$$

in the latter. Eddington paid less attention to this equation than he might have done. It is worth noticing that it can be integrated to give

$$\frac{dr}{ds} = r \cdot \sqrt{\left(\frac{\lambda}{3}\right)} \tag{31}$$

if we assume that a particle has negligible 'velocity' on leaving the origin. Here we should have had a first theoretical 'velocity-distance relation' (1922). But before making further conjectures of this sort we must indicate the important special assumptions made by Eddington, which make the step to equation (31) invalid. From the equation for the geodesic (whose first integral corresponds to (20)) equation (31) was

---

[66] loc. cit.
[67] *M.T.R.*, section 70. They will scatter, he said 'unless their mutual gravitation is sufficient to overcome this tendency'. Using the Newtonian analogue of λ, we may loosely say that gravitational attraction being absent in an empty de Sitter universe, the (massless?) test particles in it will scatter under cosmical repulsion.

obtained only 'for a particle at rest'. In our notation, putting $(d\dot{\theta}/ds) = (dr/ds) = 0$ in (20) and (21) we then have

$$\beta = 0, \quad \alpha^2 = \cos^2(r_s/R). \tag{32}$$

Differentiating (20) with respect to s, and replacing α and β by these values, we obtain

$$\frac{d^2r}{ds^2} = \frac{1}{R}\tan(r_s/R). \tag{33}$$

Changing to the variable r_e, (30) follows, bearing in mind de Sitter's relation between λ and R.

We can now understand the dissatisfaction with Eddington's result which was voiced by several of his contemporaries. A consistently accelerating particle cannot remain perpetually at rest, and it was soon to be shown that inserting general expressions for radial and angular velocities in the expression for (d^2r/ds^2), very different results were to be obtained. We shall not follow up the analysis any further, but merely indicate some of the results. The correctness of the prediction that particles should scatter was denied, for example, by L. Silberstein:[68] Eddington's particle, he explained, may certainly recede at first, gathering speed until it reaches the mid-point between its starting point and its polar. Henceforth it will recede, but with decreasing velocity. He calculated that the ratio of the volumes of the two regions within which the two kinds of behaviour were found (for any origin) were in the ratio $(\pi - 2):(\pi + 2)$, and that 'the celestial objects endowed with a gathering tendency should, therefore, be [about] five times as numerous as those showing a tendency to scatter'.[69] Some years later this very property of the de Sitter world was to detract from its value in interpreting astronomical data. Writing in 1922 Eddington had been very puzzled by the asymmetry of his result: 'In the full formula [for (d^2r/ds^2)] there are no terms which under any reasonable conditions encourage motion towards the origin. It is therefore difficult to account for these motions, even as exceptional phenomena.'[70] As Silberstein and others showed, it was all too easy. We shall shortly return to Silberstein's work in connexion with the de Sitter effect.

Another and more complete account of the 'tendency to scatter' was given by E. T. Whittaker in a memoir referred to elsewhere in connexion

---

[68] Eddington's conclusion was said to be 'hasty and inexact'. See *Phil. Mag.*, xlviii (1924) 619. See also *The Size of the Universe* (London 1930), p. 117.
[69] loc. cit. [70] *M.T.R.*, p. 162.

with the definition of distance in curved space.[71] The paper sets out very clearly the results to be expected, although by this time interest in de Sitter's world was more or less academic. Whittaker began by defining the 'proper distance' (d) between two points (x_0, y_0, z_0) and (x_1, y_1, z_1) by the equation

$$\cos d = \frac{(1 + x_0 x_1 + y_0 y_1 + z_0 z_1)}{\sqrt{(1 + x_0^2 + y_0^2 + z_0^2)} \sqrt{(1 + x_1^2 + y_1^2 + z_1^2)}} \qquad (\pi > d > 0). \tag{34}$$

(This definition of the invariant d was suggested by the Cayley-Klein geometry with Absolute

$$x^2 + y^2 + z^2 + 1 = 0 \tag{35}$$

in which the above expression is the usual one for the distance function.) The 'spatial distance' (Δ) between two points was then defined by

$$\Delta = \frac{R \sin d}{\cos (\gamma + d)} \tag{36}$$

(γ is a constant and R is the radius of curvature of the space) and it was shown that 'a freely moving star and a freely moving observer cannot remain at a constant spatial distance apart'. The 'nearest resemblance to a universe whose stars have a constant configuration' was taken to be the case of *constant proper distance*. Within such a scheme—for which the only possible justification would be an *a posteriori* one—Whittaker proceeded to explain the sequence of observations which would be made on a star at a constant proper distance. These were as follows:

(1) The first opportunity of seeing a star is when its spatial distance is equal to the radius of curvature of the universe.

(2) The spatial distance is subsequently given by

$$\Delta = R \sin d \cosh (\tau_0/R).$$

(τ_0 is the 'proper time of the observer'.)

(3) Δ has a minimum value of $R \sin d$, after attaining which the spatial distance increases indefinitely. (As the observer approaches the point where his world line terminates on the Absolute, τ_0 increases indefinitely.)

[71] *P.R.S.* (A), cxxxiii (1931) 93.

Although Silberstein and Whittaker had ostensibly shown that Eddington was only partially correct in deducing from the simple de Sitter theory a *universal* recession, yet, from the time of Eddington's writing, more and more evidence had been accumulating which seemed to support his conclusion. V. M. Slipher of the Lowell Observatory prepared for Eddington's book as complete a list of the 'radial velocities' of spiral nebulae as was then (February 1922) possible, and the indications were that the vast majority recede from us.[72] Only five from a list of forty-one spirals appeared to be approaching, and their approach could have been explained away if more had been known of the Galactic rotation.

We have so far discussed two separate questions in connexion with the de Sitter effect: What will be the relation in general between the displacement of the spectrum of a source and its kinematic behaviour? How will a test particle move in general? The next problem, and that which probably aroused most interest, concerns the relation between the spectra and distances of light-sources in a de Sitter world. For ten years the discussion of this problem was carried out at a polemical level. Authors professed not to understand any working which led to an unwelcome result. In the rest of this section we shall therefore use the results obtained at the beginning, in an endeavour to reconcile some of the more notable conclusions reached at the time.

De Sitter himself derived an expression for spectral displacement as a function of his radial coordinate r_s:

$$1 + \delta = \sec(r_s/R). \tag{37}$$

This was valid only on the restrictive assumption that the source remained with fixed r_s coordinate relative to the observer. In terms of our previous working, we write $(d\theta/ds) = (dr/ds) = 0$. From the second it follows that $(dr/dt) = 0$ $((ds/dt)$ is finite), wherefore, combining equations (32) and (29), we have de Sitter's result.

Eddington discussed in general terms the shifts to be expected in the spectra of sources accelerated as he had predicted. Had he given an explicit (δ, r) relation he would presumably have made de Sitter's assumptions (see the discussion following equation (31)), and yet these imply a radial acceleration. It seems that de Sitter's assumption of constant r_s implies the imposition of some sort of *constraint*. Would it not be more satisfactory if we could find a radial coordinate $\bar{r}$, such that

[72] *M.T.R.*, p. 162.

the condition $(d\bar{r}/ds) = 0$ implied that the corresponding acceleration vanished? This, in effect, is what Weyl did very shortly after Eddington's book appeared.[73]

Weyl began by supposing that the stars (or nebulae, in later contexts) lie on a pencil of geodesics diverging from a common event in the past. (This supposition, which has often been called 'Weyl's Principle', played an important part in later cosmology.) The stars were then found to be in an 'undisturbed state', the meaning of which may most easily be seen as follows. If, in the set of differential equations for the geodesics, the 'spatial-velocity components' $\left(\dfrac{d\bar{r}}{ds}, \dfrac{d\psi}{ds}, \dfrac{d\theta}{ds}\right)$ are zero, and if the Christoffel symbols are given the values calculated from the Lemaître-Robertson form of the (de Sitter) line-element, then it is easily shown that the corresponding accelerations vanish.[74] These stars of the Weyl model are therefore at rest in terms of $\bar{r}$, ψ, and θ, whatever the origin of the coordinates. The way in which the 'distances' r_e and r_s are related to $\bar{r}$ shows at once that they will change in time even though $\bar{r}$ remain constant.[75] Insertion of the differential coefficient (dr_s/dt) into the formula for spectral shift will then provide the desired expression. This expression will be found to differ from that originally obtained by Weyl himself, namely,

$$\delta = \tan(d/R), \qquad (38)$$

where d is the 'naturally measured distance of the star in the static space *at the moment . . . at which the observation takes place*'.

As we shall see later, the relation between $\bar{r}$ and r_s is

$$\bar{r} = \tan(r_s/R)\exp(-t/R). \qquad (39)$$

Differentiating, and bearing in mind the postulated constancy of $\bar{r}$, we find that

$$(dr_s/dt) = \sin(r_s/R) \cdot \cos(r_s/R). \qquad (40)$$

Now $\beta = 0$ (since, for a given nebula obeying Weyl's Principle, θ is constant) and combining (20), (22), and (40) we find that the constant α for the particles of a Weyl-de Sitter model is unity. In other words, α

---

[73] *Phys. ZS.*, xxiv (1923) 230; *S.T.M.*, 5th ed. (not available in translation). Cf. *Phil. Mag.* (7), ix (1930) 936.

[74] Such a simple account was not available to Weyl in 1923, but he later admitted that Robertson's coordinates 'are by far the most appropriate here' (loc. cit.).

[75] See the following chapter.

does not vary from particle to particle as is usually the case. We can now combine (40) and (29) to give

$$1 + \delta = 1/\{1 - \sin(r_s/R)\}, \tag{41}$$

a result which agrees with Weyl's only for small values of (r_s/R), if d and r_s are identified. In terms of r_e, this becomes

$$1 + \delta = 1/(1 - r_e/R). \tag{42}$$

In terms of Robertson's coordinates, the relation

$$\bar{r}/R = \exp(-\bar{t}_1/R) - \exp(-\bar{t}_2/R) \tag{43}$$

may be obtained by integrating the expression for light velocity. The transformation equation for the time coordinate is

$$\bar{t} = t + R \log \cos(r_s/R), \tag{44}$$

which with (39) gives

$$R \sin(r_s/R) = \bar{r} \exp(\bar{t}/R). \tag{45}$$

Equations (41), (43), and (45) together give (introducing suffixes 1 and 2 as before)

$$1 + \delta = \exp\{(\bar{t}_2 - \bar{t}_1)/R\}, \tag{46}$$

a result which is well-known (and which may be obtained much more simply in other ways)—a fact which serves to support the somewhat tortuous derivation of equation (41). The three (δ, r) relations can all be written in the form

$$\delta = r/R \tag{47}$$

when higher powers of (r/R) may be neglected. To this equation we shall return in the next chapter.

In using (45) we gave $\bar{t}$ the suffix 1. Weyl, as we saw, was claiming in effect that he originally used the suffix 2. If this is done, Weyl's result is not confirmed, even with the help of his later explanations. Nor does Weyl arrive at the equivalent of our equation (46) (see equation (50) below). The crux of the matter is at the equivalent of our equation (43),[76] which he differentiates in order to obtain $\Delta\bar{t}_1/\Delta\bar{t}_2$ (i.e. $1 - \delta$). There seems to be no justification for this step. Soon after its first announcement, Silberstein took exception to Weyl's Principle and claimed to reject it in determining an alternative expression for the

[76] op. cit., p. 941.

relation between spectral displacement and distance in the de Sitter world.[77] Silberstein's were probably the most important contributions to this subject, although whether he himself appreciated their relation to the work of others is obscured by his polemical style. There were references to Weyl's 'perfectly gratuitous assumption' and his 'sublime guess'; and yet Silberstein, who purported to have dropped all such assumptions himself, has somehow acquired a reputation for having taken Weyl's Principle and actually added to it. We shall see that this is at best a half-truth. It should by now be clear that in order to obtain a determinate (δ, r) relation some restrictive hypothesis must be made in order to decide α (and possibly v) in equation (29). Silberstein's merit was that he made no such hypothesis at the outset and that he discussed the orbits to be expected of particles in the de Sitter world. Tolman later gave a more complete account,[78] but most of his results were implicit in Silberstein's second paper. (Tolman gave several results using different expressions for the parameters of the orbit. Strangely enough he did not consider Weyl's Principle as such, although he considered 'a rather arbitrary assignment of values to the parameters' which amounts to the same thing.)

Before comparing Silberstein's result with ours of (29) we will combine it with a value for (dr/dt) obtained from (20) and (22). The resulting equation is:

$$1 + \delta = \alpha \sec^2(r/R)\,[1 \pm \sqrt{\{1 - (1/\alpha^2)\cos^2(r/R)} - (\beta^2/\alpha^2)\cot^2(r/R)\}]. \quad (48)$$

Now Silberstein's first paper gives a result for purely radial motion: this is hardly a severe restriction, for motion once radial will continue so without constraint. Rather than quote this result we will give that of his second paper (of which the other is a special case):

$$1/(1+\delta) = \alpha \sec^2(r/R)\,[1 \mp \sqrt{\{1 - (1/\alpha^2)\cos^2(r/R)} - (\beta^2/\alpha^2)\cot^2(r/R)\}]. \quad (49)$$

(Our notation. The upper signs of (48) and (49) correspond to recession.) This equation 'for the rigorous value of the Doppler effect, for a star and an observer endowed with any inertial motion' will only agree with (48) if V and r are related in a very special way. That Silberstein imposed such a complex hypothesis as is necessary is hardly credible, and is

[77] *Phil. Mag.* (6), xlvii (1924) 907. (Weyl's reply: xlviii (1924) 348) and xlviii (1924) 619.
[78] *Ap. J.*, lxix (1929) 245.

certainly not what later writers objected to. We find the discrepancy between (29) and (49) stems from the fact that Silberstein (like Weyl) took *the source* as the origin of coordinates and by r signified 'the distance of the observer from the star at the instant of taking its spectrogram'. The two equations do not agree, for example, under Weyl's Principle. It is easily found that (49) reduces to

$$\delta = \sin(r/R), \qquad (50)$$

which agrees with Weyl's formula ((38) above) only for small (r/R). It is now hoped that, as far as possible, the results of the several authors have been reconciled.

As to the contention that Silberstein adopted and extended Weyl's Principle, this is somewhat unfair. In deriving the formula (49) he clearly did no such thing, and the insertion of alternative signs (which, as he thought, saved him from Weyl's fate of being 'flatly contradicted' by the handful of nebulae with spectra shifted towards the violet) was analytically perfectly proper. By reason of his having done so, however, it was wrongly claimed that he tacitly added to Weyl's diverging pencil of lines a converging pencil. (The claim was Weyl's; it was repeated by Robertson and many later writers.) It is an idle claim unless it can also be shown that he took the value of α to be unity. His most heretical utterance appears to be this: 'Thus far the rigorous spectrum-shift formula. Turning to practical applications, we can at any rate confound $\gamma \: [= \alpha] \: . \: . \: .$ with unity.'[79] Why does he say this? Because at perihelion, r_s^2 and v^2 may be neglected by comparison with unity 'in all cases of actual interest' (and from this it can be formally shown that $\alpha = 1$). There seems to be no point in going further into this controversial issue, save to add that the opinions of Weyl and Silberstein were not so discrepant as each appears to have thought.

When Whittaker's expression for the shift in the received spectrum of light from a source in the de Sitter world is examined, it appears that he, like de Sitter and Weyl, inserted an hypothesis as to the relative motion of observer and source; for, as we saw, he took the case of constant 'proper distance' of star and observer.[80]

At first the indications were that astronomy supported Silberstein in finding both red- and violet-shifts. As more information accumulated, however, instances of the latter sort were seen to be an insignificant few and these were interpreted as an indication of random movements and of Galactic motion, both of which, at great distances, were obscured

[79] op. cit., p. 624.   [80] *P.R.S.* (A), cxxxiii (1931) 104.

by other effects than de Sitter's. This will be discussed in the next chapter. Silberstein's papers are noteworthy in more respects than one, however, for he was one of the first persons to realize that 'doubts as to the physical meaning' of a parameter (in this case r_s) can be legitimately avoided if the parameter can be eliminated between formulae, leaving behind a formula which is readily interpreted to the astronomer's practice.[81] Lest this should be thought a very obvious procedure it must be added that the usual alternative was to interpret the radial coordinate very freely and even to argue that this was of no matter, since astronomers' estimates of the distance of an extragalactic object were so widely discrepant. (A factor of 100 was not unusual between different authorities.) Tolman and Milne were both ahead of their time in taking a stand similar to Silberstein's on the matter.

In Chapter 10 we shall have more to say on the subject of Weyl's Principle and the way in which it appears to restrict the Principle of General Covariance.

7. Material Content

What of the physical conditions within the Einstein and de Sitter worlds? If values of the Christoffel symbols calculated from the expression for the line-element are inserted into Einstein's gravitational field equations, and if, in addition, values for the components of the energy-momentum tensor appropriate to a system consisting of a perfect fluid are taken, expressions are obtained for the pressure and density of the system. If the system is to be homogeneous, both the proper pressure (p) and proper density (ρ) must at all places take the same value.

This condition can be shown to be equivalent to the condition

$$(\rho + p) \frac{d}{dr}(f) = 0, \qquad (51)$$

where f, a function of r alone, is the coefficient of dt^2 in the expression for the line-element. (Using the expressions for p and ρ obtained from the field equations, it can be shown that this is equivalent to the condition $(dp/dr) = 0$.) This was not Einstein's approach, as we saw, and yet the condition was obviously satisfied in his case, since he took f

---

[81] In his case he eliminated r_s between his 'Doppler formula' and de Sitter's parallax formula leading to $\delta \cdot p = \pm a/R$, a result which would have been of great interest had de Sitter's model been otherwise acceptable.

to be constant. To show that the de Sitter line-element is a consequence of the remaining possibility, namely of the assumption that

$$(\rho + p) = 0, \tag{52}$$

it is necessary to add the condition that the line-element reduces locally to that of Minkowski space-time. Tolman showed this clearly,[82] but the matter had been discussed long before from a somewhat different point of view.

In 1919 Levi-Civita took Einstein's newly modified field equations and metric form. Assuming that matter (apart from a central sun) was uniformly distributed, he found the following conditions to hold outside the sun:[83]

$$\lambda = 3/R^2 - \kappa\rho \tag{53}$$

$$\kappa(p+\rho) = 2/R^2. \tag{54}$$

If p be assumed to vanish, then these are exactly equivalent to Einstein's expressions for ρ and λ. De Sitter, prompted by a letter from Lorentz, gave much the same argument later in the same year. He found that with his own form of the line-element the corresponding conditions were[84]

$$\lambda = 3/R^2 - \kappa\rho, \tag{55}$$

and $$p + \rho = 0. \tag{56}$$

These, in turn, reduce to his previous expressions for ρ and λ, if negative pressures are ruled out—as de Sitter, needless to say, chose to do. There was later much talk of the possibility of feeble cohesive forces, but after 1930 it was widely recognized that a plausible model with finite density could be obtained without such forces.

In contrasting the models of Einstein and de Sitter it should now be obvious that the questions of greatest moment to cosmology concerned spectral displacement and material content. Einstein's model could not explain the numerous red-shifts without auxiliary hypothesis, whilst de Sitter's was devoid of matter. It is worth repeating that, on the second issue, de Sitter shows by his use of the words 'inertia' and 'gravitation' that he was not unduly worried by the absence of matter

[82] *P.N.A.S.*, xv (1929) 297. It will be remembered that this was Einstein's starting point.
[83] *Rend. Lincei*, xxvi (1917) 530–1. Cf. Schwarzschild's solution of 1916 (*Berlin Sitz.* (1916) 424) for the line-element within a sphere of uniformly dense incompressible fluid. Schwarzschild used the field equations before the λ-term was added.
[84] deS. (iv) section 5.

from his 'world'. He regarded his metric as describing a 'field of pure inertia', without gravitation, an idealization which he insisted would never be useful in practice unless suitable allowances were made for the deviations of the g_{mn} from their ideal values. As we saw, in the case of the Einstein world he maintained that there was matter introducing 'inertia but not gravitation'. In supporting these distinctions it is not always clear whether or not we should accuse de Sitter of not having fully entered into the spirit of Einstein's General Theory, in which 'inertia' and 'gravitation' are not distinguished.

On the subject of the material pervading the Einstein world, both Einstein and de Sitter pointed out that, together with what de Sitter called 'straightforward astronomical data', it implied the existence of interstellar matter of far greater density than had generally been supposed. This was not the only drawback. The homogeneous and invisible substance seemed to many to be dangerously like those etherial media which still haunted good relativists. An important source of confusion was the failure to take the idealizations of Einstein and de Sitter at their face value. It was as though, like Plato, so many of them wished to hold to the reality of the form, but at the same time to complain that the form was a very poor copy of reality.

8. The Geometry of the de Sitter World

Over and above astronomical interest in the rival hypotheses—interest which was mild enough at first—the geometrical aspects of the whole problem aroused some enthusiasm. Responsibility for this was partly to be found in an urge to visualize de Sitter's four-dimensional manifold, which abounded in curious features. Neglecting the coordinates ψ and θ of the line-element (B), for example, it was seen that (only one spatial dimension being retained) the space-time could be represented by the surface of a hyperboloid of one sheet. The null-geodesics of this (the 'light tracks') would be its generators. That the ordinary geodesics do not coincide with the space-partitions (if r and t are 'space' and 'time' coordinates) corresponds, for example, with the result that two test particles in the de Sitter world do not remain relatively at rest.[85]

Transformations of the line-elements (A) and (B) which were frequently quoted led to the following expression for both:

$$ds^2 = -\{(dx^1)^2 + (dx^2)^2 + (dx^3)^2 + (dx^4)^2\} + (dx^5)^2 \qquad (57)[86]$$

[85] Cf. *M.T.R.*, pp. 164–5.   [86] Cf. Einstein's paper in Berlin Sitz. (1917).

with the condition

$$(x^1)^2 + (x^2)^2 + (x^3)^2 + (x^4)^2 - (x^5)^2 = R^2.$$

Starting with Klein and de Sitter, many early writers used the idea of the equivalence of the de Sitter world and the four-dimensional surface of the hyper-hyperboloid (or 'hyper-pseudosphere') embedded in five-dimensional Euclidean space. It is doubtful whether the burden on the intuition was not greater with these transformations than it had been without them: certainly few traces remained of the practical nature of the original problem. Geometrically these methods were very interesting and (from the point of view of popular exposition) at times very suggestive, but it is unlikely that they advanced de Sitter's cosmology very far.

Just as there are complete, or partially complete, representations of non-Euclidean geometries so, as we have pointed out, there are five-dimensional representations of the space-times of Einstein and de Sitter. Just as the Cayley-Klein metric was first applied to the two-dimensional non-Euclidean geometries, so was it now seen to be possible to regard de Sitter's geometry as having a Cayley-Klein metric with Absolute of the form

$$(x^1)^2 + (x^2)^2 + (x^3)^2 + (x^4)^2 - (x^5)^2 = 0. \tag{58}$$

Klein, in 1918, briefly mentioned the idea but, as we have seen, Edmund Whittaker was the first to work consistently within these terms of reference in his paper of 1931. For the calculation of physical effects, these methods are cumbersome. They are capable, even so, of providing a simple and attractive geometrical transcription of the more usual analytical (Riemannian) account.[87]

Einstein, in choosing between the different hypotheses which can be made as to connectivity and the identification of points with a given differential metric form, decided in favour of the (spatially) 'spherical' alternative. He was probably unaware of Klein's paper of 1890, for in a letter to de Sitter he later agreed that it was preferable to adopt an elliptical space 'for the physical world'.[88] The idea of a spherical world with its 'ghost images', over and above the opportunities which it seemed to offer for looking and seeing 'what was in the position now

---

[87] A further account of de Sitter's world of some geometrical interest, but otherwise of slight value, was given by P. du Val in terms of Weierstrassian homogeneous coordinates. *Phil. Mag.* (6), xlvii (1925) 930.

[88] See deS. (iii).

occupied by your head 6000 000 000 years ago', later gave immense scope to writers on the professional fringe.[89] De Sitter acknowledges (apart from his debt to Klein's synthesis of non-Euclidean geometry) the importance of the work of Newcomb,[90] in which the distinction between the elliptical and spherical contingencies was drawn. This was a distinction neither Riemann nor any of his early followers (such as Beltrami) seems to have appreciated. De Sitter also made use of Schwarzschild's paper of 1900, where the distinction is clearly made.[91] The characteristics of the elliptic plane were in fact implicit in some work of Cayley's, of 1859, but this was not at the time linked with non-Euclidean geometry.

Reference to papers by Whittaker and H. P. Robertson already cited will show that with their chosen definition of 'spatial distance', the de Sitter world is not spatially finite.[92] Robertson was well satisfied with having removed the de Sitter singularity from the finite region. Now shortly after the publication of de Sitter's first papers, both Klein and Einstein criticized them, this being the occasion of the third paper of the series. It will be recalled that (in reply to Einstein) de Sitter admitted the discontinuity at $r = \pi R/2$ ('at a finite distance in space'), and pointed out that this discontinuity was 'physically inaccessible'. Klein, in a private letter of 19 April 1918, now made a further objection. He argued that if we were to travel along a straight line of de Sitter space we should return to our starting-point, but with the positive direction of motion reversed. De Sitter conceded that if we travelled along a straight line intersecting the polar line of the starting-point this would be true, but he added that this motion would be 'physically impossible'. Klein, who as early as 1890 had remarked upon the result that the Newtonian potential becomes infinite at the antipodal point in spherical space[93] (and hence that elliptical space was to be preferred), now pointed to the ambiguity of the sign of the potential in elliptical space. The ambiguity is concerned, however, with physical interpretation. Once again de Sitter pointed out that no ambiguity can arise with the interpretation he proposed: the circuit which Klein assumes is physically impossible, and therefore 'no physical experiment could ever lead to contradiction'.[94]

---

[89] P. du Val (op. cit.) was surely wide of the mark when he remarked that 'the elliptic case is certainly more amusing than the spherical'.
[90] *Crelle's J.*, lxxxiii (1877) 293.   [91] *Viert. Astr. Ges.*, xxxv (1900) 337.
[92] Whittaker, *P.R.S.* (A), cxxxiii (1931) 93. See pp. 97–8 *supra*. Robertson (*Phil. Mag.*, v (1928) 835; &c.) used $l = r \exp\{t\sqrt{(\lambda/3)}\}$.
[93] *Math. Ann.*, xxxvii (1890) 557–8.   [94] deS. (iii) p. 1310.

Between 1920 and 1930 there was something of a stalemate in cosmological discussion. Although some important contributions not concerned with the Einstein and de Sitter models were made during this period, a surprisingly large number of them passed almost unnoticed for several years. Even the interest aroused in the de Sitter world was as much geometrical as physical. In 1930 Silberstein could write that 'there are in our days but two serious rival theories abroad'. The many remaining problems of relativity promised more immediate returns, whilst much that seemed fundamental had still to be defended against hostile criticism. Added to this the Quantum Theory of 1926 stole much of the limelight. Einstein, moreover, was seen to have abandoned his own model which, after about 1922, was seriously entertained by only one astronomical writer of wide repute, namely Hubble.[95] New observations of any relevance accumulated slowly and there could not invariably be any great confidence in their accuracy. But in 1930 interest in theoretical cosmology quickened with Eddington's recognition of the significance of a paper published some years before by Lemaître. Only now was the importance of an even earlier paper by A. Friedmann appreciated. The year 1930 marks a second renaissance of modern cosmology.

[95] *Ap. J.*, lxiv (1926) 321.

CHAPTER 6

THE EXPANDING UNIVERSE: THE FORMAL ELEMENT

IN CHAPTER 2 we saw numerous demonstrations of inconsistencies in the notion of an infinite Newtonian system. We saw that, at least in the first quarter of this century and before, most of these demonstrations were founded upon the assumption that the universe is, as a whole, static. This tacit hypothesis was by far the greatest barrier to the progress of cosmology, Newtonian and relativistic, in the early century. When it was abandoned, fresh attempts to formulate a Newtonian cosmology would eventually follow; but it was not on Newtonian grounds, nor on the grounds of any consideration which could clearly be labelled 'physical', that the cosmologist's world was first conceived to be systematically changing. The move was made within the realm of what may be called 'applied mathematics', on the logical status of which there has always been a good deal of disagreement. As in so much cosmological writing, one suspects that the authors in question cared little whether or not the field equations and so forth (which provided their starting point) had a sound empirical justification: the formal argument was what really mattered. Even so, one of the most striking features of the scene is that on several occasions arguments of a very general kind led to conclusions which were only supported by observation at a much later date. De Sitter's model was a case in point, but even more remarkable instances were provided by Friedmann and Lemaître, whose theories were lifted from obscurity after the relevant astronomical findings were published by Edwin Hubble.

On the other hand it would be a mistake to characterize the introduction of a non-static hypothesis into cosmology as amounting to no more than the introduction of a new mathematical variable into the coefficients of the metric. The new variable referred to *time*, and whatever else time was, it was not conceived as something purely mathematical. In this chapter (with such a qualification as this) we shall consider the formal element in the early theory of the expanding universe. We shall later see that astronomy was not entirely unprepared for the ideas involved in this new conception, but that liasons between astronomer and mathematician were weak. They would have been weaker still but for the outlook of such writers as Eddington, de Sitter, and Tolman (the first two were practised astronomers). The two streams of thought eventually converged at the end of the 1920s.

1. THE NON-STATIC LINE-ELEMENT

The metric in a non-static form made no sudden and dramatic appearance. De Sitter seems at times to have been on the point of presenting it, and it was in giving a transformed version of de Sitter's line-element that Cornel Lanczos first introduced the idea. As it turned out, de Sitter's three solutions ((A), (B), and (C) *supra*) for the line-element of a *static* gravitational field exhausted the possibilities for a static and homogeneous configuration (under the General Theory of Relativity). This was proved by Alexandre Friedmann[1] and, more simply, by R. C. Tolman some years later.[2] Now it is apparent that new coordinates could be taken as functions of the old, in terms of which the line-element would no longer be of static form. At times de Sitter seems to have been on the verge of effecting the transformation, being restrained only by his confidence in the more or less static world of astronomy. Comparing his own metric (B) with Einstein's (A), he observed that 'it is always possible, at every point of the four-dimensional space-time, to find systems of reference in which the g_{mn} depend only on one space variable (the 'radius vector') and not on the "time" . . .'[3] On the other hand, speaking of (B), he added *'there is no essential difference between the "time" and the other three coordinates. None of them has any real physical meaning.'* (In (A) the time-variable was essentially different from the rest: Einstein's model is obviously curved only with respect to the three spatial dimensions.)

De Sitter was not as willing as he might have appeared to regard all coordinates on an equal footing, and he was not the first to discard the hypothesis of a static field. When Lanczos transformed the de Sitter line-element to a non-static form[4] he introduced variables in terms of which the line-element assumed the form

$$ds^2 = -dt^2 + \cosh^2 t \cdot (d\phi^2 + \cos^2 \phi \cdot d\psi^2 + \cos^2 \phi \cos^2 \psi \cdot d\chi^2), \quad (1)$$

'with t the time variable and the others of a spatial character'. (He later took $r = \cos \phi$ as the 'radius vector'.) Lanczos was greatly indebted

---

[1] *ZS. Phys.*, x (1922) 377.
[2] *P.N.A.S.*, xv (1929) 297. Here 'static' signifies the absence of the time-variable from all g_{mn}. Weyl (*S.T.M.*, p. 241, for example), following Levi-Civita (*Rend. Lincei*, xxvi (1917) 458), uses the different convention that (with the usual notation) there shall be the additional requirement that $g_{14} = g_{24} = g_{34} = 0$. They do not appear to distinguish between 'stationary' and 'static'.
[3] deS. (iv) p. 11.   [4] *Phys. ZS.*, xxiii (1922) 539.

112   THE EXPANDING UNIVERSE: THE FORMAL ELEMENT

to Klein for a geometrical mode of presentation which he extended by incorporating the parameter t.[5] As Lanczos remarked, 'It is interesting to observe how one and the same geometry can appear with quite different physical interpretations . . . according to the interpretations placed upon the particular coordinates.'[6] This remark was not then the platitude it has since become.

If 't' is given a temporal interpretation. Lanczos's metric is obviously not static in the senses used above. Unfortunately each of the words 'stationary' and 'static' has been used in several senses and the two have been used interchangeably. As before, we use the word 'static' only when the g_{mn} are all independent of the time coordinate. As for 'stationary', it is quite impossible to give anything resembling a universal meaning. Lanczos later said quite explicitly what he meant by the word '*stationär*'.[7] A universe, he explained, is stationary 'when its metrical coefficients are independent of the time, provided we have a coordinate system in which the mean measurements are constant'. The proviso is essential, 'for time is not in itself an invariant concept but is only significant as proper time'. He pointed out that neither de Sitter's 'statical' solution, nor that of Schwarzschild, could be called 'stationary' in his sense. (It will shortly be seen that this terminology stands in sharp contrast with that proposed by Robertson.) Lanczos added that there is no general stationary solution which is at the same time spherically symmetrical (in four dimensions). In other words, at the time of writing, Einstein's model was the only 'stationary' one.

Lemaître[8] and Robertson[9] independently effected a similar transformation of de Sitter's solution (B) by introducing coordinates leading to the line-element

$$ds^2 = c^2 dt^2 - \exp\{2t\sqrt{(\lambda/3)}\}(dx^2 + dy^2 + dz^2). \tag{2}$$

The equations of transformation which it is necessary to apply to (B) are:

$$\bar{r} = R \tan(r/R) \exp(-t/R); \ c\bar{t} = ct + R \log \cos(r/R). \tag{3}$$

(Robertson actually presented a second coordinate transformation which changed the Schwarzschild line-element into a non-static variety, but this led to results which, on the scale of the solar system, were for

---

[5] Klein, *Gött. Nachr.* (December, 1918). Cf. *Gesammelte Mathematische Abhandlungen* (3 vols., 1921–3) i, p. 604.
[6] Lanczos, op. cit., p. 539.   [7] *ZS. Phys.*, xxi (1924) 73.
[8] *J. Math. Phys.*, iv (1925) 188.   [9] *Phil. Mag.* (7), v (1928) 835.

practical purposes indistinguishable from Einstein's and Schwarzschild's.)[10] The fact that the 'spatial' part of the metric contained an expression involving time was given the interpretation that 'natural processes are not reversible'. Robertson added that there was a sense in which the properties of the metric may be regarded as independent of the new time-variable. For, taking a different time origin and a suitable transformation in the radial coordinate ($\bar{\bar{t}} = \bar{t} - t_0$ and $\bar{\bar{r}} = \bar{r}\varepsilon^{kt_0}$), the form of the line-element remains unchanged. For this reason Robertson later proposed to call it 'stationary' rather than merely 'static'.

We recall that in the last chapter we derived several of the properties of the de Sitter world in conjunction with the transformation equations provided by Robertson and Lemaître. We shall later see that the new form of the line-element allowed the de Sitter world to be classed as a limiting case of a much more general 'expanding universe'. Bearing in mind that particles obeying Weyl's Principle in a de Sitter universe are in an 'undisturbed state' (see section 6 of the last chapter) when $\bar{\bar{r}}$ is constant, it seems that the line-element given by (2) is somehow implicit in Weyl's work.

2. The Work of Friedmann, Lemaître, and Robertson

Almost simultaneous with the paper by Lanczos we find the first really important extension of theoretical cosmology since de Sitter's papers of 1917. In 1922 Alexandre Friedmann of St. Petersburg drew attention to the possibility of a model with space curvature dependent only on the time,[11] whilst in 1924, prompted by his friend Tamarkine, he investigated the cases of both stationary and non-stationary worlds with negative curvature.[12] Much of his theory followed closely the assumptions made by Einstein and de Sitter, especially as regards the gravitational field equations (taken with the λg_{mn}-term) and the velocity and state of matter. In consequence of this assumption,

---

[10] The Schwarzschild line-element, as modified in the light of the later field equations, had been known since 1922 when J. Chazy (*C.R.*, clxxiv, (1922) 1157) and E. Trefftz (*Math. Ann.*, lxxxvi (1922) 317) obtained it independently. The metric can be written

$$ds^2 = c^2\{1 - (2m/r) - (\lambda r^2/3)\}\,dt^2 - \{dr^2/(1 - 2m/r - \lambda r^2/3)$$
$$+ r^2 d\theta^2 + r^2 \sin^2\theta\,d\phi^2\}$$

If we set $m = 0$ this reduces to a form of no other than the space-time of de Sitter. Robertson reproduced the argument for the time-dependent forms of the two expressions for the line-element (op. cit.).

[11] *ZS. Phys.*, x (1922) 377. It has been pointed out by Robertson that A. Calinon had long previously hinted at the possibility that our space may not be quite Euclidean and that the discrepancy might vary with the time. (*Rev. Philos.*, xxvii (1889) 588).

[12] *ZS. Phys.*, xxi (1924) 326.

namely that matter is incoherent and exhibits relative velocities which are small by comparison with the velocity of light, simple expressions were available for the components of the energy-momentum tensor.[13]

Friedmann's point of departure from the earlier work was in regard to the assumed geometrical character of the model. We have seen how Einstein and de Sitter approached the problem without any very definite precept to guide them. Friedmann, in much the same position, rather than simply transform their versions of the line-element as Lanczos had done, merely decided on the grounds of generality that at any instant of time the model should represent a space of constant curvature. He wrote the line-element as

$$ds^2 = R^2(dx_1^2 + \sin^2 x_1 dx_2^2 + \sin^2 x_1 \sin^2 x_2 dx_3^2)$$
$$+ 2g_{14}dx_1dx_4 + 2g_{24}dx_2dx_4 + 2g_{34}dx_3dx_4 + g_{44}dx_4^2 \quad (4)$$

(R dependent only on x_4). Now g_{14}, g_{24}, g_{34} may be chosen so that time and space are orthogonal—just as Einstein had previously supposed.[14] Friedmann remarked that 'there are no physical or philosophical grounds for thinking so'.[15] Nevertheless he took them to be so, without stating his reasons very clearly. Robertson later derived a similar line-element and also stated explicitly that he made the hypothesis of a cosmic time with space orthogonal to it.[16] In contrast with these methods, Tolman gave an account which, set against this postulational approach, seems more satisfactory. The orthogonality condition was not assumed at the outset, but was shown to follow from a carefully chosen set of hypotheses—notably that of spatial isotropy—which most people would admit to be more plausible.

Friedmann took, therefore, the following expression for the line-element:

$$ds^2 = R^2(dx_1^2 + \sin^2 x_1 dx_2^2 + \sin^2 x_1 \sin^2 x_2 dx_3^2) + M^2 dx_4^2, \quad (5)$$

where R is a function of x_4 and M may be a function of all coordinates. (The coordinates are subject to the restrictions $0 < x_1 < \pi$, $0 < x_2 < \pi$, $0 < x_3 < 2\pi$.) Using the field equations together with the chosen form of the energy-momentum tensor, it follows that

$$R'(x_4) \cdot (\partial M/\partial x_1) = R'(x_4) \cdot (\partial M/\partial x_2) = R'(x_4) \cdot (\partial M/\partial x_3) = 0, \quad (6)$$

[13] As before, $T_{mn} = 0$ unless $m = n = 4$,
and $T_{44} = c^2 \rho g_{44}$.
[14] Cf. (iv), p. 82 *supra*.   [15] op. cit., p. 379.
[16] *P.N.A.S.*, xv (1929) 822. Robertson supposes that Friedmann, in deducing a normal form of the line-element, had not realized that he was going beyond his stated assumptions.

whence either R is independent of x_4, or M depends only on x_4. In the former case the line-elements of Einstein, de Sitter and the Special Theory of Relativity were shown to exhaust the so-called 'stationary' possibilities: de Sitter's demonstration of this has been mentioned above, whilst it will be seen below that Robertson showed there to be no further 'stationary' solutions, in his own sense of the word. As to the latter case, in which M depends only on what was referred to as the time-coordinate (x_4), 'dies soll die nichtstationäre Welt heissen'. Through an appropriate choice of x_4, M can be chosen unity, and the final form of the line-element becomes, retaining Friedmann's notation[16*]

$$d\tau^2 = -\frac{R^2(x_4)}{c^2}(dx_1^2 + \sin^2 x_1 dx_2^2 + \sin^2 x_1 \sin^2 x_2 dx_3^2) + dx_4^2. \quad (7)$$

Again using the field equations, together with this expression for the interval, it follows that

$$(R'^2/R^2) + (2RR''/R^2) + (c^2/R^2) - \lambda = 0 \quad (8)$$

and

$$(3R'^2/R^2) + (3c^2/R^2) - \lambda = \kappa c^2 \rho \quad (9)$$

$$(R' = (dR/dx_4), \quad R'' = (d^2R/dx_4^2)).$$

Integrating the first of these equations, it follows that

$$\frac{R}{c^2}(dR/dt)^2 = A - R + (\lambda/3c^2)R^3 \quad (10)$$

and therefore

$$t = (1/c)\int_a^R \sqrt{\left\{\frac{x}{A - x + (\lambda/3c^2)x^3}\right\}}dx + B, \quad (11)$$

where a, A, B, are arbitrary constants and t is used for x_4, to facilitate comparison with later writers. Equations (9) and (10) together give

$$\rho = (3A/\kappa R^3). \quad (12)$$

The relation (11) contains an elliptic integral, and in the absence of independent information it had to be assumed that λ could take any value whatever. The three cases of interest, with the corresponding inferences drawn from (11) and (12), are set out below:

(i) $\lambda > (4c^2/9A^2)$. R is an increasing function of t. There are no restrictions on the initial value (R_0) of R $(= a$, above) other than that it should be positive. The model is 'monotonic of the first kind'. It should be noticed that the critical value of λ is that

---

[16*] He explained that the change from s to τ was made on account of its having the dimension of time.

which corresponds to an Einstein model. This can be readily verified using (7) (with $dR/dt = 0$) together with Einstein's relation between λ_e and R_e.

(ii) $0 < \lambda < (4c^2/9A^2)$. The initial value of R can, in principle, lie in three ranges. One can be ruled out as involving imaginary time. One gives a model 'monotonic of the second kind'. The third range makes R a periodic function of t.

(iii) $0 > \lambda$. Here again imaginary time is avoided only if R_0 has less than a certain value, x_0. In this case R turns out to be a periodic function of t, with period t_π('*die Welt-periode*'), where

$$t_\pi = \frac{2}{c} \int_0^{x_0} \sqrt{\left\{\frac{x}{A - x + (\lambda x^3/3c^2)}\right\}} \, dx \qquad (13)$$

In this case, as in the periodic model under (ii), R varies between 0 and x_0; t_π increases with λ and tends to infinity as λ tends to the value $(4c^2/9A^2)$; whilst for small λ, $t_\pi \simeq (\pi A/c)$. (We shall return to the question of the 'cyclical universe'.)

In his second paper Friedmann emphasized the essential difference between stationary worlds of positive and negative curvature: the model of negative curvature will not allow of a positive material density. This case 'must therefore be ruled out as physically impossible'. There is, in other words, no analogue of Einstein's model, although there is one of de Sitter's (i.e. one with constant density). Friedmann found two cases, as before, stationary and non-stationary. The line-element he took as

$$d\tau^2 = -(R^2/c^2)\left(\frac{dx_1^2 + dx_2^2 + dx_3^2}{x_3^2}\right) + M^2 dx_4^2, \qquad (14)$$

the analysis being very similar to that of the previous paper as far as the division into stationary ($R' = 0$) and non-stationary ($R' \neq 0$, $(\partial M/\partial x_1) = (\partial M/\partial x_2) = (\partial M/\partial x_3) = 0$) is concerned. The former lead either to

(i) $\lambda R^2 = 3, \quad \rho = 0$

or to (ii) $\lambda R^2 = 1, \quad \rho = -2/\kappa R^2$.

Case (i) bears comparison with de Sitter's model, whilst case (ii)—the possibility of which had previously been remarked upon in conversation with W. Fock[17]—was obviously of no practical use.

[17] *ZS. Phys.*, xxi (1924) 330, footnote.

THE EXPANDING UNIVERSE: THE FORMAL ELEMENT 117

For the non-stationary model, with M a function of x_4 alone (and ultimately taken to be unity, as before) the 'Friedmann equation' turns out to be

$$(R'^2/R^2) + (2RR''/R^2) - (c^2/R^2) - \lambda = 0, \qquad (15)$$

which differs from (8) only in the sign of the third term. The integration of this is not very different from that already given for (8). (In section 5 of this chapter we shall summarize the several possible cases, and therefore not go into further details here.) The density in such a model is found to be (cf. (12))

$$\rho = (3A/\kappa R^3). \qquad (16)$$

For the first time the possibility of a non-stationary world of constant negative curvature and with a positive material density had been shown. Friedmann emphasized, moreover, a point which was often overlooked: space may quite well be curved and yet not finite.

It is hard to explain why these two papers by Friedmann were ignored. Although they have been of late criticized for their lack of rigour, they are more rigorous than a great deal of original work. Perhaps the mathematical channels they explored seemed to offer up too many difficulties. Perhaps they seemed unlikely to lead to results of any astronomical interest. Einstein considered them briefly in two half-page notes. In the first he criticized Friedmann's findings, only to point out his own mistake—an arithmetical error—in the second.[18] It is notable that neither Friedmann nor Einstein related the model of variable radius to the large class of nebulae then known by astronomers to be receding. Friedmann's theory was noticed again some years later by V. Fréedericksz and A. Schechter who calculated formulae for aberration and parallax in both this theory and the static theories.[19]

Friedmann's work was not to be widely known until nearly ten years after its publication, and then only as a result of interest stimulated by the work of Lemaître and Robertson. By 1927 Lemaître, working in ignorance of Friedmann's two memoirs, had developed the relativistic theory in a way remarkably similar to Friedmann's. Again independently, by 1928 Robertson, as we have explained, had used the coordinate transformation which gives the de Sitter model a non-static

---

[18] *ZS. Phys.*, xi (1922) 326 and xvi (1923) 228. He thought that Friedmann's working had shown conclusively that the world-radius was constant.

[19] ibid., li (1928) 584. The expressions they obtained were shown to agree, approximately, with the customary expressions.

form. Unlike Friedmann and Lemaître, Robertson was regarded by his contemporaries and himself as having given little more than a mathematical variant of the de Sitter model. Later the usefulness of his mode of presentation was better appreciated.

On the question of interpretation, Robertson is fairly typical of his contemporaries. Whilst he was aware of the logical difficulties in interpreting 'time' and 'distance' variables, he offered no criterion by which his own particular choice could be justified. (For 'distance' he used de Sitter's radial coordinate (r_s), whilst for the 'proper time' of a clock at the origin he used his own variable ($\bar{t}$, above). As far as the use of the term 'velocity' is concerned, he contrasted the 'radial velocity' which follows, given 'the ordinary interpretation of the Doppler effect', with the 'velocity which would be measured' (namely dr_s/ds). The old problem persisted: '... we cannot, from the Doppler-effect alone, distinguish between proper motion and [the] distance effect'.[20] In the last chapter we found the theoretical (δ, r)-relation for the de Sitter model and saw the several hypotheses under which δ and r were simply proportional. Robertson's simple conclusion (based on Weyl's Principle and surprising only to those unfamiliar with the work of Weyl, Silberstein and the rest) was that 'we should ... expect a correlation

$$v \simeq (cl/R) \tag{17}$$

between assigned velocity v, distance l and radius of the observable world R'. Robertson's conclusion—which, as we shall see, was at the time 'roughly verified' and within a year confirmed by Hubble—was eclipsed by the 'discovery' of the work of Friedmann and Lemaître.

Although in point of time Lemaître's first great contribution to cosmology[21] preceded this paper by Robertson, we have reserved mention of it until now because it marks the beginning of a new phase in cosmological thought, just as Robertson's paper marks the termination of what had gone before. Lemaître saw plainly the advantages and disadvantages of the models of Einstein and de Sitter. In the de Sitter case both space and space-time are separately homogeneous and yet 'the partition of space-time into space and time disturbs the homogeneity'. Here spoke Lemaître the mathematician, for whom symmetry was nearly as important as truth. Next he reaffirmed his dislike of the horizon properties of the de Sitter model. We recall that both he and Robertson had found a way round what they conceived to be these

---

[20] *Phil. Mag.*, v (1928) 844. Cf. section 6 of the last chapter.
[21] *Ann. Soc. Sci. Brux.*, xlvii (A) (1927) 49; *M.N.*, xcl (1931) 483.

THE EXPANDING UNIVERSE: THE FORMAL ELEMENT 119

difficulties. More than this, however, he wanted a closed model 'intermediate between that of Einstein and de Sitter'—having, that is to say, both material content and definite spectral displacement. Lemaître found what he wanted, proceeding in a way very similar to Friedmann's: in fact he reached differential equations equivalent to Friedmann's, if we ignore the pressure term. These equations he wrote as follows (taking units such that $c = 1$)

$$(3R'^2/R^2) + (3/R^2) = \lambda + \kappa\rho \tag{18}$$

and $\qquad (2R''/R) + (R'^2/R^2) + (1/R^2) = \lambda - \kappa p. \tag{19}^{22}$

At this stage the physical assumptions of the two authors diverged, Lemaître supposing his model to comprise a mixture of incoherent matter (i.e. exerting negligible pressure) and radiation (the pressure of which was taken, as usual, to equal one third of its energy density). He took a principle of energy conservation which amounted to saying that the variation of total energy—of radiation and of mass—plus the work done by radiation pressure is zero. (In symbols: $d((\rho_m + 3p) \cdot R^3)dt + pd(R^3)/dt = 0$, where ρ_m is the material density. This may be confirmed by substitution from (18) and (19), bearing in mind that $\rho = \rho_m + 3p$.) But he made the assumption that 'the mass of the universe is constant'. ($\rho_m R^3 = $ constant.) Restricting himself to this example, which was tractable and yet, at the same time, likely to be thought sufficiently plausible, he was led to the equation

$$t = \int \frac{dR}{\sqrt{\{(\lambda R^2/3) - 1 + (\alpha/3R) + (\beta/R^2)\}}}. \tag{20}$$

From this equation de Sitter's solution is recovered if α and β are set equal to zero. Einstein's solution is found when R is made constant and β zero.[23] If, on the other hand, R is allowed to vary with time, whilst β is made zero, we have Friedmann's solution, of which Lemaître was unaware. In Friedmann's case the total pressure is zero. It would be

[22] For future reference, taking the possibility of open models into account (as did Friedmann) and inserting the terms in the velocity of light, we may write

$$\lambda + \kappa\rho c^2 = 3(R'^2 + kc^2)/R^2$$
$$\lambda - \kappa p = (2RR'' + R'^2 + kc^2)/R^2.$$

where k is the 'curvature constant' ($k = +1, 0$ or -1). (Cf. Friedmann's equations (9) and (8) above, having put $p = 0$.) We notice that p and ρ are functions of t alone, in keeping with ideas of spatial homogeneity. It should be noticed that it is not necessary to suppose λ positive and R real in order to ensure non-negative pressure and density. Some had previously argued on these grounds that space must necessarily be closed.

[23] Note that the R here is not the same as Einstein's.

misleading, however, to call Lemaître's a 'generalization' of the earlier model, for in this not only is the mass conserved, but also the energy (owing to this circumstance of zero pressure). One should also remember that the earlier writer had also considered the possibility of open as well as closed models. Lemaître's integral, which he himself did not discuss very fully, was carefully examined by de Sitter not long afterwards,[24] and a more complex integral, taking into account not only both open and closed models but also positive material pressures, was carefully discussed by O. Heckmann.[25]

Lemaître, like Friedmann, concluded that the universe must have a 'radius increasing without limit'. Unlike Friedmann, he derived for it 'an apparent Doppler effect', where 'the receding velocities of extragalactic nebulae are a cosmical effect of the expansion of the universe'. The apparent nebular velocity (v) was given in the way which now is usual, by

$$\frac{v}{c} = R(t_2)/R(t_1) - 1 \equiv (R_2/R_1) - 1, \qquad (21)$$

where t_1 and t_2 are the times of emission and observation of the light from the nebula. (Cf. (39) below.) If r is the distance of the source, then it was found that the approximate relation

$$v = cr/R_0\sqrt{3} \qquad (22)$$

holds.[26] Here was another prediction of 'Hubble's Law', a prediction which was not, however, fitted to the meagre data available. At least as much of this as was contained in Eddington's *Mathematical Theory of Relativity* must have been known to Lemaître. As we saw, Robertson did try to confirm an almost identical relation in the following year. He found

$$R = 2 \times 10^{27} \text{ cm } (= 6 \cdot 5 \times 10^8 \text{ pcs}),$$

which was of the same order of size as Hubble's estimate of $R_0(2 \cdot 7 \times 10^8 \text{ pcs})$. With the determination of R_0 as the end in view, Lemaître quoted only Hubble's figure, calculated from the mean density by the use of Einstein's relation. (Cf. Chapter 5, section 3.)

While Lemaître's work remained unnoticed, Robertson wrote another paper in which was set out an approach to the cosmological

---

[24] *B.A.N.*, v (1930) 211 and vi (1931) 141.   [25] *Gött. Nachr.* (1932) 97.
[26] Here R_0 is referred to as the 'initial radius'. This obscures the way in which R_0 was introduced, namely, as defined by $\lambda = 1/R_0^2$—this being simply the Einstein relation.

problem which writers have tended to follow to the present day.[27] His concern was with the formulation of principles governing homogeneity and isotropy, and with the formulation of all line-elements compatible with these principles. Claiming that Friedmann and Tolman had both previously introduced unwarranted assumptions concerning the energy tensor, Robertson now carefully stated his two geometrical hypotheses. The first concerned the possibility (from a large-scale point of view) of separating space-time into a space and a time orthogonal to it. The second was that space-time should be spatially isotropic and homogeneous (once again, when regarded on a large scale).[28] The line-element is found to be of the form

$$ds^2 = dt^2 - \exp(2f) h_{\mu\nu} dx^\mu dx^\nu, \tag{23}$$

where f is an arbitrary real function of the time t, and the coefficients $h_{\mu\nu}$ are functions of 'spatial' variables alone, such that the differential form

$$h_\mu dx^\mu dx^\nu \tag{24}$$

is positive definite and defines a three-space of constant curvature.

Robertson next argued in the usual way that it must be possible to find components T_{mn} of the energy tensor so that Einstein's field equations are satisfied. These were duly found and it was shown that the only stationary models satisfying the chosen geometrical requirements are those of Einstein and de Sitter. We recall that for Robertson a 'stationary' model was one whose intrinsic properties are independent of time. A static manifold, namely one the coefficients of whose line-element are independent of time, is evidently stationary, but the converse is not necessarily true. For a non-stationary model, coordinates can be chosen so that the first metric can be written

$$ds^2 = dt^2 - \exp(2f) \left(\frac{dr^2}{1 - r^2/R^2} + r^2 d\theta^2 + r^2 \sin^2\theta d\phi^2 \right). \tag{25}$$

Finally, the appearance of the spectra of distant sources was predicted to be one which will be 'attributed to a velocity of recession

$$v = c \tanh \{f(t) - f(t_0)\}'. \tag{26}$$

Here t and t_0 are the times of reception and emission of the light from the distant sources in question.

[27] *P.N.A.S.*, xv (1929) 822.
[28] For the full statement see ibid., p. 823. Robertson quotes some mathematical results first obtained by Fubini (*Annali di Mat.* (3), ix (1904) 64) and others.

Robertson's outlook greatly influenced Tolman and, indirectly, Milne. Above all, the search for new solutions of those equations which express the conditions for isotropy in an expanding model—especially of a set of equations first proposed by A. G. Walker[29]—soon became a mathematical pastime.[30]

3. Eddington and the Lemaître Model

Despite Robertson's contribution and its influence on American astronomy through Tolman, it was Lemaître who taught the majority how to avoid the dilemma involved in adopting either de Sitter's or Einstein's model. At a British Association meeting in 1931, de Sitter said that 'the way out of the dilemma has been shown by Professor Lemaître, whose brilliant discovery, the "expanding universe", was discovered by the scientific world about a year and a half ago, three years after it had been published.' This is rather hard on Friedmann and Robertson, but it expressed a general feeling. The channel by which the 'brilliant discovery' became known was a paper by A. S. Eddington, published in 1930.[31]

Eddington had been working with G. C. McVittie, at that time his research student, on the problem of the 'instability of Einstein's spherical world'. Before their work was complete they came across Lemaître's paper. 'Although not expressly stated,' wrote Eddington, 'it is at once apparent from his formulae that the Einstein world is unstable . . . [a point which had] not hitherto been appreciated in cosmogonical discussions.'[32] Eddington's argument begins with Lemaître's expressions for the density and pressure in terms of λ and the function $R(t)$.[33] It was seen at once that (our notation)

$$3R''/R = \lambda - \kappa(\rho + 3p)/2. \qquad (27)$$

Now Lemaître had concluded his paper of 1927 with the remark: '*Il resterait à se rendre compte de la cause de l'expansion de l'univers*',

---

[29] *Q.J.M.*, vi (1935) 89.

[30] See, for example, V. V. Narlikar and D. N. Moghe, *Phil. Mag.* (7), xx (1935) 104, (cf. Moghe, *J. Univ. Bombay*, xi (1942) 6); McVittie, *M.N.*, xciii (1933) 325; Tolman, *P.N.A.S.*, xx (1934) 169; Einstein and E. G. Strauss, *Rev. Mod. Phys.*, xvii (1945) 620; H. Bondi, *M.N.*, cvii (1947) 410; G. C. Omer, *Ap. J.*, cix (1949) 164.

[31] *M.N.*, xc (1930) 668.   [32] op. cit., p. 668.

[33] See footnote 22, p. 119, *supra*. It is worth remarking that p and ρ are both functions of t alone, in keeping with the ideas of spatial homogeneity. It can also easily be seen that it was no longer necessary to suppose λ positive and R real in order to ensure non-negative pressure and density. It had previously been argued that, on these grounds, space must necessarily be closed.

going on to draw attention to the work done by the pressure in expanding the universe. (That $(d(\rho V)+pdV)$ is identically zero, where V, proportional to R^3, is the volume, is easily verified as a consequence of Lemaître's equations for pressure and density.) Eddington was careful to point out that it would be wrong to speak of the pressure as the *cause* of expansion, for there may be expansion even in the case of zero pressure. In such a case

$$3R'' = R(\lambda - \kappa\rho/2), \tag{28}$$

and if
$$\lambda = \kappa\rho/2 \tag{29}$$

Lemaître's model reduces to an Einstein universe in equilibrium. A slight disturbance such that ρ is reduced below the value of $2\lambda/\kappa$, however, and the sign of R'' suggests that there will be an expansion which will continue, as its effect will be to decrease ρ yet further: and similarly for contractions. The conversion of matter into radiation he would not allow as a likely inducement to instability, for the increase of pressure entailed would obviously lead to contraction, and this possibility was of little interest to an astronomer.[34]

As Eddington noticed, the proper mass of the universe is not a conserved quantity, since mass is lost in the conversion of matter to radiation (the rest mass of a light quantum being zero). Nor, according to Eddington, is the mass of the universe 'relative to axes at rest' conserved, for a decrease in the kinetic energy of random motions is to be expected as the universe expands. These two 'insignificant complications' were surprisingly not referred to when Eddington later discussed the relation between M, the present mass, and M_e, the mass of the Einstein state. Taking the equations

$$\lambda + \kappa\rho = 3(R'^2 + 1)/R^2 \quad \text{(Lemaître's equation—(18) } supra\text{)} \tag{30}$$

and
$$M = 2\pi^2 R^3 \rho, \tag{31}$$

an equation emerges similar to Lemaître's, and to Friedmann's before it, namely

$$R' = \sqrt{\{(\lambda R^2/3) - \eta(M/R)\}}, \text{ where } \eta = \kappa/6\pi^2. \tag{32}[35]$$

---

[34] Cf. Tolman, *P.N.A.S.*, xvi (1930) 320. Tolman had been of a different opinion as will be seen.

[35] Differentiating this equation with respect to the time variable, and writing $M' = 0$, we arrive at Friedmann's second order differential equation for R. (Equations (8) and (15) *supra*.)

There are three cases to consider (notice that M is presumed constant):

(i) $M > M_e$ Continuous expansion.[36] R' has a minimum when $(R/R_e) = (M/M_e)^{1/3}$.

(ii) $M_e > M$ R has either a single maximum or minimum value.

(iii) $M = M_e$ As $R \to R_e$, $R' \to 0$ and hence 'the time that the radius remains in the neighbourhood of (R_e) is logarithmically infinite'.[37] This case can be integrated without using elliptic functions.

These results are essentially the same as those published by de Sitter at the same time. In two papers to which reference has already been made, de Sitter showed that if matter were to be conserved the integration of Lemaître's equation (18) could be directly effected if the expansion began from a static Einstein state.[38] The expansion is away from this state and towards a final empty de Sitter state, in both cases asymptotically. The effect of radiation pressure on the rate of expansion was shown to be very slight and the case of zero pressure, which Lemaître had studied, was evidently representative enough of any model of this sort, regardless of the proportions of matter and radiation which it contained.

Eddington's 'insignificant complications' concerning the non-conservation of 'the mass of the universe' seem far from insignificant when one considers that Eddington's personal preference was for case (iii), the case of *exact* equality of M and M_e. In this case, evolution is allowed 'an infinite time to get started, but once seriously started its time scale of progress is not greatly different from (i)'. Considering the emphasis which he placed on this third case, both here and subsequently, it is surprising that he should have glossed over the fact that mass is not conserved within his model.

Tolman, at much the same time, formally discussed the time-variation of mass within the expanding model.[39] He singled out for treatment two cases involving the conservation of mass. In general, however, he found that there would be some change in the proper mass associated with the matter of the model and, like Eddington, he

---

[36] Suffix 'e' means that the corresponding Einstein value is intended. Note:
$$(2M_e/\pi) = R_e = (1/\sqrt{\lambda}).$$

[37] op. cit., p. 672.   [38] See p. 119. Cf. also *R.T.C.*, p. 409.
[39] *P.N.A.S.*, xvi (1930) 409.

observed that this conformed with current ideas on light emission and on the source of stellar energy.

The reason for Eddington's preference was that he thought it 'necessary if the universe is to have a natural beginning'.[40] The case of continuous expansion 'seems to require a sudden and peculiar beginning of things'. Case (ii) meant that it was difficult to find 'a natural starting point for the universe on the locus mapped out for it'.[41] He added—and this is the crux of the matter—that 'with any reasonable estimate of the present degree of expansion, the date of the beginning of the universe is uncomfortably recent'. Here he seems to have fallen into an obvious fallacy, namely that of taking the length of the subtangent to the $R - t$ curve as the age of the universe. More will be said on this score, but there is no doubt that the last quotation explains Eddington's wish to have a logarithmic eternity on which to fall back, if the need should arise. Choosing a suitable 'time-scale' was to become the nightmare of the cosmologist of the 1930s. Eddington argued here for a period of about 10^{10} years between the present time and the Einstein state. In as many years from now the magnitudes of the stars will, he predicted, be fainter by a factor of ten. A time-scale of billions of years he thought unlikely, as astronomers then would have to 'count themselves extraordinarily fortunate that they are just in time to observe this interesting but evanescent feature of the sky'.[42] This is a second kind of fallacious argument which is not uncommon in cosmology. Omer, for example, points out that certain hypotheses 'would indicate that the 100-inch reflector has already surveyed a large fraction of the existing universe. This might be true, but it seems philosophically repugnant.'[43] Both appear to rely on a Principle of the Improbability of Good Fortune. But good fortune is in the eye of the beholder, and it seems odd to suppose that the beholder can affect the probability of external occurrences. On the matter of the time-scale, however, there were nearly as many opinions as authors.

4. The Stability of the Einstein World

Over the next ten years or so, cosmology tended to divide into two streams. One of them began with the question of the stability of the Einstein universe, a question which might, it was thought, throw light on the beginning of the universe—and even, perhaps, on its

---
[40] op. cit., p. 678.
[41] There are, of course, well-known instances in the physical sciences where the *entire* locus is not given a significance.
[42] op. cit., p. 677.   [43] *Ap. J.*, cix (1949) 164.

'cause'. These questions turned to matters of purely astrophysical interest—condensations in interstellar gas, cosmic radiation, and the synthesis of the chemical elements—and to speculation on the nature of the so-called 'Primaeval Atom'. Those who followed the other stream of thought argued from geometrical and kinematic premisses. They tended (for example, in their appeals to symmetry) to a rigid idealization of the situation. Often this was not accompanied by any but the slightest reference to astronomical practice. One who was always conscious of the need to connect theory and practice was Tolman. His paper on the estimation of distance in a universe with a non-static line-element is notable for giving an early definition and examination of the concepts of luminosity-distance, distance-by-angular-extension, rod-distance and coordinate-distance, not to mention their interconnexions. De Sitter also took great pains not only to point out the pitfalls of the astronomer ('how do we know that the "brightest star" in a nebula is not a cluster . . . ?'), but also to relate the theoretically connected 'radial velocity' and 'distance' with the ways in which both quantities are determined by the observer.[44] The concern which Tolman and de Sitter showed for these matters was unusual, and not until after the second world war were theory and observation properly merged.

Turning to the problem of instability of the Einstein world, the study of which Eddington began, there were obviously many questions left unanswered. Why should the universe expand rather than contract? How did the initial change in density come about? Eddington had given no firm answer but hinted that 'the formation of condensations might suffice'. The task of discovering processes which would induce instability was now undertaken by McVittie, W. H. McCrea, and Lemaître.

McCrea and McVittie[45] began by determining the exact metric for a condensation of mass at the origin. It was found that the equilibrium state of a model of given total proper mass in which a condensation has formed is one of *smaller* proper volume than the equilibrium state for a uniform distribution. McVittie next gave a general method of deriving an approximate metric where an arbitrary number of particles (of small mass) is randomly distributed throughout the Lemaître model.[46] He now decided that the volume of the model must be *greater* than that of the corresponding Einstein model. Lemaître tried to resolve the paradox,

---

[44] *P.N.A.S.*, xvi (1930) 474 (May). The style of much of this paper shows Lemaître's influence.
[45] *M.N.*, xci (1931) 128.   [46] ibid., p. 274.

arguing that the formation of condensations and the degree of concentration have no effect whatever on the equilibrium of the model. He concluded that an expansion from the Einstein state could be started by a reduction in pressure, an effect which he termed 'the stagnation of the universe'.[47] The formation of condensations affects the stability of the model, according to Lemaître, only in so far as it is accompanied by a change in pressure.

Certain features of both analyses were unsatisfactory, but despite their disagreement with Lemaître, McCrea and McVittie eventually reached conclusions not very different from his.[48] They had previously overlooked singularities at infinity exhibited by the metrics of their earlier papers. (When these are properly allowed for, both metrics agree in showing that the presence of particles does not affect the volume of space to the first order.)[49] The presence of a condensation may or may not lead to an expansion or a contraction. Their method of determining first order perturbations could not decide the question. The paper ended, however, with an interesting remark on the 'mass of the universe'. 'There must', they wrote, 'be some "condition of permanency" such as the constancy of total proper mass, to ensure that we are dealing throughout all changes with the same universe.' Variation of the total proper mass remained, nevertheless, an implicit property of most relativistic models.

Over the next few years many other results were forthcoming. Herbert Dingle proved the homogeneity of an expanding model to be stable and that of a contracting model to be unstable.[50] He was sceptical of the chance occurrence of a *simultaneous* and *universal* condensation of matter into galaxies, a chance occurrence which Eddington seems to have favoured as having inaugurated the expansion from the Einstein state. Dingle later proposed that the universe as a whole may be static, with recession a purely local phenomenon.[51] This virtually unassailable position has so far commanded little support, perhaps because it has appeared to be insufficiently constructive. It is a position taken up by R. G. Giovanelli, however, who has recently put forward a model in which density fluctuations can occur on all observable scales. In particular, what is visible to a given observer need not be representative

---

[47] *M.N.*, xci (1931) 490.
[48] McCrea and McVittie, *M.N.*, xcii (1932) 7; McVittie, ibid., 500; Lemaître, *Ann. Soc. Sci. Brux.*, liii (1933) 51.
[49] McCrea and McVittie, op. cit., p. 8.
[50] *M.N.*, xciv (1934) 134.
[51] *J. Wash. Acad. Sci.*, xxvi (1936) 183.

of the universe as a whole. There will be cells of both expansion and contraction.[52]

In 1934 Tolman found that instability of the Friedmann model with zero pressure would be a likely consequence of a disturbance's taking the form of either a condensation or a rarefaction.[53] N. R. Sen, however, found (for the same model) that although a condensation would increase in intensity yet a rarefaction would at first encounter resistance.[54] This does not indicate stability to rarefaction, for his later article showed that when the pressure was *not* zero, certain deformations might be stable and others not. R. O. Lifshitz went further and showed that for deformations of a very general kind, perturbations of the gravitational field and of the distribution of matter in the relativistic expanding universe either decrease with time or increase slowly—so slowly, in fact, that 'they cannot serve as centres of formation of separate nebulae or stars'.

From the few instances quoted here,[55] it should be clear that different mathematical treatments of a group of what appeared to be the same physical hypotheses led to results which were in some respects incompatible. The reason for this discordance is not to be found in the general method, which in all cases involved supposing a perturbation to occur in the model and inferring the model's subsequent behaviour (given various forms of discontinuity in the pressure, the density and the metric itself). The discordance was due, as W. B. Bonnor later observed, to the use of different boundary conditions; but the key to the subject was not, it seems, found until O'Brien and Synge and, independently, André Lichnerowicz, discovered some very general boundary conditions for surfaces of discontinuity in the General Theory of Relativity. The whole subject was finally reduced to some sort of order, notably by Bonnor and A. Raychaudhuri.[56]

The large measure of agreement now to be found on the subject of the perturbed Einstein model does not mean that all aspects of the subject have been equally cogently developed. Raychaudhuri, for example, found that the density differences in a perturbed relativistic model

---

[52] *M.N.*, cxxvii (1964) 461. The principle assumption made by Giovanelli, 'aimed at achieving a self perpetuating universe on a statistical rather than a steady-state basis' is that 'if the matter in the universe were smeared out uniformly, the gravitational potential would be uniform throughout; there would be no gravitational field, and the body-force would be zero'. Probably his most difficult problem will be that of satisfactorily resolving the de Chéseaux-Olbers paradox.

[53] *P.N.A.S.*, xx (1934) 169.   [54] *ZS. Ap.*, ix (1935) 215 and x (1935) 291.

[55] Cf. the more recent paper by H. Bondi which considers single condensations involving large deviations from the Einstein universe. (*M.N.*, cvii (1947) 410.)

[56] See Appendix, Note VI.

change too slowly for the formation of the nebulae to be explained in these terms. (This, it will be noticed, is a conclusion resembling that arrived at by Lifshitz.) Again, if a single condensation led to a process of universal expansion from the Einstein state, why do we observe no condensation with unique properties? The 'point-source' models have their own special difficulties and, speaking generally, they can be said to have left many of their adherents with the uneasy feeling that 'the vanished brilliance of the origin of the worlds' (Lemaître) might well have been over-rated. The problems associated with 'the origin of the worlds' have in fact been amongst the principal obsessions of cosmology ever since the rediscovery of the work of Friedmann and Lemaître. In order to discuss them more profitably when the time comes, we shall now therefore take stock of those individual relativistic models which had been introduced by the mid-1930s, outlining at the same time a very general classification of the possible time-variation of these and other models. This was separately presented by de Sitter, Heckmann and Robertson.

5. Types of Relativistic Model

We have already seen certain conditions under which Lemaître's equation

$$t = \int \frac{dR}{\{\sqrt{(\lambda R^2/3)} - 1 + (\alpha/3R) + (\beta/R^2)\}} \tag{33}$$

can be directly integrated. We have seen how, as a single illustration of the wider theory with which Lemaître began, a model could expand from the Einstein state in the infinite past, to the empty de Sitter state in the infinite future. Bearing in mind the instability of the Einstein state, it is clear that if once a start to an expansion from this state can be justified, and this at an arbitrary time in the past, then there can be no claiming that the time allowed to independently known phenomena is too short. This appears to be the sole argument of any weight which was offered in favour of this sort of expanding model by the writers of the period.

When expanding models which do not start with a stationary state are considered, there is no simple solution to Lemaître's integral. We have seen that Friedmann had already qualitatively described the form of the function $R(t)$. So likewise did de Sitter and Heckmann, afterwards, by numerical methods. Now the number of distinct possibilities is large. For example, should we classify our models by the three values of k and the three significant ranges and two significant values of λ (namely 0 and λ_e), fifteen cases would result. This number of distinct types is increased by varying the boundary conditions of the differential equation. This all supposes that we have agreed on the precise form the differential equation shall take. Actually, in supposing that we have reached the point of writing down a differential equation in R and t, we are presupposing some restrictive physical assumption. This is necessary because, given the Friedmann-Lemaître equations for the density and pressure, we are faced with a mathematical situation in which the number of dependent variables (R, ρ, p) is in excess of the number of equations. We have seen how different authors removed this indeterminacy by further assumption involving (usually) p or ρ or both. (Restrictions on λ and k are irrelevant.) The most important of these, historically speaking, are:

(Einstein) $R = \text{const.}$; $p = 0$. (Also $k = +1$, $\lambda \neq 0$.)
(de Sitter) $\rho = 0$; $p = 0$. (Also $k = +1$, $\lambda \neq 0$.)
(Friedmann) $p = 0$.

(Lemaître) $\dfrac{d}{dt}\{(\rho_m + 3p)R^3)\} + p\dfrac{d}{dt}(R^3) = 0$; $\rho_m R^3 = \text{const.}$

(Also $k = +1$.)

We shall shortly refer to a paper by Einstein and de Sitter which introduced the conditions $p = 0$, $k = 0$, and $\lambda = 0$.

The number of possible relativistic models is thus large, but from our point of view it is not necessary to provide an exhaustive account of them. For the purposes of later discussion, however, we list seven distinct forms for the function $R(t)$:

(i) Contracts to minimum, after which it expands monotonically to a de Sitter state. (Models exist for $k = +1$ and $\lambda_e > \lambda > 0$.)

(ii) ('Eddington-Lemaître model') Expands asymptotically from the Einstein state to a de Sitter state. ($\lambda = \lambda_e$, $k = +1$.) De Sitter investigated two cases, one with matter and no radiation, the other with radiation and no matter, and found them to differ only slightly from case (ii).

(iii) Einstein state. ($\lambda = \lambda_e$, $k = +1$.)

(iv) Expands monotonically from singular state. Point of inflexion. (Cases for $\lambda > 0$ and $k = 0$ or < 0; also $\lambda > \lambda_e$, $k = +1$. This last was Lemaître's favoured model, namely that beginning with what he called a 'primeval atom'.)[57]

[57] *Revue des questions scientifiques* (1931) p. 391.

THE EXPANDING UNIVERSE: THE FORMAL ELEMENT 131

(v) Expands monotonically from singular state but the curve has no point of inflexion. (Cases for $\lambda = 0$ and $k = -1$ or $k = 0$. The latter is the 'Einstein-de Sitter model'.)

(vi) Limiting case, separating examples of case (iv) (monotonic) from examples of case (vii) (oscillating). ($\lambda = \lambda_e$, $k = +1$).

(vii) Oscillatory models. (Cases for $\lambda < 0$, $k = -1$ or 0 and $\lambda_e > \lambda$.)

The seven types of model are illustrated in Fig. 1.

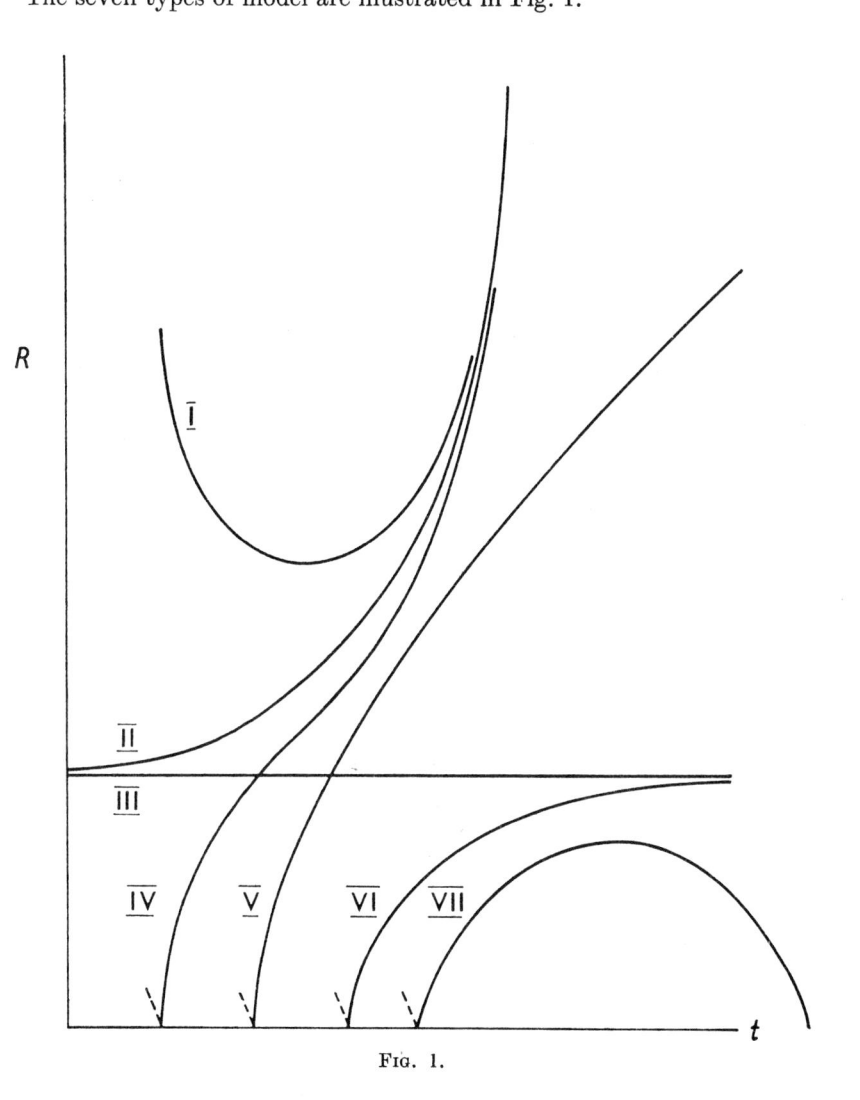

Fig. 1.

The periodic models (vii) are of some interest and have even been favoured by one or two recent writers. The first examples were given by Friedmann in his paper of 1922. One of these, namely that with zero cosmological term, was advocated by Einstein in 1931.[58] We recall that Friedmann showed that if λ lies in the interval $(-\infty, 0)$ whilst $R_0 < x_0$ (see p. 116 *supra*.), the equation (11) relating R and t involves an imaginary integral. In consequence he held that such a space is impossible. If, on the other hand, λ lies in the interval $(-\infty, \lambda_e)$, whilst $R_0 > x_0$, then R is a periodic function of the time, with '*die Welt-Periode*' t_π given by equation (13) and with R varying between 0 and x_0. This world-period increases with λ, tending to infinity as λ tends to the positive limit. For small values of λ the period is approximately $\pi A/c \, (= 2\pi/(3\sqrt{\lambda}))$. With non-zero pressures the analysis is essentially similar.

Accepting Friedmann's analysis, Einstein was extremely pleased to find that it allowed him to dispense with λ: his reasons for wishing to do so have already been explained. The solution of Friedmann's integral is in this case readily found: it is, a cycloid in the (R, t)-plane. At the time of this brief note by Einstein the indications were that, accepting Hubble's figures, the time which had elapsed since the onset of the current expansion was very short by comparison with the ages of the stars themselves.

The cycloidal (Friedmann-Einstein) universe was further considered by Tolman from the point of view of its thermodynamic properties.[59] Any periodic solution must needs satisfy the requirement of thermodynamic reversibility, for otherwise it would be impossible for the model to return to an earlier state. (Tolman had previously laid down an important part of the thermodynamics of the General Theory of Relativity,[60] an unexpected outcome of which was the prediction that *irreversible* thermodynamic changes take place at a finite rate with a continued increase in entropy, that is to say, without its reaching a maximum upper value.) Einstein's expansion and contraction were now shown to be accompanied by no increase in entropy and 'hence could presumably be repeated over and over again'. This solution may be called quasi-periodic, for the model returns to zero proper volume without the analytical conditions for a minimum holding at the lower limit. Strictly periodic solutions were ruled out by Tolman's

---

[58] *Berlin Sitz.* (1931) 235.
[59] *Phys. Rev.*, xxxviii (1931) 1758.
[60] *Phys. Rev.*, xxxvii (1931) 1639; xxxviii (1931) 797; xxxix (1932) 320.

principles.[61] Whether for this or some other reason, even the quasi-periodic solutions have never been widely advocated, despite Tolman's addition of an interesting example to their number. Tolman examined the Friedmann-Einstein case afresh[62] before considering an intermediary between this and another extreme. The former example corresponds to a universe filled with incoherent particles of matter, exerting no pressure and associated with no radiant energy. Tolman had dealt with the other extreme, namely a model containing blackbody radiation alone, in an earlier paper.[63] He now showed that not only did this represent a permissible quasi-periodic solution, but that so also did another model, interpreted as a perfect monatomic gas in equilibrium with black-body radiation.[64]

In 1932 Tolman wrote a paper jointly with Morgan Ward in which the behaviour of a closed model, more general than the Friedmann-Einstein example, was investigated.[65] This involved the possibility of irreversible changes in a 'perfect fluid unable to withstand tension and having at some initial time a finite volume and a finite rate of expansion'. Starting from Robertson's form of the line-element it was shown that such a model would expand to a finite maximum volume in a finite time, afterwards contracting to zero volume, again in a finite time. As in Einstein's example, the mathematical analysis unfortunately fails to indicate how the model will continue after reaching the lower limit of contraction (the conditions for an analytical minimum not being satisfied). Einstein was aware of this difficulty and suggested that the idealization on which the analysis is founded (the assumption of homogeneity and isotropy, for example) may be without justification in the vicinity of the lower limit of contraction.[66] Tolman and Ward, however, thought that 'from a physical point of view it is evident that contraction to zero volume could only be followed by renewed expansion'. (In his book, Tolman compares the example with that in which the differential equations of motion fail to predict the bouncing of an elastic ball.)[67] Had this strange model caught the imagination we might well have been given a theory of recurrent 'primeval atoms'.

---

[61] Cf. R. Zaycoff, *ZS. Ap.*, vi (1933) 128 (on the impossibility of periodic solutions of the gravitational field equations). An example of a cyclical universe had already been given by Tokio Takéuchi. (*Proc. Phys-Math. Soc. Japan*, xiii (1931) 166.)
[62] *Phys. Rev.*, xxxviii (1931) 1765-7. [63] ibid., xxxvii (1931) 1639.
[64] ibid., xxxviii (1931) 1797. [65] ibid., xxxix (1932) 835.
[66] *Berlin Sitz.* (1931) 235.
[67] *R.T.C.*, p. 439. Bonnor has since suggested a way in which a Friedmann model could pass from contraction to expansion without passing through a singular state. (*ZS. Ap.*, xxxv (1954) 20.)

Successive cycles of expansion and contraction would, of course, be physically equivalent only if the processes taking place in the fluid were reversible.

What of the open models? 'It is fairly certain that our space is finite, though unbounded. Infinite space is simply a scandal to human thought . . . the alternatives are incredible.'[68] Cosmologists were at this time so preoccupied with closed models that only one example of an open model was widely noticed, and then, presumably, only on account of its great simplicity. This example was presented by Einstein and de Sitter in 1932.[69] Its simplicity is achieved by the neglect of the cosmological term and the pressure. The line-element was taken to have the form

$$ds^2 = - R^2(dx^2 + dy^2 + dz^2) + c^2 dt^2, \qquad (34)$$

and space itself was represented as being without curvature and of infinite extent. The expression for the density, now easily derived from the field equations,[70] was

$$(\kappa \rho / 3) = (R'^2 / R^2). \qquad (35)$$

The expression for the pressure would have been

$$\kappa p = - (2R''/R) - (R'^2/R^2), \qquad (36)$$

but with $p = 0$ this can be written

$$2RR'' + R'^2 = 0, \qquad (37)$$

which has the integral

$$R = (at + b)^{2/3} \qquad (a, b \text{ constant}).[71] \qquad (38)$$

By the early 1930s the idea that the cosmological constant should be, if not zero, at least positive, had become no less a part of the accepted scheme than the idea of positive spatial curvature. Friedmann's theory was an exception, but not until Heckmann provided a timely

---

[68] Bishop Barnes, British Association, 1931. Discussion on the Evolution of the Universe.

[69] *P.N.A.S.*, xviii (1932) 213.

[70] See Lemaître's equations, footnote 22, p. 119 *supra*.

[71] As a simple extension of the Einstein-de Sitter example a model can be obtained in which the radiation exerts a pressure of 1/3 its energy density. This leads to the function $R = (at + b)^{1/2}$ (cf. *R.T.C.*, p. 416). In 1938 P. A. M. Dirac introduced a model, similar to these in some respects, in which R can be expressed as $(at + b)^{1/3}$. (*P.R.S.* (A), clxv (1938) 199.)

reminder (acknowledged by Einstein and de Sitter) were the alternatives thoroughly investigated.[72]

The logical advantages of an open model were generally thought to be fewer than those of a model with positive curvature. The latter involved, it is true, the need for a decision between spherical and elliptical space; but it avoided, on the other hand, the concept of an 'actual infinite'. This was treated lightly enough by most writers on cosmology. Robertson, for example, remarked that whether or not the world is infinite in extent hardly matters, for 'the only events of which we can be aware must occur within a sphere of constant radius'.[73] The problem will be taken up in a later chapter.

6. The Relativistic Models: Further Points of Comparison. The Expanding Universe

Although our survey of cosmological models is not complete, yet it is sufficiently so to allow of an examination of many of their more important features. Despite discrepancies between the underlying theories, the temporal behaviour of most later models finds an analogue in this early work.

We saw that in Lemaître's case the matter in the universe was taken to be conserved, whilst the pressure it exerted was supposed negligible. This is clearly unsatisfactory since, as already observed by Eddington, proper mass is continually being lost as radiation. Of the fractional rate of decrease in the proper mass little was known. De Sitter made a rough estimate of it,[74] but even given an accurate figure for the present epoch there is no reason to suppose it constant: indeed, on most 'evolutionary' models there are good reasons to suppose it to change appreciably with time. Friedmann had ignored not only material pressure but that due to radiation. In this case not only energy but the total proper mass of the material is conserved. Heckmann's treatment was far more general than Lemaître's, in so far as he did not ignore the pressure of matter. Again these assumptions were scarcely ever discussed. The justification for ignoring either the material or radiation pressure (an indication of the non-material energy density) by comparison with the material density, might well be acceptable for the present epoch; but in the cases discussed it is not at all obvious

---

[72] Cf. *Gött. Nachr.* (1931) 126.   [73] *Phil. Mag.*, v (1928) 837.
[74] *B.A.N.*, v (1930) 211. De Sitter also showed that in the case of a closed expanding model the pressure and density of radiation would, after a certain time, decrease steadily. The theory thus 'explains the apparent loss of stellar radiation'.

that the justification could be extended to the early stages of an expansion. For in these one might imagine the average condition to have approximated to that of the interior of a star.

It was difficult to imagine how these conservation issues were to be settled, for pure contemplation on the Mass of the Cosmos was unlikely to lead to any sound conclusion. We have already indicated how special assumptions as to the values of pressure and density made it possible to avoid this sort of problem, but none was really satisfactory. Tolman was probably the first to point out that the expressions for pressure and density might reasonably be supplemented by an 'equation of state'.[75] With the three equations, and with assumptions as to initial conditions and the values of the sign of the curvature and the cosmological constant, the problem could, in principle, be solved. This kind of approach to the problem was committed to so many uncertainties that it found no following.[76] There was instead, amongst adherents of the relativistic theories, a gradual movement in favour of deciding, on the grounds of observation, parameters which had often been previously regarded as decidable only on logical or aesthetic grounds— the cosmological term providing the most obvious example. At first the most important relations associated with observation were those involving magnitude and spectral displacement.

We saw that there was some controversy in regard to the correct version of the 'Doppler formula' in connexion with the de Sitter model. This was not so in the case of the non-static models. Admittedly, to begin with, Friedmann gave only the vaguest indications of the sort of empirical test he had in mind. 'Electrodynamic investigations' might lead, he suggested, to a value for λ, whilst it was 'possible that the Causality problem and the problem of Centrifugal Force'[77] might illuminate the task of deciding between the various plausible models.[78] Later, however, Lemaître, Robertson, Eddington, Tolman, de Sitter,

---

[75] That is to say, an equation of the form $p = p(\rho, R(t))$. If changes were to be irreversible the equation would have to include derivatives of $R(t)$ as well.

[76] Even less definite, but illuminating from the point of view of a classification of models, was what may be called the 'method of inequalities of state'. Robertson (*Rev. Mod. Phys.*, v (1933) 62) adopted the inequalities $p \geqslant 0$ and $\rho \geqslant 0$ and J. L. Synge (*Trans. R.S. Can.*, xxx (1936) 165) added to them the inequality $\rho \geqslant 3p$ (for the material density is presumably positive). Together with the equations for p and ρ, these now limit the regions of the (x, y) phase-plane in which relativistic models can be placed, where x and y are suitable functions of R and t (say R'^2 and R^2). They also imply that the gradient of the phase-plane at a point must lie between the gradients of two curves, each of a different family of 'guiding' curves. The classification of models which is made in this way does not appear to have been greatly valued by later cosmologists.

[77] Mach's problem, presumably.   [78] See *ZS. Phys.*, x (1922) pp. 385 ff.

THE EXPANDING UNIVERSE: THE FORMAL ELEMENT

and others all agreed that, allowing for various degrees of approximation, one might reasonably begin with a formula which may typically be written

$$(1 + \delta) = (R_2/R_1)(1 + u_r)/\sqrt{(1 - u^2)}. \tag{39}$$

Here u is the velocity of the source as measured by a local observer at rest with respect to the space-coordinates; u_r is the radial component of u. The times t_2 and t_1 are those of reception and emission of the light. The working is much the same as that which we gave in deriving a general expression for the de Sitter effect. Beginning with Robertson's form of the metric:

$$ds^2 = dt^2 - \{R/(1 + kr^2/4)\}^2 \cdot d\Sigma^2$$
$$(d\Sigma^2 = dr^2 + r^2 d\theta^2 + r^2 \sin^2 \theta d\phi^2), \tag{40}$$

the motions of free particles are found by integrating the equations of the geodesics of (40). Just as before it was the case only for the Robertson-Lemaître coordinates, so now it is the case for (r, θ, ϕ), that if the components of coordinate-velocity (dr/ds), $(d\theta/ds)$ and $(d\phi/ds)$ are once zero, they will remain so (i.e. it follows formally that the corresponding accelerations will be zero). As before, the particles are to conform to Weyl's Principle. It is now becoming customary to go no further in deriving the Doppler formula $(1 + \delta = R_2/R_1)$, as Weyl's Principle is a restriction on most accounts of relativistic cosmology which is accepted at the outset. In the more general case, the geodesic equations contain the following:

$$(d^2t/ds^2) + RR' d\Sigma^2/(1 + kr^2/4)^2 = 0, \tag{41}$$

which may obviously be written

$$(d^2t/ds^2) + \frac{R'}{R}\{(dt/ds)^2 - 1\} = 0. \tag{42}$$

This equation has the integral

$$(dt/ds)^2 = \{1 + (\alpha^2/R^2)\}, \tag{43}$$

where α^2 is a positive constant of integration characterizing the individual particle. Now the local proper-velocity of the particle can obviously be written

$$u = \sqrt{\{1 - (ds/dt)^2\}} \tag{44}$$

(bearing in mind that intervals of proper-time and proper-distance are the values of ds when we put, respectively, $d\Sigma^2$ and dt^2 equal to zero in the equation for the line-element). Eliminating (dt/ds) between (43) and (44) we then have

$$R^2 = \alpha^2(1 - u^2)/u^2, \tag{45}$$

showing how the velocities of particles in an expanding model will decrease in time.

Considering next a radial light-ray, its coordinate velocity is given by

$$(dr/dt) = \pm (1 + kr^2/4)/R. \tag{46}$$

Integrating this equation between the coordinate times of emission and reception of the light (t_1 and t_2) we have

$$\int_0^r \frac{dr}{(1 + kr^2/4)} = \int_{t_1}^{t_2} \frac{dt}{R}. \tag{47}$$

As in the case of the de Sitter effect, we now obtain the relation

$$V \cdot \Delta t_1/(1 + kr^2/4) = \Delta t_2/R_2 - \Delta t_1/R_1 \tag{48}$$

between the (coordinate) time-intervals separating, for example, successive light waves as emitted and received. Here V is the radial component of coordinate-velocity (dr/dt) of the source at emission. Relating Δt_1 to Δs_1 and Δt_2 to Δs_2 by (40), we have

$$(\Delta s_2/\Delta s_1)(= 1 + \delta) = \{(R_2/R_1) + R_2 V/(1 + kr^2/4)\}/\sqrt{(1 - u^2)}. \tag{49}$$

Since u_r is equal to $R_1 V/(1 + kr^2/4)$ (cf. the derivation of (44): $d\theta = d\phi = 0$) this can be written in the form of (39).

The term $1/\sqrt{(1 - u^2)}$ of (39) may be compared with the factor expressing, in the Special Theory of Relativity, the influence of velocity upon the rate of a moving 'clock'.[79] The last term in the numerator of the equation is in practice usually ignored, as we shall explain. It is the first term, that in the ratio of the radii, which is of greatest interest. It was this term, taken together with the discovery of a preponderance of red-ward spectral shifts, which properly justified the acceptance of the idea of the 'Expanding Universe'.

Now (39) is not analagous to the very general expression relating δ and r for a freely moving source in the de Sitter model: to find the analogue of this we must express u and u_r in terms of r and t. To do so we must supplement equation (45), giving u as a function of R, with a further equation for (dr/ds), obtainable by integrating the geodesic

---

[79] Sometimes called the 'transverse Doppler effect'.

equation. As before we shall have two constants of integration (α and one other), characterizing each particle. The result will obviously be very complicated—it is in any case not of direct astronomical application—and we shall not pursue the matter further. If we assume only free radial motion, however, we can combine (39) and (45) to give

$$1 + \delta = \frac{R_2}{R_1} \sqrt{\left\{\frac{\sqrt{(\alpha^2 + R_1^2)} \pm \alpha}{\sqrt{(\alpha^2 + R_1^2)} \mp \alpha}\right\}}. \tag{50}$$

Weyl's Principle has not yet been introduced and without such an assumption we can go no further in this direction. Weyl's Principle in this context amounts to setting $u_r = u = 0$ (and thus $\alpha = 0$). The particles then have fixed coordinates and the interval along the geodesic of each is equal to the coordinate t.

With very few exceptions writers were careful to avoid interpreting the spectral displacements by the use of the classical Doppler formula. They avoided what would have amounted to a naïve definition of a velocity of recession, choosing to speak merely of spectra which would *appear* to indicate velocities of such-and-such an amount. The caution which was usually displayed was a fortunate asset for another reason. The law of the new cosmology which was most exposed to the public gaze and from which the greatest number of 'philosophical' conclusions were deduced was without any doubt 'Hubble's Law'. It was obviously of the greatest importance that not only the concept of velocity, but that of distance should be handled carefully. In the early 1930s 'distance' was usually synonymous with 'proper-distance', and 'velocity' was defined accordingly. We shall later refer to other concepts of distance more in keeping with the sorts of operations traditionally used in the determination of distances. For the time being, however, we may briefly indicate the theoretical support for Hubble's Law. We must first distinguish two questions:

What justification is there for taking spectral shifts as indicative of velocities?

What is the expected form of the 'velocity-distance' relation, and can we expect a simple proportionality?

The first question has been answered to some extent in this section. From (39) we see that if R is constant and u may be neglected in higher powers than the first, $\delta = u$ as in the classical case. Rejecting this implausible approximation and accepting, say, Weyl's Principle, we have simply

$$1 + \delta = R_2/R_1. \tag{51}$$

There is no sign of a velocity here until we recall that (for example) the proper-distance of the source at a fixed r coordinate is proportional to R. With this example it follows that

$$\delta = (l_2 - l_1)/l_1, \tag{52}$$

from which it is very tempting to pass to the classical Doppler formula. But once again the step is only valid for nearby objects. In the early work of Hubble and his collaborators, however, the meaning of 'velocity' was determined solely by the classical formula. The answer to the first question must in fact be: A knowledge of δ for a single source, granted a prior knowledge of the function $R(t)$ may, in principle, be used to derive the local velocity of the source; but without this knowledge it can tell us neither the local velocity of the source nor the rate of expansion of the universe for any time (except in the impracticable case of a local source, and then Weyl's Principle must be assumed). In particular it is worth remarking that there is no reason why a 'velocity of recession' corresponding to $\delta = 1$ should be taken as equal to the velocity of light.

The second question can best be answered in terms of these theoretically predicted formulae (Weyl's Principle being assumed):

$$(dl/dt)_1 = (R_1'/R_1)l_1 \text{ and } (dl/dt)_2 = (R_2'/R_2)l_2, \tag{53}$$

$$V_D = \{(2R_2' - R_1')/R_2\}D, \tag{54}$$

and $$V_\xi = (R_1/R_2)\xi. \tag{55}$$

(Here V_D and V_ξ are the velocities corresponding to luminosity-distance (D) and distance by apparent size (ξ), being defined, respectively, as $\underset{\delta t_2 \to 0}{\text{Limit}} (\delta D/\delta t_2)$ and $\underset{\delta t_2 \to 0}{\text{Limit}} (\delta\xi/\delta t_2).$)[80] It appears that there is a proportionality of a sort. One can hardly claim 'linearity' for the relations, however, considering the time variation in the proportionality.[81] (We might add, further to the last paragraph, that from (51) it should be clear that there is no simple connexion between any of these 'velocities' and δ.)

Now having found that the velocities and distances of nebulae are likely, according to relativistic cosmology, to be related in a simple way, what have we shown? That Hubble's results were to be expected? This cannot be evident until a δ-distance relation is found (and, strictly

---

[80] See *G.R.C.*, section 8.5.

[81] One could of course take an exponential form for R, together with (53), to give a linear relation, but this would amount to special pleading.

speaking, not even then; for distances are known only indirectly and with the intrusion of theoretical constructions). Are we to conclude that the universe originated at a point at a distance in time equal to the reciprocal of 'Hubble's constant'? At the very least, such an argument would be mathematically naïve in the absence of evidence for the constancy of the Hubble factor. Must we merely suppose that the 'general picture' of a proportionality of velocity and distance is correct? There is no doubt a psychological importance to be attached to this sort of 'general picture', but the example appears to have had far too much weight thrown upon it for purposes of popular exposition. On the other hand, whatever simple picture we may have of the relation between the velocities and distances of distant nebulae, it is scarcely likely to do justice to the relevant concepts of relativistic cosmology.

As far as the astronomer is concerned, none of the formulae of this section is of direct application. In Chapter 9 we shall turn to the development of relativistic cosmology and shall consider the gradually increasing awareness of the importance of directly interpretable formulae. In the next chapter, however, we shall explain how astronomers were led, independently of these theoretical considerations, to believe that there might be a correlation between the velocities and distances of nebulae.

CHAPTER 7

THE EXPANDING UNIVERSE: THE ASTRONOMER'S CONTRIBUTION

IN THIS chapter we propose to discuss the astronomical antecedents of Hubble's Law, according to which the velocities of recession of the nebulae are proportional to their distances. We shall then outline the way in which improved astronomical techniques subsequently led to a revision of the 'Hubble factor', the constant of proportionality in the law. We shall not, for the moment, question the validity of Hubble's interpretation of his results.

1. THE K-TERM AND HUBBLE'S LAW

In Chapter 1 it was seen that not until 1899 were spectrograms published which were sufficiently convincing as evidence of the stellar constitution of spiral 'white nebulae' (in this case, M.31). Spectra of many other nebulae were subsequently shown to be of a similar nature. By 1912 V. M. Slipher of the Lowell Observatory was able to publish the radial velocity of M.31, no mean achievement in the face of the difficult problem of obtaining a sufficiently clear spectrogram.[1] Slipher, whose assistance Eddington acknowledged in his *Mathematical Theory of Relativity*, had so made this subject his own that in 1925 when Gustav Strömberg discussed the known radial velocities of globular clusters and 'non-galactic nebulae',[2] of the forty-five nebular velocities quoted all but five had been derived by Slipher. (Ten of Slipher's were independently verified.) Slipher's results were not without interest to astronomers, for it was generally hoped that they would allow a derivation of the solar motion. Truman and others obtained a value for this motion with respect to neighbouring nebulae,[3] but as information accumulated it became increasingly clear that the recorded nebular velocities could be accounted for only if, in addition to the solar velocity, there were some further systematic effect. For this reason, C. Wirtz introduced the 'K-term' in 1918:[4] it was supposed at this time to represent a

---
[1] Slipher used a fast short-focussed camera in conjunction with a low-dispersion spectrograph attached to a 24-inch refractor. The mean of four spectrograms indicated a velocity of *approach* of about 300 km/sec. See *Lowell Obs. Bulletin*, no. 58 (1914).
[2] *Ap. J.*, lxi (1925) 353 = *Mt. Wilson Cont.*, no. 292.
[3] *Pop. Astr.*, xxiv (1916) 111. Cf. Slipher, *Proc. Am. Phil. Soc.*, lvi (1917) 403.
[4] *Astr. Nachr.*, ccvi (1918) 109.

constant velocity to be subtracted from the apparent nebular velocity before evaluation of the solar motion. (The name 'K-term' had already been used in the evaluation of the solar motion with respect to the B-type stars. Here it was estimated at about 4 km/sec.) In 1920 and 1921 K. Lundmark and Wirtz successively confirmed Wirtz's earlier estimate of the value of the term, using Slipher's steadily accumulating data, and astronomers had to become reconciled to their astounding result: the term was of the order of 800 km/sec and was much larger than the resultant solar velocity.[5]

Wirtz hesitantly pointed out that the K-term might represent a systematic recession of nebulae from the Sun.[6] The scatter in the residual velocities of the nebulae was not at first easily explained, but with the eventual acceptance of the de Sitter effect it was recognized that the K-term might be a function of the distance. Wirtz appears to have been the first to investigate the possibility,[7] finding a general increase in velocity with reduced apparent diameter.[8] (Apparent diameter was the only available indication of nebular distance.) This, the first rough correlation of velocity and distance, was surprisingly overlooked by the theorists, who instead singled out another of Wirtz's findings which treated of an apparent connexion between velocity and nebular concentration. This was variously explained—most frequently as resulting from a 'gravitational red-shift'. This was later seen to be a selection effect: of the more distant nebulae, those which were highly concentrated tended to be selected simply because they were the *brighter* nebulae.

Wirtz's claim that the K-term may be a function of distance was at once examined by Lundmark[9] and Strömberg, who used nebular luminosities as an additional criterion of distance.[10] Strömberg found no correlation between radial velocities and distance from the Sun. 'The only dependence fairly well established is one that is a function of position in the sky'—and in fact this seemed to indicate a solar motion

[5] Lundmark (20 nebulae) *Stock. Acad. Hand.*, Bd. lx, no. 8 (1920); Wirtz (29 nebulae) *Astr. Nachr.*, ccxv (1921) 349.

[6] *Astr. Nachr.*, ccxvi (1922) 451.

[7] 'De Sitter's Kosmologie und die Bewegungen der Spiralnebel', *Astr. Nachr.*, ccxxii (1924) 21; *Scientia*, xxxviii (1925) 303.

[8] He gave a linear relation between velocity and the *logarithm* of the diameter—not quite the de Sitter effect.

[9] *M.N.*, lxxxiv (1924) 747.

[10] Strömberg, op. cit. Harlow Shapley in 1918 had noted that 'the speed of spiral nebulae is dependent to some extent upon apparent brightness indicating the relation of speed to distance or, possibly, to mass'. (*Ap. J.*, xlvii (1918) 20 = *Mt. Wilson Cont.*, no. 161.)

which was in close agreement with the motion as calculated from the asymmetrical velocity-distribution of stars in our neighbourhood (300 or 400 km/sec in direction $\alpha = 315°$, $\delta = 62°$). Lundmark found a 'not very definite' relation. Not only was he uncertain as to distances but, dealing as he was with the nearer spirals, their 'peculiar motions' (of perhaps 100 or 150 km/sec) must have represented a relatively large fraction of the recorded 'velocities'. In 1925 he replaced K by the series

$$k + lr + mr^2,$$

where r was the distance.[11] (As unit he chose the distance of M.31.) From data concerning forty-four nebulae he found that this could be written

$$(+ 513 + 10\cdot 365\, r - 0\cdot 047\, r^2) \text{ km/sec.}$$

This 'reaches its maximum value (of 2250 km/sec) at some 110 Andromeda units, which, according to results given later on, corresponds to a distance of 10^8 light years'.[12] He went on to add that 'as the peculiar velocities of the spirals seem to be smaller than 800 km/sec, one would scarcely expect to find any radial velocity larger than 3000 km/sec among the spirals.[13]

Few astronomical discussions have been so exhaustively treated when so little information was available as here. Slipher determined nebular velocities no longer, for there was little point in doing so until more reliable estimates of the nebular distances were forthcoming. It must be remembered that a large body of opinion still believed the 'white nebulae' to be close satellites of the Galaxy—Island Universes perhaps, but of much smaller dimensions than the central body. In our first chapter we saw the reasons which ultimately led to the rejection of this belief. We saw also how the Cepheid method of estimating the distances of the spirals eventually supplemented the nova method. After 1925 the investigation of the K-term as a function of distance could therefore be reasonably carried through, but only granted a telescope sufficiently powerful to carry out measurements on novae and Cepheids in distant spirals. Only one telescope in the world was in fact suitable, namely, the 100-inch reflector at Mount Wilson.

[11] *M.N.*, lxxxv (1925) 865.
[12] ibid., p. 867. Lundmark, as early as 1919, quoted the distance of M.31 as $1\cdot 7 \times 10^5$ pcs. (*Astr. Nachr.*, ccix (1919) 369.) This was obtained by the nova method—see p. 12 *supra*. The result was in order-of-magnitude agreement with Hubble's $2\cdot 75 \times 10^5$ pcs, obtained nine years later.
[13] loc. cit. It is noteworthy that Lundmark believed there to be a relation between the radial velocities and 'the stage of nebular evolution', as also insisted upon by Wirtz.

Hubble in 1929 knew the velocities of forty-six objects, although the distances for only eighteen of these, and that of the Virgo cluster, were available. This, however, was the basis of his famous 'velocity-distance relation', the basis, in other words, of a representation of the K-term as a linear function of distance. It is to be emphasized, in passing, that the paper in which Hubble's Law was announced (the paper in which he 'establishes a roughly linear relation between velocities and distances among nebulae for which velocities have been previously published and . . . [which] appears to dominate the distribution of velocities')[14] was presented without his knowing that Friedmann and Lemaître had published theories which it could be interpreted as supporting. The discovery which Hubble announced was not, however, made in a theoretical vacuum. Not only had the K-term been continually discussed for many years, but Hubble mentioned as the 'outstanding feature' of his discovery the possibility that the velocity-distance relation might represent the de Sitter effect. Robertson's attitude has a bearing upon this remark, for it is likely that he influenced Hubble strongly. We recall that eight months before he had proposed a linear relation between the assigned velocities and distances of the extragalactic nebulae. Using data provided by Slipher and Hubble, he held that this relation was 'roughly verified'. Hubble's Law could therefore equally well be ascribed to Robertson, not to mention Wirtz or Lundmark. Robertson was not the first to obtain a linear velocity-distance relation, but he was probably first to state it explicitly in connexion with an underlying theoretical explanation.[15] We are reminded, however, that his 1928 paper, rather than belonging with the Friedmann-Lemaître theories of an expanding universe, sets a seal to the physical development of the models of Einstein and de Sitter.

2. Minor Revision of the Recession Factor

The task of making observations of the kind used by Hubble was resumed in 1928 by M. K. Humason at Mount Wilson. (Slipher had used only a 24-inch refractor.) Within a year he had recorded the velocity of N.G.C. 7619 (a cluster in Pegasus) at 4000 km/sec—more than double the largest velocity previously known.[16] By 1936 he had reached the

---

[14] *P.N.A.S.*, xv (1929) 168. (Communicated 17 January 1929).
[15] We recall the supposed advantages of the transformation he gave for the de Sitter line-element viz., that the radial coordinate lay within an infinite range and the de Sitter singularity was 'removed from the finite region'.
[16] *P.N.A.S.*, xv (1929) 167. Notice that Hubble's paper follows this.

limits of the 100-inch reflector and his lists of velocities[17] contained one of 42000 km/sec.[18] During these eight years of patient and intensive observation there was no drastic revision of Hubble's first recession factor.[19] In 1930 Shapley revised the 'zero-point' of the Cepheid period-luminosity curve,[20] and the figure of 500 km/sec per Mpc, originally quoted for the recession factor, was now increased to 558 km/sec per Mpc.[21] There were several small revisions of this sort during the 1930s, owing to frequent adjustments of the Cepheid curves. There was revision too in the form of a correction to the apparent magnitudes of the nebulae, for reasons of the diminution of brightness as the result of the displacement of the spectrum. The correction formula applied by Hubble[22] can be justified only if the model is first specified, for example, in the sign and magnitude of the curvature. This, unfortunately, is the very matter over which one hopes to be guided by the observations. A more critical reduction of the observations was to come later.

In more recent times it has been necessary to make adjustments to the Cepheid curves owing to the changes in the value assigned to the recession factor. Shapley and his fellow Harvard astronomers concentrated upon the classical Cepheids, which are abundant in the Magellanic Clouds, and by 1940 had arrived at several slightly different period-luminosity curves for these objects.[23] In the last resort, of course, any calibration of the period-luminosity curve relied upon a knowledge of the parallaxes of Cepheids within the Galaxy.[24] In 1940 the best available work was that which R. E. Wilson had undertaken on the statistical parallaxes of the classical Cepheids.[25] The assumption

[17] *Ap. J.*, lxxiv (1931) 35 = *Mt. Wilson Cont.*, no. 426 (46 objects, 9 observed by F. G. Pease); *Ap. J.*, lxxxiii (1936) 10 = *Mt. Wilson Cont.*, no. 531 (100 objects).

[18] This was for a member of U.Ma.Cl. no. 2.

[19] By 1934, such was the confidence placed in the linearity of the velocity-distance relation that it was used by Hubble and Humason as a datum from which they could deduce the intrinsic luminosity of nebulae, given their velocities, and from which they were thus able to draw up a luminosity function for the external nebulae. (See, for example, *P.N.A.S.*, xx (1934) 264.)

[20] *Star Clusters* (1930), p. 189. The Large and Small Magellanic Clouds were now put at distances of 26 200 and 29 000 pcs respectively. (Recent figures vary between about 40 and 70 kpc.) This meant a reduction in the unit of nebular distance of about 11 per cent.

[21] Hubble and Humason, *Ap. J.*, lxxiv (1931) 43 = *Mt. Wilson Cont.*, no. 427.

[22] *Ap. J.*, lxxxiv (1936) 158, 270, 516.

[23] *Harvard Reprints*, no. 207 (1940); *Harvard Circular*, no. 439 (1940); *P.N.A.S.*, xxvi (1940) 541.

[24] The function was eventually taken in a non-linear form. There also appears to be a scatter of about unit absolute magnitude in the luminosity corresponding to any given period. There are, too, some unexplained systematic differences between the two Magellanic Clouds.

[25] *Ap. J.*, lxxxix (1939) 218. For later work see A. Blaauw and H. R. Morgan, *B.A.N.*, xii (1954) 95. The accuracy of this work, which is obviously of the greatest importance, still falls far short of what is needed by cosmologists.

was originally made that the classical Cepheids of the (smaller) Magellanic Cloud are identical with those of the Galaxy, although the truth of the assumption has been since doubted.[26] Shapley also assumed continuity in the sequence of classical and type II Cepheids and R.R. Lyrae stars; that is to say, he believed that different types of star with the same period were of the same absolute magnitude. In 1952 W. Baade showed that this was unwarranted. For some time it had been known that the average absolute magnitude of the globular clusters of M.31 differed appreciably from that of our own globular clusters (-5 as against $-7 \cdot 5$). The former was evaluated in terms of the distance of M.31, this having been assessed on the strength of the known periods of its classical Cepheids. In 1944 Baade proposed that the stars in galaxies should be divided into two groups—Population I (containing the classical Cepheids) and Population II (containing the RR Lyrae and type II Cepheids).[27] The 200-inch telescope had not been long in operation when Baade was able to deny the continuity which Shapley assumed. In short, he was unable to detect RR Lyrae stars in M.31—a result perfectly compatible with a distance decided from novae, globular clusters, and other criteria not generally believed valid. He concluded that a classical Cepheid was about 1·5 magnitudes brighter than a type II Cepheid of the same period, and also that there is no continuity between the curves for these three types of variable star.[28] From these considerations sprang a drastic revision of the value assigned to the recession factor, which, from 550 km/sec per Mpc, was altered to about 180 km/sec per Mpc.[29]

If other methods of measuring distances have been ignored in this short account, this is not because they are unimportant, but because they stand or fall very largely on the success of the Cepheid methods. Novae and supergiants have been and are still used as distance indicators, but all methods alike ultimately converge on the problem of ascertaining absolute magnitudes for the galaxies themselves. In 1929

[26] W. Buscombe et al., *Aus. J. Sci.* (suppl.), xvii (1954) 3.

[27] *Ap. J.*, c (1944) 137. The two types of stellar population had been recognized amongst the stars of the Galaxy by J. H. Oort as early as 1926. He showed that there the high velocity stars (type II) are essentially different from the slow-moving stars (type I) which predominate in the solar neighbourhood. (*Groningen Publications*, no. 40 (1926).) Baade was able to announce that photographs (taken with the 100-inch on red-sensitive plates) had now, for the first time, resolved into stars the two companions of M.31—M.32 and N.G.C. 205—and the central region of M.31 itself. Baade's paper threw a great deal of light on the question of nebular evolution.

[28] See Appendix, Note VII.

[29] Had the 'time scale' which derives from these figures always been so generous as it is now often imagined to be, it seems unlikely that the supposed problem of finding an age for the universe would have so obsessed the cosmologists of the 1930s.

Hubble relied, first, upon the distances of twenty-four extragalactic nebulae, assuming a value of approximately −6·3 for the absolute magnitude of their brightest stars; and second, he assumed that the nebulae themselves are of a 'definite order of absolute luminosity exhibiting a range of 4 or 5 magnitudes about an average value $M_{(vis)} = -15\cdot2$'.[30] In 1944 Baade showed that the luminosities which Hubble recorded for the (classical) Cepheids in M.31 are systematically too bright.[31] As the distance modulus of M.31 was the key to the figure of −15·2, Hubble's errors were of some importance. The brightest members of the Local Group[32] had for long been carefully studied, and individual stars were even thought to have been found in members beyond the Group (for example, M.81, M.87, N.G.C. 4321) yielding a more reliable value for the absolute magnitudes of the intrinsically brightest galaxies. These are now thought to lie within the range −19 to −20, although any alteration in the Cepheid calibration will mean that these figures, and hence the recession factor, must be altered.[33] More recently A. R. Sandage has discussed the use of the 'brightest-star' criterion for the distances of galaxies beyond the Local Group, concluding that Hubble almost certainly identified as brightest stars in galaxies beyond the Local Group, 'knots' which are really H II regions. Observations on M.100 suggest to Sandage that what are indeed the brightest stars are 1·8 magnitudes fainter than these knots. In adjusting Hubble's moduli for galaxies in the Local Group and correcting the ambiguity in the information dealing with 'brightest stars', Sandage derived a new recession factor of about 75 km/sec per Mpc.[34] But the end of the story is not in sight, and more recent figures suggest a value between 80 and 115 km/sec per Mpc.

[30] *P.N.A.S.*, xv (1929) 168. His views had changed little by 1936 when the Mount Wilson programme was virtually complete. In *Ap. J.*, lxxxiv (1936) 270 (= *Mt. Wilson Cont.*, no. 549) he gives the brightest cluster nebulae as $M = -16\cdot9$. The luminosity function for cluster nebulae 'may be provisionally described as a normal error-curve with a dispersion of 0·53 mag. and an average range of about 5·5 mag'. (p. 289.)

[31] *Ap. J.*, c (1944) 137.

[32] S.M.C., L.M.C., M.31 and M.33 together with the Galaxy. There are nearly 30 less luminous members.

[33] For the accumulated information of more than 20 years work at Mount Wilson, Palomar, and Lick Observatories, see Humason, N. U. Mayall, and A. R. Sandage, *A.J.*, lxi (1956) 97. Sandage, rather optimistically, takes M to two places of decimals (−19·82).

[34] *Ap. J.*, cxxvii (1958) 513. i.e. $(1/h_1) = 1\cdot3 \times 10^{10}$ years. Sandage suggested an uncertainty of factor 2.

CHAPTER 8

KINEMATIC RELATIVITY AND THE REVIVAL OF NEWTONIAN COSMOLOGY

IN CHAPTER 4 we saw that, for the General Theory of Relativity, the concept of gravitational force is superfluous. In an excess of zeal, several authors, including Eddington and Max Jammer, have held that anyone requiring such forces does so because he 'mistakenly' takes the metric to be Galilean. A field of force, according to Eddington, must needs be taken by such a person only in order to represent the discrepancy between the 'natural geometry', which is 'true', and the geometry 'arbitrarily' assumed to hold—in most cases Euclidean geometry.[1]

Such an account as this is of value perhaps, should one wish to make an approximate interpretation of the newer language into the old. It is written, on the other hand, by one who has fully imbibed the later language and who has accepted it as something final and irrevocable. It represents an unsatisfactory scientific attitude in so far as it would have us rule out, *a priori*, all theories of cosmology which accept the notion of gravitational force, and, by implication, those which require an Euclidean metric. In the two chapters which follow, several important contributions to cosmology will be evidenced, none of which would have been developed had Eddington's views prevailed.

1. THE CONCEPTUAL BASIS OF MILNE'S THEORIES

The first theory we consider, namely that due to E. A. Milne, is one of great originality.[2] Partly for this reason it proves difficult to grasp the trend of Milne's argument without first attempting to discern his attitude towards the subject of scientific explanation. His investigation, he tells us, progressed along a line which granted the initial ideas, was 'inevitable' and which proved to be in 'through command of the phenomena'.[3] When Milne said that he believed his work to have an interest

---

[1] *M.T.R.*, section 16.
[2] The reference '*W.S.*' is to Milne's book *Relativity, Gravitation and World Structure* (C.P., Oxford, 1935). This extended his first important statement of the theory which is found in *ZS. Ap.*, vi (1933) 1–95, and a draft of which was first presented in the form of lectures at Aberystwyth in 1933.
[3] *W.S.*, pp. 3–4.

independent of the extent to which it reproduced 'observable characteristics of the universe',[4] he spoke with the pride of the mathematician: thus elsewhere he referred to his work as being 'devoted to the mathematical consequences of general notions derived from experience'.[5] It is worth remembering, however, that Milne, so often accused of rationalism, submitted that the final interest in his results lay in their application to astronomy.[6] He was seemingly torn between the precepts of an early intellectual training in applied mathematics and a genuine feeling for astrophysics, a subject to which he made important contributions. He appears to have been attracted by the positivistic doctrine that an unverifiable proposition is meaningless.[7] Unfortunately, in an argument purporting to show that physical space-time must be finite, he immediately interprets this doctrine as saying, in effect, that any proposition which in terms of *his own* theory is meaningless is unverifiable. He appears to wish to endow his theoretical arguments with some sort of necessity, and in doing so undermines the conventionalism which he affirms elsewhere. Milne's treatment of physics led to several attacks on his epistemological position. He defended himself with a form of conventionalism which, indeed, appears to have been largely responsible for the methods he used from the very first.

It will later be seen that A. N. Whitehead had previously argued, like Milne, that Einstein's arguments must be rejected because epistemologically unsound. Milne was on much safer ground than Whitehead in that he preferred to rest his case on the principle for which S. R. Milner and others have argued: 'it is always open to the observer to choose the flat static space of Euclid in which to embed the events the observation of which constitute his field of phenomena'.[8] We shall later (in Chapter 13) have reason to qualify this statement, but as a heuristic device it has much to recommend it. The relativistic view that 'space, time, and material phenomena are interdependent' and that 'no one of them can be assigned without affecting the character of the others',[9] he very properly described as 'a rule of procedure, not an assertion about some preconceived entity "space" '.[10] For some purposes the rule may, he believed, be very convenient, but it is not so 'for visualizing the phenomena in the space we habitually use in other branches of physics and in ordinary life'.[11] In saying as much he was

[4] *W.S.*, p. 267. [5] ibid., p. 140, note. [6] ibid., p. 19.
[7] ibid., p. 131. [8] ibid., p. 10. [9] McVittie, *Obs.*, lvii (1934) 63.
[10] *W.S.*, p. 11. [11] loc. cit.

taking issue with a considerable body of opinion, according to which there exists in some sense a real entity—'the space of nature'—which in the fullness of time we shall come to know more intimately.[12] In Part II we shall consider some of the arguments for and against the various doctrines of conventionalism thus preached by Milne and others.

Milne's conventionalism was, on occasion, over-zealously applied. That the observer is at liberty to 'choose his space arbitrarily' gave occasion for criticism of Einstein's methods wherein freedom of choice was said to have been overlooked. Milne at times appears to have believed that to be free to choose a geometry is to be compelled to make a single and specific choice of metric. Misquoting Sir Thomas Browne, one might say of him that he liked flat, but not flexible, truths. He offered, however, a timely reminder that none of the spaces used in physics was possessed of 'objective reality'. The idea of spatial relation was suggested, he said, by tactual experiences. 'But no one has constructed a physics out of tactual data.'[13] Milne sought a firm foundation for physics. The concept of the transport of rigid bodies he held to be unreliable because one does not know with what objects the idealized rigid body may be identified. Time, on the other hand, was 'on a totally different footing from space', the passage of time being 'an undeniable constituent of our consciousness'.[14] The coincidences of events which are used in measuring time all occur *at the observer*, and if spatial and temporal measures could be reduced to nothing but such coincidences, some conceptual simplification might result. It was, more or less, Milne's claim to have brought about such a simplification. He showed how to define the concept of 'rigid length scales at a distance',[15] without invoking the transport of either 'rigid scales' or 'invariable clocks' (a phrase of Eddington's). Length coordinates and distance coordinates were 'combinations of time measures'.[16] This all involved laying down a procedure for comparing clocks at a distance, and here we first encounter Milne's notion of 'equivalent particle observers'. For some obscure reason he preferred to speak, not of coordinate transformations, but of transformations from one observer to another. Where the General

---

[12] Milne quotes from the works of de Sitter and Bertrand Russell, but he could have found many others. A quotation from de Sitter, more explicit than that provided by Milne, is to be found in *B.A.N.*, vii (1933–6) 205.
[13] *W.S.*, p. 14.   [14] ibid., p. 14.
[15] We notice that the definition, when analysed, involves the notion of simultaneity and varies according to the frame of reference with respect to which the simultaneity is reckoned.
[16] *W.S.*, p. 17.

Theory of Relativity was said to begin with 'coordinates . . . as a possible mode of description of events' which are then 'translated into the possible observations which they imply', he chose 'to begin with observations, and any coordinates used will be constructs out of observations'.[17] The fact is that Milne did not begin with observations but with hypothetical observations. His 'particle observers' are hypothetical observers. The two kinds of observation which they are supposed to make are never made. Milne's first book virtually ignores, for example, photometric measurements, and he supposes, for *conceptual* purposes, that the 'clock' is the astronomer's sole instrument. He believed it necessary to consider only a single instrument, as 'two observers might be equivalent in regard to one piece of apparatus and not in regard to others'.[18]

By the word 'clock' Milne intends any means of 'correlating with the real numbers the totality of events' in the experience of an observer.[19] To what extent can spatial measurements be reduced to combinations of clock readings? The comparison of path times in interferometry is nothing new, and both here and in the many 'echo-sounding' techniques we have the reduction of the measurement of distance to that of time.[20] Bondi has gone further and maintained that 'only time standards are true primary standards'.[21] His grounds are that 'the size of our measuring rods is determined by atomic interactions, which themselves are fully characterized in principle by the atomic *frequencies*'. This is reasonable enough if it is simply intended to indicate that the rigid body must be defined, if at all, in terms of chronometry. But the difficulties inherent in the idea of a standard of length which has a guarantee of rigidity, however it may be moved, haunted physicists for far too long. It usually originates with those who, wishing to provide theories with what they imagine to be a sound conceptual basis, believe that coordinates must be 'operationally defined', or that they must at least be provided with a hypothetico-operational definition. We shall revert to this matter in Chapter 16. The belief that any theory must be provided with a 'sound conceptual foundation' is, again, often closely allied to the doctrine of operationalism, coinciding with that variant form according to which *every* concept must be explicitly 'operationally defined', even if only hypothetically so. In other cases, the belief may be prompted by the desire to avoid a multiplicity of concepts (such as

[17] *W.S.*, p. 22.   [18] ibid., p. 25.   [19] loc. cit.
[20] The mean equatorial parallax of the Sun has recently been found by the radar technique, applied to Venus.
[21] *Rep. Prog. Phys.*, xxii (1959) 105.

the several concepts of distance). This can be done by affecting a reduction, part formal, part semantic, whereby the complex or unfamiliar concept is defined in terms of the elementary or familiar. Milne had the elementary concept—as, for example, that of distance chrometrically determined—but failed to connect it with, for example, the *customary* methods for determining distances. At the same time he made full use of Hubble's results. Milne claimed, nonetheless, that his two kinds of possible observation were not far removed from astronomical practice, and before deciding whether this claim was justified, the types of observation will be described.

The first consists of an observer (A) noting the times (t_1 and t_2 respectively), by his own arbitrarily graduated clock, at which a signal is sent to a second observer (B) and at which the reflected (or immediately returned) signal is received. The determination of t_2 as a function of t_1 determines, in Milne's view, the description by the first observer of the motion of the second in the line of sight.[22]

The second kind of observation consists of the observer A noting the reading on the clock of the other (t'_B) as seen when the echo-signal first returns. A can thus, by determining t'_B as a function of t_1 or t_2, investigate the apparent behaviour of B's clock.

B can carry out this sort of observation on A. (The corresponding epochs are denoted by t'_3, t'_4, t'_A.)

That such an interchange of signals is closely connected with astronomical procedure is said to follow from the fact that 'the Doppler effect from B observed by A will be found to be simply the differential coefficient dt_2/dt'_B, once the clocks have been graduated so as to be comparable'.[23] Milne submitted that the whole of his analysis could be carried out by starting with Doppler effect observations and subsequently integrating them.[24] Although his analysis is more realistic than it might at first have appeared, it still overlooks the exigencies of astronomical practice in the determination of distances. Granted Milne's procedure, however, the way in which he defines the coordinates, epoch, and distance, of the reflection event (E_B) is clearly explained. These are to depend not on t'_B, but on t_1 and t_2 alone, owing to their immediacy. The epoch must (if 'earlier' and 'later' are to retain their customary meanings) lie between t_1 and t_2. Regraduating A's clock by the addition of a constant t_0, we should require the epoch of E_B to increase by the same amount; and should E_B occur at A we should require its epoch to agree with t_1 and t_2. With these requirements it can

[22] *W.S.*, p. 27.   [23] loc. cit.   [24] ibid., pp. 34–6.

be shown that the epoch-coordinate of E_B must be of the form

$$\frac{1}{2}(t_1 + t_2) + \psi_1(t_2 - t_1). \tag{1}$$

When the constant t_0 is introduced we should require the distance coordinate of E_B to be unaffected; whereas when E_B occurs at A its distance should be found to vanish. The distance coordinate of E_B must then be of the form

$$\psi_2(t_2 - t_1). \tag{2}$$

Clearly $\psi_1(0) = \psi_2(0) = 0$. Simplicity dictates that ψ_1 be identically zero, whilst ψ_2 be a (positive) constant multiple of its argument. Milne thus writes as the final expressions for the coordinates,

and
$$\left. \begin{array}{l} T_B = \dfrac{1}{2}(t_2 + t_1),^{25} \\[2mm] R_B = \dfrac{1}{2}c(t_2 - t_1). \end{array} \right\} \tag{3}$$

All observers are to construct coordinates in the same way using the same positive number c.

Now the observer A can determine both R_B and t'_B as functions of T_B,[26] and B can determine the corresponding functions obtained by making observations on A. If their graphs are identical, A and B are said to be *equivalent observers*. Milne showed that there was a sense in which both could be said to possess *identical clocks*. He suggested that we might immediately recognize whether two 'atomic' clocks (atoms) are identical or not, and showed how the apparent relative running of two clocks in terms of the relative motion is equivalent to an evaluation of the Doppler effect, as already indicated. He showed, moreover, that if the Doppler shift of light from an equivalent observer B were constant in time $(s = (dt_2/dt_B) = \text{constant})$[27] the time-distance function (ϕ) would be given by

$$\phi(\xi) = \{(s^2 - 1)/(s^2 + 1)\} \cdot \xi, \tag{4}$$

from which it follows that (dR_B/dT_B) would be constant. In other words, what A would term 'the radial velocity of B'' would be constant, and in fact equal to $c(s^2 - 1)/(s^2 + 1)$.[28] For Milne there was no question

---

[25] This agrees with Einstein's definition in the Special Theory of Relativity.
[26] Milne writes $R_B = c \cdot \phi(T_B)$ and $t'_B = f(T_B)$.   [27] In our notation $s = 1 + \delta$.
[28] *W.S.*, p. 36. Denoting this by V it follows that $s = \sqrt{\{(1 + V/c)/(1 - V/c)\}}$, a well-known formula of the Special Theory of Relativity.

as to the 'true nature' of the shifts in nebular spectra. That they imply non-zero velocities in the nebulae themselves follows immediately from a more general result which can be shown, namely, that whether s is constant in time or not, V is non-zero *so long as the nebulae are equivalent to our own, in Milne's sense.*

One of the commoner criticisms of Milne concerned his dictum that 'the famous "postulate of the constancy of the velocity of light" is at bottom a convention'.[29] Recalling equations (3), the quantity $R_B/(T_B - t_1)$, which A would clearly designate the mean outward velocity of his signal, is constant and equal to c. The same result is found for the mean return velocity, and if B is an equivalent observer it is easily shown that B arrives at exactly the same results. This quantity the observers will term the 'velocity of light'. It is obvious that the 'convention' as to its constancy for all the observers was really made, first, when it was specified that the function ψ_2, for reasons of simplicity, should be taken as constant; second, when equivalent observers were defined as constructing coordinates in the same way. William Band has criticized Milne's claim that the velocity of light is constant by convention, believing that this cannot be so whilst the 'flatness or otherwise of the continuum is also a matter of convention'.[30] No explanation is given and Band is almost certainly mistaken. For the decision to choose a Euclidean geometry seems to be in no way connected with the decision to construct distances out of observations in any particular way, and it is this latter decision (or 'convention') which makes the constancy of the velocity of light an analytic truth.

As is well known, in formulating the Special Theory of Relativity, Einstein spoke of the velocity of light (and thus the simultaneity of distant events) as a convention.[31] His treatment differed from Milne's only in the sort of observations made fundamental. Time was measured in terms of the constant velocity of light, and distance measured by rod. A. A. Robb objected to the definition of simultaneity which this implied, believing that the only 'really simultaneous' events are those which occur at the same place.[32] The objection is not serious, for Robb appears to have failed to appreciate the need for such 'mere conventions'

---

[29] ibid., p. 39.   [30] *Phil. Mag.* (7), xxxvii (1946) 554.

[31] In later life he wrote: 'the type of critical reasoning which was required for the discovery of this central point was decisively furthered, in my case, especially by the reading of David Hume's and Ernst Mach's philosophical writings'. (ed. Schilpp. op. cit., p. 53.) It is interesting to note that Milne was also influenced by writers in the same tradition.

[32] A. A. Robb, *The Absolute Relations of Time and Space* (Cambridge, 1921), p. 13.

in the natural sciences—conventions for which Robb's own system would have been richer. There are, of course, good and bad conventions, and this particular one would be unsatisfactory should the velocity of light not be constant when measured in terms of a unit of time (in Milne's case, distance) *other than* what might be called 'rod-time' (in Milne's case, 'distance by light-time'). This other unit might be one which is of practical convenience or to which, for some other reason, physicists and astronomers are strongly attached. One might yet save appearances by distinguishing between the two concepts of time (or distance), although this complication would be widely regarded as unsatisfactory.

(One should remember that Einstein modified his original hypothesis when laying down the General Theory. Here the velocity of light depends upon the distribution of matter, and (even should the model be supposed homogeneous) as long as its metric is non-static the time taken by light to travel finite distances will depend upon the epoch. Although usually explained as the result of an increasing separation of the galaxies, this could equally be explained in terms of a changing velocity of light.)

2. Milne's Cosmological Principle

Milne proceeded to derive one-dimensional Lorentz formulae for the coordinates of any event in line with two equivalent observers, and corresponding formulae where the event was not in line but where the two observers were equivalent and in uniform relative motion.[33] The case where the equivalent observers were in uniform relative acceleration was solved by Milne and G. J. Whitrow some years later.[34] In his earlier book, however, Milne explained how, assuming rectilinear 'cross-country' light paths, he had been led to a set of four equations, each of which was incompatible with the other three. Rather than reject Euclidean geometry, he chose to abandon the principle of rectilinear cross-country light paths.

Next, as a preliminary to a three-dimensional scheme, he described a one-dimensional universe of discrete and equivalent particles 'satisfying Einstein's Cosmological Principle'. We have already encountered this in connexion with the Robertson-Walker metric. In this section we shall make a preliminary aside on the subject of Milne's views on the

[33] *W.S.*, pp. 40–51.   [34] *ZS. Ap.*, xv (1938) 344.

principle which in one form or another has appeared in every cosmological theory of note in the present century.

Milne, whose name is most frequently associated with the Cosmological Principle, refers his readers to Einstein's *Alle Stellen des Universums sind gleichwertig*.[35] Milne was more explicit than Einstein and proposed the principle that 'not only the laws of nature, but also the events occurring in nature, the world itself, must appear the same to all observers, wherever they be, provided that their space-frames and time scales are similarly oriented with respect to the events which are the subject of observation'.[36] This principle he referred to as the 'extended principle of relativity', a description which, although later disclaimed,[37] is far more instructive than his later characterization of it as a 'principle of selection', a 'rule of extrapolation' and a 'definition'.[38] Before discussing Milne's second thoughts we should add that Einstein was not, in the passage quoted, concerned with quite the same principle as Milne, for Einstein added the words '*im speziellen soll also die örtliche gemittelte Dichte der Sternmaterie überall gleich sein*'. Milne, nevertheless, afterwards referred frequently to 'Einstein's cosmological principle'.

Milne's second statement of the principle was more explicit than the first. It is satisfied, he said, by any system of *equivalent* particle-observers who offer identical descriptions of the system as a whole.[39] This was not to say that any pair of observers will describe the motion of a third in the same way, but that (by way of example) all will arrive at the same functional dependence of the velocity of a distant particle-observer upon its position vector and upon the time.[40] It should be clear that this is a (*restricted*) form of none other than the Principle of Covariance, and it is in this sense that Robertson and Walker accepted the principle. The 'totality of observations that any observer can make' is the same for all observers, according to Walker.[41] Later, speaking of the equations of a certain set, he added, rather more precisely, that they may be regarded as the description by an observer A of the set of free paths 'and can be considered as constructed out of observations made by A'.[42] 'From the cosmological principle', he continues, 'it follows that any other observer A' will obtain an identical set of equations for his own coordinate system', and the equations in question 'must be invariant under the transformation from A to A''. To speak of a

---

[35] *Berlin Sitz.* (1931) 235. See *W.S.*, p. 24, note and *ZS. Ap.*, vi (1933) 3–4.
[36] *ZS. Ap.*, loc. cit. By 'the world' Milne means 'the totality of the flux of events'.
[37] *W.S.*, p. 125.   [38] *W.S.*, pp. 20, note, 60, 125–6; *K.R.*, p. 5.
[39] ibid., p. 24.   [40] ibid., pp. 73, 174.
[41] *P.L.M.S.* (2), xlii (1936–7) p. 94.   [42] ibid., p. 99.

universe thronged with 'observers', even of the hypothetical kind, is a trifle odd; but allowing that we are here concerned with a comparison between descriptions in terms of two coordinate systems (each at the discretion of a single observer) rather than with the recorded observations of distinct individuals, we have a straightforward assertion of general covariance. On the other hand, just as Walker spoke of agreement as to the 'totality of observations' and Milne of 'the world itself . . . [which] must appear the same to all observers', clearly something more than this was intended. As Milne later wrote, the principle 'is an extension of the principle of the uniformity of nature from the independence of *laws of nature* as to the locality and epoch to the independence of the *structure* of the universe as to point of observation, where the word "structure" includes the complete history of the motions'. In Part II we shall return to this question of ambiguity.

Milne found that a one-dimensional universe of discrete and equivalent particles all describing the system in the same (functional) way was fully determined, given the relative motion of any two members. It was found that, were this motion uniform, the 'observed velocity' of the nth nearest member would be given by

$$c \left\{ \frac{(1 + V/c)^n - (1 - V/c)^n}{(1 + V/c)^n + (1 - V/c)^n} \right\}, \tag{5}$$

where V was the velocity of the nearest member.[43] This tends to the limit c, as n tends to infinity.

3. THE SIMPLE KINEMATIC MODEL. MILNE ON 'GRAVITATION'

Milne now approached the problem of deriving the 'simple kinematic world-system', namely that system which describes a set of equivalent particles in motion such that the Cosmological Principle is satisfied.[44] Any observer will, given an enumeration of the velocity- and position-vectors of all other equivalent particles, be able to summarize them by some such formula as

$$\boldsymbol{v} = f(\boldsymbol{r}, t),$$

where f is the same function for all observers (Cosmological Principle). Two different observers, O and O', following the motion of a particle P of the system, will write, respectively (Newtonian time):

$$\boldsymbol{v} = f(\boldsymbol{r}, t) \tag{6}$$

and
$$\boldsymbol{v}' = f(\boldsymbol{r}', t). \tag{7}$$

[43] *W.S.*, p. 57.   [44] ibid., ch. 4.

If a suffix '0' indicates a vector of O' relative to O, then clearly

$$v_0 = f(r_0, t). \tag{8}$$

Using 'Newtonian relativity', i.e. the simple vector addition rule, it follows that (7) can be written

$$f(r - r_0, t) = f(r, t) - f(r_0, t). \tag{9}$$

Hence f is a linear vector-function of the vector r, and we can therefore write

$$f = T(t) \cdot r. \tag{10}[45]$$

From the isotropic tensor character of $T(t)$ it follows that the inner product on the right of the last equation can be written as $r \cdot F(t)$. In other words, the velocity vector of a *local* particle P is along and proportional to the position vector OP, for all origins of reference O. Here was Hubble's Law, explained in a beautifully simple way, without appeal to a dynamical argument. The restriction to local particles stemmed from Milne's doubts about the legitimacy of using 'Newtonian time' for all observers. What of the form of $F(t)$? Milne said that 'observation is not yet sufficiently precise to disclose any day-to-day or year-to-year variation in the observed Doppler shifts'. Taking the motions to be uniform 'within our present accessible range of time-differences', $F(t)$ was written as $(t - t_0)^{-1}$, with t_0 some constant.[46]

Since at each position in this model there is a unique velocity, the motion, according to Milne, is of a 'hydrodynamical character'. Adding, therefore, the hydrodynamical equation of continuity, an expression is easily found for the particle density $(n(t))$.[47] This, in keeping with the assumption of homogeneity, is independent of position. The expression is as follows:

$$n(t) = N \exp\left(-3 \int_T^t F(t) \, dt\right), \tag{11}$$

where N is the particle density at any time T. For the case of a general recession, $F(t)$ is positive, and the system thus becomes progressively diluted. Milne therefore suggested that the apparent outward increase in density, which Hubble believed himself to have detected, favoured

---

[45] $T(t)$ is a Cartesian tensor of rank 2. Milne took it to be isotropic, but there seems to be no reason why it should not be otherwise. Dividing it then into symmetrical and skew-symmetrical parts, the former would correspond to an expanding universe, the latter to a rigid rotation of the entire universe about the observer. Milne would have objected to the 'absolute space' implied by this rotation.
[46] W.S., p. 75. [47] ibid., p. 77.

his system. In fact, he regarded Hubble's evidence as supporting the equivalence of the nebulae, but there was no attempt to show that, were the nebulae *not* equivalent, results which were qualitatively similar would not follow.

In the years immediately preceding the first publication of Milne's work, it was not infrequently pointed out by those who worked at the cosmological problem that the relativistic theory was incapable of predicting the expansion, as opposed to contraction, of the system of galaxies. In one of the simplest and yet most suggestive of arguments, Milne demonstrated that expansion seemed to be ultimately inevitable.[48] The argument involves the selection of a finite cluster of particles in, say, the interior of a sphere, at the centre of which is an observer O. Any particle which moves with constant velocity v and which at time t_0 lies at a distance r_0 from O will, at time t, lie at a distance r, where

$$v(t - t_0) - r_0 < r < v(t - t_0) + r_0. \tag{12}$$

It follows that

$$(r - r_0)/(t - t_0) < v < (r + r_0)/(t - t_0), \tag{13}$$

and thus, for sufficiently large $(t - t_0)$,

$$v \simeq r/(t - t_0). \tag{14}$$

Whatever the original velocity distribution, there will be a segregation of particles according to their velocity.[49] Moreover, at any epoch, velocities tend to become simply proportional to distance.

It is hard to conceive of a simpler argument for Hubble's Law, and yet several of its features must be underlined. In the first place the particles have been 'constrained to move with uniform velocities' (to use Milne's phrase). We shall later see to what extent this is in keeping with Milne's views of gravitation. It would certainly please neither a Newtonian nor an Einsteinian should the particles be identified with the nebulae. In the second place, the particles do not satisfy the Cosmological Principle, those at the edge of the cluster having a perspective quite different from those near the centre. A third point is that the time variable is a cosmic time, conceptually imposed from without. (This is of course no drawback where, as here, the motions are postulated to be

---

[48] *W.S.*, pp. 79-80.

[49] That there is a tendency for two particles ultimately to recede from one another, whether they were originally approaching or not, has been used by G. J. Whitrow (*The Natural Philosophy of Time* (Nelson, 1961) p. 11) as an argument which 'automatically reveals the irreversibility of time'.

uniform and the entities purely kinematic.) The model was not, however, seriously proposed except as affording 'insight into the origin of the features exhibited by the nebulae'. The insight was to be realized in the later sections of Milne's work.

Consider a system of particle-observers 'constrained to move with uniform velocities' according to one, and therefore all, members of the system. Milne proved that if each were to find not only the same velocity statistics for the system, but that these statistics were the same at all times, and if also the Lorentz formulae were to be used (and these he had derived for transformations between the coordinates assigned by equivalent observers),[50] then the number of particles with velocity-components lying between u and $u + du$, &c., can be written

$$f(u, v, w)\, du\, dv\, dw = \frac{B\, du\, dv\, dw}{c^3\{1 - (u^2 + v^2 + w)/c^2\}^2}, \qquad (15)$$

where B is an arbitrary positive constant.[51] The total number of particles is infinite, their density increasing as their velocity, which cannot exceed c. There is no preferential velocity frame, nor is there a velocity-centroid. Now the earlier arguments had suggested that it would be of some interest to examine the idealized system of particles which results when the velocity law

$$u = x/t,\ v = y/t,\ w = z/t, \qquad (16)$$

is imposed, x, y, z being coordinates assigned by any observer at their origin, and t the epoch which he assigns (t is no longer 'Newtonian time', but the variable which Milne so thoroughly explained at the outset of his work). It may be verified that the density distribution

$$\frac{B\, t\, dx\, dy\, dz}{c^3\{t^2 - (x^2 + y^2 + z^2)/c^2\}^2} \quad \left(\text{or,}\ \frac{B\, t\, r^2\, dr\, d\omega}{c^3(t^2 - r^2/c^2)^2}\right) \qquad (17)$$

satisfies the hydrodynamical equation of continuity. Verifying also that all equivalent observers must arrive, not only at the same velocity distribution, but at the same density distribution for the system, it follows that the Cosmological Principle is satisfied and all particles 'possess identical world-views'. This result provides, in a sense, the culmination of Milne's earlier work. What follows is an elaboration of the theme, a defence of the method which underlies it, an attempt to justify it as rendering a fair account of the world of the nebulae, and

---

[50] *W.S.*, pp. 40–52.   [51] The solution is found at *W.S.*, pp. 349–50.

an extension of this kinematic relativity to subsidiary systems. But the first goal had been reached, and Milne had 'constructed a system satisfying the Cosmological Principle in flat space'.[52]

Some of the features of the model need to be clarified. On the nature of the 'constraints' by which the members of the system were compelled to move with uniform velocities, it was sufficient for Milne to point out that a test particle released from a member of the equivalence cannot be separated from it, 'for to do so would be to select a preferential direction, and no such selection is possible'.[53] The free particle continues to move, therefore, with no acceleration relative to members of the system. The constraints on these members may thus be removed and each will follow the paths of free particles with which it originally moved. Each member of the system must therefore remain unaccelerated—a circumstance which might be expressed as a total 'action' of zero on each member of the system by the remaining members. Milne objected to this way of speaking, maintaining that no appeal to the ordinary dynamical (as opposed to kinematical) concepts—force, mass, energy, momentum, and the like—was necessary.[54] 'Gravitation' was said to be no more than 'a name for describing the motions that actually occur'.[55] The motions which Milne investigated were said to be compatible with *any* 'law of gravitation' the formulation of which was 'relativistic'.[56] This highly ambiguous term in the present context means 'capable of leading to a description by equivalent observers in equivalent terms'. With this single qualification, 'we may', wrote Milne at one point, 'impose any kind of interaction between the particles we like'.[57] Again, he termed the law of gravitation 'irrelevant', but elsewhere he made out a stronger claim. He held that the systems studied depended 'on no specific *formulation* of a law of gravitation', and that they therefore suggested that 'the same is true for any system'.[58] According to Milne, this property of being independent of a specific formulation of a law of gravitation 'constitutes an advance, however small, on Einstein's theory', which even obscures 'the inevitability of the motions here disclosed'.[59]

Since many of these remarks are tinged with extravagance it will be as well to consider the extent to which they are justified. Milne showed, it is true, that a *relativistically* formulated law of gravitation makes no difference to the motion of *equivalent* particle-observers or to 'free test-particles' *released* (as opposed to projected) from the members of

[52] *W.S.*, p. 91.   [53] loc. cit.   [54] ibid., p. 6.   [55] loc. cit., cf. also p. 60.
[56] ibid., p. 96.   [57] loc. cit.   [58] ibid., p. 7.   [59] ibid., p. 8.

the equivalence. The accelerations of free test particles, on the other hand, which are *projected* with velocity vector V was calculated by Milne.[60] The resulting formula for the initial acceleration, as reckoned from another particle of the system, is

$$(dV/dt)_0 = (P - V \cdot t)(Y/X) \cdot G(\xi), \tag{18}$$

where P is the position vector of the particle from which the test particle is projected, and where

$$\left. \begin{aligned} \xi &= (Z^2/XY), \, X = t^2 - (x^2 + y^2 + z^2)/c^2, \\ Y &= 1 - (u^2 + v^2 + w^2)/c^2, \\ Z &= t - (ux + vy + wz)/c^2. \end{aligned} \right\} \tag{19}$$

Here G is a function of ξ so far undetermined, but one which, if the particle is to have the customary Newtonian acceleration, must reduce locally to $-(4\pi\gamma mB)/(3c^3t)$ (m and B are constants and γ is the Newtonian 'constant' of gravitation).[61] This acceleration formula can obviously be regarded as a law of gravitation in so far as it describes the 'effect' of the given particles on the test particles—but this is cause and effect with a difference, for it derives from purely formal considerations which begin with the Cosmological Principle.

Milne proved that without specifying the form of the function $G(\xi)$, the equations (18) and (19) can be integrated completely and the form of the trajectories found.[62] It was shown that a particle outside the sphere $|P| = ct$ never crosses to the inside, and conversely. Also, since the motions of the particles inside and outside this sphere depend on different ranges of values of ξ, there was said to be 'no interaction whatever between phenomena outside the sphere and phenomena inside it'.[63] We recall that Milne's positivism in any case forbade him to consider phenomena outside these 'impenetrable barriers' surrounding each observer. More important than this barrier, however, was the remarkable discovery that the integration shows a slowly moving test particle in the vicinity of a fundamental particle to be *attracted towards*

---

[60] ibid., pp. 92–5, 350–2.

[61] The acceleration was seen to be 'directed towards (or away from) the centre of symmetry of the whole system in that frame in which the free particle is momentarily at rest' (*W.S.*, p. 100). Implicit in Milne's derivation is the postulate that the expression for g contains no dimensional constants. The problem of writing the acceleration equation in the form of a principle of least action was first solved by Walker (*P.R.S.* (A), cxlvii (1934) 478) although a proof that a solution must exist was given shortly before by Milner. He showed that, whatever geometry be chosen, the motion of a particle must be capable of expression in the form of a principle of least action, granted that a certain 'weighting' function of invariants may be introduced. Walker showed that the paths may be considered as geodesics of a four-space. This is a Finsler, not a Riemannian, space.

[62] *W.S.*, ch. 8.   [63] ibid., p. 144.

the latter. The trajectory of a projected test particle appears curved from all but the fundamental particle from which it was projected, whilst its acceleration goes on until the velocity of light is reached.

This, in outline, accounts for the motion of a test particle; but the practical distinction between members of the equivalence and test particles was never made very clearly. In one passage we are told that the distinguishing feature of the latter is that they have no influence upon the other particles. Perhaps because it is difficult to overcome the temptation to associate influence with mass we find it hard to accept that, as regards influence, the fundamental particles of intergalactic matter qualify with the nebular nuclei. 'What do we mean by a particle?', asked Milne in the section devoted to the statistical system. 'Do we mean an atom, a dust particle, a star or a cluster of stars?' He conceded that it seemed best to give no specific answer to the question, but to 'regard our analysis as giving the broad outlines of the distribution and trajectories of matter in motion'.[64] Even the most ardent supporter of Kinematic Relativity must agree that this is vague. Again, if not all fundamental particles are representations of nebular nuclei—and the latter, after all, form a system of *discrete* objects—how is Milne to reconcile the fact with his formulation of the Cosmological Principle? Are we to supplement the nebulae with hypothetical nebulae? It is clear that Milne himself would have liked to describe a three-dimensional model comprising a system of discrete particles, obeying the Cosmological Principle, but that he believed there to be a formal difficulty which could not be overcome.[65] (In Appendix Note IX we refer to a paper published in 1950 by H. S. M. Coxeter and G. J. Whitrow, which shows that Milne was mistaken in this belief.)

It is almost tautologous to say, with Milne, that a 'relativistic law of gravitation' cannot affect a uniform motion equivalence: it would not be 'relativistic' in Milne's sense if it were able to do so. It is easy to forget that a uniform motion equivalence is not the only formal possibility, and indeed Milne's arguments for it are usually only of a suggestive kind. The velocity-distance proportionality was said to be 'deep-lying', since it 'turns up' in several contexts—in the simple (local) kinematic theory, in neo-Newtonian, and in relativistic cosmology. As far as the general kinematic model was concerned (i.e. that which was not restricted to local particles) the method followed was to discover the formal consequences of postulating a simple proportionality of velocity and distance (which turns up so often elsewhere) and to

[64] *W.S.*, p. 201.   [65] ibid., p. 357.

justify the hypothesis *a posteriori* in the light of accumulated observational results. This is a perfectly legitimate procedure which would, on the other hand, have been stigmatized as *ad hoc* more frequently than it was in fact, had Milne not himself discovered such beautifully simple arguments for the (local) velocity-distance law, on the assumption of the Cosmological Principle. Despite these simple arguments, however, we are reminded that in relativistic cosmology there are many who have thought it preferable to allow for an accelerated recession, if only on the grounds that it is wise to be prepared for the worst.

4. The Development of Kinematic Relativity by Milne and Whitrow

In 1936 Whitrow showed that the assumption of uniform relative motion is superfluous, being no more than a convenient time-scale convention.[66] He explained how observers could invariably regraduate their clocks in such a way that the 'assumption' becomes a necessary truth. The broad outline of his method, and Milne's use of it, will now be explained. Whitrow's result does not, of course, prove the motions to be uniform in terms of commonly accepted kinds of measurement, but merely raises a new question for the observer to answer: What regraduation is necessary if I continue to use my present standards of time and distance? (To remove one source of the apparent rationalism in Milne's account, we must deny what at times appears to be his tacit assumption that all concepts of distance are interchangeable. The truth of this proposition is neither mathematically nor theoretically guaranteed.)

As the result of work by Whitrow and Milne, the foundations of Kinematic Relativity were, between 1935 and 1938, established more rigorously than hitherto, the theory of groups being liberally used.[67] The ideas of *signal functions* and of *linear equivalences* with their corresponding *generating functions* were important to this work. Briefly, the signal functions are of two kinds. The first (denoted by θ_{ab}) gives the time of reception of a signal by B, taken on his own clock, as a function of the time of its emission by A, as noted by A on his clock—this being made congruent to the first (see section 1 of this chapter). A 'signal function of the second kind' (denoted by θ_{ba}) is simply the functional inverse of the first: it gives the time of emission of a signal by

---

[66] *ZS. Ap.*, xii (1936) 47.
[67] Whitrow, *Q.J.M.*, vi (1935) 249; *ZS. Ap.*, loc. cit., and xiii (1937) 113; Milne, *P.R.S.* (A), clviii (1937) 324; clix (1937) 171, 526; and (jointly) *ZS. Ap.*, xv (1938) 263, 342.

A, noted on A's clock, as a function of the time of its reception by B as noted on his own clock. Given a set of observers such that any three remain collinear and that any pair can graduate their clocks so as to make them congruent,[68] Whitrow showed that any pair of the signal functions commute $(\theta_{pq}\theta_{rs} \equiv \theta_{rs}\theta_{pq})$. The general solution of these commutation identities Whitrow found to be of the form

$$\theta_{pq}(t) = \psi\alpha_{pq}\psi^{-1}(t), \qquad (20)$$

where ψ is any monotonically increasing function, and α_{pq} is a positive real number. A *linear equivalence* was now defined as a set of collinear observers with signal function of this form. The function ψ characterizes the whole equivalence and α_{pq} characterizes any pair of its members, being capable of taking all positive values. It was found that there is an infinite number of linear equivalences containing two given observers, but that three observers determine a linear equivalence uniquely. Such an equivalence remains linear on clock regraduation and 'all apparently different linear equivalences, generated by different functions ψ, are merely different descriptions of the same kinematic entity'.[69]

The first example of a linear equivalence was that which had already been investigated, namely the *uniform motion equivalence* given by

$$\psi(t) \equiv t. \qquad (21)$$

The formulae for the transformation of coordinates were found (much more simply than before, owing to Whitrow's calculus of signal functions) to be the Lorentz formulae. Once again the epoch-distance relation was derived, showing that the aggregate of all particle-observers moves with uniform relative velocity, having separated at the common epoch $t = 0$.

The next example which may be quoted is one discovered by Leigh Page, namely an *accelerated* equivalence.[70] Milne and Whitrow found the corresponding generating function to be[71]

$$\psi(t) \equiv t_0/\log(t_0/t) \qquad (t_0 > 0), \qquad (22)$$

and investigated its formal properties in some detail. There is, of course,

---

[68] In the previous notation $t'_B = \theta(t_1)$, $t_A = \phi(t_3)$, $\theta = \phi$. Milne and Whitrow proved that a graduation could always be effected so that this held (*ZS. Ap.*, xv (1938) p. 270.)

[69] *K.R.*, p. 25.

[70] *Phys. Rev.*, xlix (1936) 254, 466. Accelerated and decelerated linear equivalences were also found by J. B. Reid, *ZS. Ap.*, xvi (1938) 333.

[71] *ZS. Ap.*, xv (1938) 342.

an infinite number of generating functions, but as a last example we quote

$$\psi(t) \equiv t_0 \log (t/t_0), \tag{23}$$

which assumed great importance in Milne's later work. The signal function is found to be

$$\theta_{ab}(t) = t + t_0 \log \alpha_{ab}. \tag{24}$$

When the transformation formulae had been found[72] it was seen that all members of this equivalence assign the same epoch to any event, and that there must therefore be an absolute simultaneity in such a system. The members of the equivalence, so defined, are also seen to be relatively stationary, with α_{ab} providing a measure of the separation of A and B.

It is not difficult now to see that if the clocks of a uniform motion equivalence are regraduated in accordance with the relation[73]

$$\tau = t_0 \log (t/t_0) + t_0, \tag{25}$$

the signal function $\theta_{ab}(t) = \alpha_{ab} t$ becomes

$$\theta_{ab}(\tau) = \tau + t_0 \log \alpha_{ab}, \tag{26}$$

which generates the relatively stationary equivalence. It is thus formally possible to distinguish two scales of time. Whether or not either may be identified with the time of Newtonian physics is a problem which could be answered only were Milne able to derive Newtonian mechanics from Kinematic Relativity. This, as it happens, he maintained that he was able to do, finding that the appropriate time variable was τ. The equation can equally well be given in terms of the t-coordinate, but not only is the Newtonian form then lost, the equation of motion of a free particle is no longer reversible in time. (Energy, on the other hand, is an invariant in the t-dynamics.)

In a similar sense the t-scale was said to be that appropriate to the Lorentz formulae and the Maxwellian electromagnetic theory which Milne derived. It is therefore often said that 'photons keep t-time whilst mechanical systems keep τ-time'. This seems to be to insist upon the validity of both Newtonian mechanics and Maxwellian electromagnetism, given the same (independent) time-measuring device for each. Yet if physicists can agree upon any one thing, it is that (given customary measuring devices) these theories are not, at the limits of

---

[72] $t = t'$, $x = x' + ct_0 \log \alpha_{ab}$. (K.R., p. 33.)
[73] M. Leontovski also independently introduced the logarithmic transformation of time (according to Milne) in an unpublished paper of 1937.

observation, satisfactory. Perhaps, with Milne, some might even be tempted to make a stronger claim. 'Contemporary physics,' he wrote, 'thus has an ambiguity running through it, inasmuch as it confuses the time-variables used in two distinct domains of investigation'.[74] Even some who have numbered themselves amongst Milne's opponents have supported this attitude. What are its implications? The investigations by Milne and Whitrow of the concept of equivalence were of the first order of importance (perhaps not only for Kinematic Relativity) and the distinction between the two time-variables was correspondingly important; yet to accept Milne's claim to have detected an ambiguity in contemporary physics is to accept Milne's derivation of a dynamics, and of a theory of electromagnetism, formally identified with the Newtonian and Maxwellian theories respectively. And to do so can scarcely be to accept anything less than the whole of Milne's cosmology.

5. The Relatively Stationary Substratum

In *Kinematic Relativity* and elsewhere, Milne gave the name 'substratum' to the 'homogeneous equivalence in uniform relative motion'. The description of this substratum was given, of course, in t-time. The results found for linear equivalences were soon generalized to three dimensions and Whitrow found that the formulae of transformation of coordinates (where (r, t) and (R, T) are coordinates for a uniform motion and a relatively stationary equivalence respectively) are

$$t - (r/c) = t_0 \exp\{(T - t_0 - R/c)/t_0\} \tag{27}$$

and

$$t + (r/c) = t_0 \exp\{(T - t_0 + R/c)/t_0\}. \tag{28}$$

It was further pointed out that an observer may reckon small spatial intervals in terms of spatial coordinates in the conventional Euclidean manner *either in r-measure or in R-measure*. If the former, that is to say if he uses the metric

$$de^2 = dr^2 + r^2(d\theta^2 + \sin^2\theta\, d\phi^2), \tag{29}$$

then the expression for a small spatial interval (dE) in R-measure will be

$$dE^2 = dR^2 + (ct_0)^2 \sinh^2(R/ct_0) \cdot (d\theta^2 + \sin^2\theta\, d\phi^2), \tag{30}$$

when use is made of a certain general formula of the theory,[74a] relating the elementary distances as measured in conformity with the two scales

---

[74] *K.R.*, p. 50.
[74a] See *K.R.*, p. 47, equations (28) and (30).

of time. The relation (30) corresponds to a hyperbolic space. If, on the other hand, the observer chooses

$$dE^2 = dR^2 + R^2(d\theta^2 + \sin^2\theta\, d\phi^2), \tag{31}$$

then, using the general formula cited above,

$$de^2 = dr^2 + c^2(t^2 - r^2/c^2)\left[\frac{1}{2}\log\{(t+r/c)/(t-r/c)\}\right]^2 \cdot (d\theta^2 + \sin^2\theta\, d\phi^2). \tag{32}$$

It was shown that the expression (30) for dE is in form and value the same for all observers. Milne called the space it represents a 'public' space in contrast with the private spaces represented by (29), (31), and (32), which assign the same form to the interval but lead to different *values* for it—given any pair of events—as between different observers. The last metric may be seen to involve the time coordinate in its coefficients.

Suppose now that the substratum previously discussed is regraduated to give a 'relatively stationary substratum' with metric (30). Clocks must be regraduated in accordance with the relation (putting $r = R = 0$ in (27))

$$T = t_0 \log(t/t_0) + t_0. \tag{33}$$

Replacing the earlier expression[75] for the density distribution of particles in the Euclidean space (29), the corresponding expression, after some reduction, shows that the spatial density in the space of metric (30) is $(B/c^3 t_0^2)$. The new density is thus constant, being independent of T and R. It also coincides with the *local* value in t-measure at the epoch $t = t_0$. Other interesting comparisons can be made. The r-distance from an observer of the 'impenetrable sphere' $r = ct$ corresponds to an infinite value of R. The 'universe' is thus identified with the whole of hyperbolic space, populated uniformly with an infinite number of particles, as before. In the T-representation the red-shift is to be explained as a secular decrease in the frequency of light. (There is no general recession, and we recall that Milne wished to keep the velocity of light constant.)

For Milne, then, the time-scale difficulty of the mid 'thirties was effectively overcome by the recognition of the two scales of time: where dynamical arguments favoured a 'long time-scale' (10^{12} years) the *infinite* τ (or T, as written here) -scale was held to be appropriate,

---

[75] See equation (17) *supra*.

whereas with Hubble's data the t-scale must be used. Despite the 'kinematic identity' of the two descriptions, Milne always chose the uniform motion substratum for comparison with observation.

6. Milne on Gravitation and Galactic Evolution

Whether or not we call the formulae governing an accelerated equivalence a 'law of gravitation'—one way or the other the decision is trivial—the fact remains that Milne's elimination of gravitation, in Newton's dynamical sense of the word, merits a place in a history of the methods of physics. Gravitation, for Milne, rested on no assumption of action at a distance, nor was it to be identified with a curvature of space which was itself dependent on the distribution of matter. At first sight it appears that the motion of Milne's particles can be said to depend not upon the rest of the universe, but merely upon a *decision* to the effect that these particles shall move in an attractive and symmetrical way. We feel that Milne's mistake was in instructing the universe in the ways it must follow. Milne is not always consistent in his own defence, but it can be said in his favour, first, that the 'inevitability' of the motions was recognized to be one which followed only if the premises were sound; second, that although at times he appeared to want to endow these premises with moral certainty, yet he frequently refers to the need to compare his results with observation.

Actually, despite all he said about the older concepts of dynamics in his first book, Milne (in his *Kinematic Relativity*) constructed a dynamics in the following sense: where a particle's motion fails for some reason to coincide with that of a free particle, he defined suitable concepts of *force* and *potential* in terms of the differences between the motions. This was presumably necessary if he was to introduce any sort of cause which would lead to deviations from the overall cosmological pattern: it was essential, therefore, for his electrodynamics. But Milne also chose to add to his earlier theory of gravitation another draft in terms of potential functions and the like. This development does not appear to be altogether in sympathy with the outlook of his earlier book, but it emphasizes this important need of any satisfactory theory of cosmology: the theory must include, or at least be compatible with, physical laws of the small scale, where the essential problem is one of predicting *departures* from the uniformity of the universe. We shall shortly outline the sort of departures from uniformity which Milne predicted.

Before referring to the extension of the simple kinematic system to the case of statistical systems, it is worth noting Milne's distinction between 'world-map' and 'world-picture'.[76] In the former, all events are mapped at a common epoch, t. The world-picture is what would be ideally observed at a given epoch. The observed events have different epochs in the reckoning of the observer. It is clearly the world picture which is to be compared more directly with the findings of the astronomer, and it was necessary for Milne to convert his previous formulae (which are for the most part appropriate to the world-map) to this purpose. It was now found, for example, that the radius of the system observed at time t must be $ct/2$ rather than ct. The formula for the number of particles at epoch t inside a solid angle $d\omega$ and between distances r and $r + dr$ now became

$$\frac{Br^2 \, dr \, d\omega}{c^3 t (t - 2r/c)^2}. \tag{34}$$

Milne preferred to express this result in terms of the Doppler shift. It then becomes

$$\frac{B(s^2 - 1)^2}{4s^3} \, ds \, d\omega, \tag{35}$$

a formula which makes possible a fairly direct comparison of Milne's system with observation, and one which shows Milne's coordinates (like all good coordinates) to be no more than a useful scaffolding.[76a]

Milne's account of his statistical systems is more concerned with the detailed structure of the universe than with cosmology in a wider sense, and we shall therefore give only the barest outline of it. One object of constructing systems was to determine the structure of a nebula ('considered as an assemblage of freely moving particles'), without making 'any specific hypothesis as to its structure'.[77] As before, the nebular nuclei were taken as equivalent, in Milne's sense of the word. The aggregate of the trajectories was to be described in the same way from each fundamental particle *on the average*. The motions were no longer taken to be of a hydrodynamical nature, but instead to be governed by a velocity distribution rather like that postulated in the kinetic theory of gases. To return for a moment to Milne's 'theory of gravitation': he found, on the assumption that the Cosmological Principle is satisfied, a distribution function of the form

$$\frac{\psi(\xi)}{c^6 X^{3/2} Y^{3/2}} \, dx \, dy \, dz \, du \, dv \, dw, \tag{36}$$

[76] *W.S.*, p. 107.   [76a] As before (*Cf.* p. 154 *supra*), $s = 1 + \delta$.   [77] ibid., p. 169.

where a relation exists between the functions $\psi(\xi)$ and $G(\xi)$. The relation in question is

$$G(\xi) = -1 - \frac{C}{(\xi-1)^{3/2}\psi(\xi)} \qquad (37)$$

(C is a positive constant of integration).[78] This was said to be a law of gravitation for the family of systems defined by $\psi(\xi)$, to the extent that it prescribes the acceleration of each member.

The statistical systems now emerge from Milne's analysis as triply infinite collections of sub-systems, each characterized by the velocity of its nucleus (identified with a nebular nucleus) which is also the mean velocity of the sub-system. The sub-systems, which comprise a triple infinity of particles, are subject to an outward expansion, but unlike that of the idealized nebulae, this expansion is accelerated. The overall system was shown to be most densely populated in the neighbourhood of the fundamental particles (of uniform velocity) and to exhibit a local structure with spherical symmetry. Not only this, but the law of density distribution in the average sub-system was calculated and found to be in rough agreement with some measurements made by J. H. Reynolds[79] and Hubble.[80] Milne predicted the arrival, and arrest, in the vicinity of each nebular nucleus of particles from other galaxies,[81] and cited as evidence for this effect the scanty existing knowledge of interstellar material. The estimated density of this he thought (in the case of the Galaxy) too great for the material to originate with the Galaxy itself. Nebular evolution he described as 'a continual interchange of partners'.[82] 'Dissolution is accompanied by re-formation, but re-formation of a new type of system.'[83] The evolutionary process never terminates, the individuals die 'but the race of nebulae survives for ever'.[84]

Milne's treatment of galactic evolution has made little impression on the subject as a whole, but at the time of its first publication relativistic cosmology could show no comparable treatment of large numbers of co-existing nebular condensations. The advantage might be thought to have been even more with Kinematic Relativity when, in his second book, Milne put forward arguments which suggested that the arms of

---

[78] For the way in which the function $G(\xi)$ was introduced, see section 3.
[79] *M.N.*, lxxiv (1913) 132; lxxx (1920) 746.   [80] *Ap. J.*, lxxi (1930) 231.
[81] There is an interchange between any galaxy and those galaxies which lie within a definite distance of it.
[82] *W.S.*, p. 264.   [83] loc. cit.   [84] ibid., p. 286.

the spiral nebulae are the loci of the present positions of particles, which themselves describe spiral orbits.[85] The arguments (which, as Milne ruefully admitted, lacked the compulsion of those in his first book) relied on the assumption of a plane of symmetry and an axis of rotation for each sub-system. Appealing to the supposed secular variation of the gravitational 'constant', he was able to deduce both an equation to the spiral arms and an expression for the number of convolutions. He was also able to predict a reversal of the sense of winding of the arms. These results were supported, to some extent, by the findings of B. Lindblad and others. But neither this apparent success, nor the intricate theories of electrodynamics, of atomic structure and of cosmic radiation, each of which he made a part of his single theme, could preserve Milne from the widespread scepticism aroused by his supposed rationalism. If anything, Milne's position was unnecessarily worsened by the strain of argument to be found in his last book, *Modern Cosmology and the Christian Idea of God*.[86] The unfortunate outcome of all this is that his entire work is often lightly dismissed, with the consequence that many of his methods are often mistakenly ascribed to others.

7. Kinematic and General Relativity

Milne's ideas became widely known soon after their first publication, and before long a firm belief arose in many quarters that his cosmology would emerge as no more than a trivial instance of the older relativistic theory. Milne's attitude towards the element of convention in selecting a geometry fostered this belief, and shortly after his first memoir appeared,[87] W. O. Kermack and W. H. McCrea contended that if space-time is merely a convenient map for the description of phenomena, then there can be no objection to linking Milne's general method with a Riemannian metric.[88] (This was very much the plan which Robertson and Walker put into effect in 1935.) Kermack and McCrea observed that, as any manifold of constant curvature can always be conformally represented on a Euclidean manifold, in any small neighbourhood a Euclidean map and a Riemannian map of constant curvature (one, for example, with any line-element of the sort provided by relativistic cosmology) must correspond without distortion. They went on to show

---

[85] In this Milne followed up a suggestion made by E. W. Brown (*Ap. J.*, lxi (1925) 111) that the spiral arms are the *envelopes* of orbits. Milne gave reasons for rejecting this and for assuming that the orbits cannot have had an envelope. The condition that this was so provided a starting point for Milne's theory.

[86] (C.P., Oxford, 1952).   [87] *Z.S. Ap.*, vi (1933) 1.   [88] *M.N.*, xciii (1933) 519.

that Milne's hydrodynamical solution is the limiting case of one of the non-static universes of the General Theory of Relativity as the density tends to zero.[89] Observers in both kinds of universe, it was claimed, 'will have the same experiences', and 'Milne is therefore studying exactly the same material phenomena as those studied by general relativity'.[90] This presupposes a great deal about the correspondence rules chosen for the two theories. The formal comparison is certainly valuable. Later we shall refer to an attempt by Weyl to reduce a theory by G. Birkhoff to Einstein's theory. In this case the theories shared little more than their end results. In the earlier case, with suitable coordinate transformations, both Milne's theory and certain of the relativistic theories could, in addition, be conveniently described in terms of the same Riemannian metric.[91] But the theories were even there not wholly equivalent, and the extent to which they compare, whilst of great interest, is easily exaggerated. Milne was frequently irritated by these comparisons, needless to say.

Kermack and McCrea found Milne's simple kinematic solution to be only a limiting case of an expanding model of the General Theory of Relativity. Not only this, but a glance at the principles upon which the kinematic system was founded shows that they are satisfied, granted a suitable interpretation, by all the relativistic models previously considered. Why then did Milne's 'kinematic' methods lead to results less general than those of the General Theory of Relativity? Part of the answer to this question is to be found in a paper written by Walker in 1935.[92] He demonstrated that the solutions of the kinematic problems which correspond to Riemannian spaces of zero and positive curvature were implicitly rejected by Milne when he took over the Lorentz transformations between observers in a space of three dimensions. Walker showed that other transformations, although more complex, are equally valid when more than one dimension is considered. He found, however, that Milne's method did not lend itself to a determination of the consequences of these transformations. Instead he offered a very general solution, which proceeds from the 'general principles laid down by Milne, without assuming any particular transformations'.[93] The solution, also obtained by Robertson, was more general than Milne's,

---

[89] ibid., pp. 522–4. The cosmological constant, λ, was assumed zero. Dispensing with the requirement of unaccelerated relative motion, λ may be non-zero and the de Sitter solution may then be considered as an instance of Milne's general solution.
[90] ibid., p. 529.
[91] Most clearly shown by Walker, *P.L.M.S.* (2), xlii (1937) 90.
[92] *M.N.*, xcv (1935) 263.   [93] Walker, *P.L.M.S.*, loc. cit.

and geometrically speaking it included all other systems previously investigated.[94] We shall explain briefly how this was so.

The earliest relativistic models all contained the assumption that space, at least, was of constant curvature. Weyl, in using a system of co-moving coordinates and requiring that the spatial configuration should remain similar to itself as time goes by, tacitly introduced another sort of uniformity. Robertson, as we saw in Chapter 6, carefully formulated a principle of general homogeneity and isotropy from which he derived the form of line-element now known by his name. Tolman obtained it more rigorously on the basis of a somewhat different set of hypotheses. Of the two, Robertson's treatment was the more consistently geometrical. Tolman, it will be remembered, made such assumptions as that the density in the model and the rate of expansion of its elements of proper volume are independent of position. Neither demonstration, however, approached in generality that given first by Walker in 1935 (May) and independently by Robertson in August of the same year,[95] that Robertson's earlier metric (see equation (23), Chapter 6) must inevitably be common to all accounts making certain very general hypotheses. Both memoirs, considering their independence, use remarkably similar methods. Both made liberal use of the theory of continuous groups and both began with Milne's Cosmological Principle. Both, after the manner of Milne, took the 'world of matter in motion' to be idealized as a system of fundamental particles, attached to each of which was supposed an observer equipped with clock, theodolite and the means of sending and receiving light signals. These authors made assumptions which were equivalent to what Walker termed 'the principle of symmetry'[96] (each fundamental observer sees himself to be at a centre of spherical symmetry), and both proved that there exists a quadratic in the differentials of the coordinates which is invariant under transformation from one fundamental observer to another. This quadratic may, as they pointed out, be regarded as a line-element defining a Riemannian space (V_4) of constant spatial curvature. Here

---

[94] L. Infeld and A. Schild used methods similar to Milne's in deriving models which are metrically, but not topologically equivalent to those of Robertson and Walker. (*Phys. Rev.*, lxviii (1945) 250; lxx (1946) 410.) Whitrow and D. G. Randall, using the methods of Kinematic Relativity, also investigated a class of homogeneous models of which Milne's simple kinematic model was one. (*M.N.*, cxi (1951) 455.)

[95] *Ap. J.*, lxxxii (1935) 284, cf. lxxxiii (1936) 187, 257. Of these the former discusses the equations of motion of a test particle (on the assumption that they depend, at most, on second differentials of the coordinates) and finds them only within a certain arbitrary function. This function is found, assuming (i) General Relativity, (ii) certain modified forms of Newtonian mechanics.

[96] *P.L.M.S.* (2), xlii (1937) p. 103.

'space' is defined by the condition τ = constant, τ being the invariant local time common to all observers whose clocks are synchronized in the way indicated by Milne. The resulting line-element was that which, in its general form, Robertson six years earlier had shown to be appropriate to the uniform models of the General Theory of Relativity. It had at last been proved essential to any account in which a Riemannian treatment, Milne's Cosmological Principle, and one or two other widely accepted physical principles were admitted. (As an instance of the latter we have Walker's adoption of Fermat's principle as a means of determining the light paths.) The solution includes all systems derived from the General Theory of Relativity, from Milne's Kinematic Relativity, and from the more recent steady-state hypotheses. The light paths in both accounts were found to be minimal geodesics, but the general form of the line-element was determined without hypothesis as to the paths followed by free particles, paths which are not necessarily geodesics of the Riemannian map.[97]

In using the Riemannian analysis Robertson and Walker were not, of course, aiming to reduce all cosmological theories to those stemming from Einstein's General Theory.[98] They did not imply that Milne had tacitly accepted the Einsteinian idea of a theoretical connexion between local deformation of the metric and the motions of all matter in the neighbourhood, nor would they have denied that geometries other than the Riemannian may be used.[99] Their analysis cannot, without further assumption, single out a particular form for the function $R(t)$ in the expression for the line-element. Ways of restricting the function—uniquely so in the last two cases—were in fact left to the set of field equations of Einstein's theory, Milne's 'dimensional hypothesis', and the steady-state hypotheses (to take only three examples).

8. The Neo-Newtonian Cosmology of Milne and McCrea

The second group of cosmological models to be considered, involving the assumption of a Euclidean metric, involves also a return to most,

---

[97] Walker found them from a variational equation; on the assumption, that is to say, that the system satisfies a principle of least action. This last principle is implied by the Cosmological Principle and the principle of symmetry taken together. (Walker, op. cit., pp. 114–15.)

[98] See Appendix, Note VIII.

[99] Consider, for example, Walker's previous investigation of the form which a principle of least action must take if the motion of a free particle is to be described in Finsler geometry. ($P.R.S.$ (A), cxlvii (1934) 478.)

but not all, of the remaining principles of Newtonian mechanics. We recall that on each previous occasion on which unacceptable conclusions were drawn from Newton's mechanics it was assumed that the universe is, on the whole, static. Once there was reason to doubt this assumption, fresh efforts were made to formulate a Newtonian cosmology. The first notable attempt was by Milne and McCrea, who offered a version for which no more was claimed than that 'the symbolic representation of Newtonian cosmology is formally identical with the symbolic representation of the cosmologies of "general" relativity, but that the latter representation requires a different interpretation in terms of observations'.[100] The precise status of their new theory was never altogether clear. According to its authors, it ignored the failure of objective simultaneity and was held to be neither 'logical' nor 'self-consistent'. It was claimed, however, that it offered 'insight into the cosmological problem' and conveyed 'an easily intelligible idea of the *dynamical* nature of the phenomenon of the expanding universe'.

The principles upon which the system rested were those of Newtonian dynamics and gravitation (augmented, if necessary, by a 'λ-term' signifying a repulsive acceleration of the form $\lambda r/3$ (> 0)), Galilean relativity, and the Cosmological Principle. Space was taken as Euclidean and the concept of universal time was accepted, although it was pointed out that 'observers who are equivalent in the Newtonian sense fail to find themselves equivalent when they employ identical procedures for calculating coordinates (in particular, epoch of distant events) out of observations. . . .'[101] The idealization was a hydrodynamical one. Recalling the way in which Einstein chose his field equations, it is not surprising that, as far as local phenomena are concerned, the two systems agree. Perhaps more surprising was that the 'Newtonian distance' (r) of an element of the new model turned out to be the same function of the 'Newtonian time' (t) as certain relativistic coordinates are of the corresponding time coordinates. This was first shown in a paper by Milne in which a 'parabolic' model was shown to correspond to the Einstein-de Sitter model.[102] Shortly afterwards Milne and McCrea found a more general result which will now be explained.[103]

It is immediately evident from the Cosmological Principle that the density (ρ) must be a function of the time alone. It is therefore legitimate,

---

[100] This, and the quotations which follow, are taken from *W.S.*, pp. 290, 70, and 299.
[101] *W.S.*, p. 301. The Newtonian universal time, in short, was not that of Milne's Kinematic Relativity.
[102] *P.N.A.S.*, xviii (1932) 213. See p. 134 *supra*.
[103] This is substantially the account to be found in *Q.J.M.*, v (1934) 73.

and it will shortly be seen to be convenient, to write the expression $(1/\rho)\,(d\rho/dt)$ as $-3F(t)$. The equation of continuity,

$$(d\rho/dt) + (1/r^2)\frac{\partial}{\partial r}(r^2\rho v) = 0, \tag{38}$$

(where $v\,(= dr/dt)$ is the velocity of the fluid at distance r) now gives, on integration, the following expression for v:

$$v = r\,.\,F(t) + G(t)/r^2, \tag{39}$$

where $G(t)$ is an arbitrary function. Now where an observer regards himself as at the centre of a distribution of matter, which in arrangement and state of motion is spherically symmetrical about him, and where 'the observer considers the material outside the sphere as having no influence on the motions inside it, in accordance with Newtonain gravitational theory',[104] the Newtonian equations of motion will reduce to

$$(1/r)\,(Dv/Dt) = -\,4\pi\gamma\rho/3. \tag{40}$$

Since the right-hand side of this equation is a function of t only, so is the left, and substitution of the expression for v in the left-hand side soon shows that the function $G(t)$ must be identically zero. 'Hubble's law of nebular velocities' was thus derived in the form

$$v = r\,.\,F(t). \tag{41}$$

It was also easily shown, without further assumption, that the integral of this (following the motion) is

$$r = f\,.\,R(t), \tag{42}$$

where f is constant for a specific particle, and R is a universal function of t such that

$$(R'/R) = F(t). \tag{43}$$

The density is then B/R^3, where B is a constant of integration. The differential equation governing the variation of R is found to be

$$R^2 R'' = -\,4\pi\gamma B/3. \tag{44}$$

Integration of this leads to the complete expression for v, namely

$$v = (8\pi\gamma\rho/3 + A\rho^{2/3})^{1/2}r, \tag{45}$$

where A is a constant, the same for all particles. (We notice that a static universe is not possible except when $B = 0$, the case of vanishing

[104] Milne's paper, p. 67.

density.) The velocity of all particles is thus elliptic, parabolic, or hyperbolic, according as A is less than, equal to, or greater than zero.

Milne and McCrea next drew attention to the somewhat astonishing correspondence between the differential equation (44) and that of relativistic cosmology,

$$(2/Rc^2) R'' + (R'/cR)^2 + (k/R^2) = 0, \qquad (46)$$

the connexion requiring the identification of k with $-AB^{2/3}/c^2$.[105] This last equation is that which describes an expanding relativistic universe of 'radius' R and curvature k/R^2, with $\lambda = p = 0$. Spaces of positive, zero and negative curvature correspond, respectively, to the Newtonian models with elliptic, parabolic, and hyperbolic velocities. (Each particle's velocity was regarded as a 'velocity of escape from the matter "inside it" as judged by any arbitrary observer situated on any particle of the system'.)[106]

Milne and McCrea now modified their Newtonian principles in one respect, namely by adding a term proportional to r to the equation of motion, which becomes:[107]

$$(Dv/Dt) = (-4\pi\gamma\rho/3 + \lambda c^2/3)r. \qquad (47)$$

It is found that v is now given by the equation

$$v = (8\pi\gamma\rho/3 + A\rho^{2/3} + \lambda c^2/3)^{1/2}r \qquad (48)$$

and that the 'Newtonian' differential equations may be reduced to

$$(2/Rc^2) R'' + (R'/cR)^2 + (k/R^2) = \lambda \qquad (49)$$

and

$$(R'/cR)^2 + (k/R^2) = \kappa\rho/3 + \lambda/3 \qquad (50)$$

(ρ still assuming its previous form). These, of course, are no other than the Friedmann-Lemaître equations of relativistic cosmology, with non-zero cosmological term, but still with zero pressure. (The λ-term in the equation of motion is precisely that which Neumann and Seeliger had introduced, nearly forty years before, in the hope of explaining how an infinite and static universe was possible.) According to Milne, the way considered was the *only* permissible way of modifying the Newtonian inverse square law which is compatible with the Cosmological Principle —'any other modification would make it impossible for every observer

---

[105] See p. 75 of the joint paper. For equation (46) see note 22, p. 119 *supra*.
[106] Milne's paper, p. 65.   [107] Joint paper pp. 76–8. Cf. *W.S.*, p. 319.

to calculate accelerations &c., as though he were central'.[108] No proof was either given or alluded to.

Before considering some of the complications which arise in connexion with the reinstatement of Newtonian methods, some indication will be given of the ways in which the Milne-McCrea theory has been developed, notably by O. Heckmann and his colleagues at Hamburg. Under the name of *'dynamische Kosmologie'*, Heckmann gave a more or less unchanged account of the previous theory.[109] His presentation made use, however, of *linearen Strömungsfelder*: the velocity-distance relation holds in the sense that each component of a particle's velocity is proportional to the corresponding coordinate, but the three coefficients of proportionality are not necessarily the same (that is to say, $v_\mu = a_{\mu\nu}(t)x^\nu$, with an obvious notation). This allowed him to discuss problems more general than the isotropic problems previously considered.[110] Heckmann indicated that when the pressure of these dynamical models was non-zero, a smaller measure of agreement with the relativistic models was to be expected than was found in the case of vanishing pressure.

In another contribution made at this time, Heckmann investigated the conditions under which the equations of motion of Newtonian mechanics are formally identical with those of the General Theory of Relativity (namely the equations of the geodesics).[111] He concluded that the only four-dimensional metric for which the identity holds is one of constant curvature. Since as a matter of principle such a metric is locally unacceptable to an exponent of the General Theory of Relativity, it seems that there is a fundamental incompatibility between this theory and Newtonian mechanics.

9. Sources of Criticism

Perhaps the most surprising feature of these neo-Newtonian contributions was that although they faced difficulties essentially the same as those encountered by Seeliger and Neumann, they remained uncriticized for so long. The criticism finally came from D. Layzer in 1954.

---

[108] *W.S.*, p. 320.

[109] *Gött. Nachr.*, iii (1940) 169. Cf. Teil I of Heckmann's book *Theorien der Kosmologie* (Springer, Berlin, 1942).

[110] See, for example, Heckmann, op. cit., p. 20. Cf. note 45, p. 159 *supra*. Lemaître has expressed admiration for Heckmann and Schücking's anisotropy hypothesis, 'not that it destroys the initial singularity, but that it makes it less awful'! He added that it might be possible to introduce the idea of a 'spin of the primeval atom'. (I.A.U. discussion, 1961.)

[111] *Hamburg Abh. Math.*, xiv (1941) 192.

Layzer's arguments are of interest because they oblige us to consider what can be meant by the phrase 'subject-matter of the universe'. Layzer's criticism serves also to underline the conceptual gulf between 'Newtonian' and 'relativistic' theories.

Milne's first paper showed that, in the special case considered, the interpretation of Einstein's law of gravitation coincides with that of Newton's law for a spherically symmetrical distribution of mass. Layzer, working within the corpus of the General Theory of Relativity, proved that the dynamical properties of a uniform pressureless system of 'gravitating particles' are the same when the system forms part of an unbounded uniform distribution of matter as when the system is embedded in empty space. These dynamical properties he showed to be given, without approximation, by the Newtonian theory.[112]

At the same time Layzer criticized Milne and McCrea, maintaining that Newtonian mechanics cannot be applied in a self-consistent way to a uniform *unbounded* distribution of matter. In the light of this criticism McCrea modified his position.[113] Taking the case of a pressure-free fluid, moving under its own (Newtonian) gravitational forces alone, he repeated that the system could be characterized completely (with respect to a Newtonian frame of reference) by a function of time ($R(t)$) and two constants, one of which determines the *'total extent of the system'*. By considering any other reference frame, moving with a particle of the fluid and not rotating with respect to the first, McCrea showed (as he and Milne had shown before) that 'every observer moving with the material sees the same motion, which is purely radial and radially symmetric with regard to himself. If he supposes that the distribution is spherically symmetrical about himself, and that the classical laws of motion and of gravitation hold good in his own frame, then he obtains a classically correct description of the motion.'[114]

There are, so far, two features of McCrea's argument which may be thought unsatisfactory. The symmetry is a purely kinematic one. It is not dynamical, in the classical sense, for the different frames are relatively accelerated. The descriptions, moreover, of the fluid system in terms of these frames are not entirely alike, for the observed boundary is centred at the origin of the first frame only.

Layzer's objection, which was not on these grounds but on those of McCrea's extension of his results to an *unbounded* system, led McCrea to modify the last part of his argument. He now supposed the extent of the

[112] *A.J.*, lix (1954) 268.   [113] ibid., lx (1955) 271. Milne had died in 1950.
[114] op. cit., p. 272.

system (representing 'all the contents of the universe') to be 'arbitrarily large' by comparison with the 'finite range of observation' of any observer.[115] He added that 'the proportion of observers who can observe any part of the boundary can be taken to be arbitrarily small' and thus 'from the standpoint of observables . . . the difference between an arbitrarily large system and an unbounded system is scarcely significant'.

Avoiding in this way regarding an unbounded (Euclidean) system as the limit of a bounded one, McCrea had to ignore any broad heuristic principle which forbids privileged observers. It is hardly likely that anyone adhering to the principle would admit as exceptions to it even an arbitrarily small proportion of the cases conforming to it. However large the system there must be observers within sight of the boundary. In this respect the theory seems to have more loose ends than are justified, even in a theory whose principal function is merely one of suggestion.

The hierarchic type of theory, which we encountered in Chapter 2, achieved its purpose by making the infinite integrals, which represent the components of Newtonian attraction at any point, convergent. If one is prepared to dispense with certain requirements of the classical Newtonian argument, however, there arises the possibility of a further 'Newtonian' theory. An instance of this was put forward by Hermann Bondi.[116] He began with a kinematical treatment very much like Heckmann's (1940), taking the principle of conservation of mass, as expressed by the equation of continuity, and the principle of the conservation of momentum, as rendered by Euler's equations of motion. The latter reduce to

$$(D\mathbf{v}/Dt) + (1/\rho)\,\mathrm{grad}\,p - \mathbf{F} = 0. \tag{51}$$

When writing the first edition of his book *Cosmology*, Bondi was not aware of the snares to which Layzer eventually drew his attention. He acknowledged them later thus: 'The full evaluation of $\mathbf{F}$ in an infinite system is a somewhat ambiguous matter, but for our purposes it is sufficient to use Poisson's equation $\mathrm{div}\mathbf{F} = -4\pi\gamma\rho$.'[117] Whether

---

[115] loc. cit.   [116] *Cosmology*, 2nd. ed., (C.U.P., 1960) pp. 75 ff.

[117] op. cit., p. 79. Contrast this with Milne's first paper (*Q.J.M.*, v (1934) p. 64) where, in writing down the equation

$$\frac{1}{2}v^2 = \gamma M(r)/r$$

($M(r)$ being the mass contained in the sphere of radius r, centred on the observer) we are not, according to Milne, 'using the notion of a gravitational potential, here inapplicable, but are employing [the equation] simply as an integral of the equation of motion with a particular value of the constant of the integration'. (op. cit., p. 68.)

Bondi's theory merits the title 'Newtonian' is a moot point, for the boundary conditions which would normally be inserted to make a solution for F unique cannot, for an unbounded system, be specified. If one is prepared to discard the classical requirement that, for a finite system, the density distribution at any one time determines uniquely the gravitational field, then one likely objection to the resulting theory —which may certainly be consistent—will be on the grounds of its indeterminacy. To a theory in which an indefinite number of distributions of matter is compatible with a given gravitational field, and conversely, it might be reasonably objected that a theory which can explain a whole range of possibilities can have little predictive power in the relevant respects.

When McCrea's modified theory is compared with Bondi's it is found that though one is for an 'arbitrarily large' and the other for an unbounded material system, they give otherwise similar results. One such result concerns the status of coordinate systems. Only *one* of these is, on either theory, an inertial system in the old sense—but as far as observed motions are concerned it emerges that any observer at local rest (and therefore accelerated with respect to the first) would offer the same account as an observer at the origin of this 'classically inertial' system of coordinates.[118] There is a sense, therefore, in which both can regard their frames as inertial. That this is unsatisfactory on the classical account is not, of course, proof against such an argument; but once more the use of the title 'Newtonian' is seen to be misleading.

On the whole, the treatment of these quasi-Newtonian cosmologies has tended to be remote from the remainder of physical theory. When the question of light propagation arose, their authors were faced with a choice between the classical conception of a theory divorced from particle dynamics, and more recent views stemming primarily from the Special Theory of Relativity. Milne and McCrea,[119] Heckmann,[120] and Bondi,[121] all took the second alternative, freely admitting that they were guided by the formulae of the General Theory of Relativity. On what grounds is this appeal to another theory to be justified and how much importance should one attach to theories which result from this sort of argument?

Milne believed that 'imperfect as the Newtonian universe is as not being in accordance with Einstein's principle of relativity, its study is valuable as affording insight into the whole subject'. McCrea argues

---

[118] See McCrea, *Math. Gaz.*, xxxix (1955) 287. [119] *Q.J.M.*, v (1934) 78–9.
[120] op. cit., pp. 26–31.   [121] op. cit., pp. 86–9.

that the work 'seems to have proved useful in affording a physical understanding of the relativistic treatment', but that Milne and he did not suggest that it 'eliminated the need for a relativistic treatment of the large scale features of the universe'.[122] Bondi insists upon the same point, representing the theory as 'a very useful picture owing to the close analogy with relativistic cosmology', and as one which makes for clarity in so far as it avoids 'the cumbrous tensor calculus'.[123] Despite these remarks he adds, rather wistfully, that 'it is hardly worth while nowadays to compare Newtonian theory and observation, since the Newtonian concepts are known to be untenable'. This being so, from the point of view of scientific cosmology it is quite irrelevant to draw attention to the identical formulae of the two theories. Resemblances can hardly be of physical significance when they hold between formulae which are differently interpreted. To see that they do indeed differ one need go no further than the term k in equations (49) and (50). In the corresponding relativistic equations k specifies the constant of curvature. In Newtonian theory it relates to the total energy of the particles. Coming down to more basic terms (still two or three stages removed from anything resembling a 'statement of observation', but the example is sufficient to make our point) the relations between the 'Doppler-displacement velocity' and 'luminosity-distance' of an object in relativistic cosmology closely resembles equation (41) of this chapter. Does this mean that the interpretations may, after all, be made ultimately the same? In fact equation (41) is an exact formula, whereas the relativistic one is only an approximation, and is valid only locally. This does not, of course, mean that a closer correspondence than this cannot be found—a more promising start might have been made by linking the 'r' of (41) with the relativist's 'proper distance'—but it does seem to suggest that any assistance offered to our physical insight is illusory.

Bondi had more to say on the importance of the quasi-Newtonian theories, for in closing his chapter he added that 'in any problem of intermediate size, such as the structure of the nebulae, general relativity cannot be expected to explain any major features in any different or better way than Newtonian theory'.[124] Now the use of the word 'explain' is certainly very indefinite and it might reasonably be said that an hypothesis, whatever its order of generality, 'explains its instances' in some way. It seems reasonable, nevertheless, to say that a *better* explanation of something is given by one set of hypotheses than by a

[122] *A.J.*, lx (1955) 271. [123] op. cit., p. 89. [124] loc. cit.

second set, if the hypotheses most nearly resembling those of highest order in the latter are subsumed under hypotheses of a still higher order in the first scheme. In this sense, most features of Newton's theory of gravitation can be said to have been explained by Einstein. Quite apart from this, bearing in mind that the quasi-Newtonian argument rests heavily on an analogy with the General Theory of Relativity, it seems very strange to claim that neither theory offers an explanation better than that provided by the other. For to say that either theory's explanation of gravitation is as good as the other's, on the grounds that the predictions of two theories are to all intents and purposes the same, is tantamount to denying the possibility of improvement in the structure of a scientific deductive system.

CHAPTER 9

THE THEORIES OF GRAVITATION OF BIRKHOFF AND WHITEHEAD

In the last chapter two groups of theories were examined, the first originating with Milne's kinematic arguments, and using mainly a hydrodynamic model, the second following upon a renewal of broadly Newtonian methods—and once again using a hydrodynamic model. In both cases a Euclidean metric was accepted, and whatever cogent objections have been raised against them, none concerns the legitimacy of this initial conceptual choice. In the present chapter we consider two further theories for which a similar choice was made. At the present time neither of these (with the possible exception of Whitehead's theory) would be judged important to cosmology, but it is of some historical interest to see how their methods compare with Einstein's.

1. Birkhoff's Theory of Gravitation

In 1927 G. D. Birkhoff, the American mathematician, offered an unusual interpretation of the new Schrödinger wave equation, suggesting that matter be regarded as a 'perfect fluid' supporting a curved Einsteinian (static) space.[1] By 'perfect fluid' Birkhoff meant a fluid which was homogeneous and adiabatic, the components of its energy tensor being given as

$$T_{ij} = \rho u_i u_j - p\delta_{ij}. \qquad (1)^{[2]}$$

In addition, the 'perfect fluid' was to have a disturbance velocity equal to the velocity of light, at all densities.[3] In 1942 Birkhoff presented to an astrophysics congress in Mexico a new gravitational theory which incorporated the same 'perfect fluid' as before, but which made use only of the flat Minkowski space-time of the Special Theory of Relativity.

---

[1] *P.N.A.S.*, xiii (1927) 160, particularly section 1 and p. 165. Most of the references to Birkhoff's work which follow are to be found in *G. D. Birkhoff. The Collected Mathematical Papers* (Amer. Math. Soc., 1950) vol. i.

[2] u_i are the components of the velocity vector dx_i/ds. The x_i are 'normal coordinates'. Birkhoff writes $x_1 = t$, $x_2 = ix$, &c.

[3] The velocity of a small disturbance from the equilibrium state of constant pressure and density was found to be $(d\rho/dp - 1)^{-1/2}$. Equating this to the velocity of light (unity, in Birkhoff's units) we find $2p = \rho$. If the velocity of disturbance were less than that of light, difficulties would arise in describing the collision of two portions of the fluid at velocities of this order.

The resulting theory of gravitation now closely resembled that which Gunar Nordström had proposed over thirty years before.[4]

The body-force vector (f_i) per unit volume is written into the equation of fluid motion in the usual way. It is assumed to be rational and integral in the velocity components u_i, of not higher than second degree (this was true of all previous theories which had met with any success), the coefficients being homogeneous and linear in the first partial derivatives of

(i) the 'atomic potential' (ψ),
(ii) 'the usual vector electromagnetic potential' (ϕ_i), and
(iii) 'the symmetric gravitational tensor potential' (h_{ij}).[5,6]

There were also to be no degenerate quadratic terms in the velocities. The full expression for the (covariant) force components is

$$f_i = \rho(\partial\psi/\partial x_i) + \sigma\{(\partial\phi_i/\partial x_\alpha) - (\partial\phi_\alpha/\partial x_i)\}u_\alpha \\ + \rho\{(\partial h_{i\alpha}/\partial x_\beta) - (\partial h_{\alpha\beta}/\partial x_i)\}u_\alpha u_\beta. \quad (2)$$

In the years that followed, Birkhoff's theory was worked out in some detail by its author and the Mexicans A. Barajas, C. Graef, M. S. Vallarta, and others, but only as concerns *gravitational* forces. Here the theory had a measure of success, suggesting as it did formulae for the advance of planetary perihelia and the deviation of light rays by large masses, formulae which agree substantially with Einstein's.[7] The expressions for the 'gravitational red-shift' differed from Einstein's,[8] although the predicted values were much the same. It will be of some interest to see how such agreement came about.

Birkhoff, like Einstein, proceeded by analogy with Poisson's equation of the classical theory of gravitation. He assumed therefore that a similar equation holds between each component of the (symmetrical) energy tensor (T_{ij}) and the corresponding (symmetrical) gravitational potential (h_{ij}), namely,

$$\sum_\alpha (\partial^2 h_{ij}/\partial x_\alpha^2) = 8\pi T_{ij}. \quad (3)$$

He deduced (in conformity with the assumptions already stated concerning the form of f_i) that the gravitational force vector must be of the form

$$f_{Gi} = \rho\{(\partial h_{i\alpha}/\partial x_\beta) - (\partial h_{\alpha\beta}/\partial x_i)\}u_\alpha u_\beta. \quad (4)$$

---

[4] *Phys. ZS.*, xiii (1912). On the comparison of Nordström's theory with Birkhoff's and Newton's theories, see Birkhoff, *Sci. Mon.*, lviii (1944) 135.
[5] *P.N.A.S.*, xxix (1943) 231.   [6] *P.N.A.S.*, xxx (1944) 324, section 11.
[7] *P.N.A.S.*, xxix (1943) 231.   [8] *Bol. Soc. Math. Mex.*, i (1944) 1.

(It was tacitly assumed that the gravitational force is proportional to the density ρ.) Not surprisingly the theory, in approximation, agrees with the Newtonian theory. The exact equations for the motion of a particle attracted by a sphere of the 'perfect fluid' are found to be

$$\ddot{x} = -(mx/r^3) - (2mx/r^3)(\dot{x}^2 + \dot{y}^2 + \dot{z}^2) + (m\dot{x}\dot{r}/r^2), \quad (5)$$

and similarly for $\ddot{y}$ and $\ddot{z}$.[9] The first term on the right is the usual Newtonian expression for a particle moving in the z-plane. The differential equation of the path is readily integrated[10] and the perihelion advance,[11] the curvature of light[12] and the gravitational red-shift[13] found. The two-body problem was relatively easily solved,[14] whilst the problem of determining the motion of several bodies was considerably more straightforward than the corresponding problem in Einstein's theory.

Despite its more tractable form, interest in Birkhoff's theory of gravitation has been confined to a fairly small group of followers. Mathematical physicists have tended either to ignore it or to demur from acknowledging it as a theory in its own right. H. E. Ives, for example, sees its 'physical significance' as nothing more than the possibility of interpreting it into Newtonian terms: he expressed Birkhoff's equations in terms of the variable t rather than s (local time), and hence showed that the mass term was in Lorentzian form, and that to the Newtonian force of attraction there was added a force perpendicular to the direction of motion.[15] Again, although Weyl contended that Birkhoff's theory was much the same as Einstein's for the case of weak gravitational fields, he was looking only at the end products of

---

[9] The dot signifies differentiation with respect to s (local time). The exact solution for the gravitational potentials is $h_{ij} = \delta_{ij}(m/r)$. See *P.N.A.S.*, xxix (1943) p. 238.

[10] *Bol. Soc. Math. Mex.*, i (1944) 1, section 4.

[11] $\varepsilon \simeq (6\pi m/a)(1-e^2)^{-1}$—precisely Einstein's formula (ibid., section 5).

[12] $\Delta\theta \simeq 4m/P$, once again Einstein's formula (ibid., section 6).

[13] $$\frac{ds_s}{ds_T} = \frac{\text{interval local time}}{\text{interval at observer}} = \left\{\frac{\phi(M)-1}{\phi(m)-1}\right\}^{1/2} \simeq 1 - \{(m/r)-(M/R)\}$$

where $\phi(M) = \varepsilon^{2(M/R+C)}$, C is a constant of integration, R and M are the radius and mass of the Earth, and r and m corresponding values for the Sun. (ibid., section 7). Cf. Einstein's formula

$$\frac{ds_s}{ds_T} = \{(1-2m/r)(1-2M/R)\}^{1/2} \simeq 1 - \{(m/r)-(M/R)\}.$$

[14] See C. W. Berenda, *Phys. Rev.*, lxvii (1945) 56. The rate of advance of periastron now becomes $\bar{\varepsilon} \simeq \{2\pi/a(1-e^2)\}\{(3m_1^2 + 7m_1m_2 + 3m_2^2)/(m_1+m_2)\}$. As $m_1 \to 0$ this reduces to the formula previously given.

[15] V^2/c^2 is the component of gravitational force at right angle to the motion. See *Phys. Rev.*, lxxii (1947) 229.

the two theories.[16] Admittedly there is a sense in which Einstein's theory is more general than Birkhoff's, and for this amongst other reasons it will usually be preferred. Similarities are naturally to be expected for the simple reason that both Einstein and Birkhoff took the Newtonian theory as their prototype; and yet, if only from the expression for the gravitational red-shift, it should be clear that the resemblance is superficial. The trajectories of test particles in Birkhoff's theory are not geodesics in any four-dimensional Riemannian space-time,[17] and indeed Birkhoff's whole method differs from Einstein's. As Weyl said, 'the connexion between metric and gravitation is dissolved'. The dissolution was, of course, deliberate. The usual arguments against it take the form, broadly speaking, of objections to an 'absolute reference system'. In their defence, Barajas, Birkhoff, Graef and Vallarta pleaded that there was a distinct advantage in retaining a 'fundamental reference system'. In Einstein's rejection of this 'advantage' they found the reason for 'the early exhaustion of all observation tests of the General Theory of Relativity' and for the fact that 'even in the simple case of Schwarzschild's solution of the one-body problem, no clear-cut physical interpretation of the Schwarzschild coordinates seems to be available'.[18] We shall later suggest that the 'absolute background' difficulty is not necessarily serious; but that the argument which concerns the interpretation of coordinates reveals a common and important conceptual error.

Whatever its theoretical basis, it seemed almost inevitable that Birkhoff's gravitational theory should be applied to cosmological problems.[19] In the paper by Barajas, Birkhoff, Graef, and Vallarta, it was hinted that 'to account for the phenomena of nebular recession on this basis, it is necessary to suppose that at some time in the past there was a nuclear distribution of matter, with a wide range of velocities'.[20] Without any supporting argument they added that 'from the physical point of view it appears to be more natural to suppose an initial nuclear distribution at low relative velocities'. Nothing further was said, apart from a suggestion that a 'cosmological constant' (K) could be readily

---

[16] *P.N.A.S.*, xxx (1944) 205 and *Am. J. Math.*, lxvi (1944) 591. Weyl asserted that the theory differs trivially from a limiting form of Einstein's theory for weak fields, obtained by writing the components of the metric tensor as ($\delta_{ik} + h_{ik}$) and neglecting higher powers of h_{ik} than the first.

[17] See A. Barajas, *P.N.A.S.*, xxx (1944) 54.   [18] *Phys. Rev.*, lxvi (1944) 138.

[19] Barajas and Fernández gave an axiomatic system for the theory and P. Kustaanheimo has investigated the sorts of theory which conform to it. (*Op. Phys-Math. Fenn.*, xvii (1955), Paper 11, 15 pp.).

[20] op. cit., p. 142.

incorporated in the theory. Again, no argument was given over and above the remark that 'from the formal point of view' a more general form of Poisson's equation

$$\sum_\alpha (\partial^2 h_{ij}/\partial x_\alpha^2) = 8\pi T_{ij} + K g_{ij} \qquad (6)$$

'needs specially to be examined'.[21] According to Birkhoff, the 'most natural possibility from the electromagnetic point of view is to take the cosmological term in the gravitational potentials to be

$$h_{ij}^* = \frac{K}{8}(t^2 - x^2 - y^2 - z^2) g_{ij} \qquad (7)$$

where x, y, z, t are any Lorentz coordinates and $x = y = z = t = 0$ is the origin in space-time. Then 'h_{ij} are regularly infinite to at most the second order, with a boundary distribution at infinity which is spherically symmetrical in a spatial sense'.

No further justification was provided, although a detailed study was promised. It is quite possible that Birkhoff would have fallen foul of the sort of objection Layzer raised against Newtonian cosmology, had this line of thought been developed. Birkhoff, however, died in 1944. Not until 1950 did anything of a cosmological nature appear in print,[22] and thus far nothing of much promise has come from Birkhoff's school. The position is somewhat like that in which Whitehead's theory of gravitation found itself for almost thirty years after its first formulation. Whitehead's theory, like Birkhoff's, uses a flat space-time and the concept of force. Even so, Whitehead's theory, in at least one respect, resembles Newton's theory more closely than does Birkhoff's: it does not involve the dynamics of a continuum. In the rest of this chapter we shall indicate the way in which Whitehead's theory of gravitation has been applied to the cosmological problem.

2. WHITEHEAD'S THEORY OF GRAVITATION AND ITS EXTENSION BY J. L. SYNGE

Not long after Einstein's earliest formulation of the General Theory of Relativity, A. N. Whitehead, in part II of *The Principle of Relativity*,[23] presented an alternative theory of gravitation which lies somewhere between the Einsteinian and Newtonian versions.[24] Whitehead's

---

[21] $g_{ij} = 1, -1, -1, -1$, for $i = 1, 2, 3, 4$ respectively; $g_{ij} = 0$ if $i \neq j$.
[22] *Phys. Rev.*, lxxx (1950) 127. (Abstract of lecture only.) See Appendix, Note X.
[23] (Cambridge University Press, 1922.)
[24] A similar point of view, but one which was never followed up, was that of C. K. V. Row. See *Nat.*, cxv (1925) 261.

theory has much to recommend it: it admits Lorentz invariance, unlike its Newtonian predecessor (but not invariance with respect to general coordinate transformations), and yet, unlike Einstein's General Theory, it does not impose the Herculean task of solving sets of non-linear partial differential equations whenever a gravitational field is sought. Whitehead himself was mainly concerned with providing a rationale of certain very general 'philosophical' ideas, and cannot be said to have developed them very far from a cosmological point of view. We shall shortly indicate the extent to which others have performed the task for him.

At the end of Chapter 3 it was seen that attempts to formulate a law of gravitation which was Lorentz-covariant all failed. Birkhoff and Whitehead were more successful. Whitehead derived the Lorentz formulae afresh, not being prepared to accept Einstein's 'principle of relativity'. For the same reason he would not generalize to a principle of general covariance. Like Birkhoff (and before him), Whitehead had shown that values agreeing with those obtained by Einstein for the advance of planetary perihelia and the deflection of light by the Sun were to be expected. Eddington derived this result from Whitehead's principles with greater accuracy.[25] As for the gravitational red-shift, Whitehead deduced a result which was (13/6) times as great as Einstein's. (His book contains the figure 7/6, but Whitehead later amended this.) In this calculation he used an over-simplified picture of the molecule. J. L. Synge, using Whitehead's theory but Einstein's method, has since obtained Einstein's figure precisely. George Temple, shortly after Whitehead's theory was first published, extended it to space-times of constant curvature.[26] Whitehead, as will be seen elsewhere, had given reasons for denying that space can be heterogeneous, but gave none which indicated that space-time must necessarily be considered flat.[27] Temple, whose exposition was not only more general than Whitehead's, but also more rigorous, showed that a lower limit to a uniform and isotropic curvature is implied by each value of

---

[25] See *Nat.*, cxiii (1924) 192. Eddington provided a transformation which carried Whitehead's line-element into Schwarzschild's.

[26] *Proc. Camb. Phil. Soc.*, xxxvi (1924) 176. (A small error is indicated by Rayner, *Proc. Phys. Soc.* (8), lxviii (1955) 944.) Cf. also *Phil Mag.* (6), xlviii (1924) 277. In this the Hamilton-Jacobi method of classical dynamics is extended to relativistic dynamics, both Einstein's and Whitehead's. Temple further studies one of the two laws of gravitation which Whitehead had suggested as further alternatives to his main law, and to Einstein's.

[27] Arguments similar to Whitehead's are to be found in C. D. Broad, *Mind*, xxxii (1923) 211, and Samuel Alexander, *Space, Time and Deity* (London, 1920).

the advance of the perihelion of Mercury. He showed, furthermore, that the effect of the curvature on the deviation of light rays grazing the Sun's disk is at the most of the order of 10^{-8} times the total deviation. It is therefore never likely to be detected by observation of this deviation.

Despite this initial activity, Whitehead's methods were soon to fall into neglect, a neglect which persisted for nearly thirty years. We have no intention of following the early history of Whitehead's relativity in more detail. Not only has it been dealt with very fully elsewhere, but only recently has the theory been adapted to the needs of cosmology. We begin, therefore, in 1952, when Synge extended Whitehead's formula for the gravitational field (hitherto only applied to the field of a single massive particle) to include the case of a static and continuous distribution of matter. The result calculated by Synge for the deflection of a light ray by a spherically symmetrical mass agrees exactly with that formerly obtained by Einstein. The values obtained for the advance of the perihelion and the periodic time of a planetary orbit differ, however, from those previously obtained, being now dependent not simply upon the Schwarzschild parameter (viz. the 'mass of the central sphere') but upon the density distribution within the central mass. The difference is slight and is of doubtful numerical importance. It stands as evidence, if any be needed, that the Synge–Whitehead theory is not simply the Schwarzschild–Einstein theory in disguise: that the former conforms to nothing more than Lorentz invariance distinguishes the two in a more general way. It is this aspect of Lorentz invariance which, in the words of C. B. Rayner, makes it 'particularly appropriate as the basis of a cosmological theory which treats the recessional motion of the extragalactic nebulae as *uniform*'.[28] (Perhaps most cosmologists would regard this as an intolerable restriction.) Rayner was, in fact, the first to apply Whitehead's ideas to cosmological problems.

Before considering Rayner's treatment of non-static systems, one or two details of the Synge-Whitehead theory will be given. Whitehead contemplated 'action-at-a-distance', but in this he is often misunderstood. His particles interacted, not after the manner of Newtonian gravitating particles, but through the retarded gravitational tensor potentials g_{mn}, given by

$$g_{mn} = \delta_{mn} - (2\gamma/c^2) \sum_i \{m_i y_m y_n (y_n \cdot dx'_n/ds')^{-3}\}. \tag{8}$$

[28] *P.R.S.* (A), ccxxii (1954) 509.

The summation contains a term for each particle giving rise to the field at an event $P(x_p)$, whilst $y_p = x_p - x'_p$. Here x'_p is the event at which the world line of the typical particle (proper mass m) cuts the null cone

$$(x_n - x'_n)(x_n - x'_n) = 0 \tag{9}$$

drawn into the past from $P(x_p)$.

(A. Schild has drawn attention to the analogy between Whitehead's theory and the Liénard–Wiechert potential theory of classical electrodynamics.[29] Schild's paper describes an infinity of new gravitational theories which are generalizations of Whitehead's, and none of which can be obtained from a single action principle. Like the simple Whitehead theory, they were found to 'violate the law of action and reaction and to predict a secular acceleration of the centre of mass in the two-body problem'.[30])

A particle moves in the field g_{mn} in accordance with the variational principle

$$\delta \int (-g_{mn} dx_m dx_n)^{1/2} = 0, \tag{10}$$

the particle whose motion is in question making no contribution to the g_{mn}. Light paths are null-geodesics for the form $-g_{mn} dx_m dx_n$. It must be emphasized that Whitehead's space-time remains that given by the Minkowskian form

$$ds^2 = -dx_n dx_n \quad (x_4 = ict), \tag{11}$$

and that for him the concept of a continuous Riemannian manifold with an associated tensor calculus is not a part of what is normally understood by 'geometry'. It is, as it happens, possible to evaluate world lines and light paths more conveniently by using Euler–Lagrange equations with a Lagrangian involving g_{mn}.[31] Synge considered the limiting case of (8) when the number of particles tends to infinity whilst their masses tend to zero in such a way as to ensure a finite density.[32] He considered only a static distribution, with results for light-ray deflection and perihelion advance which have already been indicated.

---

[29] ibid., ccxxxv (1956) 202.

[30] See G. L. Clark, *P.R.S. Edin.*, (A), lxiv (1954) 49. The possibility of detecting such an acceleration (in the direction of the major axis of the orbit, towards the periastron of the larger mass) in a binary star had been discussed by Levi-Civita in 1937. Clark gave instances where it might be detected in less than a century. Cf. Robertson, *Ann. Math.* Princeton, xxxix (1938) 65.

[31] See Synge, *P.R.S.* (A), ccxi (1952) p. 310.   [32] op. cit., p. 311.

3. The Rayner–Whitehead Cosmology

It was now left to Rayner to extend the argument to cover continuous but non-static distributions[33] and thus to open the way to cosmological applications. The equation (8) was now replaced by

$$g_{mn} = \delta_{mn} + (2\gamma/c^2) \int\int\int \frac{y_m y_n \rho_0 dx'\,dy'\,dz'}{c(t - t')\{y_n(dx'_n/ds)\}^2}, \qquad (12)$$

where $\rho_0(x'_p)$ represents the proper density of the distribution at the event $P(x'_p)$. Under conditions of spherical symmetry it was found possible to simplify this considerably, and thus to determine the gravitational field in a uniformly expanding model. (Rayner's method of reaching this result was, on the other hand, fairly involved. Some months afterwards Synge obtained an equivalent expression for the field in a much simpler way.)[34] The fundamental form of the Whitehead–Rayner universe can be written

$$g_{mn}\,dx_m\,dx_n = (1 + A/R)dx_n\,dx_n + (4A/R)\,dR^2, \qquad (13)$$

where

$$A = (8\pi\rho\gamma R^3/9c^2), \qquad (14)$$

R being the Minkowskian separation of an event from the centre of symmetry ($R^2 = -x_n x_n$).

Rayner used his equivalent expression of this form to investigate the motion of test particles (using equation (10)) and to calculate the red-shifts in the spectra of fundamental particle-sources.[35] He found, for example, that light paths are limiting cases of the orbits of particles, and that they correspond to a projection velocity which cannot be exceeded. He found that any freely moving particle comes to rest with respect to the matter in its locality, whatever its velocity of projection. The system is thus stable in the sense that slight disturbances will die away. The velocity of light waves, however, increases with time, tending to the limit c as the density of the distribution tends to zero. The model possesses what Rayner called a 'visual barrier', through which neither particle nor photon can pass; but the barrier recedes and no part of the system can remain beyond it for more than a finite period. There is also a 'critical epoch' ($3A/c$, reckoned in the proper time of a fundamental observer) at which all particles and photons are at rest with respect to local fundamental particles.

[33] See Rayner, op. cit., pp. 511–12. [34] *P.R.S.* (A), ccxxvi (1954) 336.
[35] op. cit., pp. 519–24.

The precise form of the expression obtained for this spectral shift was

$$1 + \delta = \sqrt{\{ct_0/(c^2t^2 - r^2)\}}, \tag{15}$$

where t_0 is the observer's time of reception of the light and t is the time of emission (in the observer's reckoning). The quantities r, t_0, and t are connected by a further integral relation, and when the quantity A/c is negligible by comparison with t_0 and t it appears that, to the second order,

$$\delta = (r/ct_0) + 3(r/ct_0)^2/2 \tag{16}$$

(with the accepted approximation, c being the velocity of light at epoch t_0). To the first order, the equation for δ coincides with the simple Hubble's Law, granted the classical interpretation of the Doppler effect.

This in itself is only a first step in the genesis of a satisfactory cosmological theory, yet it suggests that the Rayner–Whitehead theory is capable of relevant elaboration. The theory lacks, at the present time, those formulae which make possible a direct comparison of theory and observation on any scale much larger than that of the planetary system, and also those formulae which make possible extrapolation to epochs far removed from the present. Indeed, the approximations which were made in deriving the last equation rested upon neglect of the term A/ct_0, a neglect which, it is implied, amounts to supposing A/c to be 'small compared with the present age of the universe'.[36] The formula (16) is said to be 'the same to the first order as the ordinary Doppler formula, and so, when we apply it to the universe, we obtain the same value for the age of the universe as is given by any other recessional theory which has the Doppler formula as its first approximation'.[37] This, however, is to invoke a host of hidden assumptions concerning the concept of an age of the universe. From the approximations made in deriving the relation (16) between δ, r, and t_0, it is obvious that the nature of the implied physical discontinuity is illusory. It would indeed be interesting to hear of the behaviour of the model for values of t_0 comparable with (A/c).

One further point which may be noticed is that the time of observation enters into the approximate law (16), and that if we are to speak of a 'Hubble parameter' it will not be constant.[38]

[36] Rayner, op. cit., p. 524.   [37] loc. cit.

[38] In a yet more recent contribution to that group of theories which sprang from Whitehead's, Rayner considers the extension of the Synge–Whitehead law of gravitation to a space-time of constant curvature. (*Proc. Phys. Soc.* (B), lxviii (1955) 944.) Beginning with a discrete set of particles he extends the analysis, as Synge and he had done previously, to systems of finite and continuous proper density.

We have now considered four theories capable of cosmological extension, the authors of which showed a desire to resist the 'natural geometry' of the General Theory of Relativity. Milne's kinematic method, as its name implies, differs from the other three in that it may in principle avoid the concept of force, just as did Einstein's General Theory. All four are usually regarded as alike in being on the side of reaction, this being most apparent in respect of their retention of flat space-time. Now reaction, in a scientific context, is not necessarily a fault. When, as here, the older methods are still very much better known than Einstein's, one might have expected to find their reactionary element regarded with favour. It is significant that this is not so, and that at the present time relativistic cosmology is still the yardstick for all others. The theories originating with Whitehead and Birkhoff are still in their infancy, whilst conceptual objections to the new 'Newtonian' theories cannot be overlooked. Against Milne it must be said that his simple models require the equivalences to be satisfied precisely, and unless a statistical treatment is adopted his methods are unlikely to be of much astronomical interest in the long run. Such impediments as these should not, however, blind us to the fact that the decision to work within the context of flat space-time does not by itself carry the blame. As we shall argue in Chapter 15, contrary to the claims of at least one recent author, it is highly improbable that the outcome of any series of observations could oblige a resolute cosmologist to abandon Euclidean geometry or, indeed, any other of comparable complexity.

Before going further we should point out, however, that such a conventionalist attitude was not uppermost in the mind of either Whitehead or Birkhoff. Furthermore, neither of them was an inflexible Euclidean. Whitehead, believing that there were some requirements of a general sort which could be placed on spatio-temporal relations as a result of *a priori* considerations, argued simply that we are required to adopt a space of *uniform* curvature: 'I should be very willing to believe that each permanent space is either uniformly elliptic or uniformly hyperbolic, if any observations are more simply explained by such an hypothesis.'[39]

In conclusion, Whitehead, Birkhoff, Milne, and the rest were concerned with something more than proving the feasibility of a conventionalist view of the geometrical structure of the world: they shared a belief that the Einsteinian theory of gravitation was unnecessarily

[39] Preface to *The Principle of Relativity*.

complex. In some cases they believed that Einstein's entire order of approach was at fault. As to the first point, whilst many may agree, few will concede any of the alternatives so far given to be entirely satisfactory. As to the supposed contrast of methods, the matter will be taken up again in Chapter 16.

CHAPTER 10

CONTINUAL CREATION AND THE STEADY-STATE THEORIES OF BONDI, GOLD, AND HOYLE

WHETHER or not the steady-state theories of Bondi, Gold, and Hoyle constitute relativistic cosmology's most important rival, a considerable number of people believe this to be so. In the first place these theories appeared to be well supported by observation, and they overcame the 'time-scale difficulty', to which we shall refer below, in a particularly simple way. When the analysis of existing astronomical data was revised shortly after the steady-state theories were first elaborated in 1948, however, this apparent advantage was lost. As time has passed, it has become increasingly difficult to adhere to the theories in their original form, and yet later developments are for the most part outside the range of this book. What must surely be of permanent interest are the conceptual problems which arise in connexion with these theories—in particular those associated with the modification of the law of conservation of mass—and the formal considerations which led to them. Questions of Creation and of the Cosmological Principles will be dealt with in a more general way in Part II. In this chapter we are chiefly concerned with the ways in which Bondi, Gold, and Hoyle added to cosmology. We begin, however, by indicating some of the ways in which the principle of mass conservation was questioned by earlier writers.

1. CONTINUAL CREATION: SOME EARLY ARGUMENTS

In 1918 and 1925, in order to remove so-called 'paradox' of de Cheseaux and Olbers, W. D. MacMillan proposed a form of continual material creation.[1] MacMillan was primarily concerned with problems dealing with the formation of planets and stars, and amongst certain 'postulates' he included two (nos. 6 and 13) which suggest that the universe maintains a steady state, and a third (no. 12) according to which the energy of a large region of the (unbounded) universe is conserved. Holding to the conversion of matter to energy in stellar interiors, he explained away the de Cheseaux-Olbers paradox in conformity with these principles by claiming that, in travelling through

[1] *Ap. J.*, xlviii (1918) 35 and *Science*, lxii (1925) 63–72, 96–9, 121–7.

empty space, radiation disappears or is dissipated—but that it reappears in the form of hydrogen atoms which constitute 'the nebulosity with which cosmogonists have always started'. (Atomic physics was not yet fitted to explain this transition.) MacMillan's theory was not so implausible as to be without supporters. R. A. Millikan, one of MacMillan's colleagues at Chicago, used it to account for the origin of cosmic rays; but by 1935, in the face of evidence provided by A. H. Compton, it was realized that processes of transmuting radiation into matter could not account for the high energies which cosmic rays were known, on occasion, to possess. The hypothesis was discarded by all concerned. It should be noted, before leaving MacMillan's theory, that it differs from the first Bondi–Gold–Hoyle steady-state theories in identifying a source from which the energy of the 'created' particle is drawn. It does not involve creation *ex nihilo*.

Tolman also contemplated a process of conversion of radiation into matter—this time as a means of explaining a hypothesis of continual creation within a de Sitter world.[2] We recall that Silberstein, Tolman, Whittaker, and others discussed the orbits of free particles within such a model, and applied the results in deriving a general formula for the de Sitter effect. Each particle, it will be remembered, having passed perihelion will move away from an observer never to return. We can, on the other hand, still see a great many nebulae and thus, according to Tolman, either these nebulae must have entered the range of observation so recently as not to have had time to disperse, or they must have been recently formed or created within that range. It was held to be possible that either process would be a continuing one, a process 'which is still acting to maintain an approximately uniform concentration of nebulae in the neighbourhood of the origin'.[3] (A clear anticipation of the 'Perfect Cosmological Principle'?) Tolman did, however, consider the alternative to a continual change, with his 'hypothesis of recent formation'. There were thus four possibilities from which to choose, and of these Tolman preferred continuous entry. He found it virtually impossible, even so, to explain away the known preponderance of red-shifts in nebular spectra without introducing too many scarcely plausible hypotheses. As to the 'hypothesis of continuous formation' he had this to say:

> This idea appears to have little inherent probability. Science has not yet provided us with any clearly worked-out mechanism by which such a formation of nebulae could take place, and an appeal to special acts of creation is not

[2] *Ap. J.*, lxix (1929) 266 ff.   [3] op. cit., p. 267.

within the province of our discussion. Nevertheless, we should not completely disregard the possibility that some such process—perhaps associated with a condensation of radiation into matter—might be taking place.'[4]

As to the known preponderance of red-shifts, Tolman thought it might be that nebulae 'tend to form at or near perihelion and then escape, or that they condense around particles which have been dislodged from our own system and hence are already escaping'. Within two years the de Sitter effect was outmoded as an explanation of the red-shifts, and such special pleadings as this became unnecessary. In any case, the idea of the creation of matter within a world ostensibly devoid of matter was clearly unsatisfactory.

We have already touched upon the non-conservation of proper mass within some of the early non-static cosmological models (cf. Chapter 6). Granted the need to allow for a one-way conversion of proper mass to radiation, it is not surprising that this non-conservation was regarded as in itself no drawback. The non-conservation of total energy is another question. Almost invariably total energy was held to be conserved (although from a given coordinate-region in most expanding models energy is lost whilst no region gains).[5] We recall Lemaître's conservation equation (Chapter 6) and the contention that 'the pressures do work in the expansion'.

Before the steady-state theories an increase in the number of particles in any region of the universe without the sacrifice of radiant energy was rarely countenanced, but one or two isolated references may be mentioned. In 1928 a conjecture (unsupported, however, by subsequent astrophysics) was presented by James Jeans. He felt that no satisfactory account of the spiral character of the arms of the nebulae had been given, and surmised that 'the centres of the nebulae are of the nature of "singular points" at which matter is poured into our universe from some other, and entirely extraneous, spatial dimension, so that, to a denizen of our universe, they appear as points at which matter is being continually created'.[6] Milne acknowledged that Kinematic Relativity 'is in general conformity with these speculations of Jeans'.[7] He also followed Jeans in appealing to the secular dependence of the gravitational

[4] op. cit., p. 271.
[5] R. Zaycoff (*ZS. Ap.*, vi (1933) 193) was an exception. He allowed for the annihilation of matter and other forms of energy. Zaycoff considers a possible basis for a covariant form of wave mechanics, and his non-static model universe is homogeneous in the distribution of atoms and radiation quanta. (The first part of this paper is to be found on p. 128.)
[6] *Astronomy and Cosmogony* (Cambridge University Press, 1928), p. 352.
[7] *K.R.*, p. 167. Cf. *M.N.*, cvi (1946) 180.

'constant' on the epoch, but on the whole he was unhappy with the idea of continual creation. By this time there were other instances of writers with similar ideas in mind. In Japan, mathematicians of the Hiroshima school[8] were working on the reduction of cosmology to what they called 'wave geometry'. They found the total mass of a de Sitter-type model to increase with time, which meant that the number of particles in the model increases with time (for the individual particles were supposed to be of constant mass).[9]

There are many reasons for the neglect, outside Japan, of the arguments of the Hiroshima group, above all their complexity, their unconcern with the problems of the observatory and the restriction of the type of model to one of two cases, both of which had for long been superseded. It is also unlikely that there was much sympathy for a movement away from the idea of a world continuum, which had become an accepted part of most cosmological thought by the end of the 1930s. The Hiroshima school was, nevertheless, led to the idea of a continual creation of particles in a natural and consequential way which contrasts with the work of one or two more recent authors who have arbitrarily superimposed upon the Friedmann model a time-variation of the total mass of the model.[10]

Another curious way in which an increasing number of fundamental particles is suggested was indicated by P. Dirac in 1937. Dirac had already, in expressing the wave equation in a form invariant under the Lorentz transformations, removed an incompatibility between the Special Theory of Relativity and the quantum theory of matter. Eddington tried to do as much for the General Theory, and in his highly involved attempts there appeared in the wave-tensor calculus a number of the order of 10^{79}. (The number was $3/2 . 136 . 2^{256} \simeq 2 \cdot 4 . 10^{79}$.) Around this number Eddington erected some very strange arguments purporting to show that it, or certain powers of it, represented the 'only

---

[8] Y. Mimura, T. Iwatsuki, T. Sibata, H. Takeno, and K. Itimaru.

[9] They replaced the Riemannian metric by a 'microscopic' metric $ds . \psi = \gamma_i dx^i . \psi$ (where the γ_i are matrices for which $\gamma_i \gamma_j + \gamma_j \gamma_i = g_{ij}$) and invoked the theory of spinors (*Hiroshima Journal of Science*, viii (1938) 193 (in five parts): cf. Sibata's article in *Tensor*, vi (1943) 62). A certain vector (u^k) (formed out of the matrices γ_i and the fundamental spinor ψ) being the tangent vector to a geodesic of the space, is next interpreted as the momentum density of a particle identified with a nebula. The mathematical restrictions upon this vector restrict in turn the cosmological model, which may be of either the Einstein or the de Sitter kind. Itimaru (part iv) showed that only in the latter case is a motion of universal recession possible, and that in this case the total mass increases with time.

[10] e.g. T. Vescan, who took the mass of the universe to be $M = M_0 + M_1 \exp(-qt)$. (*C.R.*, ccxxv (1947) 278.)

possible values' of certain theoretical terms of cosmological significance. Dirac on the other hand, took a radically different view in making similar 'naturally occurring' numbers of this order functions of the epoch.[11] Dirac's argument has had little influence on cosmology as a whole, although it was responsible for drawing attention to the ease with which the mathematics might lead one into a theory involving the continual creation of matter. The 'age of the universe' as derived from cosmological theories was at this time put at 2.10^9 years. In terms of a commonly used unit of time, provided by atomic constants (namely e^2/mc^3), this age is of the order 10^{39}. Now the ratio of the 'mass of the universe' to the proton mass had been estimated as of the order of 10^{78}, whilst the ratio of the electrical force between electron and proton, to the gravitational force between the same particles, was well determined as nearly 2.10^{39}. This, according to Dirac, suggests that both ratios are functions of the age of the universe and hence of the time, the number of protons and neutrons in the universe increasing as the square of the time, and the 'gravitational constant' being inversely proportional to the time.[12] Needless to say, the second conclusion would not have followed had the charge or the mass of electron and proton been supposed functions of the time.

It is difficult to discover who was the first to appreciate the numerical coincidences from which Dirac worked, but J. Q. Stewart of the Princeton University Observatory wrote on the subject as early as 1931.[13] The six universal constants e, h, c, γ, m_e, and m_p determine 'three presumably independent pure numbers', namely (hc/e^2), $(e^2/\gamma m_e^2)$, and (m_p/m_e). Trial and error showed Stewart that the simplest formula for a quantity of correct dimensions and of the same order of size as the currently observed quantity (c/h_1) was $(e^6/h\gamma m_e^3 c^3)$. Using this expression, the ratio of the calculated value of the Hubble factor to Hubble's figure was $1\cdot21:1$. The expression corresponds, however, to no formula in the Dirac–Jordan development of the subject.

Discounting the uneasy mixture of conceptual schemes, and allowing for recent revision of the Hubble factor, the coincidences are very striking. And if they are more than mere coincidences, some sort of

---

[11] Letter of 20 February 1937: *Nat.*, cxxxix (1937) 323.

[12] A more complete, but somewhat modified argument is provided in *P.R.S.* (A), clxv (1938) 199. The unit of time (e^2/mc^3) Dirac chose as being roughly the geometric mean of some half-dozen alternatives. It had the additional merit of being, in reciprocal, much the same as the accepted value of the Hubble factor! He conceded that if two numbers seem to be of the order of 10^{39} one must allow that their ratio may be as much as a few powers of ten.

[13] *Phys. Rev.*, xxxviii (1931) 2071.

explanation is called for. The interpretation given of them, on the other hand, was too simple. If the reciprocal of the Hubble parameter is interpreted as the 'age of the universe' then, as we shall see, R must be a very special function of 'time'. It is not that there is something inherently wrong in the acceptance of this function, but that if it is to be accepted at all, it should be on stronger grounds than these. Assuming, nevertheless, that the ratios are *some* function of the time, using a suitably transformed time variable Dirac found it possible to 'arrange for the laws of mechanics to take their usual form'.[14] It is likewise possible to effect a (different) transformation of the time variable in order to avoid the consequence of an increasing number of particles, but this possibility was not mentioned. It is also possible, without any such transformation, to suppose that both the mass of the proton and the mass of the universe are functions of time. Why the former should be constant and the latter not, is unclear. The only argument which it seems might have been offered is that since the number of particles must by hypothesis change, to suppose their mass to change in time would require a mechanism of continuous mass-redistribution, over and above that of particle-creation—and this regardless of whether or not the total mass changes. (Dirac, in the paper referred to below, contemplated that the ratio of the mass of a proton to that of an electron might be required by 'future developments . . . to vary slowly with the epoch'.)

Dirac modified and enlarged upon this first contribution in the following year,[15] and now assumed that matter is conserved when expressed in terms of proton or neutron mass as unit. He made it clear that he would assume three general principles: the first was to the effect that one must allow a preferred time-axis defined by the 'natural velocity of matter at each point'. The second was a cosmological principle in Milne's sense. The third, since known by Dirac's name, was that 'any two of the very large dimensionless numbers occurring in Nature are connected by a simple mathematical relation, in which the coefficients are of the order of magnitude unity'.[16] The distance between nebulae, expressed as a dimensionless number, $f(t)$, multiplied by Dirac's chosen atomic standard of length, was then to be the same

[14] Dirac suggested that Milne's logarithmic transformation be used. Milne approved much of Dirac's work, relating it to his own, but he repudiated the idea of continual creation. (*Nat.*, cxxxix (1937) 409.)

[15] *P.R.S.* (A), clxv (1938) 199. Meanwhile A. Haas had published similar conclusions, see *Phys. Rev.*, liii (1938) 207.

[16] Dirac, op. cit., p. 201.

for all nebulae except for an arbitrary constant factor. Expressing the density of matter, averaged throughout large finite regions of space, in terms of the number of protons per unit atomic volume, produced a figure of the same order of size as the Hubble factor.[17] In terms of $f(t)$ this can be expressed

$$f(t)^{-3} \simeq f'(t)/f(t), \qquad (1)$$

whence $\qquad f(t) \simeq (t + a)^{1/3}. \qquad (2)$

By 'a suitable choice of the zero from which we measure time' this was reduced to

$$f(t) \simeq t^{1/3}. \qquad (3)$$

Recalling the dangers of interpreting this formula lightly, we notice that Dirac emphasized that 'Hubble's constant' changes with epoch and that the time from the 'natural origin of time' is only one-third of 'the value we gave it on the assumption of a constant velocity of recession'.[18] The value indicated (7.10^8 years) was certainly on the small side, 'being less than the age of the earth as usually calculated from data of radioactive decay'. But Milne had unwittingly set the fashion for explaining away 'the time scale difficulty' in terms of differing rates of time-keeping processes, and Dirac avoided the conflict between his theory and the astronomical evidence by maintaining that 'a thorough application of our present ideas would require us to have the rate of radioactive decay varying with the epoch and greater in the distant past than it is now'.[19]

To return, however, to the conservation of the number of protons and neutrons which Dirac was now assuming at the outset of this second paper, 'apart from processes involving the transformation of the rest-energy of these particles to or from another form'.[20] He realized that the contrary hypothesis could not be expected to be supported on the grounds of direct evidence, either terrestrial or astronomical, and he held to a principle of conservation simply because the 'spontaneous creation or annihilation of matter is so difficult to fit in with our present theoretical ideas in physics as not to be worth considering, unless a definite need for it should appear, which has not

---

[17] Previously Dirac had simply accepted that this varied inversely as the time. Here his argument is slightly more general, the Hubble factor being taken as a quotient of nebular velocity and distance.

[18] Dirac, op. cit., p. 203. [19] op. cit., p. 204. [20] loc. cit.

happened so far, since we can build up a quite consistent theory of cosmology without it'.[21]

No explanation will be given here of the way in which Dirac tried to link his ideas with General Relativity: his attempts to do so are in many respects unsatisfactory, and they add nothing of interest to the present context.

One of the few writers on whom Dirac's methods made a lasting impression was Pascual Jordan.[22] Shortly after Dirac first broached the subject, Jordan gave a résumé of the aims of both Dirac's and Eddington's work, listing some of the dimensionless ratios.[23] He rejected two possibilities which Dirac had considered, namely that of a varying rate of radioactive decay, and that of a variation in the ratio of proton to electron mass.[24] Jordan took much from Dirac's earlier draft, adhering to the proposition which Dirac abandoned, namely that the 'number of fundamental particles in the universe increases as the square of the age of the universe'. Jordan thought that Dirac might have been influenced by 'fear of contradicting the principle of the conservation of energy'.[25] Whether or not this was so, Jordan made the conservation of energy an important feature of his theory, a theory in which a steady increase of the matter in the universe occurs by the creation of *stars* and even entire *galaxies*, 'whose negative gravitational energy balances the new mass energy'.[26]

In order to see Jordan's methods more clearly we may collect together his elementary arguments.[27] He distinguishes two groups of

[21] loc. cit. For other features of Dirac's theory see Bondi, *Cosmology*, p. 161; C. Gilbert, *M.N.*, cxvi (1956) 684; *Obs.*, lxxvii (1957) 57. Gilbert considered the theory in the context of General Relativity. More recently he has linked Dirac's results with a highly ingenious theory of his own. (*M.N.*, cxx (1960) 367.) The most novel feature of this theory is his use of a quadratic, rather than a linear, action principle. The cosmological model which results has some resemblance to Milne's kinematic model, and in a local gauge the average field has the same metric as an Einstein–de Sitter universe.

[22] Jordan is remembered as having collaborated with Heisenberg and Born in the invention of matrix mechanics.

[23] *Naturwiss.*, xxv (1937) 513; xxvi (1938) 417. Cf. Haas, *P.N.A.S.*, xxiv (1938) 274. A fifth dimensionless constant is defended, viz. the ratio of the number of neutrons in the universe to the number of protons. Haas gives a general argument for the equality of the two numbers should the current empirical value of the ratio $(M_p/M_e) \cdot (e^2/hc)$ be too large by 1 part in 200.

[24] *ZS. Phys.*, cxiii (1939) 660. The former would imply a complex form of displacement in the spectral lines of distant nebulae. The latter, according to Jordan, is ruled out by evidence from the discoloration haloes of radioactive minerals which, he claimed, show that nuclear forces (and hence α-particle ranges) have been constant over times of the order of 10^9 years.

[25] *Nat.*, clxiv (1949) 638.

[26] *Ann. Phys.* Lpz., xxxvi (Sept. 1939) 64.

[27] Works cited and his book *Schwerkraft und Weltall* (Braunschweig, 2nd ed., 1955).

terms, cosmological and microphysical, namely

(i) c, κ, h_1, A (the age of the universe), ρ (its mass density), R (the radius), and M ($\propto \rho R^3$, the mass of the universe) and

(ii) e, $l(= e^2/m_e c^2$, the 'electronic radius'), $\tau(= l/c)$, m_e, m_p, m_n, m_m (mesonic mass), &c.

There are clearly dimensionless ratios $h_1 A$, R/cA, and $\kappa \rho c^2 A^2$ to be formed from the first group. This is not surprising: it is true for any dimensionally consistent interpretation of the terms. But that all three should (as Jordan held to be the case) turn out to be of the order unity was accounted very strange. Accordingly, three laws were suggested, namely

$$h_1 A = 1, \qquad (4)$$

$$R = cA, \qquad (5)$$

and $\qquad \kappa M \simeq R$ (using (5)). $\qquad (6)$

Now the first of these (which, we are told, means that space 'must once have been very small')[28] has a positively tautologous appearance. The usual figures quoted for A are simply reciprocals of h_1. And even where a less naïve interpretation is given, it is almost always from a knowledge of h_1 that A is inferred. In other words, Jordan must of necessity become involved in a prior cosmological theory—and one which, presumably, his theory is aimed at superseding. Similar remarks may be made in regard to (5) and (6), and indeed Jordan retained the '$\simeq$' sign in (6) in order to make a comparison with the corresponding equation for the Einstein universe,

$$\kappa M = 4\pi^2 R. \qquad (7)$$

The surprising discovery that the 'empirical value' of $\kappa M/R$ is of order unity can surely be no more than the discovery that the accepted value of R was in fact derived from formula (7) using an estimate of the value of the mean density of matter. The surprise, in short, was assured the moment astronomers quoted values for R which were (for want of better information) arrived at in this way. There would have been less of a surprise had Einstein's formula contained a constant differing greatly from unity. In any case, the derivation of a value for R (for which there are several plausible interpretations) depends very intimately upon the particular theory which is used. The best modern evidence,

[28] *Nat.*, clxiv (1949) 638.

used within the relativistic scheme, seems to suggest a negative value for k;[29] and the sign of the curvature can hardly be ignored, for it occurs as a multiplier of the only relevant term (apart from the density term) in the nearest equation equivalent to (6) of *general* validity in relativistic cosmology (namely Friedmann's equation for the density). In any case, the two are so different as to virtually guarantee the *falsity* of (6) should it be used to derive a value for R from observation. It is unfortunate that Jordan does not appear to realize the extent to which he was indebted to the cosmological theories which had preceded Dirac's. It was as though he imagined that each 'physical constant' is the common property of all theories in which its name and symbol occurs.

This facility for accepting certain features of a given theory, only to reject that theory later, is even more in evidence when one considers the use to which Jordan put equation (6). The resemblance to Einstein's formula was, in a sense, unfortunate; for was Einstein's universe not static? Since R must be supposed to vary, so then must κ or M, or both. The reasons for supposing both to do so were precisely those already given in Dirac's first paper. (At this point Jordan brings in the second group of physical quantities and the numerical coincidences between dimensionless ratios which we have already encountered.) Must the principle of the conservation of energy be denied? Jordan was indebted to Haas for indicating a way in which the relation (6) might be regarded as an expression of this very principle. 'The relation', wrote Jordan, 'can be written in the form $\gamma M^2/R \simeq Mc^2$, which means that the negative potential energy of gravitation for the whole universe is equal to the sum of the rest-energies of the masses of the stars', and that it is therefore possible for the total energy of the universe to be exactly zero. 'Exactly' might appear to be the wrong word, more especially as the factor 8π has been dropped.

On the question of spontaneous creation, with which we began, Jordan was much more explicit than Dirac—who, in any case, abandoned the idea. In order that the negative gravitational energy should exactly balance the material energy, the new matter must enter the universe in a highly condensed form, and not as a diffuse aggregate of simple particles. These large condensations will, he later argued, first appear as supernovae. The rate of creation of supernovae is, as might be expected, related to Dirac's number 10^{39}. The number of stars in each galaxy and the number of galaxies in the universe are both said to be

[29] Jordan seems to have believed it necessary that space should be of positive curvature if the de Chéseaux–Olbers paradox is to be avoided. If so, he is wrong.

proportional to the one-quarter power of this number (there are grounds for believing that the former is not far from the truth), and this is interpreted as meaning that each is proportional to the one-quarter power of the time—since the 'age of the universe' in Dirac's units was supposed to be of the order of 10^{39}. The predicted average rate of occurrence of supernovae turns out to be rather more than one per galaxy per year, a figure which is two or three hundred times too great. Jordan failed to explain the unsuitable chemical composition of novae and supernovae considered as a means of supplying the material for the remaining parts of the observed universe.[30] Despite its inadequacies in these respects, Jordan has aimed at providing a 'physics of creation', although he obviously felt uneasy at the idea of a recurrent spontaneous creation; for he writes of isolated spaces, external to the 'large universe', which by 'the gradual unfolding of time' and by the expansion of the large universe, may coalesce with it ('when the gravitational constant is the same for each'). These 'little worlds' were said to be the embryonic states of stars or supernovae.[31] This notion of the intrusion of matter from another dimension is not, of course, new: it was a mathematician's source of amusement a century ago. It seems not to have been developed further. It shows, perhaps, that at least one supporter of a theory of continual creation was unhappy at the idea of a fragmented First Cause.[32]

2. Energy Conservation. The Bondi–Gold Theory

Little notice was taken of the idea of a process of continual creation until Bondi and Gold made it a part of their steady-state theories of cosmology. Once they had committed themselves to the Perfect Cosmological Principle[33] ('the universe presents on the large scale an unchanging aspect'), and hence to a universe in a 'stable, self-perpetuating state', they were—in holding to an expanding universe—obliged to maintain that matter is continually created within it, in order to keep the density constant.[34] Once again the problem of energy conservation arose, and although Bondi and Gold admitted that they had to 'infringe the principle of hydrodynamic continuity', yet they

[30] But see Jordan, *Naturwiss*, xl (1953) 407 where P. W. Merrill's detection of the lines of Technetium I in the spectra of S-type stars (*Ap. J.*, cxvi (1952) 21) is claimed as support.
[31] *Astr. Nachr.*, cclxxvi (1948) 193; *Nat.*, clxiv (1949) at p. 639.
[32] See Appendix, Note XI.   [33] *M.N.*, cviii (1948) 252.
[34] op. cit., p. 256. From estimates of the mean density and the rate of expansion, the rate of creation was thought to be at the most one particle of proton mass per litre per 10^9 years (later amended to 5×10^{11} years).

maintained that it was 'by no means clear whether we can regard any principle of conservation of energy as infringed'.[35] The distinction between the two principles is a useful one which is seldom made because it is seldom necessary. Although hydrodynamic continuity is clearly abandoned, yet in the steady-state theory an observer will in principle always be able to observe a finite amount of matter, which will be constant in time. 'At great range', according to Bondi and Gold, 'matter is drifting into an unobservable state by approaching the velocity of light, and without a process of creation this would not allow any principle of the conservation of energy to be applied to the sum total of all observable matter.'[36] In this theory, matter is conserved in any given *proper volume* of space. Clearly no conservation principle can be laid down in a non-finite model which does not specify the kind of region within which matter or energy is constant. In the relativistic models *coordinate volume* is generally specified.

If co-moving coordinates are used, and the matter per given coordinate volume is taken as constant, the average density over the observed region (or proper volume) must diminish in time. 'It may well be considered correct', wrote Bondi paradoxically, 'to speak of conservation of mass in the steady-state model rather than in relativity, since proper volume is more fundamental than coordinate volume.'[37]

Does the ambiguity in the definition of distance not transfer itself to that of volume when it becomes necessary to offer meta-scientific arguments for or against one of these conservation principles? There are, after all, other definitions of distance as acceptable as that of proper distance, each carrying with it a possible conservation principle. Again, the region 'which it is possible to observe' might seem to be more fundamental than any of these other sorts of volume in the sense that different authors might agree in ostensively defining it. But the traditional enunciation of the energy conservation principle speaks of a 'system of bodies', and the observable limits do not, in our case, follow the system of bodies as they move outwards. (Co-moving coordinates have this advantage.) It is, of course, immaterial which of the several alternatives is accepted, in phrasing conservation principles for cosmology, so long as conclusions follow which are otherwise acceptable. For the moment we are simply pointing out that it would be naïve to

---

[35] loc. cit.

[36] op. cit., pp. 258–9. A controversy followed over the phrase 'observable universe' which was useful if only because it led to a clarification of the 'horizon' terminology. See p. 274 *infra* and Appendix, note XVI.

[37] *Cosmology*, p. 144.

dismiss the Bondi–Gold theory—as did some when it was first proposed—on the grounds that it violates some inviolable Principle of the Conservation of Energy.

Whilst on this subject, it is worth remembering that a conservation law usually relates the change with time of a quantity, integrated through the interior of a volume, to the integral of a second quantity over the surface of that volume. The great value of conservation laws is that they permit of our making assertions about complete systems without an intimate knowledge of the properties of their contents. In this way they may be of use in theoretical cosmology. Most modern theories of cosmology have followed Einstein, who in turn followed the Newtonian example,[38] in making an energy conservation principle an hypothesis of high order. The difficulty for the orthodox Theory of General Relativity has been that in a general Riemannian space conservation laws are not easy to formulate, for the addition of tensors at different points of a finite region does not lead to a tensor. In constructing a conservation law for the tensor T_{mn}, Einstein saw that this could only be done outside curved regions of Riemannian space. Constructing the components T_{mn} accordingly (that is, as in the flat space of Special Relativity) we have already seen how he related them to geometrical quantities which are automatically conserved (that is to say, have vanishing divergence) and thus laid down the field equations. Containing as they do the conservation laws of classical dynamics, these field equations have, as we have seen, only been tested in a very indirect way—for example, in so far as they lead to the laws of planetary motion.

An alternative to Einstein's first approach is to formulate laws in Riemannian space in terms of non-tensorial quantities which, unlike tensors, are themselves strictly conserved. As they are not likely to be transformed so easily as tensors they are likely to be difficult to interpret. The earliest example was provided by Einstein in the form of the 'pseudo-tensor', but many others have since been investigated. The use of pseudo-tensors allows the equations of mechanics to be written in the form of an ordinary divergence with a classical analogue. All formulations of an energy conservation principle in cosmology are, however, alike in being equally remote from the sort of considerations which were once believed to point inevitably towards their truth. According to Max Planck for example, 'every single experiment thus far undertaken has shown that . . . there is no reason to deny that the

---

[38] Thomson and Tait claimed to trace the principle to the finishing sentences of the Scholium to Lex III.

law of the conservation of energy is an absolutely and universally valid law of Nature'.[39] But the observation which would detect a uniform intergalactic medium such as Bondi and Gold contemplated has yet to be devised. Such a medium could not be expected to have any appreciable scattering effect on the light received from distant galaxies, nor would it alter sensibly the peculiar motions of any celestial object.[40]

As explained already, the starting-point of the theory of Bondi and Gold was the so-called 'Perfect Cosmological Principle'. This will be discussed again in Part II. For the present we notice that whereas Milne's Cosmological Principle required the large-scale aspect of the universe to be independent of the position of the observer, it is now made also independent of the time of observation. Thus, for example, not only must the average density of both matter and radiation remain constant, but also the age-distribution of the nebulae must be unchanging in time: as the older nebulae separate with the general expansion, new nebulae are formed in the intervening spaces out of newly created matter. The maximum entropy paradox is thus obviated by the introduction of a continued creation process: radiant energy, of high entropy, is lost continuously through the Doppler effect of the expanding universe, whilst material creation supplies energy at low entropy.

The Bondi-Gold theory owed much, not surprisingly, to the writings on Kinematic Relativity, in particular to the kinematic derivation by Robertson and Walker of the line-element

$$ds^2 = dt^2 - R^2(t)\ (dr^2 + r^2 d\theta^2 + r^2 \sin^2\theta d\phi^2)/(1 + kr^2/4)^2. \qquad (8)$$

Here $R(t)$ is simply an arbitrary function of the time. It will be recalled that apart from Weyl's Principle and certain properties concerning light propagation, the ordinary Cosmological Principle was used in this derivation. Imposing the additional requirement of a steady-state (that is, 'strengthening' the Cosmological Principle) it is said to follow that $k = 0$, if R is not to be a constant function of time (as in an expanding universe). The reason given is that the curvature of the three-space (r, θ, ϕ) is k/R^2, and that this 'is an observable quantity since it affects the rate of increase of the number of nebulae with distance, and must hence be constant by virtue of the perfect cosmological principle'.[41] There is here an implicit assumption that the nebulae are distributed uniformly with regard to proper volume. (The proper distance between the individual nebulae will of course, vary in time.)

---

[39] *A Scientific Biography* (London, 1950), p. 175.
[40] See, for example, F. Hoyle, *Nat.*, clxv (1950) 68.   [41] *Cosmology*, p. 145.

Once more applying the Perfect Cosmological Principle, it follows that the Hubble parameter (R'/R) must remain constant (say $1/T$), whence $R = \exp(t/T)$. The resulting metric is none other than the de Sitter metric in the form first provided by Robertson and Lemaître, that is to say

$$ds^2 = dt^2 - (dr^2 + r^2 d\theta^2 + r^2 \sin^2\theta \, d\phi^2) \exp(2t/T). \qquad (9)$$

When null geodesics are identified with light rays, the Doppler shift and light intensity formulae of this theory are exactly those of de Sitter's model, namely

$$\delta = r/T, \qquad (10)$$

and
$$\text{(intensity)} \propto 1/\{4\pi r^2 (1 + r/T)^2\}. \qquad (11)$$

As a consequence of the fact that in relativistic cosmology the nebulae are regarded as uniformly distributed with respect to coordinate volume, whilst here it is proper volume which matters, any number-count formulae will differ between the two theories. We shall refer to this question again in the following chapter. By and large the steady-state theory is in its consequences the simplest cosmological theory of repute produced so far this century. It requires, for example, that the *average* nebular luminosity be independent of epoch, whereas in almost all other theories this varies with time. The Bondi–Gold theory is in such respects as this more specific, with fewer parameters to enable the 'appearances' to be 'saved'. This property of the theory will be considered again in Chapter 14.

3. Hoyle's Steady-State Theory and McCrea's Interpretation of It

The second important contribution to a theory of a steady-state universe was made shortly after that of Bondi and Gold. Hoyle, as we have seen, was prepared to abandon the principle of the conservation of energy, but he preferred methods less radical than those used by Bondi and Gold.[42] Rather than work from such a very general principle as the Perfect Cosmological Principle, he followed Einstein's precedent in modifying the earlier field equations, $R_{mn} - R g_{mn}/2 = -\kappa T_{mn}$. Where Einstein included the cosmological term (λg_{mn}) on the left hand side, thus preserving the energy conservation principle and the tensor

---

[42] *M.N.*, cviii (1948) 372; *Nat.*, clxiii (February, 1949) 196; *M.N.*, cix (1949) 365 (justification of his method and reply to criticism). Much of Hoyle's paper was, I believe, written before that by Bondi and Gold, but the latter was published first.

character of the equation, Hoyle included a symmetrical tensor C_{mn}, preserving, as he first thought, only the latter.[43] He was guided by Weyl's Principle (which, however, he modified) according to which there is a preferred velocity at each point of space-time. Now in previous expositions, this principle—which might be rigidly interpreted as positively denying a general covariance—had only been taken into account once the complete field equations were laid down (and solved). These equations had been formulated in such a way as to uphold general covariance. Weyl's Principle seems to cancel this, saying in effect that not all observers, but only certain fundamental ones, see the universe in the same way. We can look at the relativistic procedure in several ways: it can be regarded as an acceptance of covariance as a local requirement, dispensed with only on a cosmic scale. (We recall that the requirement of covariance was retained at the time of the inclusion of the cosmological term, a term of no significance on a local scale.) The difficulty now is in knowing where to draw the line between what is local and what cosmic. A second alternative would be to accuse those relativists who use Weyl's Principle of inconsistency: general covariance was, after all, originally introduced as a methodological desideratum which should not be cast aside lightly, even when it cannot be followed to the letter. To reject it outright would be to sever the very roots of the General Theory of Relativity.

A third and more satisfactory point of view might be summarized as follows: general covariance is (for reasons which we have discussed elsewhere) desirable; and thus our most basic equations, with an application to very many domains—from pendulum clocks to the universe of nebulae—should conform to it. But the requirement of general covariance is a formal one. It is stipulated that all laws are to have the same form, not that all observers should see the contingent distribution of matter in the same pattern of motions. One is presumably at liberty to interpret Weyl's Principle as of the same character as this; but that was certainly not the intention of those who first used it in the 1920s. For them it was a restriction on the motions which nebulae were allowed to have. It would not, perhaps, be too much to call it a 'law of motion', but it would certainly be misleading to treat it as a law of laws of motion. If, over the next thirty years, the assumed character of the principle changed, this was probably as the result of the fashion which Milne introduced, whereby 'observer' and 'nebula' were virtually taken as synonyms.

[43] See p. 376 of his first paper.

According to Hoyle's use of Weyl's Principle, particles in local condensations were not held to conform to it. He qualified the principle so that only at the time of its creation does a particle follow a geodesic of the diverging bundle.[44] He did not require that it should continue to do so: 'the particle will subsequently move along a path not passing through O if it should be perturbed by an electromagnetic field'. (O is the point of divergence of the geodesics.)[45] This might have been said of ordinary particles by anyone upholding the principle in its original form, but as yet electromagnetic fields were virtually outside the province of cosmology.

We shall now briefly explain how Hoyle's 'creation field' was introduced.[46] A quantity C is defined as a scalar function of position, such that the difference between C at two points is proportional to the difference between the geodesic distances of these points from O.[47] The first derivative of C defines a field of vectors, C_m, directed always along the geodesics through O and of constant length $(3c/a)$. The second derivative of C, C_{mn}, defines a symmetric tensor field. Introducing the tensor C_{mn} in the same way as Einstein had introduced λg_{mn}, the new field equations were now written

$$R_{mn} - \frac{1}{2} R g_{mn} + C_{mn} = -\kappa T_{mn}. \qquad (12)$$

It was shown that any line-element which is a solution of these equations tends asymptotically to a metric of the de Sitter variety in Robertson–Lemaître form:[48]

$$ds^2 = c^2 dt^2 - (dx_1^2 + dx_2^2 + dx_3^2) \exp(2ct/a). \qquad (13)$$

(The Bondi–Gold scale-function T corresponds to Hoyle's a/c, the metrics being otherwise alike.) The density of matter in the corresponding model was not, as in de Sitter's case, zero.[49] The constant

---

[44] *M.N.*, cviii (1948) p. 380. Cf. Bondi and Gold, op. cit., p. 266: 'The identification of the preferred direction defined by the creation of matter with the preferred direction defined by the motion of matter is compelling'. They were, however, prepared to contemplate a random distribution of initial velocities about the preferred direction (i.e. non-zero temperature).

[45] Hoyle added that 'the nature of the point O remains mysterious'. (op. cit. p. 381.)

[46] This follows Hoyle's first account: the second differs from it slightly.

[47] This, a slightly more general form than in Hoyle's first account, allows O to be infinitely remote.

[48] Any fluctuation from this solution was found to be exponentially damped in a time interval of order a/c ($= 1/h_1$). McVittie has found conditions under which this is not true.

[49] It was given by $3/\kappa a^2$ for the limiting case. The C_{mn} term, whose role resembles that of the cosmological term in the de Sitter model, is different, as Hoyle pointed out, in that there is no contribution from the C_{00} component, making a solution possible with non-zero density.

value derived for it suggested that since the line-element represents an 'expanding universe' there must be a continual influx of matter. Hoyle's theory was therefore yet another involving continual creation.[50]

Energy, for Hoyle, was not conserved, since $-\kappa(T^{mn})_{,n}$ is equal to $(C^{mn})_{,n}$, and this expression was not intended to be identically zero.[51] McCrea, however, later showed how Hoyle's form of the field equations may be interpreted as not requiring any modification of Einstein's equations, namely by incorporating the additional term in the T^{mn}.[52] He also suggested that although Einstein's field equations should be adaptable 'to describe phenomena in Riemannian space', a space which they had previously been regarded as themselves determining, yet what was usually required was the relativistic analogue of some specified classical problem. The General Theory of Relativity provides no essential way of formulating the stress-energy tensor: this task is completed with the help of a non-relativistic physical theory, or perhaps of the Special Theory of Relativity. Once the tensor T^{mn} is specified, so, consequently, is the metric. Now when Einstein's theory suggested to Hoyle a phenomenon without classical counterpart (namely the creation process), the temptation was to regard it as having nothing to do with the formulation of the T^{mn}. McCrea, however, chose to work with the Einstein equations, rewriting T_{mn} as $(S_{mn} + \kappa^{-1}C_{mn})$, where S_{mn} is Hoyle's (that is the conventional) form of the stress-energy tensor. Since $(T^n_m)_{,n} = 0$, there is obviously a sense in which energy may be said to be conserved. One is reminded of the dictum that the principle has survived because energy is defined as that which is conserved!

McCrea now sought a physical interpretation for his T_{mn}. He found it in obtaining a correspondence between the results of this relativistic treatment and that 'Newtonian' account first presented by Milne and himself in 1934.[53] The standard Friedmann–Lemaître expressions for the density and pressure in relativistic cosmology were now interpreted as, respectively, 'an equation of continuity which includes the mass equivalent of the work done against the stress of the medium in the expansion process' and a 'first integral of the equation of motion of the medium under its own gravitational attraction, this attraction including

---

[50] The new theory, according to Hoyle, arose out of a discussion with Gold, who had insisted that it must be possible to reconcile expansion with constant density by postulating continual creation. Hoyle's first paper, p. 372.
[51] To see this, take the divergence of (12).
[52] *P.R.S.* (A), ccvi (1951) 562.   [53] *Q.J.M.*, v (1934) 73.

a gravitational effect of stress disclosed by general relativity theory'.[54] Now Hoyle's steady-state solution has

$$\left.\begin{array}{rl} \rho &= 3/(\kappa a^2), \\ p &= -(3c^2)/(\kappa a^2) \\ \text{and} \quad \sigma &= -6/(\kappa a^2) \text{ (density of gravitational mass),} \end{array}\right\} \quad (14)$$

where a is Hoyle's constant, having the dimensions of length, such that $R'/R = c/a$. (As in the Bondi–Gold case, this means that the observed rate of expansion is independent of the epoch.) These were essentially the results found by de Sitter and rejected by him owing to the negative pressure. Introducing the cosmological term, de Sitter found the sole solution $\rho = 0$, $p = 0$. But de Sitter, and all who followed, were interpreting the stress p as the kinematic pressure of the matter together with the radiation pressure. McCrea questioned the classical interpretation, for 'in classical mechanics the absolute value of the stress does not enter into the equations of motion. The latter depend only upon the gradient of the stress, so the stress may be deemed indeterminate to within an arbitrary constant (actually a constant tensor in the general case). In relativity theory, on the other hand, the absolute value of the stress does become significant by virtue of . . . [Einstein's field equations], that is to say, the arbitrary constant has to be determined.'[55]

Using Hoyle's interpretation of the stress-energy tensor (written here S_{mn}) the pressure vanishes. For McCrea it is negative. He adds that as there is no pressure gradient 'this is not an observable difference'. (The theorist who postulates uniformities in his idealization does not usually expect their correlates in the real world to be observed as strict uniformities. McCrea is presumably not emphasizing the absence of pressure gradients in the real world, but rather that their absence from his model and Hoyle's is—through the medium of the underlying theory—capable of ensuring that all predictions capable of being confirmed will be alike on both theories.) It should be noticed that the gravitational mass (in this case $\sigma = \rho + 3p/c^2$) is also negative, owing to the contribution of the negative pressure; but it was shown to be unlikely that this will have any detectable effect upon the gravitational field of ordinary massive bodies. Perhaps the most interesting of all McCrea's results, however, was his calculation of the work done by the negative pressure (per unit volume per unit time) in the expansion of the universe. This $(3\rho c^3/a)$ reappears in the mass density at the rate

[54] *P.R.S.* (A), ccvi (1951) p. 569.   [55] ibid., p. 571.

($3\rho c/a$) units of mass per unit volume and unit time, thus accounting for the constant density despite expansion. McCrea now speaks of Hoyle's tensor C_{mn} as a 'zero-point stress' which 'it is only natural to suppose . . . would be in the form of an isotropic pressure (positive or negative)'.[56] 'If so,' he continued, 'at each point in space-time it will have only one definite principal direction . . . [and therefore] will be associated with a unique time-like vector at each point of space-time.' This provided a highly satisfactory link with Hoyle's vector field and the tensor field associated with it.

Before leaving this subject it is worth adding that McVittie has shown that in many of the older models of relativistic cosmology the conversion of stress-energy into mass, and conversely, was a commonplace—although not always recognized as one.[57]

4. Creation as a Physical Process

In one of the more recent presentations of his work, Hoyle incorporated and extended McCrea's point of view, in so far as he now wrote the energy-momentum tensor in four parts:[58]

$$T^{mn} = T^{mn}_{(g)} + T^{mn}_{(e)} + T^{mn}_{(c)} + T^{mn}_{(n)}. \tag{15}$$

These parts give, respectively, the ordinary gravitational field, the electromagnetic field, the creation field and the nuclear fields. 'In this way the creation field plays an entirely similar role to the other fields', wrote Hoyle, adding that from 'a cosmological point of view the important terms are $T^{mn}_{(g)}$, $T^{mn}_{(c)}$', and that 'ironically it is $T^{mn}_{(e)}$, $T^{mn}_{(n)}$ that are studied extensively in macroscopic physics'.[59]

Hoyle, as we saw, conceived himself to be disregarding the requirement of covariance, 'not by a departure from tensor equations, but through a preferred direction being defined by geometrical considerations at each point of space-time'.[60] He gave reasons for discarding two wholly covariant systems, in which $(C^{mk})_{,k}$ was related to the density (ρ) and the velocity vector of matter (V^m), first by the equation

$$(C^{mk})_{,k} = \text{const} \cdot \kappa \rho V^m \tag{16}$$

and secondly by the equation

$$(C^{mk})_{,k} = \text{const} \cdot \sqrt{(\kappa \rho)} V^m \tag{17}$$

---

[56] op. cit., p. 573. [57] See Appendix, Note XII. [58] *M.N.*, cxx (1960) 256.
[59] op. cit., p. 261. No satisfactory form for $T^{mn}_{(n)}$ is at present known.
[60] *M.N.*, cix (1949) p. 368.

(suggested by Bondi). The former was shown to require all bodies to double their masses in times under 10^9 years—a rate of increase ruled out by planetary theory. The second, although avoiding this objection, was rejected because the universe was said to be of 'a markedly non-invariant structure' wherein the observed motions of galactic recession, as described by Weyl's Principle, make necessary 'severe restrictions' on several equations, amongst them the field equations. Because no explanation of the significance either of these restrictions or of the unwanted solutions of the equations seemed to be forthcoming, a non-covariant formulation of the law of the 'creation process' was adopted. Hoyle insisted on its being recognized that the non-covariance applied locally only to the creation properties: 'Once matter has been created the full invariance of Einstein's General Theory of Relativity is required to describe the states occurring in local condensations'.[61] Bondi and Gold, in their original paper, had criticized Hoyle's forthcoming account on the grounds that it was *not entirely non-invariant*.[62] They were prepared to restrict the Principle of Relativity to observers in 'similar environments'. Hoyle remained unconvinced, and in his paper of 1960 he presented a covariant law for the 'creation field' of matter. This paper makes an important distinction between two possible procedures. Originally Hoyle's technique was to show that, with an arbitrary distribution of material, his theory predicted the eventual development of isotropy and homogeneity. To this end he found it necessary to adopt a creation law which was not covariant. He now attempted the problem on the assumption that the model describes, for all times, an isotropic and homogeneous state, and derived a covariant law capable of maintaining such a state. The tensor C_{mn} was now obtained by differentiating twice a scalar field ϕ, where ϕ is related to the mass density ρ through the equation

$$\Box \phi = 3c^2 \rho. \tag{18}$$

Hoyle suggested that 'matter be thought of as radiating a creation field',[63] and expressed the hope that microscopic physics would throw some light on the process.

In the earlier theory it turned out that Hoyle's 'creation rate' was more or less independent of the mass density ρ.[64] This, averaged over large spatial volumes, was thus found to be maintained at a more or less

---

[61] op. cit., p. 369.   [62] ibid., pp. 269–70.   [63] ibid., p. 259.
[64] *M.N.*, cix (1949) p. 369. There is a slight connexion through the metric tensor g_{mn}.

constant value. The form of the line-element which represents the completely homogeneous steady-state solution is in fact asymptotically approached by a model containing localized condensations.

To the problem of the physical characteristics of the newly created material within the context of their steady-state model, Bondi and Gold devoted a great deal of thought. They decided that the local rate of creation could not be a function of the local density, for the new material would have to be ejected from the stars (in which it would in this case be mainly formed) rapidly enough to supply intergalactic space with material sufficient to provide new galaxies at a great enough rate.[65] The necessary rates of emission would have required a far larger number of stars of the character of novae or supernovae than is observed. These authors consequently favoured accretion as the most important feature of the processes of nebular and stellar evolution, and as for the distribution of the new matter, it seemed 'most reasonable to assume that it is created in a random manner'. Two points may be made here: first, the objection to a creation which is a function of the density appears to militate against Hoyle's covariant creation law. Secondly, one might imagine that, against all steady-state theories with an infinitely long time scale, whatever the 'law of creation', there must be galaxies of indefinitely large mass formed by indefinitely prolonged accretion. Although these may be sufficiently old to have disappeared over our horizon, they should be visible in principle to some observer, and were this so the Cosmological Principle would not be obeyed. This objection can be disposed of by anyone who is prepared to admit D. W. Sciama's ideas on galaxy formation.[66] Sciama believes that by virtue of its motion through intergalactic gas, a galaxy collects material in its wake, and that gravitational forces will form this material into a daughter galaxy. A critical condition is found for the parent to break away from the daughter. After separation the process is repeated. Should the two fail to separate there is a chance that a cluster of galaxies will result, although other processes will encourage the 'evaporation' of cluster-members. Evidence for Sciama's theory would provide important evidence of intergalactic hydrogen of the requisite density.

We have seen that Hoyle believed the most probable initial motion of the material to be that of the background material. (In his theory, since $(C^{\alpha k})_{,k} = 0$ ($\alpha = 1, 2, 3$), it follows that the matter has no momentum in terms of co-moving coordinates.) As for its electrical

[65] See *M.N.*, cviii (1948) pp. 266–8.   [66] *M.N.*, clv (1955) 15.

charge, it was concluded that if there is any average charge-excess in newly created matter, the average charge-mass ratio must be very small (less than $10^{-19}e/m$). The repulsion of excess charge would otherwise take place at a speed greater than the speed of expansion of the galaxies as a whole. All this suggested the creation of either protons or electrons (separately created but at identical or nearly identical rates), neutrons or hydrogen atoms. 'For simplicity and definiteness' the last alternative was adopted, but even the first alternative would ultimately give rise to hydrogen in the ground state.

Hoyle at first added little to these suggestions, although later he made many contributions of a kind which provide a suitable starting point for those theories of stellar evolution which have been his chief concern. At first he contented himself with drawing attention to the result that although 99 per cent of all material appeared to consist of hydrogen yet the rate of its conversion to helium was probably as much as 0·1 per cent per 10^9 years.[67] This apparent anomaly could be explained, provided that the new material consisted of either hydrogen or neutrons, the former perhaps resulting from the latter by subsequent disintegration. For some years to come little was written on the subject, but there was general agreement that the creation process must yield fundamental particles which, in turn, must produce hydrogen. In 1958 Gold and Hoyle drew from a tentative theory of the condensation of interstellar matter into galaxies the conclusion that the new matter might first have appeared in the form of neutrons at a temperature of the order of $10^{9\circ}$K.[68] The radiation from a gas at this temperature would be small and would be quickly absorbed by the C, O, N, and Ne in a galaxy.

Speculation on the 'creation process' continues, and one most interesting suggestion is that, on the grounds of symmetry, one would expect statistically equal amounts of matter and anti-matter (positrons or anti-protons: the former predicted by Dirac in 1928 and found experimentally in 1932; the latter found in 1955) to be produced. G. R. Burbidge and Hoyle have shown, however, that regardless of which kind predominates, a galaxy cannot be expected to contain the two kinds of matter in a ratio greater than about $1:10^7$.[69] Clearly, if it is not the case that each kind originates almost solely in the presence of like matter, there must be some highly involved process of segregation

[67] *Nat.*, clxiii (1949) 196.
[68] *I.A.U. 1958 Paris Symposium on Radio Astronomy* (Stanford University Press, 1959) p. 583.
[69] *A.J.*, lxii (1957) 9 and *Nuovo Cim.*, iv (1956) 558.

at work.[70] McCrea has very recently outlined a model in which the creation process depends upon the presence and physical state of existing matter.[71] He suggests that the presence of either matter or anti-matter leads only to creation of the corresponding kind; and furthermore, that any particular galaxy is effectively composed of either the one kind or the other. The virtue of these hypotheses is that they obviate the need for a mechanism of segregation.

In Part II we revert to the question which the steady-state theories claim to answer: Is it possible to bring the problem of creation within the scope of physical inquiry? We have now outlined the conceptual framework in terms of which (or so, it is held) the 'problem of creation' is aptly treated. This chapter now ends with some short remarks by way of comparing the formal aspects of the two principal steady-state theories.

Both theories share a metric which is unaltered if we effect the transformations
$$t \to t + t_0 \tag{19}$$
and
$$x \to x/\exp(t_0/T). \tag{20}$$

This shows that the geometry is essentially the same regardless of time origin. The metric is therefore highly appropriate to a steady-state model, and it is easy to see that the result would not have followed had k been equal to $+1$ or -1. We can now see that Robertson's criterion for applying the word 'stationary' (which we are here reproducing) covers the current use of the phrase 'steady-state'.

Again, using an old result of Eddington's, it should easily be seen that in a de Sitter model (at least with zero pressure), since material is accelerated away from an observer, the field equations of Einstein's theory cannot be fitted into the steady-state metric taken together with

---

[70] Speculation on the existence of large tracts of anti-matter is not, of course, solely the province of the steady-state theories. Light relief has been provided by A. Goldhaber who, with Lemaître's theory in mind, suggested that early in its existence the universe divided into two independent parts, one of matter, one of anti-matter, with a large relative velocity. (*Science*, cxxiv (1956) 218.) See also the prophetic remarks of Sir William Schuster, p. 37 *supra*.

[71] *M.N.*, cxxviii (1964) 335. An alternative formulation of continual creation, with somewhat similar properties, is that given by Hoyle and J. V. Narlikar. (See, for example, *P.R.S.* (*A*), cclxxiii (1963) 1; ibid., cclxxviii (1963) 465.) In their so-called '*C*-field cosmology', matter is created out of a negative energy field *C*. When the strength of this field at any point satisfies a certain condition, matter is created there, and at the same time a pulse of *C*-field is emitted. There is, in other words, a feed-back process.

a constant recession-rate. McCrea's means of harmonizing the two without zero-pressure requires the concept of 'zero-point stress'. This, on occasion, has been criticized on the grounds that it is 'inconceivable'. Without for the moment entering into the question of the folly of this type of general argument from inconceivability, we may refer forward to an interesting model which McCrea has offered of an expanding universe with zero-point stress.[72]

The steady-state theories differ from the evolutionary models perhaps most conspicuously in that they involve no singular state, either in the past or in the future. If we are prepared to admit that the physical interpretation of such singularities is as difficult a problem as any in cosmology, then the supporters of the steady-state theories may be thought fortunate to have side-stepped the issue. But in its place they have the problem of continual creation. The two alternatives are again considered in our last chapter.

We have already referred to the highly specific character of the Bondi–Gold theory (i.e. in regard to number counts and average luminosities). Hoyle's theory was even more specific, for it provided a quite definite value for the mean density of matter in the universe, whereas the former theory did not. In one sense, however, Hoyle's theory was more specific than it need have been, for the de Sitter metric is only one of many solutions of Hoyle's modified relativistic field equations.

We now pass to a more general comparison of the theories of the steady-state with those others whose history has been followed in previous chapters.

[72] See Chapter 14, last section.

CHAPTER 11

THE RECEDING GALAXIES: APPEARANCE AND REALITY, THEORY AND OBSERVATION

THIS book is primarily concerned with the language and methods of cosmology, and few references have been made either to past or to current astronomical data. It is hardly possible to make good this deficiency in a short space. Indeed, it is hardly possible to fulfil such a task at all, for the number of different interpretations of the meagre and uncertain data available at any one time during our period is formidable. As for the present situation, which is rather more stable, almost every item in the very extensive literature of the subject is given over in part to collating the best available information.[1] We can, however, consider the ways in which a comparison of theory and observation is made. We can compare the kinds of predictions made from different theories, and in doing so we shall be comparing the theories in the way most relevant to their scientific worth. As a means of introducing this comparison we begin by discussing a controversy closely related to it, namely that which concerns the 'reality of the expansion of the universe'. The reasons for questioning this reality were not entirely philosophical or theoretical. To a large extent they were prompted by the so-called 'time-scale difficulty', which might even be regarded as the mainspring of the cosmology of the late 1930s. We begin, therefore, by explaining the difficulty.

1. THE TIME-SCALE DIFFICULTY

We have already alluded to some of the reasons for refusing to accept, at least within relativistic cosmology, any simple linear 'velocity-distance relation' naïvely interpreted as indicating a point-origin for the universe, $1/h_1$ years ago. Although de Sitter was careful to avoid this way of speaking, and although he considered that the beginning of the expansion and the beginning of the universe cannot be identified, it does not appear that he thought it logically improper to do so, but merely that he interpreted various properties of the nebulae as supporting the hypothesis that they had existed in some form or other

[1] To no author do we owe more in this respect than to G. C. McVittie. See especially his article in *Handbuch der Physik*, iii (Springer, Berlin, 1959) p. 445.

long before approaching to a minimum separation some 10^9 or 10^{10} years ago.[2] De Sitter was unusual: with the great majority there was an uncritical acceptance of the bare mathematical concepts involved. A great many examples could be quoted. Eddington, for instance, was typical in frequently referring to 'the date of the beginning of the universe' (although the Eddington–Lemaître universe does not begin with a point-singularity). Again, most early discussions of the Einstein–de Sitter model refer to the lapse of time between the present epoch and the instant when $R(t)$ was supposedly zero as 'the age of the universe'.[3] For the Eddington–Lemaître model the present rate of expansion suggested an origin of 'uncomfortably recent' date, forcing Eddington to 'allow evolution an infinite time to get started'.[4] For the Einstein–de Sitter model the age of the universe was even more uncomfortably recent.

Without, for the moment, discussing further what can be meant by the phrase 'age of the universe', one can discover the cause of the uneasiness. If it is possible to speak of the age of the universe at all, it will be agreed that this cannot be less than the age of any constituent part, or the duration of any single constituent process. (It is not necessary to go as far as Robertson, who took the 'age of the universe' to be 'a significant parameter because other considerations . . . require that it lies within certain reasonably well defined limits'.)[5] Now when the reciprocal of the 'constant' in Hubble's Law was first compared with estimates of the age of the Earth—then, as now, taken to be of the order of 10^9 or 10^{10} years—the similarity was striking.[6] Unfortunately there was a firm belief that Jeans had established the age of the Galaxy at 10^{12} or 10^{13} years,[7] basing his arguments on the dynamics of star systems (extended moving clusters and binary systems). Here was a paradox under which the evolutionary theories laboured for many years. How can the Galaxy be older than the universe itself? Milne avoided the paradox by assigning different scales of time to the two phenomena, but the problem was resolved to the satisfaction of most cosmologists only with the discrediting of Jeans's arguments. This was gradually brought about during the 1930s, largely by reason of a

[2] *Proc. Acad. Amst.*, xxxv (1932) 596.
[3] R was usually given as $(at)^{2/3}$, making the time lapse in question $2/(3h_1)$, in our previous notation.
[4] *M.N.*, xc (1930) 678. [5] 'Cosmological Theory', *Jubilee R.T.*, (1956).
[6] The papers by Hubble (*P.N.A.S.*, xv (1929) 168) and M. L. Humason (*P.N.A.S.*, xv (1929) 167) suggested a figure of a little under 2.10^9 years.
[7] See *Astronomy and Cosmogony* (1929) which contains references to Jeans's earlier work (1922).

hypothesis of galactic rotation proposed by J. H. Oort and B. Lindblad. The differential galactic rotation seemed to indicate that the disruption of the less dense clusters would be faster than Jeans had said.[8]

It is of interest to notice that here, as elsewhere in astrodynamics, there has always been an uneasy truce between Newtonian methods and those of the General Theory of Relativity. This is evident, for example, in the writings of Lemaître, S. Chandrasekhar, B. J. Bok, and F. Zwicky on the formation and stability of nebular clusters— writings which are also aimed at providing a lower limit to the 'age of the universe'. The method, briefly, is to assess the numbers of members and their masses, the velocity-dispersion, and radius, of a cluster of nebulae, and thereby to deduce a half-life for the cluster. With a few plausible assumptions as to its origin, an upper limit to its age might also be estimated. Taking the well-known cluster in Virgo, it was found that, assuming the 'short time-scale' of 10^9 or 10^{10} years, agreement with the Newtonian theory of gravitation could only be maintained if the mass of the cluster were of the order of 10^{14} solar masses. Sinclair Smith, in 1936, tentatively arrived at such a figure,[9] but in the general opinion it was too large by a factor of 10 or 100. Almost all the information used has since been drastically revised, but at least as recently as 1944 the age of the Virgo cluster was quoted as 10^{11} or more years.[10] Such arguments as were given, however, were very loosely bound to the main stream of cosmology, ignoring almost entirely the question of the formation and evolution of the nebulae themselves, as referred to a particular model universe. Investigations of the latter subject were not wanting—between 1931 and 1937 McVittie, Lemaître, Synge, V. V. Narlikar, D. N. Moghe, and N. R. Sen made valuable contributions to it—and yet even Lemaître, who began with relativistic cosmology, eventually fell back upon a Newtonian dynamics to which was added a field of 'cosmical repulsion'.[11] (His early work was, however, orientated in the direction of the cataclysmic explosion of the primeval atom and, by the nature of his argument, the age he assigned to the

---

[8] See Appendix, Note XII.

[9] *Ap. J.*, lxxxiii (1936) 23. A mistake which was frequently made in this connexion was to identify the average mass of the members of a group of galaxies with the mass of some galaxy judged to be 'average' on other grounds. An inherent difficulty is that the luminosity per unit mass varies considerably between different types of star, much of the mass of a nebula being in the form of non-luminous matter. Cf. Shapley on population (*P.N.A.S.*, xix (1933) 389) and Humason on radial velocities and their dispersion (*Ap. J.*, lxxxiii (1936) 10).

[10] See Appendix, Note XIII for references to work by Chandrasekhar (10^{11} years). Cf. Tuberg, *Ap. J.*, xcviii (1943) 501 (10^{12} years).

[11] See Appendix, Note XIV for references to work by these authors.

nebular clusters was derived from considerations of the 'age of the universe' rather than conversely.)[12]

A further 'lower limit' argument (probably the strongest, historically speaking) in favour of the 'long' time-scale was founded upon the apparent equipartition of stellar energy. In 1911 Halm argued that the kinetic energy of each star was approximately the same.[13] Seares, eleven years later, independently discovered the same rule, although he noted that B-type stars were to be excepted, having only half the usual energy. Disregarding these exceptions and assuming that the state of equipartition was produced by stellar encounter, an age of 5.10^{12} years or more was deduced for the Galaxy. Jeans regarded this argument as suggesting that, as the universe cannot (as he supposed) have been expanding for more than 10^{11} years, there must have been a previous period of contraction.[14] As we have seen, this was much the same as de Sitter's view. Jeans's argument has one major drawback: if the state of affairs was achieved by stellar encounters then it should be a state in which *all* stars are on an equal footing, and we have seen that this is not so. Since it was first announced, many other objections have been made to it.[15] Other processes than direct encounter of star with star are probably involved—for example, a process whereby interstellar material is accreted more rapidly by slow-moving than by faster stars. The dynamical arguments for the age of the Galaxy are not generally thought capable of inspiring much confidence. A conclusion which seems unavoidable is that the various 'ages' quoted in many of the works cited would have been quite different had it not been for the influence of the all too simple interpretation of Hubble's Law.

Three more important ways which have been used to assign 'a lower bound to the age of the universe' remain, all closely related. The first of these is concerned with the Earth or the planetary system. The age of terrestrial rocks and meteorites, as determined by the relative abundance of radioactive materials; the age of the atmosphere and of the oceans; an upper limit to the age of the Earth-Moon system, estimated from a knowledge of the effects of tidal friction: the series of values derived for all these quantities has shown a remarkable consistency.

The second way draws upon the theory of stellar evolution, a theory which was insufficiently advanced before the post-war period to be of much interest to cosmology. Evidence as to the ages of the stars of the

---

[12] See Appendix, Note XV, for views on the status of the nebular clusters.
[13] *M.N.*, lxxi (1911) 634.
[14] See, for example, *P.R.I.*, xxix (1936) 65 and *Nat.*, cxxxvii (1936) 17.
[15] A. N. Vyssotsky, *Ap. J.*, xcvii (1943) 381 and xcviii (1943) 187 (with Williams).

Galaxy came only with the recognition of one of the two principal sources of stellar energy (namely the carbon-nitrogen cycle, the cyclic nuclear reaction converting H to He) by H. Bethe[16] and C. F. von Weizäcker[17] in the late 1930s. The ages of the majority of stars now appear to be of the same order as the age of the planetary system, although once again the stellar masses upon which the evolution depends must be increased, to some extent, by the continual (but so far undetermined) accretion of interstellar matter.

The third way of assigning a lower bound to the age of the universe relies upon those theories whose aim is to account for the history, distribution and relative abundance of the elements. These theories might be said to have originated with an article written by W. D. Harkins in 1917. He drew attention to the connexion between the relative abundance of the elements and the stability of atomic nuclei.[18] Tolman followed with a study of the thermodynamic equilibrium between H and He, concluding that if equilibrium were to be attained in a reasonable time, the temperature and density of the system must be high.[19] S. Suzuki independently arrived at much the same conclusion some years later,[20] and by the early 1930s this vast subject ('equilibrium theory') was under way.

Equilibrium theory soon encountered difficulties. We do not intend to go far into a discussion of this subject, which is more properly a branch of astrophysics, yet it is interesting to note the way in which cosmology, and especially many of Lemaître's ideas, affected it.[21] It was found impossible to account for the observed abundances of the elements in terms of an equilibrium state at a single density and temperature.[22] Now although general agreement could be obtained by making density and temperature suitable functions of position, there remained the difficulty of explaining the *distribution* of the elements throughout space. The outbursts of novae seemed an unlikely means, for they chiefly involve the ejection of material from the surface, where the proportions of the elements are different from those of the majority of stars. Supernovae, involving the disruption of the whole star, also appeared to be unsuitable. When it seemed unlikely that these

[16] *Phys. Rev.*, lv (1939) 434.   [17] *Phys. Rev.*, xxxix (1938) 633.
[18] *J. Am. Chem. Soc.*, xxxix (1917) 856. 'The evolution of the elements and the stability of complex atoms'.
[19] *J. Am. Chem. Soc.*, xliv (1922) 1902.
[20] *Proc. Math. Soc. Japan*, x (1928) 166; xi (1929) 119; xiii (1931) 277.
[21] Cf. Chapter 11.
[22] R. A. Alpher and R. C. Herman give a very full account of the entire subject. (*Rev. Mod. Phys.*, xxii (1950) 153.)

difficulties would be resolved, astrophysics looked to the early, highly condensed, high temperature states of the expanding universe for conditions favourable to the formation of the heavier elements. 'Non-equilibrium theory' was first investigated in the late 1930s and has since met with some success.[23] According to Alpher and Herman its most important failure was in the range of atomic weights from 5 to 8. Its early success was largely due to the failure of its rival to find regions in which the temperature was sufficiently high (10^9 °K or higher) to account for the origin of the elements beyond helium. It is now known that the presence of neutrons in the nuclei of stars at the red-giant stage of evolution can explain the transmutation of light into heavier elements at lower temperatures.[24] The whole subject is mentioned here as one of the few branches of astrophysics to have taken into account, even to have integrated itself with, theories of cosmology. This it was bound to do if it was to specify a suitable 'pre-stellar state of the universe'. Even so, as far as their assisting in the choice of an appropriate lower bound to the age of the universe is concerned, both equilibrium and non-equilibrium theory alike are of little help, being themselves largely influenced by the age limits obtained in ways already outlined. They may be said to support these findings to the extent that the processes which they assume to take place can usually be accommodated within a time of 10^9 or 10^{10} years.

The oppression of the time-scale difficulty was not completely thrown off until 1958, although it was certainly never felt so seriously as in the two or three years during which Jeans's figure of 10^{12} years was almost universally accepted. Now considering that observations relating to spectral displacement and distance were in the 1920s made for the most part without reference to theoretical cosmology, the most surprising feature of the situation is that those immense difficulties of technique which had stood in the way for half a century were removed at the very time when cosmology called upon the observations in question. This coincidence perhaps explains the severity with which the time-scale difficulty was felt. There was a lack of scepticism shown towards the findings of extra-Galactic astronomy in the 1930s which can perhaps be explained in part by the fact that there was only one telescope in the world really capable of being used to this end. Whatever the

[23] G. Gamow and E. Teller, *Phys. Rev.*, lv (1939) 654. Further early investigation was undertaken by Gamow, Teller, Walker, von Weiszäcker, E. Fermi, W. S. Smart, and especially by Alpher and Herman.

[24] E. M. and G. R. Burbidge, W. A. Fowler, and F. Hoyle, *Rev. Mod. Phys.*, xxix (1957) 547.

explanation, criticism of the empirical value of the Hubble factor was not often forthcoming, and it appears at times that astronomers were more concerned with bringing the rest of astrophysics into line with the parameter in Hubble's Law than the other way about.

2. THE REALITY OF THE EXPANSION. ALTERNATIVE EXPLANATIONS OF THE RED-SHIFTS

Is the expansion of the universe real? The question, commoner twenty or thirty years ago than now, is doubly confusing; for 'expansion' may refer either to a general increase in the distance separating the nebulae, or to a certain kind of dependence of the metric upon time. Moreover, to question the reality of the expansion may be to ask whether one theoretical explanation of what we interpret as an expansion is not better replaced by another; or it may be to question our rules of correspondence—to ask, in short, whether we are not using the words 'distance' and 'velocity' in unreasonable ways.

Now one very obvious way of dispelling the time-scale difficulty, and a way not infrequently resorted to, was to assert that the observed radial velocities were spurious. Indeed, no sooner was Hubble's paper of 1929 published than Zwicky followed it with just such a claim.[25] He dismissed explanations of the observed spectral shifts both in terms of the influence of the galactic gravitational field and of a true Compton effect with free electrons in interstellar space. He pointed out that the former (which is likely to be too small for measurement) would not be related to the distance of the observed galaxy, whilst the number of collisions which would be required on the second hypothesis, would render interstellar space 'intolerably opaque'.[26] He proposed instead a new phenomenon, a supposed gravitational analogue of the Compton effect. The possession of inertial and gravitational mass by light quanta means, he said, that in passing a gravitational material they will not only be deflected, but will transfer momentum and energy to this material. The loss of energy being accompanied by a lowering of frequency, with a few plausible assumptions concerning the distribution of gravitating matter in intergalactic space a law not unlike Hubble's was, to Zwicky's satisfaction, explained.[27] Observations of some importance to Zwicky's hypothesis were those which allowed a correlation

[25] *P.N.A.S.*, xv (1929) 773 and *Phys. Rev.*, xxxiii (1929) 1077.
[26] Even were it not opaque the galactic images would be much too widely diffused. With heavier particles than electrons the position would be much worse.
[27] $\Delta\nu/\nu = (1{\cdot}4\pi\gamma/c^2\rho D) \times L$ see Zwicky, op. cit., p. 778. D gives a measure of the distance over which the drag operates.

of the distances of objects *within the Galaxy* with their recession-velocity. On the recessional hypothesis the velocities should be too small to be determined—some would have maintained that there was likely to be no systematic recession of objects within a galaxy at all—but according to Zwicky's hypothesis an appreciable effect was to be expected. E. von der Pahlen and E. Freundlich of Potsdam had correlated the 'radial velocities' of globular clusters with Galactic latitude some years before, and now P. ten Bruggencate argued for an increase in these velocities with decreasing latitude,[28] thus supporting Zwicky.

Zwicky held to this form of explanation of the shifts in galactic spectra. In 1935 he discussed statistical means of treating relevant spectroscopic information;[29] and then, and again in 1954,[30] he emphasized that if his hypothesis were correct the variation of emission line width with nebular distance would differ from that obtaining with absorption lines. Discrepancies do appear to exist—emission lines seem to be sharper than absorption lines—but the evidence is very inconclusive and the actual spectral broadening does not appear to be sufficiently great for Zwicky's argument.[31]

A progressive interaction between light and matter was by no means the only alternative to the theories of an expanding universe. Within a few years of Hubble's announcing the 'velocity-distance relation', several other interpretations were forthcoming. J. Q. Stewart, for example, proposed on intuitive grounds that light quanta may be subject to 'fatigue' during their journey. He drew the conclusion that the frequency of light which has travelled a distance d bears to the frequency with which it is emitted the ratio $1/\exp(d/H)$.[32] This is not altogether different from another recurrent theme, namely the attribution of a secular variation to one or more of those physical quantities usually regarded as constant. P. I. Wold, for example, investigated the properties of a wave travelling in a medium in which the velocity of

[28] *Phys. Rev.*, xxxiii (1929) 1077 (Abstract).
[29] *Phys. Rev.*, xlviii (1935) 802.
[30] *Helv. Phys. Acta.*, xxvii (1954) 481.
[31] Some experimental evidence was presented in 1936 that the observed spectral displacements were not caused by light traversing a highly ionized gas with free interstellar electrons. (R. J. Kennedy, W. Barkas, *Phys. Rev.*, xlix (1936) 449.) See also E. Schrödinger, *Nuovo Cim.* (10), i (1955) 63. H. S. Shelton recently proposed an argument not unlike Zwicky's (*Obs.*, lxxiii (1953) 159, 243. Reply to first paper by R. d'E. Atkinson ibid., p. 159.) He suggested that in a collision between photons and elementary particles energy may be given to the particle only in the direction of motion of the photons. No evidence was given in support of such a process.
[32] *Phys. Rev.*, xxxviii (1931) 2071. Cf. H. E. Buc, *J. Franklin Inst.*, ccxiv (1932) 197. H resembles our c/h.

light is a function of the time, and showed that a displacement towards the red of the lines in nebular spectra is a consequence of the hypothesis.[33] For simplicity he supposed the variation to be of the form $c = c_0(1 - Qt)$, c_0 and Q being constants, but he pointed out that the decrease would be unlikely to continue indefinitely and that the variation may prove to be periodic.[34]

Wold's hypothesis was actually little more than old wine in new bottles, for it was already realized that the non-stationary line-element

$$ds^2 = c^2 dt^2 - \exp(\eta t)(dx^2 + dy^2 + dz^2) \tag{1}$$

may be interpreted as requiring that the velocity of light diminishes with time.[35] The line-element, in a form approximating to that for our own locality at times not far removed from the present, can be written

$$ds^2 = c^2 dt^2 - (1 + \eta t)(dx^2 + dy^2 + dz^2), \tag{2}$$

the velocity of light being then ($ds = 0$)

$$c(1 - \eta t/2). \tag{3}$$

A knowledge of the present value of the Hubble factor (h_1) was sufficient to give Tokio Takéuchi, for example, a value for η of 10^{-10} or 10^{-16} yr^{-1}. He too added that the velocity of light need not tend to the limit zero, 'for the universe may be cyclic'.[36] G. E. J. Gheury de Bray[37] and V. S. Vrkljan[38] had pointed out four years earlier that the velocity of light appeared to be diminishing with time, and quoted as evidence the various determinations of the velocity over a period of fifty years.[39] Vrkljan, seemingly aware of only the Einstein and de Sitter line-elements, had remarked that the General Theory of Relativity could account for the decrease if it were supposed that the g_{mn}, at least in the region surrounding the Earth, were functions of the time.

Considering the expression for the energy of a photon (hc/λ), it is clear that apart from a reduction in the value of c a loss of the energy

---

[33] *Phys. Rev.* (2), xlvii (1935) 217. Cf. F. L. Arnot, *Time and the Universe* (Sydney, 1941) p. 40.

[34] Wold presumably disliked the idea of light coming to rest altogether! Hubble's data would require $Q \simeq 1 \cdot 81 \times 10^{-17}$ sec^{-1}.

[35] Cf. Takéuchi, *ZS. Phys.*, lxix (1931) 857; *Proc. Phys.-Math. Soc. Japan*, xiii (1931) 178. The idea is discussed in less explicit terms by earlier writers.

[36] See p. 133, note 61.

[37] *Astr. Nachr.*, ccxxx (1927) 449, cf. *Nat.*, cxxvii (1931) 522, 739, 892.

[38] *ZS. Phys.*, lxiii (1930) 688.

[39] Cf. Thomson and Tait, *Natural Philosophy* (1874), i, p. 403. The quoted figure was of the order of 8 km/sec per century, considerably greater than that indicated by Takéuchi and Wold.

of the photon can be accounted for in terms of a decrease in time of h, the quantum of action. This solution was proposed first by J. A. and B. Chalmers[40] and later by S. Sambursky.[41] Again, a secular increase in the wavelength of the light would secure the same conclusion, and in a sense Milne and Whitrow were arguing for this when they showed that it was possible to explain the shifts in the spectra of extragalactic sources in terms if the divergence of t- and τ-scales of time over long periods of light travel.[42] Lemaître, some years before, had put aside any explanation of the expansion which relied upon a change in the atomic constants arising from 'some artificial change of gauge'.[43] In a passage which owed something to Eddington he pointed out that 'expansion of the universe is in some sense relative: it is relative to the whole set of essential properties of matter being assumed to be constant'.

The hypotheses so far considered all involve a progressive degeneration in the energy of the quanta of radiation; but, with the exception of the last, they were only loosely connected with particular theoretical schemes. Each could have been added, without much difficulty, to almost any of the theories which predicted a 'true nebular recession'. That they were not, on the whole, well received, was perhaps due in part to the fact that the astronomer already had more undetermined parameters than he could evaluate: in part it was that they were generally acknowledged to be *ad hoc* and even, perhaps, perverse.

Against the arguments for a progressive reddening of the light from distant nebulae stand those which support the hypothesis of a general nebular recession. Those which proceeded from the General Theory of Relativity had the inestimable good fortune of being first, but those originating in the work of Whitehead, Milne, Bondi, Gold, and Hoyle involve, not complete originality, but at least a measure of self-sufficiency. By contrast with these, there is little overall merit to be found in the so-called 'classical' explanations of the 'recession phenomenon'. We have already referred to the hypothesis of a small repulsive force in addition to Newtonian attraction.[44] This is only one of many classical explanations. Consider, for example, a proposal by R. Gunn

---

[40] *Phil. Mag.*, xix (1935) 436, supplement.
[41] *Phys. Rev.*, lii (1937) 335. D. H. Wilkinson maintains that Planck's constant could change by 1 part in 10^{12} per year, at the most.
[42] *ZS. Ap.*, xv (1938) 263.
[43] *Nat.*, cxxviii (1931) 704, reprinted in Lemaître, *The Primeval Atom* (Van Nostrand, 50) p. 80.
[44] G. Armellini investigated the form of the nebulae under this hypothesis (*Rend. Lincei*, xviii (1933) 342).

to the effect that 'the asymmetrical radiation of mass from a star which has broken up' would provide the star with momentum, whilst 'a secular and statistical increase in the kinetic energies of all the stars implies an expansion of the system'.[45] 'The mechanisms developed and applied to the Galaxy are, with certain restrictions, applicable to nebulae', he added, but he did not explain himself further.[46] Consider also the much more sophisticated explanation put forward by M. S. Eigenson of Leningrad.[47] He assumed a distribution of galaxies isotropic about the Galaxy and moving under mutual Newtonian attractions. Each of the galaxies was assumed to radiate mass and, as a result, it was found that each galaxy must recede from our own with a velocity proportional to its distance. The rate of loss of galactic mass calculated from the observed rate of expansion was found to be of the same order of magnitude as the rate of loss of mass by radiation from objects within the Galaxy. Eigenson used the results of a paper by Karl Pilowski (on the influence of a secular loss of mass on the dynamics of stellar systems)[48] which, in turn, rested heavily upon Jeans's work.[49]

In favour of the modified Newtonian argument it could have been said to 'explain' the nebular recession, however badly, even though not originally laid down with this in mind. But this ignores the conceptual difficulties which Newtonian theories encounter, not to mention the failure of the Newtonian account on both planetary and microphysical scales. Those theories of cosmology which have been most influential, and rightly so, are those which have been concerned with the solution of problems regardless of scale. It is wrong to regard theories of natural cosmology as being formulated solely with an explanation of the redshift phenomenon in mind, as did many of the accounts described at the beginning of this section.

How are we to decide between the alternatives? Several means of doing so have been suggested, although most of them are too specific. Zwicky's line-broadening criterion, for example, cannot decide the issue between theories which hold to a general recession and theories of a progressive reddening of the light from nebulae. Hubble and Tolman, and more recently Whitrow, have, on the other hand, proposed means of deciding this very broad issue. The earlier method depended upon the number-magnitude relation,[50] and distinguished only the expanding

[45] *Phys. Rev.*, xliii (1933) 764 (letter).
[46] I have failed to locate 'the complete paper which should appear shortly' (1933).
[47] *ZS. Ap.*, iv (1932) 224.   [48] *ZS. Ap.*, iii (1931) 53.
[49] *M.N.*, lxxxv (1925) 2, 912. Also *Astronomy and Cosmogony* (1929) p. 298 &c.
[50] See pp. 245 sqq. *infra*.

models of the General Theory of Relativity and the static Einstein universe, the latter with an unspecified means of providing a progressive reduction in the wavelength of light. Whitrow's analysis was still more general, presupposing no particular world-model. We shall discuss the earlier criterion first.

3. The Decision: Empirical Considerations

(For convenience, the hypotheses of expansion and of progressive reddening will be referred to hereafter as A and B. They are not, of course, mutually exclusive.)

In 1935 Hubble and Tolman considered the theoretical (m, N) and (m, θ) relations as derived from the hypothesis A or B.[51] These relations are discussed later in this chapter. Unfortunately it was impossible to make critical observations of sufficient accuracy. For an extended object such as a nebula it was difficult to pronounce upon the relation between photographic and apparent bolometric magnitude;[52] nebular sizes were themselves uncertain, for the decrease in surface brightness from the centre outwards was gradual. Under these circumstances there are, in any case, many different ways of specifying size.[53] Number-counts to a given limiting magnitude were, of course, subject to the uncertainties of the magnitudes themselves, but the real difficulty here was in deciding the theoretical significance of a large scatter in the values of the *absolute* magnitudes. Hubble and Tolman decided that the observations, whilst agreeing with hypothesis B, agreed with A only if space has a positive radius of curvature of about $4 \cdot 7 \times 10^8$ L.Y.—a relatively small figure which implied that most of the universe had already been telescopically explored.

Having failed to draw a definite conclusion, Hubble and Tolman recommended as a means of coming to a decision that measurements be taken on 'the surface distribution of nebular luminosity', adding that 'the results of such a test should be less dependent on the assumed homogeneity of the model than those derived from nebular counts'.[54] This interesting suggestion does not appear to have been pursued, presumably as a result of the obvious difficulty of making the necessary

---

[51] *Ap. J.*, lxxxii (1935) 302 (= *Mt. Wilson Cont.*, no. 527).

[52] The so-called K-term of the magnitudes was later found to be in error by as much as $0^m \cdot 6$.

[53] See C. W. Allen, *Astronomical Quantities*, 2nd. ed., (University of London, 1963) p. 9. It is worth noticing that C. Hazard has recently found that the ratios of optical to radio diameter of certain spiral galaxies (Sb or Sc) is not constant. (*M.N.*, cxxvi (1963) 489.)

[54] *Ap. J.*, lxxxii (1935) 336.

measurements: the images of distant galaxies are of a size comparable with the photometer aperture and the grain of the photographic plate.

The numerical analysis contained in these papers by Hubble and Tolman was criticized by Eddington and McVittie. Briefly, Hubble found (in the last-named paper) that between (m_{pg}) $18^m\cdot 47$ and $21^m\cdot 03$ the number of nebulae brighter than photographic magnitude m (corrected) is given by

$$\log_{10} N = 0\cdot 501 m - 7\cdot 371. \qquad (4)$$

Now had the nebulae been uniformly distributed in an otherwise void Euclidean space then, according to an inverse square law of photometry, the coefficient of m would have been $0\cdot 6$. The discrepancy was explained as a consequence of the increase in wavelength of the received light. Its relation to the first coefficient in a (δ, m) or 'velocity-distance relation' was calculated, but was found to differ from that which he had obtained in the more direct way. He concluded that his error was in assuming hypothesis A. (More properly, it should be said that he had apparently been wrong in assuming hypothesis A *alone*.)

Now in the relation (4), m must be corrected for spectral displacement, and the amount of the correction made was $2\cdot 94\delta$.[55] The theoretically evaluated corrections corresponding to hypotheses A and B are, however, respectively $4\cdot 0\delta$ and $3\cdot 0\delta$, and thus B was clearly favoured by Hubble's working.[56] Eddington was quick to point out that in deriving the coefficient $2\cdot 94$ Hubble had ignored the large probable dispersion of the absolute magnitudes of the nebular sources (an omission which, as it happened, proved insignificant) and had overlooked a probable error of at least $0\cdot 56$.[57] Neither criticism was wholly merited,[58] but both served to underline the uncertainties involved in this type of argument.

McVittie at first confined himself to a criticism of Hubble's method of extrapolating from the observations—the empirical (δ, m) relation was established only as far as $m = 17$, whilst (4) held down to $m = 21$.[59] Hubble remained unconvinced. Yet even using Hubble's method of extrapolating the (δ, m) relation, McVittie found (as he had found before) indications of a hyperbolic rather than a closed universe.[60]

---

[55] The amount of the correction (Δm) was calculated as follows: assuming that Δm is a linear function of distance, $\log \Delta m = 0\cdot 2\,(m - \Delta m) + b$, and the corresponding number-magnitude relation for a uniform distribution is given by $\log N = 0\cdot 6(m - \Delta m) + c$. Using the last equation, together with (4): $\Delta m = 0\cdot 165 m + (7\cdot 371 + c)/0\cdot 6$. Now the 'ideal' velocity-distance relation is $\log \delta = 0\cdot 2(m - \Delta m) - 4\cdot 239$; and hence $\Delta m = 2\cdot 94\,\delta$.

[56] For reference to method see Appendix, Note XVII.   [57] *M.N.*, xcvii (1937) 156.
[58] See Hubble, ibid., p. 506.   [59] ibid., p. 163.   [60] *ZS. Ap.*, xiv (1937) 274.

This was so, however, only so long as the average galactic mass was no greater than 2×10^{10} solar masses: increasing this value by a factor of 10 led to the possibility of a spherical space. Once again the uncertainties of the observations were seen to be of the greatest significance.

The controversy continued for the next three years, and out of the discussion came a far greater respect for quantities which had previously been 'omitted in approximation'.[61] Perhaps the most important contribution to it was that in which McVittie demonstrated that Hubble's 'semi-empirical' formulae (for δ and N as functions of m) implied arbitrary assumptions which restricted the choice of model.[62] Writing the equations

$$\log_{10}\delta = (m - \Delta'm) + \alpha_2 \tag{5}$$

$$\log_{10}N = (m - \Delta m) + \alpha_4 \tag{6}$$

(where $\Delta'm$ and Δm are the corrections), Hubble seemingly made the following four assumptions:

(i) $\Delta'm = 3\delta$.

(ii) Equation (5) is valid down to $21^m\cdot 03$.

(iii) $\Delta m \propto$ nebular distance. This was expressed as:

$$\log_{10} \Delta m = 0\cdot 2(m - \Delta m) + \text{const.} \tag{7}$$

(But, as McVittie remarked, 'Hubble does not enter into the thorny problem of distances'.)

(iv) $\Delta'm = \Delta m$.

Using these, together with the results of his observations, Hubble calculated the coefficients first in (5) and then, discarding the value 3δ for $\Delta'm$, in (6) and (7). He also deduced that

$$\Delta'm = \Delta m = 2\cdot 94\delta. \tag{8}$$

The assumption (iv) was probably Hubble's most serious mistake, for its intuitive appeal was so strong. It is very natural to assume that there is a single 'corrected apparent magnitude' which is to be used in the separate equations relating the recorded magnitude, first to δ and then to N; but as McVittie showed, this implies that h_1 and h_2 are related by the equation[63]

$$h_1^2 = h_2. \tag{9}$$

---

[61] See W. Fricke, ZS. Ap., xvi (1938) 11 and McVittie's reply at p. 21.
[62] Proc. Phys. Soc., li (1939) 529.
[63] See also McVittie, G.R.C., pp. 172-3.

This restriction upon the model was totally unnecessary, and McVittie has shown two other ways of reducing the data of observation whereby it is avoided.[64] Hubble, realizing that the small radius and high density of the universe could be avoided if δ appeared in the first rather than the second power in the function for the rate of reception of energy from a distant source, made the amendment on grounds which are acceptable in the kinematic models but not in those of the General Theory of Relativity.[65] These examples illustrate well the dilemma of the observing astronomer who was now confronted with a range of theories each with its own means of reducing the observations. This has never been an obvious situation in the other natural sciences, and it was less obvious still in astronomy. The astronomer, until thirty years ago, conceived his task to be primarily that of discovering new phenomena and of evaluating the parameters of a theory which, except in its detail, was almost universally accepted. By the end of the 1930s, in cosmology at least, the intrusion of theory into the realm of fact was slowly becoming appreciated.

At the time of these early attempts to decide the nature of the redshifts, the distances of the nebulae examined were not such as to merit concern over any change with time in the energy they radiate. E. Schatzman was one of the first to draw attention to yet another uncertainty in the relation between nebular counts and luminosities, namely that the luminosities are likely to decrease with time, perhaps as a result of the evolution of giants and dwarfs.[66] It has more recently been realized that a plausible theory of galactic evolution is of the first importance in deciding between steady-state and evolutionary models.[67] Hubble and Tolman, it is true, made a passing reference in their 1935 paper to the probable significance of galactic evolution, but up to 1939 it was the exception rather than the rule for the physical properties of

[64] op. cit., pp. 170–2. McVittie, in his paper of 1939, apparently made an unwarranted assumption about the coefficients of the expansion in δ of the K-term. Corrected, this would affect his inference to a negative curvature. Heckmann supported the findings of Hubble and Tolman, but pointed out that the result was very sensitive to small observational errors. (*Theorien der Kosmologie* (Berlin, 1942) pp. 60 ff.)

[65] McVittie and other supporters of the latter theory criticized him on these grounds. He was also criticized by Milne and Whitrow (*ZS. Ap.*, xv (1938) 263) for his apparent belief that a density distribution which appears to increase radially outwards must entail that the observer is in a favoured position.

[66] *Ann. astrophys.*, x (1947) 14. On calculating the decrease with time see H. A. Bettie, *Phys. Rev.*, lv (1939) 434. On the giant-dwarf hypothesis see M. A. Greenfield, *Phys. Rev.*, lx (1941) 175.

[67] See W. Davidson, *M.N.*, cxix (1959) 54.

the galaxies to be taken into account in cosmological work. (Of perhaps equal significance is the observer's selection bias which might be called an 'apparent evolutionary effect': of two sources at an equal distance the observer is more likely to include the brighter in his counts.)

Yet another instance of the way in which one might decide between the hypotheses A and B (the hypotheses of expansion and progressive reddening) was provided by S. N. Milford, who suggested that a decision as to the nature of the red-shift could be made if the relation between this and the rate of decline of the light-curve of distant supernovae could be found.[68] Milford's idea was a natural extension of one taken from a paper written many years before by McCrea, who observed that any periodic phenomenon taking place within a galaxy—for example the fluctuation in Cepheid luminosity—must show the same change in frequency as the radiation itself, as long as this change in frequency is a velocity effect.[69] This had not been proposed as a means of deciding between the different hypotheses about the red-shifts, but in 1939 O. C. Wilson suggested that a time dilation in the declining light curves of galactic supernovae might easily provide such a test. Milford, unaware of Wilson's suggestion, now revived the idea and added one or two variants. In one of these he found a rather complicated expression which compared, in terms of an expanding relativistic model, the apparent rate of variation in luminosity of two similar sources at different distances from the observer. In order to use this expression to decide between hypotheses A and B it is necessary to be able to recognize a class of (necessarily bright) objects which are 'reasonably likely to have identical intrinsic light curves'.[70] Milford proposed Type I supernovae (Cepheids lacked the necessary brightness). He then enlarged upon what was orginally McCrea's idea, deriving the relation

$$F_2/F_1 = (1 + \delta_1)/(1 + \delta_2), \tag{10}$$

where F_1 and F_2 are the frequencies at two places of 'a particular type which is the same at all times and all places', and δ_1 and δ_2 are the red-shifts in the spectra of the nebulae in which they are situated.[71] The relevant observations must of necessity be imprecise, as can be seen from the list of periodic effects visible at great distances—the frequency of occurrence of novae and supernovae and of irregularities in their light curves.[72] Milford believed (and there seems to be no counter

[68] *Ap. J.*, cxxii (1955) 13.   [69] *ZS. Ap.*, ix (1935) 290.
[70] Milford, op. cit., p. 19.   [71] loc. cit.
[72] Perhaps it will shortly be possible to add quasar fluctuations to this list. See p. 248 *infra*.

evidence) that these tests would be free from the evolutionary objections which were raised against the principle of galaxy-counts.

One of the most valuable contributions to the problem of interpreting the shifts in the spectra of extragalactic nebulae was made by Whitrow in 1954.[73] Assuming no particular world model, he pointed out that whilst the hypothesis of progressive reddening requires only a decrease in the apparent energy if each quantum of radiation, yet the hypothesis of recession requires, in addition, a decrease in the rate at which quanta are received, as well as an increase in the solid angle in which they are emitted. (Although, ostensibly, those theories which attribute the shifts to a secular variation in one of the 'constants' of nature fall into the 'progressive reddening' category, yet they are excluded from consideration under this heading since it is often possible to change them into recession theories by a suitable transformation of the scale of measurement and since, also, many of them are time-dependent.)

The analysis which follows is much simplified by the result (then established between λ 3400 and 6600 to an accuracy of within 1 part in 100 by A. E. Lilley and E. F. McClain[74] and to an accuracy of 1 part in 560 by R. Minkowski and O. C. Wilson)[75] that δ is independent of the received wavelength. In addition, such assumptions are made (for the strictly non-recessional models) as that the time-scale is such that the system of galaxies is static; that their spectra are time-independent; that distances obey the linear addition law; that δ is a function of the distance travelled by the received light; that if one observer were to send out light of the same (single) wavelength as that which he received from another, then the light received from both sources by an observer beyond the second would appear to be of the same wavelength, all three being on the same light path (this amounts to saying that red-shifts possess the group property); and that δ is a continuous function of a continuous variable (the distance (r) of the source). From these simple assumptions Whitrow found the following law:

$$\delta = \exp(\eta r) - 1 \quad (\eta \text{ const.}). \tag{11}$$

[73] *M.N.*, cxiv (1954) 180.
[74] See *Ap. J.*, cxxiii (1956) 172. Lilley and McClain, however, thought that they had found the 21 cm H line in Radio Source Cygnus A to be displaced in keeping with the optical shifts. This was later shown to be a mistake: this line is not observable in the spectrum of Cygnus A.
[75] *Ap. J.*, cxxiii (1956) 373.

Writing δ as a series in r:

$$\delta = Ar + Br^2 + \ldots, \tag{12}$$

it can be seen that this requires A^2 to be equal to $2B$. There is no need to recount the list of coefficients which the different recessional models require, for it is sufficient that the recessional hypothesis is known to be compatible with *any* relation between the coefficients. A particular theory may, of course, require A and B to be simply related. Milne's (original) uniformly expanding model, for example, requires that $2B = 3A^2$. The usefulness of Whitrow's analysis[76] must rest, therefore, upon its ability to rule out the progressive reddening hypothesis (given suitable empirical evidence). Clearly it can neither prove the latter to the exclusion of the recession hypothesis, nor can it rule out all of the large variety of (δ, r) relations which are compatible with the hypothesis of recession. It is incapable of giving the former proof since there may well be a theory which, arguing from recession, reaches the conclusion $A^2 = 2B$. But is the method capable of ruling out the strictly non-recessional progressive reddening hypothesis? If Whitrow's five postulates are by themselves a necessary condition of this hypothesis then the answer is Yes, given suitable empirical evidence. With a non-exclusive progressive reddening hypothesis it would appear not, for then it is always open to a supporter of this hypothesis to combine it with a theory of recession. But he might refuse to do so and instead object to Whitrow's restrictions on the concept of distance; for the distances commonly used in cosmology do not all obey the linear addition law.[77] With a non-evolutionary cosmology it would, however, be difficult to sustain this argument.

The sort of analysis which Whitrow gave provides a valuable paradigm for the general problem of comparing the structure of two or more groups of theories. It proceeds from the selection of a small number of hypotheses common to the theories of one group, to the simple differentiae of the groups. It may hope to succeed in ruling out one or more of certain widely accepted hypotheses, but this will always be difficult so long as some supporter is willing to extend, modify or reinterpret them.

[76] The rest of Whitrow's paper is concerned with (*a*) a recessional theory in which the red-shifts are independent of time (as in Kinematic Relativity but not in General Relativity). The difficulty here is that it is necessary to consider a time-independent correlation of shift and *velocity*, and velocities do not obey the linear law of addition. (Whitrow uses Einstein's velocity addition formula.) (*b*) The relaxing of the assumption that δ is independent of λ. In view of the findings of Lilley and McClain, (*b*) is only of academic interest; (*a*) of interest only in illustrating Whitrow's unusual methods.

[77] See Ch. 15.

This matter of interpretation is perhaps the most difficult. Consider, for example, E. Schrödinger's remark that the alternative explanations of the red-shift are not so very different from those based on an expanding universe because, in this case, the shift arises 'by gravitational interaction with the matter which supports the geometry'.[78] Judging by Whitrow's five postulates, this would certainly not be counted a progressive reddening hypothesis (although at one time the de Sitter effect was spoken of in this way). Again, consider a static cosmological model proposed by D. K. Sen,[79] which uses Lyra's modified form of Riemannian geometry.[80] The model resembles the original Einstein universe: it does not introduce the cosmological constant but represents a concentration of matter of finite density. It exhibits the displacements within nebular spectra as a consequence of a property inherent in the geometry and quite independently of any expansion. In this respect it bears comparison with the de Sitter model. There seems to be no reason, in fact, for supposing that the hypotheses of expansion and of progressive reddening (in its usual sense) must always be looked upon as covering all possibilities. Bearing in mind such oddities of phrase as Schrödinger's, however, it seems that the best way of judging the issue is to consider the various explanations (which, we repeat, are not mutually exclusive) so far as possible within a single theory. Here, with observational help, their relative importance can be assessed. (As for the observations, those of the kind which Milford advocated would seem capable of providing an invaluable supplement to the galaxy-counts.) With this in mind, the remaining part of this chapter is devoted to the development of those formulae which lend themselves to empirical test. It will be found that quite regardless of the incorporation of any progressive reddening hypothesis into relativistic cosmology, the undetermined parameters in these formulae are already as many as can be tolerated, given existing observational techniques. It will soon be evident that these formulae are not, on the whole, adapted to deciding the relative importance of hypotheses of expansion and progressive reddening within a given model, but are primarily of value in making possible the choice of a suitable model—usually ignoring progressive reddening entirely.

[78] *Nuovo Cim.* (10), i (1955) 63.   [79] *ZS. Phys.*, cxlix (1957) 311.
[80] G. Lyra, *Math. ZS.*, liv (1951) 52. A summary of Lyra's geometry (which introduces Weyl's gauge function into affine space) is given in Sen's article, pp. 314–17.

4. The Theoretical (δ, m) Relation in Relativistic Cosmology and the Stebbins-Whitford Effect

In Chapter 5 we saw the dangers, when using the parlance 'velocity-distance relation', of regarding velocity as the cause of a classical Doppler effect, of ignoring time dependence in any 'linearity' which may be predicted, and even of presupposing a Newtonian time-variable and the older methods of defining 'distance'. Gradually this sort of inconsistency became recognized—a few writers had certainly borne it in mind from the first—and, by the mid-1930s it had become usual to seek formulae which contained (apart from undetermined parameters) nothing but 'observed quantities'. And under this heading, 'velocity' and 'distance' were certainly not to be included. Tolman made many important contributions in this direction, and although many of his formulae contained terms corresponding to 'coordinate position' and 'luminosity',[81] yet he was clear as to the way in which the first could be eliminated in favour of 'astronomically determined distance' or even of nebular counts. He related the latter, amongst other things, to apparent angular diameter, and proposed some useful means of testing the (relativistic) hypothesis of nebular recession. One such test, for example, was that of determining the constancy or otherwise of the expression

$$\Delta\theta/\{(1 + \delta)^2\sqrt{l}\}, \qquad (13)$$

where l is the luminosity. As yet insufficient attention had been paid to the theory of correcting photographic magnitudes—as we have seen —and, as Tolman realized, 'the test would be complicated by the difficulties of handling the data on diameters'.[82]

One of Tolman's more important innovations was an expansion of the function $R(t)$ as a power series in t around the present epoch,[83] followed by an expression for δ as a function of coordinate position and the coefficients of this power series.[84] The resulting expression

$$\delta = k\bar{r} - l\bar{r}^2 + (k/6R_0^2 + kl/3 + m)\bar{r}^3 + \ldots, \qquad (14)$$

together with an 'empirical' (δ, r) relation, is in principle capable of yielding information as to the evolution of the line-element. In practice only the first coefficient could be given a value—although upper limits could be assigned to the (absolute) values of the next two.

---

[81] See, for example, *P.N.A.S.*, xvi (1930) 511; *R.T.C.*, pp. 462–77.
[82] *R.T.C.*, p. 481. [83] *P.N.A.S.*, xvi (1930) 409.
[84] *R.T.C.*, p. 472. The constants k and l are respectively equivalent to our h_1/c and $(h_2 - h_1^2)/2c^2$.

A satisfactory single equation relating δ to a series in m was not forthcoming, however, until 1938, when McVittie derived it.[85] He had previously criticized W. Fricke, who had taken an inexact form of the (δ, m) relation in support of Milne's Kinematic Relativity.[86] It seems that many people thought a high degree of theoretical accuracy to be unimportant whilst the astronomical evidence was subject to great uncertainties. Heckmann, however, extended McVittie's methods,[87] and these have since been acknowledged as providing the most suitable way of comparing theory and observation.[88] Of late, discussions of the (δ, m) formulae have turned on a possible evolutionary effect in the power of the sources[89] and on the correction of the measured magnitudes. Broadly speaking, McVittie now proceeds as follows: expanding an expression for δ in powers of the luminosity-distance D, as far as the term in D^2, he obtains[90]

$$\delta = (h_1 D/c)[1 - \{(h_1^2 + h_2)/2h_1^2\}(h_1 D/c) + \ldots] \quad (15)$$

Luminosity-distance (D) and apparent magnitude (m) are connected, according to McVittie, by the formula[91]

$$\log_{10} D = 0 \cdot 2\{m - K - (M_0 + \Delta M)\} + 1, \quad (16)$$

where

$$K = K_1 \delta + K_2 \delta^2 + \ldots \quad (17)$$

and where M_0 is the absolute magnitude of a galaxy in the neighbourhood of the observer, judged to be similar to the galaxy under observation.

The most controversial term in this expression is ΔM, the 'Stebbins-Whitford correction', whose origin will be explained. J. Stebbins and A. E. Whitford believed in 1948 that they had found a reddening in the spectra of distant elliptical galaxies in excess of that found in nearby galaxies, when due allowance was made in both cases for the shifts in the spectral lines.[92] Two explanations were offered which relied, respectively, on assumptions of selective intergalactic absorption and

---

[85] *Obs.*, lxi (1938) 209. Cf. *Proc. Phys. Soc.*, li (1939) 529.
[86] Fricke, *ZS. Ap.*, xvi (1938) 11; McVittie, ibid., p. 21.
[87] *Theorien der Kosmologie* (Springer, Berlin) 1942.
[88] See, for example, *G.R.C.*; Robertson, *P.A.S.P.*, lxvii (1955) 82; W. Davidson, *M.N.*, cxix (1959) 54.
[89] Milne was probably the first writer to include in his formula a function which took into account the evolution of nebular sources. (*ZS. Ap.*, vi (1933) 90; *W.S.*, p. 354.)
[90] McVittie, op. cit., p. 163. Cf. the equation (14).
[91] ibid., p. 157. For further remarks on the K-term see Appendix, Note XVII.
[92] *Ap. J.*, cviii (1948) 413. Cf. Stebbins, *M.N.*, cx (1950) 416.

of evolution of the E-type nebulae, perhaps as a result of the fading of red supergiants. The consternation to which this announcement gave rise[93] was to some extent reversed when, eight years later, Whitford and A. D. Code showed that the observed effect arose perhaps entirely from a mistaken method of observation.[94] They now held that at the very most it was only a tenth as important as previously estimated.

By the time of Whitford's second announcement, McVittie had offered an explanation of the effect in terms of the astronomer's natural but mistaken choice of comparison sources. Two separate definitions can be given of a source at a luminosity-distance of 10 pc from the observer, of a definite area and spectral distribution function, and without spectral displacement. The two definitions of the standard differ in the assigned temperature. If the distant nebula has a temperature $T(t)$ and if t_1 is the time at which light is emitted, to be received at time t_0, then the temperature of the standard may be defined as either $T(t_1)$ or $T(t_0)$—and it is the standard with a temperature $T(t_0)$ which an observer would arrive at from an examination of the light sources in his neighbourhood, whereas the physical condition of the distant nebula corresponds to a standard with temperature $T(t_1)$. The term ΔM, which was necessary to correct for this mistaken choice of standard, was evaluated in terms which involved the temperature of the nebula—an unknown function of time.[95] It was shown on general grounds that it must be possible to write ΔM in the form

$$W_1\delta + W_2\delta^2 + \ldots, \tag{18}$$

where W_1 and W_2 are constants.[96]

The various results were now collected to give, after some working, the following important relation:[97]

$$0 \cdot 2m = \log_{10}[(c/h_1 A)\,\delta + (c/h_1 A)\{(h_1^2 + h_2)/(2h_1^2) \\ + 0 \cdot 4606(K_1 + W_1)\}\delta^2], \tag{19}$$

where

$$A = 10^{1-0 \cdot 2M_0}\,\text{pc}. \tag{20}$$

This expression, which does not differ radically from that which McVittie gave in 1938, provides one of the most important points of

---

[93] See Bondi, Gold and Sciama, *Ap. J.*, cxx (1954) 597 and de Vaucouleurs, *M.N.*, cxiii (1952) 134.
[94] Whitford, *A.J.*, lxi (1956) 352.   [95] McVittie, op. cit., pp. 156–67.
[96] op. cit., p. 164.   [97] ibid., p. 165.

contact between relativistic cosmology and astronomical observation. It is far more important than those theoretical expressions of, for example, the 'distances' D, ξ, and so on, in terms of the 'velocity' $c\delta$. Such, if not necessary stages in the derivation of relations like that just quoted from McVittie, are as often as not merely concessions to a strong desire to translate unfamiliar ideas into familiar ones. Unfortunately for most of us there is absolutely nothing familiar about the concepts in question.

McVittie chose a (δ, m) relation with m the dependent variable, since the apparent magnitudes are less accurately measured than the redshifts. The coefficients of δ and δ^2 being determined from measurements on δ and m, the values of h_1 and h_2 may be found; h_1 so long as M_0, K, and W_1 are known.[98] One corollary of this is that it should be possible to decide between an accelerated, uniform, or retarded expansion without any knowledge of K_2, W_2, or even the space-curvature constant.

This treatment of the problem has recently been criticized by Davidson who points out that the name 'Stebbins-Whitford correction' is misleading: for even were the Stebbins-Whitford effect absent, the correction (ΔM) could be non-zero.[99] 'This is', explained Davidson, 'because the observations of Stebbins and Whitford would detect only one evolutionary change in the relative intensities at different wavelengths and not an evolutionary change in the absolute intensities.'[100] This is an extremely important point: not until one possesses information which will lead back to the absolute values of the power of the sources at any epoch—information which Whitford's later seven-colour photoelectric analysis does not provide—can one state with any confidence values for h_1 and h_2. It now seems that such information may be drawn from the counts of galaxies. The type of observation which Stebbins and Whitford hoped to make and which Whitford and Code have since obtained is, however, by no means superfluous. It might well provide evidence, for example, unfavourable to the steady-state theories.

5. The Theoretical (N,m) Relation

Of even greater importance than the shift-magnitude relation is the relation between nebula-counts and magnitudes. Its importance lies in

---

[98] Most nearby galaxies appear to have absolute magnitudes lying between -19.9 and -20.9. The values of W_1 and W_2 must be based upon a theory of galactic evolution. See *H.M.S.*, (1956) where an argument is given for taking $W_1 = 0$ (photographic magnitudes).

[99] *M.N.*, cxix (1959) p. 469. [100] ibid., p. 60.

the opportunity it offers for taking much fainter galaxies into account. Einstein appears to have been the first to refer to the opportunities of using expressions involving nebula-counts to test one of the models of cosmology: he did so in 1932 in connexion with the Einstein–de Sitter model.[101] Tolman and Hubble took up the idea and applied it, as we have seen, to the problem of deciding the nature of the red-shift. Their work led to a very valuable discussion on the best means of reducing the observations. One error may be mentioned which, although noted by Eddington when discussing the distribution of nebulae, remained otherwise uncorrected until 1946 when A. Fletcher drew attention to it.[102] Both Hubble and Shapley assumed in their work that the absolute magnitudes of the nebulae are distributed in Gaussian manner about a known mean value M_0, with dispersion (standard deviation) σ. Counts of all galaxies with apparent magnitude less than a definite value favour the intrinsically bright galaxies. The mean absolute magnitude of the observed galaxies must, as Fletcher showed, be corrected not by an amount of $1\cdot 382\sigma^2$ (as Hubble, following K. G. Malmquist,[103] had believed), but by an amount of only $0\cdot 691\sigma^2$ (mag.). This correction meant that current estimates of the number of galaxies per unit volume had to be revised (from about 6 per mpc^3 to about 12).

Surprisingly little attention has been paid to the luminosity-function of the galaxies. The great difficulty here is that observing in an unconsciously selective manner will lead the observer into the mistake of favouring sources which are progressively more luminous the greater their distance. J. H. Bigay has proposed an alternative to the limiting magnitude criterion, namely one whereby galaxies are selected only if they can be resolved into stars.[104] Bigay has carried out the difficult measurements on the magnitudes of sixty-four resolvable galaxies, and has provisionally decided in favour of a Gaussian distribution. These methods obviously cannot be extended to very distant galaxies, and the limiting magnitude criterion is bound to remain. The contributions of J. Neyman and Elizabeth L. Scott are therefore of great value, for they show that one may draw important conclusions from counts made to a single limiting magnitude which need not be accurately known.[105] Neyman and Scott also deserve mention as having introduced

[101] See *R.T.C.*, p. 469.
[102] *M.N.*, cvi (1946) 121.
[103] *Arkiv. Mat. Astr. Fys.*, xvi (1922), no. 23.
[104] *Ann. astrophys*, xiv (1951) 319; *Publ. Obs. Lyons*, v (1952).
[105] *Ap. J.*, cxvi (1952) 144 and, with C. D. Shane, *Ap. J.*, cxvii (1953) 92.

statistical methods into the theory of the spatial distribution of nebulae.[106]

Neyman and Scott's single limiting magnitude need not be accurately known: but most of the immense practical difficulties remain. Shane and others investigated Hubble's measurements of twenty years before and found that not only were the limiting magnitudes different for different types of galaxy, but that they differed from plate to plate and even from one place to another on any given plate.[107] These authors estimated the mean galactic extinction towards the poles at 0·46 magnitudes—nearly twice Hubble's value.

We have hinted at the fact that the (N,m) relation is of great potential value in deciding between various models. How this comes about may be briefly explained. The relation between N_m, the number of nebulae per square degree of sky having apparent magnitudes not greater than m, and m, was given for relativistic cosmology by McVittie in the form

$$N_m = (B_0 + B_1 x + B_2 x^2)x^3, \qquad (21)$$

where $x = 10^{0 \cdot 2m}$, and where the terms B_0, B_1, B_2, if determined from the numerical data, are capable of yielding in succession α/R_0^3 and $1/h_1$, where α is the number of sources per unit volume. Either R_0''/R_0 or the sign of the curvature may be found, given that the other is known.[108] This knowledge can only be acquired if the following assumptions are granted:

(i) All nebulae of the model have the same intrinsic luminosity.
(ii) A common absolute magnitude of these sources can be decided.
(iii) The number of the sources in unit coordinate volume is constant.
(iv) The coefficients K_1, K_2, W_1, W_2, are known.

---

[106] C. D. Shane and C. A. Wirtanen had found that most galaxies must be regarded as members of clusters (*A.J.*, lix (1954) 285). Neyman and Scott introduced the idea into their theory and assumed that the centres of the clusters, rather than the galaxies themselves, were randomly distributed. Uniformity of spatial distribution was now interpreted in terms of the Robertson–Walker metric. If the number of sources of apparent magnitude smaller than a given figure is to be proportional to the cube of any distance, it must be 'distance by volume' (Ξ) which is taken, and this, as McVittie pointed out, 'in a formal sense only, makes N_m increase at the instant t_0 *as if* the galaxies were uniformly distributed and at rest in a Euclidean space wherein Ξ measured distance'. (*G.R.C.*, pp. 173-4 (my italics). The assumption of relativistic cosmology is that the number of nebulae per unit coordinate volume is constant.) The (classical) relation $\log N = 0 \cdot 6m + \text{const.}$ is nevertheless sometimes carelessly referred to as a criterion of a uniform distribution of sources by those whose point of view is ostensibly non-Euclidean.

[107] *A.J.*, lix (1954) 285; lxi (1956) 292; lxiv (1959) 197.

[108] See *G.R.C.*, pp. 169-73.

Bearing in mind the imprecision of the optical counts, the information likely to be yielded must seem slight. It was shown, on the other hand, that if the value of h_1 which is provided from the (δ,m) relation data is used in the theoretical expression for B_1, then $(K_1 + W_1)$ can be determined (assuming that (iv) is ignored, of course). Thus the galaxy-counts (especially using radio-methods) are likely to be of very great importance in distinguishing between the merits of steady-state and evolutionary theories.

Of the first three assumptions much remains to be written. The first two serve to remind us that in cosmology statistical methods were for long neglected where they were needed most. The application of the methods of radio astronomy to cosmology falls outside our period, but it should be added that a form of (iii), namely the assumption of a constant number of radio sources in unit coordinate volume, may very well be unfounded. As W. Davidson has pointed out, the question of the nature of the strong radio sources is highly relevant to a decision one way or the other. Before stating how this is so, one or two remarks may be made on the recent history of the question.

For a time it was widely believed that a large proportion of the most powerful Class II (extragalactic) sources represent a collision between two galaxies. It was rarely possible to identify these sources optically, but many of those which had been identified showed some peculiarity—for example, a double nucleus, an apparent ejection of matter or a collision. On the other hand, the most powerful Class I source, Cygnus A, is inconspicuous to the visual observer, a fact which does not favour the collision hypothesis. In 1962 Hoyle and W. A. Fowler were working on the evolution of stars within galaxies.[109] They decided that clouds of galactic material would normally condense and fragment into smaller clouds, each of which might form a star. But this fragmentation might occasionally fail to occur, in which case a large part of the cloud would contract as a whole, forming a central star of enormous size. This 'quasi-star', or 'quasar', would proceed to grow by gravitational accretion, increasing its mass and hence its ability to attract yet more material from its surroundings. The implosion of this material might well lead to the ejection of jets of material at speeds approaching that of light—an immensely powerful source of radio waves.

At much the same time, Cyril Hazard, using a new technique, found a set of highly accurate coordinates for the radio star 3C 273. These were sent to Palomar, where Maarten Schmidt found photographic

[109] See *M.N.*, cxxv (1963) 169.

evidence of the object, together with a faint wisp of gas to one side of it. The distance of the object was estimated from its spectral shift, and hence its brightness and size were found. A hundred times brighter than the average galaxy, and only a twentieth the size, here was the first object which might qualify as one of the Hoyle–Fowler quasars. Over a score have since been found. Not for the first time in astrophysics we have the observer a very short, but flattering, step behind the theorist.

The Hoyle–Fowler theory is hardly likely to be the last word on the subject, and until the issue is fully resolved the radio-counts must be treated with reserve. The reason for this was given by Davidson in 1959: if most collisions occur within clusters, and if these do not expand with the universe, then assumption (iii) holds (even if the identification of Class II sources with colliding galaxies is made). On the other hand, the formation of clusters might be a recent phenomenon, and it might then be unreasonable to use the assumption for the earlier phases of an evolutionary model.[110] Davidson derived an alternative nebula-count formula for collision-type sources in an evolutionary model. The steady-state (N,m) relation will, of course, be unchanged.

The (N,m) relation for the steady-state theories is particularly simple, for as an immediate consequence of the Perfect Cosmological Principle the number of nebulae in unit proper volume must be constant in time. As Bondi and Gold showed,[111] the number of nebulae with radial coordinates between r and $r + dr$ from which light reaches the origin at $t = 0$ is

$$4\pi r^2 n \, (1 + Hr)^{-3} dr. \qquad (22)$$

It is worth repeating that the factor $(1 + Hr)^{-3}$ would not occur were proper volume replaced by *coordinate* volume, as in relativistic cosmology. The steady-state model is therefore not an exact analogue of the de Sitter model. The steady-state theory must naturally introduce magnitude corrections in the form of the K-term, or some similar expression, and Davidson has given suitable formulae.[112]

Finally, the radio-source counts have proved to be by far the most widely discussed feature of recent cosmology. The controversy began when Martin Ryle and his colleagues presented what they believed to be conclusive evidence from this quarter against the steady-state

[110] *M.N.*, cxix (1959) 665 at p. 675.
[111] *M.N.*, cviii (1948) p. 261. Cf. Bondi, *Cosmology* (2nd ed.) p. 147.
[112] op. cit. p. 677.

theory.[113] Hoyle and Narlikar,[114] and Hanbury-Brown[115] next offered several criticisms of Ryle's conclusions, paying attention to several points of detail in the hypotheses accepted. What about clustering, which Ryle ignores?—and so on. All Ryle's critics made the assumption, however, that the sources under discussion were mainly extra-galactic. D. W. Sciama points out that many of the sources may lie within the Galaxy, in which case Ryle's data are compatible with the steady-state theory.[116] Sciama has described a way in which the necessary Galactic distribution of sources may have orginated.[117] One feature of his model has a somewhat artificial appearance, involving as it does a local deficit in the concentration of Galactic sources. Why should powerful radio sources systematically avoid the Sun? Sciama's is unlikely to be the last word on the interpretation of Ryle's data.[117*]

6. The Theoretical (N,δ) Relation

Neither Milne, nor (so far as we can determine) any other writer, has provided a straightforward (N,m) relation for Kinematic Relativity.[118] McCrea, however derived an expression for Milne's theory relating the difference between the spectral displacements $(d\delta)$ of nebulae at the two boundaries of a spherical shell, at the centre of which the observer is situated, and the number (dN) of nebulae within the shell.[119] This may be written[120]

$$(dN/d\delta) = \pi\alpha(2 + \delta)^2\delta^2/(1 + \delta)^3 \qquad (23)$$

$$= 4\pi\alpha(\delta^2 - 2\delta^3 + 13\delta^4/4 \ldots). \qquad (24)$$

McCrea had previously derived the corresponding formula for relativistic cosmology.[121] Expressing r, the radial coordinate of the shell of the nebulae, as a function of δ, and showing that (with the usual

---

[113] P.R.S. (A), ccxxx (1955) 448; P.R.S. (A), ccxlviii (1958) 289; M.N., cxxii (1961) 349.
[114] M.N., cxxiii (1961) 133 and cxxv (1962) 13.
[115] ibid., cxxiv (1962) 35.   [116] ibid., cxxvi (1963) 195.
[117] ibid., cxxviii (1964) 49. Sciama's explanation involves the rotational instability of gravitationally contracting low mass stars, with the conversion of rotational energy into magnetic energy and the energy of high-velocity particles.
[117*] Notice Davidson's discussion of evolutionary hypotheses in conjunction with Ryle's findings, and their implications for several models. (M.N., cxxiii (1962) 425 and cxxiv (1962) 79.)
[118] But see W.S., p. 354, note 7.   [119] W.S., p. 109.
[120] First found by McCrea, ZS. Ap., ix (1935) p. 307. Cf. M.N., xciii (1933) 668 where Milne gave the relation between dN/dr and r. McCrea's formula is reprinted in W.S., p. 109. Milne's 'Doppler shift ratio' (s) is $(1 + \delta)$. McCrea uses 'D' for the same quantity and 'σ' for our 'δ'.
[121] ZS. Ap., ix (1935) 290.

homogeneous and isotropic form of the metric, using co-moving coordinates)

$$(dN/dr) = 4\pi\alpha r^2/(1 + kr^2/4)^3 \tag{25}$$

(α = the number of nebulae per unit coordinate volume), after much calculation he finally obtained the following formula connecting the $dN/d\delta$ and δ:[122]

$$(1 + \delta) \, dN/d\delta = (4\pi\alpha c^3/R_0'^3) \{\delta^2 + (2/R_0'^2)(2R_0 R_0'' - R_0'^2) \delta^3$$
$$+ (1/12 R_0'^4)(11 R_0'^4 - 46 R_0'^2 R_0 R_0'' + 45 R_0^2 R_0''^2$$
$$- 10 R_0' R_0''' R_0^2 - 4c^2 k R_0'^2) \delta^4 + \ldots\}. \tag{26}$$

It is worth noticing that, to a first approximation,

$$dN/d\delta = (4\pi\alpha c^3/R_0'^3) \delta^2, \tag{27}$$

and that this expression is independent of k, the sign of the curvature, and of the general form of the function $R(t)$. It was, of course, only to be expected from the assumption of homogeneity that for small shifts, when δ is proportional to proper distance, N would be proportional to δ^3. To this order of approximation the formula agrees with that derived by Milne. It must be admitted, however, that the relation is likely to be of little practical value, for in order to decide the coefficients of δ^2 from optical measurements—and hence to deduce a value for $(R_0 R_0''/R_0'^2)$—it is necessary to face all the uncertainties of extrapolating the empirical (δ,m) relation beyond the largest apparent magnitude used in its determination, in order to eliminate m between it and the empirical (N,m) relation. Added to this there are all the uncertainties of the galaxy-counts themselves. With refinement the method may prove useful in ruling out some of the many alternative theories. Here, as elsewhere, the steady-state theory of Bondi and Gold, like Kinematic Relativity, is relatively inflexible: it predicts a relation between N and δ of the form[123]

$$N = (4\pi\alpha c^3/H^3)\{\log_e(1 + \delta) - \delta(2 + 3\delta)/2(1 + \delta)^2\}, \tag{28}$$

giving

$$dN/d\delta = (4\pi\alpha c^3/H^3)(3\delta - 8\delta^2 + 15\delta^3 \ldots). \tag{29}$$

7. Other Criteria

An assumption which has been made in deriving all those expressions here quoted involving galaxy-counts is that the number of sources within a given coordinate- or proper-volume (whichever is appropriate)

[122] This was checked by K. K. Mitra; for the formula see McCrea, op. cit., p. 303.
[123] Bondi and Gold, *M.N.*, cviii (1948) 261.

is constant. Some such theory of galaxy formation as Sciama's may explain how, with continual creation, this state of affairs may be brought about; but in other theories, expanding systems will presumably pass a stage at which the mean intergalactic density is too small to allow galaxy formation, and sources in all regions must then gradually become invisible. As the lifetime of a star is generally supposed comparable with the reciprocal of the Hubble parameter, galaxies at very great distances should nevertheless appear fewer (and younger) than those close at hand. Although the discovery of such an effect would tell against the steady-state theories, taking it into account adds at the same time yet another intricacy to the already complex rival theories which lack the hypothesis of continued galaxy formation. The effect would tend to offset the apparent increase in congestion with distance which all such systems predict.

This last-named effect (which might most clearly manifest itself in the increasing proximity of clusters of galaxies, rather than of galaxies, with distance) provides yet another criterion for comparing the merits of the various theories. Hubble and Tolman gave some thought to the method of relating the observed angular diameters of galaxies to the spectral shifts, but considered it too difficult to apply.[124] Galaxies, as they pointed out, are not conveniently spherical and well-defined, and their photographic images continue to increase in size with exposure.

If a well-defined procedure is to be laid down, the phrase 'apparent radius' must be interpreted in some such way as 'radius to a point at which the intensity has fallen to a given fraction of the central intensity'. A specific class of galaxies—elliptic perhaps—would be chosen, and presumably the 'galactic radius' in the direction of their greatest apparent extension would be averaged over the members of the class within a given cluster. Alternatively, the well-known 'nth brightest' method might be used. It seems unlikely that the relationship between redshift and apparent galactic diameter will assume any great importance, at least as long as radio frequency-shifts remain unobtainable. Hoyle, nevertheless, has discussed the $(\Delta\theta,\delta)$ relation for steady-state cosmology, together with the corresponding relation for the Einstein-de Sitter model.[125] Davidson, it seems, was first with the relevant formula for the steady-state model.[126] This may be written

$$\Delta\theta = \bar{d}\{(1+\delta)/\delta\}H/c \qquad (30)$$

[124] *Ap. J.*, lxxxii (1935) 302 (= *Mt. Wilson Cont.*, no. 527).
[125] *Paris Symposium on Radio Astronomy*, 1958 (Stanford University Press, 1959) 529.
[126] Unpublished Ph.D. thesis (London) 1958.

($\bar{d}$ = mean proper linear diameter or greatest extension, at the epoch of emission, of galaxies at a common distance). The two models lead to radically different forms for the relation. In the steady-state, $\Delta\theta$ clearly tends to a finite lower limit $(\bar{d}H/c)$, whilst in the Einstein–de Sitter universe Hoyle found that $\Delta\theta$ decreases to a minimum value (when $\delta = 5/4$), after which it increases. Yet another kind of variation is entailed by Milne's theory, in which $\Delta\theta$ decreases monotonically to the value $(2\bar{d}H/c)$ at the observable horizon.[127] Each of the three models predicts, in this respect as in so many others, a closely circumscribed mode of behaviour which contrasts with that implied by the most general relativistic cosmology. This, as might be expected, involves a $(\Delta\theta,\delta)$ relation in which $R(t)$ occurs, together with its derivatives (that is, $h_1, h_2, \ldots$). It should presumably allow also for the variation in the proper size of galaxies with epoch. Davidson gave it in the form[128]

$$\log_{10}(\Delta\theta \cdot \delta) = \log_{10}(h_1 \bar{d}_0/c) + 0\cdot 217\{3 - h_2/h_1^2 \\ - 2(\dot{\bar{d}}_0/h_1\bar{d}_0)\,\delta + O(\delta^2)\}. \quad (31)$$

Should reliable measurements ever become available the numerical determination of the coefficients might well provide valuable information concerning evolutionary trends in the galactic dimensions, assuming an independent knowledge of h_1 and h_2.

W. A. Baum has recently presented a few preliminary measurements, connecting the angular dimensions of galaxies with their apparent magnitudes,[129] whilst Davidson has given the corresponding theoretical relations for both the steady-state and evolutionary theories.[130] Into these relations are introduced correction terms not altogether unlike McVittie's K- and W-terms,[131] the values for which still cannot be stated with any confidence. Although observations of this kind are not, therefore, likely to be of much assistance in determining the several parameters of relativistic cosmology in the near future, Davidson's analysis makes it clear that even if the form of evolution taken by the individual sources remains unknown, the observations might serve to distinguish the merits of various specialized cosmological theories— Einstein's, de Sitter's, Kinematic Relativity, and the steady-state, for example. As another means of doing this (or at least of deciding between the steady-state theories and theories which assert, on the contrary, that a distant region is unlikely to appear so sparsely populated as our

[127] See Davidson, *M.N.*, cxx (1960) 283.
[128] op. cit., p. 278.
[129] *A.J.*, lviii (1953) 211.
[130] Davidson, op. cit.
[131] See *M.N.*, cxix (1959) 54, 665.

own), P. S. Florides and McCrea have recently suggested the use of a relatively simple criterion.[132] This involves measuring the 'congestion of the universe' at a certain distance. This is defined as (Δ^*/θ^*), the ratio of the average value of the angular diameters of clusters of galaxies (Δ^*) to the average value of the angular separations of any such cluster and its closest neighbours (θ^*). (These neighbouring clusters must be subject to the expansion, that is to say, must not be 'gravitationally bound'.) The steady-state theory is not compatible with any dependence of (Δ^*/θ^*) on Δ^*. $((\Delta^*)^{-1}$ is clearly a measurement of the distances between objects, assumed to be approximately the same for all. But other theories would be expected to imply different forms for the dependence of the ratio upon, for example, spectral shift. Florides and McCrea discovered the surprising result that 'Newtonian' cosmology, the cosmologies of the Special Theory of Relativity, Kinematic Relativity and General Relativity all lead to the same result, namely that (θ^*/Δ^*) is proportional to $(1 + \delta)^{-1}$. The relation is exact whether or not θ^* is a small angle. The effect is a first order effect in δ, and therefore 'no other observational means of discriminating between the steady-state and other theories can demand less accuracy'.[133]

Whether or not the required observational accuracy can be attained is uncertain; there are, in any case, two theoretical difficulties which must be faced.[134] The first of these concerns the possibility that the intrinsic sizes of clusters might depend upon their age at the time of emission, and hence upon their distance from the observer. This would make it difficult to refute the non-steady-state theories, whilst even if the observations seemed to support the steady-state theory it would be possible to hold that the evolutionary trend was compensating for the change in the factor $(1 + \delta)$ to within the errors of observation. Florides and McCrea point out that there are, nevertheless, certain 'size parameters' (as, for example, the distance between the components of a binary galaxy) which are unlikely to vary much with age. The second difficulty concerns the possibility of galactic clustering of the second (and perhaps even higher) order, which might mean a rate of expansion which varies from place to place. They remark, however, that even supposing a higher order of clustering of galaxies, giving rise to an appearance of non-uniformity in the distribution of matter, the diffuse matter distributed between galaxies and clusters may redress the balance.

[132] *ZS. Ap.*, xlviii (1959) 52.   [133] ibid., p. 64.
[134] Ryle has pointed out that the method does not readily lend itself to radio techniques.

8. Cosmology and the Formation of Galaxies

Using the most powerful optical telescopes the displacements of lines in the spectra of distant nebulae can scarcely be determined beyond the nineteenth magnitude. Beyond this figure approximate magnitudes and galaxy-counts can be obtained out to, perhaps, magnitude 21 (Mount Wilson) or 23 (Mount Palomar), although the shapes of nebulae beyond magnitude 18 or 19 can rarely be appreciated. It is unfortunate that to discriminate between any two current theories one must rely (with the exception already noted) upon effects of second or higher order, effects which fall near the limits of both optical and radio telescopes. Once again, whilst it may well be possible to rule out some of the alternatives, it is highly probable that a succession of theories will accumulate between which it will be impossible to decide by means of the (δ,m) and (N,m) relations alone Evidence of a less precise nature is also called for, and one example of the sort of evidence which is likely to assume an increasing importance relates to nebular condensation. To what extent are the various theories able to accommodate a plausible mechanism for the genesis of stars, galaxies, and even clusters of galaxies?

The uncertainties of the investigations by such writers as Lemaître, McCrea, and McVittie, into the effects of condensation on the stability of the Einstein state, were no greater than the uncertainties in the problem of explaining the formation of condensations, such as we now see, in the context of an expanding universe. If the Einstein state persisted for a long period, the temperature and density must have rendered any perturbing condition an improbable affair. With expansion and consequent decrease in density, however, the circumstances would have become even less favourable. How, in this model, are condensations after the first to be explained? Lemaître later held that three stages must be recognized in the evolution of the universe:[135]

(i) Progressive fragmentation of the primeval atom giving radiation of very high energy and particles of too great a velocity for condensations to occur.

(ii) A period of deceleration. 'Gravitational attraction' and 'cosmical repulsion' approach equilibrium and conditions favour condensation. The distribution deviates from homogeneity on a local scale.

---

[135] See, for example, *C.R.*, cxcvi (1933) 903, 1065; *P.N.A.S.*, xx (1934) 12.

(iii) Renewed expansion. The probability of forming new condensations diminishes.

Apart from Lemaître, few at first made this problem their concern, although some interesting results were obtained by Synge,[136] N. R. Sen,[137] and D. N. Moghe[138] in connexion with the expansion or contraction of clouds or spiral nebulae in an expanding universe.

The essential difficulty with all relativistic theories in which λ is positive is that of accounting for the formation of condensations in terms of gravitational instability; for, to use the 'force' metaphor, the present expansion indicates that the forces of cosmic repulsion exceed those of gravitational attraction.[139] This is not likely to disturb the stability of systems (such as the Galaxy) of high average density, but it is likely to prevent new condensations in regions of low density. With this difficulty in mind, Gamow and Teller[140] and later Gamow,[141] Alpher and Herman[142] have attempted to infer the conditions obtaining at the time of formation of the galaxies, and before, from present observations on the relative abundances of the elements. Their treatment is not so specific as Lemaître's, in the sense that their choice of a particular evolutionary model is left rather uncertain, but it cannot be denied that their occasional successes have been thought to reflect favourably on Lemaître's model.

Evidence from the observed relative abundances of the elements is never likely to play a decisive part in the selection of a cosmological theory, but it may well be instrumental in the rejection of such a theory. Some opinions are that recent developments have already made Lemaître's model untenable. For this reason a few chronological remarks in this connexion may be of interest.

As mentioned at the beginning of this chapter, attention was first drawn to the connexion between the relative abundances of the elements and the stability properties of atomic nuclei by W. D. Harkins in 1917.[143] On the other hand, the nature of the constituents of the system is not the only important factor: temperatures and densities must be sufficiently high for equilibrium to be established in a reasonable

[136] ibid., p. 635. [137] *Bull. Calc. Math. Soc.*, xxix (1937) 185.
[138] op. cit., xxxi (1939) 19. (A criticism of Sen, followed by Sen's reply.) See also Appendix, Note XIV.
[139] See, for example, Gamow and Teller, *Phys. Rev.*, lv (1939) 654. [140] ibid.
[141] *Phys. Rev.*, lxxiv (1948) 505 (letter); *Nat.*, clxii (1948) 680.
[142] ibid., p. 774; *Phys Rev.*, lxxiv (1948) 1577, 1737; lxxv (1949) 1089; *Rev. Mod. Phys.*, xxii (1950) 153, 406 (erratum).
[143] *J. Am. Chem. Soc.*, xxxix (1917) 856.

time. We referred to the fact that Tolman, contributing to what became known as 'equilibrium theory', published in 1922 a study of the thermodynamic equilibrium between H and He in which he came to the surprising conclusion that it is not possible to understand the observed abundance ratio at a temperature below 10^6 °K.[144] S. Suzuki, independently, put the minimum temperature at 10^9 °K.[145] The subject had begun to attract the attention of many astrophysicists, and by the time of Gamow's attempt to connect it with cosmology, it had acquired an extensive literature.

The opinion of Alpher and Herman, summarizing the work done before 1950, was that equilibrium theory is not capable of reproducing the observed relative abundances so long as it is applied to an assembly at a single density and temperature, but that general agreement may be obtained if density and temperature are suitable functions of position.[146] For this reason there arose the ideas of element formation within an 'isothermal material prestellar body embedded in a sea of radiation' or in 'dehydrogenized, collapsing, rotating stars' of various temperatures and densities. Each hypothesis, however, led to the prediction of only a narrow range of atomic weights. Another difficulty was, as we saw, in explaining the distribution of the elements throughout space. In novae, for example, only the surface material appears to be ejected, and the elements in the interior are likely to remain there. Convection within the star was not ruled out, but here was another complication: equilibrium ratios would change with radial position. The disruption of the entire mass of supernovae might have provided an answer had not difficulties arisen here too.

In order to avoid these difficulties, Gamow suggested that the formation of the elements took place in an early stage and that the process was arrested by the rapid expansion of the universe, a process which the theory must obviously be prepared to specify.[147] Gamow, Alpher, Herman, Smart, and others have developed the idea quantitatively. (The theory is known variously as 'non-equilibrium theory' and the 'α—β—γ theory'—the latter being a play on the names of three prominent authors.) In the first seconds of the expansion, we are told, the density decreases so rapidly that the universe comprises no more than a neutron gas. The neutrons subsequently decay, and the resulting

---

[144] *J. Am. Chem. Soc.*, xliv (1922) 1902.
[145] *Proc. Phys.-Math. Soc. Japan*, x (1928) 166, xi (1929) 119, xiii (1931) 277.
[146] *Rev. Mod. Phys.*, xxii (1950) 153. (This memoir of 59 pages contains a bibliography of 180 references.)
[147] *Phys. Rev.*, lxx (1946) 572.

protons and neutrons by capture, with β- disintegration, then produce the heavier elements.

As Alpher and Herman pointed out, the most serious difficulty was the absence of stable nuclei of mass either 5 or 8. In another commentary on the rival theories, D. ter Haar remarked that the equilibrium theory seemed to offer the better solution.[148] More recently the supernovae have been reinstated as likely centres of nuclear transmutation. As A. G. W. Cameron has shown, neutrons would appear in their nuclei and, even more important, in the nuclei of the much commoner red-giants; and these neutrons would quickly transmute light into heavy elements.[149] E. M. and G. R. Burbidge, W. A. Fowler, and Fred Hoyle, using Cameron's idea, have since accurately accounted for the observed relative abundances of the elements,[150] whilst it has also recently been shown that nuclei heavier than helium could not have survived that early phase of the expanding universe of high temperature and density which Gamow presupposed. The actual abundance of helium is still uncertain, however, and it may eventually be necessary to invoke some such explanation as Gamow's.

The hypothesis that the formation of galaxies is a continuous process would almost certainly have obtained more support could the difficulty of accounting for some form of gravitational instability have been overcome. The application of Jeans's formula for gravitational instability requires large density fluctuations such as might be explained in terms of turbulence.[151] S. Chandrasekhar discussed the astrophysical roles which a theory of turbulence might be expected to play,[152] but it was von Weizsäcker who really introduced a theory of turbulence into cosmogony.[153] The way in which he did so may be simply explained. A sheet of gas with axial symmetry in which is situated a central mass will comprise successive ring-elements which, moving in circular orbits with appropriate Keplerian angular velocities ($\propto 1/r^2$), will be in relative motion. Turbulence will follow, viscous forces will then perturb the motions and, as von Weizsäcker concluded, matter within a certain critical radius will fall to the centre whilst the rest will be dissipated. The theory was applied to both stellar and galactic problems and an

---

[148] *Rev. Mod. Phys.*, xxii (1950) 142.   [149] *Ap. J.*, cxxi (1955) 144.
[150] *Rev. Mod. Phys.*, xxix (1957) 547.
[151] *Phil. Trans.* (A), cxcix (1902) 1. See p. 49. Cf. *Astronomy and Cosmogony*, pp. 345–7. L. Spitzer questions the application of the criterion to a rotating system, but Chandrasekhar (*Vistas*, I, 344) lent support to the general application of Jeans's formula. See *J. Wash. Acad. Sci.*, xli (1951) 309.
[152] *Ap. J.*, cx (1949) 329.
[153] *ZS. Ap.*, xxii (1944) 319; xxiv (1947) 181; *ZS. Naturforsch.* (A), iii (1948) 524.

attempt was even made to work into the theory a repulsive force such as might account for the dispersing of the galaxies themselves.[154]

Sciama, as already explained, has accounted for the continuous formation of galaxies within a steady-state universe by gravitational means.[155] W. B. Bonnor, on the other hand, reviewed the problem as it is seen in the context of relativistic cosmology. He claimed that although closed models are more favourable than open models for the occurrence of condensations, yet the density fluctuations required to account for the present state are much greater than classical statistical theory would lead one to expect.[156] Bonnor believed that the difficulty might be avoided using either the Eddington–Lemaître, or perhaps an oscillating model.

Sciama's findings have recently been further qualified by M. Harwit, who finds that unless matter is preferentially created in the presence of existing galaxies, a stationary galaxy cannot gravitationally attract a sufficient amount of matter from the extragalactic medium to double its own mass within a time of $1/3H$—as the steady-state theory requires.[157] He also finds that a moving galaxy is incapable of accreting matter at the required rate. It follows that either the hypothetical creation process is strongly localized, or there are forces other than gravitational which give rise to galaxy formation. Gold and Hoyle have welcomed this result, having previously proposed that pressure gradients are necessary for producing primary condensations.[158] They maintained that the pressures exerted by the intergalactic gas (at temperatures of 10^7 or 10^9 °K) might be sufficient to compress locally cooled regions of gas to form new galaxies. It was shown that the local cooling can take place in times of the right order ($1/3H$) provided that the density of the gas is not less than about 10^{-27} g cm^{-3}. Hoyle's ideas are, for various reasons, not readily extended to other systems of cosmology, which are thus clearly at a disadvantage until they are provided with some satisfactory alternative.

It is worth adding that although the steady-state theories do not suffer from the old time-scale difficulty, it would have been virtually

---

[154] Gamow adopted the idea of a 'supersonic' and 'primordial turbulence', the present distribution of the galaxies being described as a 'fossilized' relic of this early (Mach 30) agitation. He did not explain the sources of the early turbulence. (*Phys. Rev.*, lxxxvi (1952) 251 (letter) and *P.N.A.S.*, xl (1954) 480.) Jeans's formula was modified to take into account the energies corresponding to the velocities of recession of the galaxies.

[155] *M.N.*, cxv (1955) 3.   [156] *Ann. Inst. Poincaré*, xv (1957) 158.

[157] *M.N.*, cxxii (1961) 47 and cxxiii (1961) 257.

[158] Hoyle, *Proceedings of the 11th Solvay Conference* (ed. R. Stoops, Brussels) p. 53, and *Paris Symposium of Radio Astronomy*, p. 583.

impossible to fit these more recent ideas on galaxy formation (the average age of a galaxy being put at between 10^9 and 10^{10} years) with the *old* value for H (which would make the average age of a galaxy $(1/3H) = 6 \times 10^8$ years).

Galaxy formation in the Kinematic Theory of Relativity was discussed in Chapter 8. Milne's account was rarely taken seriously, especially when he predicted the interchange of particles between galaxies. Oddly enough, it has recently been announced by the American J. Linsley that a cosmic ray particle has been detected with such a high energy and from such a direction that it must come from another galaxy.[159] It seems unlikely, even so, that the main body of Milne's work will ever be revived.

Before closing this chapter some mention must be made of the simplest observation of all having cosmological significance: the sky is dark at night. In a previous chapter we ascribed the resulting paradox to Olbers and de Cheseaux, but it was Bondi above all who emphasized its importance for modern cosmology. It is a paradox no longer, for there are several different sorts of solution, although Bondi has drawn attention to flaws in the solutions offered by Olbers and de Cheseaux themselves. It is now generally recognized that the existence of nebular red-shifts is a sufficient condition for the resolution of the paradox, and this is so whatever interpretation is offered for the shifts. Milne was probably the first to be aware of the need to show that the total flux of radiation at all points of his chosen model is finite.[160] Since then the greatest interest has been shown in the behaviour of both contracting and expanding models, and the results of several investigations have recently been published.[161] Thus Bonnor finds that the reduction in the intensity of light received due to the obscuration of distant galaxies by those nearer to us is furthered by expansion. All the common models— for example, the Einstein–de Sitter, cycloidal and steady-state models— yield approximately the same results. Contracting models are seemingly more discrepant: a steady-state contracting model with sufficiently rapid contraction may have an infinite light-flux, whilst even when the contraction is most rapid, the cycloidal model may exhibit a finite flux.

[159] *The Times*, 2 March, 1963.
[160] *ZS. Ap.*, vi (1933) 1. Discussed by McVittie and Wyatt, *Ap. J.*, cxxx (1959) 1, and Whitrow and Yallop, *M.N.*, cxxvii (1964) 301.
[161] McVittie, *Phys. Rev.*, cxxviii (1962) 2871; Davidson, *M.N.*, cxxiv (1962) 79; Bonnor, *M.N.*, cxxviii (1964) 33; Metzner and Morrison, *M.N.*, cxix (1959) 657 (this paper is criticized by Whitrow and Yallop, op. cit.); see also the joint papers by McVittie and Wyatt.

To say merely that the sky is dark at night is not likely to satisfy an astronomer, and, as long ago as 1901, Newcomb was attempting to measure accurately the background radiation of the sky by visual means.[162] Photoelectric photometry was brought to bear on the problem in 1937, and observations from satellites working above the Earth's atmosphere may soon overcome the greatest obstacle to present progress—namely, the zodiacal light and airglow. It does seem, however, that no theory currently held is likely to be ruled out on these grounds for many years to come.

It is probably not too much to say that, despite their conflicting predictions, none of the theories here considered can be categorically dismissed; for the number of undetermined factors which it is possible to introduce is still far in excess of the number of different sorts of observation which it is possible to make. Success in the future appears to rest with the application of radio-astronomical techniques, with number-counts and flux-densities in particular, and with the optical measurement of the variation of mean angular diameter with distance. Measurements of the change with distance in the physical properties of galaxies, and especially of their average ages, might also be decisive in excluding the steady-state theories.

[162] *Ap. J.*, xiv (1901) 297. Later work is discussed by Whitrow and Yallop, op. cit.

PART II
PHILOSOPHICAL ISSUES

He gave man speech, and speech created thought,
Which is the measure of the universe.

SHELLEY, Prometheus Unbound

CHAPTER 12

'FACT' AND THE 'UNIVERSE'

'. . . he hath strange places cramm'd
With observation, the which he vents
In mangled forms'.
                            As You Like It II, vii, 37.

CRITERIA for the application of the word 'cosmology' have so far been deliberately avoided. No doubt there will be those who would wish to assert that the first part of this work was not concerned with cosmology at all, but merely with extra-galactic astronomy. This chapter is no philological digression, but is devoted instead to a claim with which most reviews of our subject begin. Cosmology we are told, is simply the study of the universe. Two questions which this statement presupposes to have been satisfactorily answered are at once prompted. Can two distinct theories be said to have the same objects of study? Can a cogent account of the concept Universe be given? These questions will be considered in turn.

1. BASIC OBJECTS

It is a commonplace that what often passes for the final presentation of a scientific theory reveals a mixture of seemingly incongruous elements. Fragments of systems which are historically anterior to the main theory, half-formalized appeals to intuition, a casual deductive manner mingled with argument by analogy: these features all appear alien to any such rules of formal procedure as those which the sciences are generally supposed to follow. This is difficult to understand if one supposes the theorist to be working on strictly deductive or inductive lines. The author of a physical theory, on the contrary, generally has his attention directed more to his explanatory purpose than to the formal structure whereby it is achieved. His theory, in addition, has to satisfy several requirements which are not purely formal. It must, for example, be possible sooner or later to make the observations which are relevant to the theory. But above all the theory must demonstrate its adequacy for explaining what is felt to be, without it, unexplained. With his attention thus divided the physicist is more likely to be satisfied with something less than logical coherence; and yet the position is not as hopeless as it might appear. To begin with, he might be working with some form of correspondence between the terms of his own theory and those of another, in which case some sort of logical form will be imposed on his theory from outside. In any case, mathematics is capable of pulling a theory into shape—and more than this, mathematics is capable of giving two theories with ostensibly different foundations very much the same ultimate form.

This, together with the fact that different theories often share a great deal of their vocabulary, gives rise to the first of the two problems discussed in this chapter. In order to deal with this problem it is first necessary to consider some of the things which theories can have in common. Two theories which differ only in regard to their notation or the order of their exposition (from either a logical or a historical point of view) must of course be counted structurally identical. It seems highly improbable that any attempt to prove the exact isomorphism of any two major cosmological theories would meet with any success, but it is often held that they may be in some weaker sense *similar*, and many examples can be seen in Part I. In each of two structurally identical theories it must be possible to find exact translations of inferences made in the other, and this might well provide a suitable definition of such an 'identity'. But it should be obvious that 'similarity' may be defined in many different ways. For example, two theories might be judged similar if they differ principally in their rules of correspondence, or as between unobservable consequences, or if well-defined *parts* of them are structurally identical. Without attempting to give a formal definition of 'similar theories', it should already be clear that theories which would be said to 'predict the same consequences' need not necessarily be judged 'similar'; for they may conceivably differ in respect of both rules of correspondence and formal structure.

The question now arises as to whether 'same consequences' can be given a wholly satisfactory meaning when syntactical identity is lacking. Of course the very notion of 'meaning' is notoriously ambiguous. Although it would be out of place to dwell long on the matter here, what will surely be generally agreed is that at least part of the significance of the terms of cosmology is decided not merely by their application (for those that can be said to have an application) but by their theoretical context. Now throughout cosmology descriptions are encountered which at first sight appear to be incompatible whilst purporting to describe the same objective state of affairs. And yet this way of talking is quite misleading, for just as outside the context of the appropriate theory it is impossible to ascribe a meaning to either description, so can neither, failing a common theoretical basis, be regarded as conflicting with the other.[1]

We have spoken here of descriptions, but what of simple nouns? Do these not have meanings independent of theoretical context? What of 'nebula' and 'star', 'telescope', and 'spectrum', words which can on the face of it be ostensively defined, granted that all percipients can agree on the resemblances between two objects 'defined' in this way. It seems clear that there must be some such words corresponding to what might be called 'basic objects'. (It is certainly not held that the four named here are in any way unusual.) It is equally clear, even so, that there are pitfalls in supposing that all common names are of this sort. Common nouns are usually assigned to things belonging to a specified class or having characteristics which are again somehow specified. The logician, for example, in giving symbolic form to language, will often replace a common noun by the (proper) noun of a class. In this case it is Platonic forms and the like which tend to be regarded as our 'basic things'. Alternatively the logician may represent a common noun as a variable ranging over a class, which may be defined in terms of variables 'denoting properties'. From this point of view it is properties which are our 'basic things'. On the whole, it seems likely that most people would be happier without either reduction or paraphrase in the case of most familiar objects. It would be misleading to stigmatize this attitude as improper, although the grammarian may find it inconvenient and the logician might point to its inconsistencies. But in a scientific context, whether one likes it or not, these 'reductions' are essential to the theory. One cannot point to electrons, and no part of the meaning of 'electron' can be given without a discussion of tendencies to move away from like objects, of differences between electrons and other objects with this property, and so on: in other words, none can be given without our first becoming well and truly immersed in a specific theory.

Of the many words which seem to function as common nouns, even in ordinary day-to-day speech, some at least appear then to have meanings which are far from common to all contexts. The discovery that a great many of the words of our language are markedly ambiguous is no occasion for surprise, but to hold that a word such as 'electron' is ambiguous as between two rival theories which use it is not likely to

[1] This lack of direct conflict does not, of course, undermine the basis of scientific disagreement, which pivots on both predictive success and relative formal advantage.

command much support. This ambiguity is unavoidable, however, just as with more obvious cases of word sequences which would quite clearly pass for descriptions. It is unavoidable simply because the meaning of a word cannot be given in isolation from others. Part of it might be given by ostension alone, but the complete meaning depends very largely on its theoretical involvement. Thus a large proportion of the individual terms common to more than one theory of cosmology must be regarded as each having different meanings which cannot be strictly compared.[2]

Why do we not say that *no* term can be compared as between two different theories? Even terms denoting what we previously referred to as 'basic objects' have a syntactical element in their meanings. Does this not rule out the possibility of a comparison of the meanings of any simple term which features in two theories? The answer seems to be that with terms like 'telescope', 'star', and so on, we are most of us prepared to ignore all but the semantical element in their meanings. This is notably true when we are in that region where scientists agree on what constitute the data from which all theories must proceed. And yet although some sorts of data are accepted in principle by all cosmologists, where data end and theories begin is often unclear—we recall the Hubble–McVittie controversy as a case in point.[3]

Now this may all seem a far cry from one of the problems with which we began, namely that of finding a meaning for the concept Universe. It seems necessary, even so, to discuss these matters first, for no one is likely to claim the universe as a datum, and the meaning of 'universe' must at least be coloured by the meanings of the names and descriptions attached to its parts. The trouble with this word is that it has never been customary to distinguish it in kind from words like 'galaxy'. We shall see that they do indeed have something in common, but that they also differ in an important way.

2. Fact

What can two cosmological theories have in common? It would generally be held that they may share a basis of fact, and that the system of facts which they characteristically study is nothing less than the universe as a whole. A more modest answer might be that the factual basis is simply the sequence of all galactic states. Now it is well known,

---

[2] Terms of two theories may admittedly be judged to have a certain consonance, and this judgement will most easily be made when the theories are structurally similar.

[3] See pp. 235 ff. *supra*.

and we assumed as much in the first section, that the meaning of a referring expression is not to be identified with the object to which it applies. If it were so then, at the very least, we should be in a quandary in deciding whether 'a galaxy', for example, is a referring expression; for such expressions as this could only have meaning if there existed objects to which they applied. It is now widely accepted that the meaning of such an expression is given by the set of linguistic conventions governing its proper use—but its proper use both as a referring expression *and* as a term of the theory in which it is featured. The former sorts of convention may be the same, even as between different theories. But the latter conventions, *ex hypothesi*, must differ. Once again we are reminded of the importance of context in the determination of meaning. To return now to the claim that two theories may share a common basis of fact: it is widely supposed that all sciences have a common factual basis simply because it would be nonsense to speak of the context of a fact, let alone of a fact changing from context to context. No one doubts his own findings more systematically than a scientific observer, and yet few scientific observers doubt that in embodying their selected findings in 'statements of observation' they are providing a part of the foundation for *any* relevant theory. They suppose themselves to be stating fact, and fact is vaguely thought to be something which will endure, by contrast with speculative and controvertible theory.

Unfortunately no such distinction is reasonable. What is supposed to be so immune from criticism is, ultimately, a *statement* of observation. What of the observation itself? What of instrumental distortions which we cannot even know about without some theory? And what of the inference from the last link to the first of a 'chain' of facts—from the photographic image to the galaxy, for example? Such inferences may be intricate, lengthy and tenuous. Far from their mirroring a 'purely factual' situation it would be a slight on an astronomer's intellectual ability to say so.

So far we have not held that the very notion of fact is at fault: we have simply pointed out that no description can be a simple reflexion of fact, for all descriptions carry contextual implications which are linguistic and artificial and can thus have no factual counterpart. 'There is no pure appearance or observation' has, on the other hand, an air of paradox, and it is not equivalent to the plausible assertion that to speak of appearance we must always, in the very act of speaking, involve ourselves in an act of theoretical interpretation. The impossibility is one of expression and not of perception.[4] Now the impossibility of *saying* what can, without doubt, be the case, is not easy to reconcile with the ambitions of the natural sciences. Fortunately, from the point of view of ordinary communication, the majority of people agree as to the way in which they shall give expression to their experience. Even in the sciences, when agreement cannot be reached either on suitable concepts or on the way in which these are to be used, the scientist who deviates from the norm is usually prepared to justify himself. What he may not do, however, is justify himself by referring to some system of objective entities, external and absolute, which compel the more perceptive members of the scientific fraternity to theorize in a certain definite way.

Once the myth of a uniquely describable objectivity is rejected, the way is left clear for an appreciation of the malleability of what is usually accepted as 'factual statement'. This might at first seem an obstacle to any definite scientific progress: how can we so much as communicate, without something stable and external to all of us? Again, it is not denied that at the basis of science there is something which might be called 'unutterable fact' as the object of attention, but merely that this can never be distilled from any scientific description. It is always possible to hold to the metaphysical belief that the several theories of science merely describe the same things from different points of view. On the other hand, a relative evaluation of two theories which purports to be based upon 'that set of facts which is described by both' is likely either to be tacitly favouring one of them or to be using a third theory as a yardstick for both. In neither case does the assumed basis of fact have the objectivity which was presumably desired.

Can it be that the word 'fact' has been taken in an unusual sense? To begin with, some would have us believe that facts may *imply*: the fact that Einstein is a man is said to imply that Einstein is mortal. But here prefixing 'the fact' is just like asserting the truth of the proposition 'Einstein is a man'. This may be seen more clearly in terms of this sort of dialogue: (A) 'This doesn't fit the facts'. (B) 'What facts?' (A) 'The fact that . . .' and so on. Here A has been obliged to *express* the facts only to draw attention to an inconsistency between propositions. It seems that the word 'fact' is either superfluous or is used in a futile attempt to get at a world behind our language and totally independent

---

[4] It is not, of course, logically impossible for people to have illusions, nor is it logically impossible for them to disagree over what is illusion and what is not. But illusions are only recognized as such by their lack of coherence with more usual experience.

of it. Facts are often held to be timeless, for example, and some have even gone so far as to define them as 'those propositions which are true'. The collision of this star and that would then be an event; that the stars collided, if true, would be called a fact. It may be the case that this is a better account of ordinary usage than ours, but this does not vitiate what we have said. However, if we accept these recommendations and then say, for example, that two facts may imply a third, it must not be supposed that we are adding to empirical knowledge. And in this case it is worth asking whether we are prepared to say that the supressed premiss in the example at the beginning of this paragraph ('All men are mortal') expresses a fact. If this (and likewise the whole gamut of scientific generalizations) is truly fact, in the present sense, by what criterion do we decide the matter? Without pursuing it further, such a concept of fact as this is clearly not calculated to assist in the discovery of a realm underlying all cosmological theories, true as it may be to ordinary speech.

3. Objectivity and the Uniqueness of the Universe

Suppose that a person were to make this claim: 'I know what it is to be an objective entity. To understand the concept Universe I need know no more, for the word "universe" signifies the aggregate of all such entities.' Would he be wrong to allow only one category of existence? Are galaxies, living creatures, electrons, and gravitational fields all to be classed together? If so, then existence is presumably to be tied to current scientific belief—and the universe becomes a rather impermanent entity.[5] If not, are we to admit different sorts of co-existing universe? To both ideas our claimant might object that his one class of entity has nothing to do with the exigencies of theory—being at the level of 'unexpressed fact' perhaps. The trouble then is that although his concept of the universe is not totally unfitted for answering the question 'What have different cosmological theories in common?' it explains no more. It can scarcely be called a *scientific* concept, for it can be understood in terms of a single mode of existence and the logic of the word 'all'. Furthermore, it seems to commit us to the view that the universe is without structure, unless in some transcendental sense.

[5] It would be wrong, of course, to use the old argument that the universe itself cannot, since things have bounds, be termed a 'thing'; for, as Riemann showed, there is nothing illogical in the idea of a finite and unbounded space. Notice, in this connexion, that the great appeal of Einstein's 'finite and unbounded universe' was that it allowed relationships between the universe and its contents to be readily conceived (cf. equations (12) and (13) p. 83 *supra*). Bishop Barnes, for example, often claimed that we could at last begin to understand the range of God's activity.

The creation of a new scientific theory, of a 'new way of looking at the world', is accompanied by the insinuation of a structure which is not of the world but of the theory itself. From the structure of any one of the many theories of cosmology few people would wish to make inferences to a unique structure for the universe: but there is a prevalent feeling that the universe is capable of being *truly* described in only one way. It is felt that although, perhaps, one may never discuss the structure of the universe without the mediation of language, the world nevertheless possesses a structure which is independent of the linguistic habits of humanity. At one time this might have been explained in terms of certain 'cognitively inaccessible but real relations at the level of the *Ding an sich*'. Since neither relata nor relations are accessible to scientific enquiry we are fortunately spared further consideration of this sort of 'world-structure'. A second approach would be to draw upon current cosmological usage and to attempt to refine the meaning of the term 'universe' in the hope that one would arrive at a hard core of significance acceptable to all parties. Yet the idea that such a concept can be formulated, common to all theories and called in question by none, is impossible to accept. We have thus still found no way of characterizing natural cosmology as 'any theory which deals with the universe', if the universe is thereby meant to be the common referent of all theories.

The possibility has not been ruled out that *some* aspects of the meaning of the term 'universe' may be common to all theoretical contexts. The most obvious aspect is in regard to uniqueness. It is invariably held that in one respect cosmology is a very unusual science, because the universe is in principle unique.[6] Whatever new item is discovered which is capable of description by a specific theory must, by definition, be accepted as part of the universe of that theory. Some of the implications of this will now be considered.

It is tempting to say that our scientific method must be a very unusual one; for the universe can be compared with no other set of objects or events, and such a comparison is often supposed to be the very essence of a scientific procedure. This sort of problem seems to have worried M. K. Munitz, who argues that, as it is unique, there can be no laws *of* the universe, although there may be 'devices for

---

[6] Notice, however, that astronomers have found it natural to talk of 'island universes', and in both this century and the last we find mentioned the possibility of two distinct universes, one of matter and one of anti-matter, flying apart under the influence of their mutual repulsion. With a few obvious changes of terminology, this convention can be accommodated in what we have said here.

comprehending the structure of the universe as a whole'.[7] He admits laws only for the constituents of the universe: the laws of galactic recession, for example, he believes to be simply laws of galactic astronomy. If by this he merely wishes to object to the idea of a universal statement having as its subject a class of objects which, by definition, can have only one member, then it is easy to agree with him. One should object to this way of talking, if only because it is superfluous to cast into universal form a law having only one instance. But, as it happens, 'laws of the universe', in Munitz's sense, are scarcely ever considered in actual cosmological writing, and there was no occasion to mention them in Part I.

Although the cosmologist occasionally holds that the universe as a whole is the object of his study, he is nowadays in the habit of excluding such things as man, the freedom of man, and the origin of evil (all of which Hegel listed amongst the chief topics of cosmology) from his investigations. Clearly the term 'universe' cannot mean (as we are so often told it does) 'the totality of existent events' in its widest sense.[8] The cosmologist, in fact, restricts himself to such things as give his own particular formalism its interpretation. His universe, loosely speaking, is nothing more than the domain of these individuals. It is nothing more than a universe of discourse, and this being peculiar to his own theory, as we have already intimated, it is necessary to look elsewhere for what makes each of several different theories properly termed 'cosmological'.

Putting aside the question 'What makes theories cosmological?' let us consider possible meanings for the word 'universe' when this is obviously intended in the sense of 'universe of discourse'. Bondi offers two meanings for the term. In one case it is said to be synonymous with 'the largest set of events that can be considered to be physically linked to us'.[9] It is suggested, on the other hand, that the word is often taken to denote the largest set of events 'to which our physical laws (extrapolated in some manner or other) can be applied'. Now to claim that an event is physically linked to us is to claim that some at least of

[7] *B.J.P.S.*, xiii (1962) 39.

[8] During the half-century under review the preoccupation with gravitational cosmology has led some to assume that cosmology is the province of no other form of scientific explanation. It will be surprising, even so, if electrodynamics does not play a much larger part in the cosmology of the next fifty years, beginning with the cosmogonical problem. (Cf. the paper by Gold and Hoyle in *Paris Symposium on Radio Astronomy*, p. 583.)

[9] *Cosmology* (2nd ed.) p. 10.

our chosen physical laws can be applied to it, and thus the two meanings are not, on this score, radically different. The second meaning presumably avoids ruling out events inferred to be beyond our 'visual horizon' whereas the first does not.[10] The distinction is worth keeping for theoretical, and perhaps even for epistemological reasons. Both meanings allow only what might be called 'permissible fact'—fact, that is to say, over which a theoretical scheme claims to legislate. In both senses, to speak of a 'theory of the universe' is to speak of the structure of such fact, such sets of events, such universes of discourse.

These two meanings do less than justice to the term 'universe' as it is used in many of the contexts of cosmology. On occasion it is used as synonymous with 'model', with 'metric', and even with 'theory'. Seldom, even so, does this sort of use give rise to serious confusion: without any loss of meaning the word 'universe' could simply be replaced by another. For example, to say that 'the universe has such and such a total curvature' might well be to assert something about the line-element of an accepted theory. To speak of 'the Einstein universe' might be to speak of a specific model, in one sense of that word (see Chapter 14). To speak of 'the expanding universe' might be to maintain that the curvature is changing with time. (To explain the meaning of this phrase it is neither necessary to perform the logically impossible feat of getting outside the universe to observe the expansion nor is it necessary to indicate that *into which* the universe expands. On both grounds it has been held that it is meaningless to speak of an expansion of the universe.) Statements concerning 'the age of the universe' are much harder to deal with, but for the moment we notice that they involve the notion of a universe given at a specific time. There are many well-known snares in the idea of a universal time[11] but even without these it will be found difficult to provide the phrase with a simple and useful equivalent. From almost all points of view there is something to be said for regarding the concept Universe as a theoretical concept, in the accepted sense of this phrase. Perhaps then we shall be less perplexed by the failure to find any existent entity corresponding to it. At the very least it should be generally acknowledged that the full significance of the term 'universe' cannot be given in terms of denotation alone.

[10] For the two kinds of horizon found to obtain with most relativistic models, see Appendix, Note XVI.
[11] No doubt it was for this reason that Bondi's definitions were given in terms of events rather than objects.

To return to an earlier question: In what respects do cosmological theories differ from the other natural sciences? The answer cannot be, in any straightforward sense, that each studies a unique entity, the universe as a whole. One might, perhaps, try to avoid the idea entirely. One might, for example, hold that a cosmological theory is simply one which is incompatible with any other theory which predicts the existence of systems more extensive than those which it does itself. (There is here, of course, the assumption that the two theories share a concept of extension.) Cosmology deals with a pattern of matter in motion which it does not admit of being instantiated more than once. This at least marks out cosmology as a comprehensive, if not all-embracing study. Compatible with this answer is a claim that we wish to make here, namely that, these respects apart, there is no essential difference between cosmology and the other physical sciences. 'Cosmological theories have in common a concern with the distribution of matter on the largest scale'. To say this, and no more, is much better than characterizing cosmology as 'the science of all there is'. It is a commonplace that other branches of science deal with a restricted class of objects. Why should cosmology be thought different? The reasons usually given are that cosmology is the science of the universe, and that the universe comprehends all things. It might further be said that since the cosmological theories discussed in Part I made mention of neither cabbages nor kings, they were not truly cosmological. 'True cosmology'—one can hear the claim—'deals with the All, and nothing less: at the very most, modern scientific cosmology has provided a rough outline so transitory that the details will probably never be filled in.' But it is of the nature of *all* sciences to have a restricted subject-matter—their finite expression sees to that. For scientific cosmology it is perhaps more succinct, but certainly misleading, to refer to this subject-matter as the universe. Rather is it the system of those galaxies and other large-scale distributions of material which are either accessible to observation, or susceptible of physical discussion, or both.[12] There is, of course, no reason why one should drop the term 'universe' entirely, dubious though some of its uses may be. Some of the ways in which it may reasonably be applied have already been indicated. Others will be considered in the chapters which follow.

---

[12] To forestall etymological objections we might add that besides signifying *order*—and hence *world-order*—the Greek term κοσμος could mean any *region* of the universe. In later Greek it could even be used to refer to the *known* world.

CHAPTER 13

THE ELEMENT OF CONVENTION IN COSMOLOGY AND SCIENCE

THE Cartesian dualism of matter and mind is not far removed from the classical opposition of nature and human convention. This may be found in Parmenides. According to Plato we owe it to Pindar.[1] Time upon time, the truth of nature is to be found contrasted with the unreliability and deceitfulness of human conventions: the idea runs through theories of aesthetics, of politics, of poetry, and of philosophy—and, of course, of science. In science the idea has become enshrined in a series of principles which have proved unpalatable to the majority. Perhaps for this reason the principles have as often as not been randomly grouped together under the title 'conventionalism', and collectively dismissed. We shall begin this chapter by drawing some distinctions between different versions of conventionalism, by rejecting some, and by answering objections to others. Much of the chapter will be concerned with exegesis, defending some adherents to these ideas from misconceived attack. A way will be indicated whereby science as a whole will benefit when one of the more modest conventionalist theses is accepted, as it is already largely accepted by cosmologists. We shall, of course, draw upon cosmology for purposes of illustration: it is doubtful whether any other branch of natural science would serve as well.

1. 'CONVENTIONALISM': AN AMBIGUOUS TERM. POINCARÉ AND HIS CRITICS

Conventionalism in the context of logic is not our thesis. Several important points may be brought out in connexion with it, none the less.

There are some necessary statements—for example those of logic and mathematics—which do not in any obvious sense record information about the world. Can it be that they are no more than expressions of linguistic agreements and prescriptions—of conventions, that is to say? First one must distinguish between two aspects of necessity which have attracted a good deal of attention, those of analyticity and the property of being *a priori*. '*A priori*' is usually retained as an epistemological predicate, and is nowadays not usually distinguished from

[1] Plato: Gorgias 482E and Protagoras 337C.

'non-contingent'; that is to say, it usually signifies independence of fact, 'come what may'. 'Analytic' is a logical predicate, meaning something like 'derivable within an uninterpreted calculus'. In one extreme sense, an analytic statement need not be true in a more general sense than 'derivable', even when the calculus is interpreted: the truth issue hangs on the truth of the basic formulae of the calculus under the interpretation, and on synonymy rules. This definition of 'analytic' is not to be confused with that which has been current since, shall we say, Frege. Although there are many minor variants of his definition, we may take his own as fairly typical: an analytic statement is said to be one which requires for its justification only the laws of logic and definitions. The extent to which this differs from the extreme definition above, depends on what attitude one has to the inherent correctness of logic itself as an interpreted calculus. For the moment all that matters is that the two predicates *a priori* and analytic be distinguished.[2]

A great many writers have been bewildered at the prospect of having to explain why the postulation of a set of conventions (in the form of a system of symbols operated upon by a series of rules) is able to generate a feeling of certainty. One virtue of making a distinction between '*a priori*' and 'analytic' is that it helps to clear up the confusion. Certainty in regard to the world has nothing to do with analyticity alone. When it comes to delineating the concept of necessity in relation to ordinary language, however, the problem is not so clear-cut; and ordinary language does, after all, have a part to play in science as we find it. It is easy enough for those who think of logic as an uninterpreted calculus to make 'necessary' a synonym of 'analytic' and to ignore interactable uses of the former word; but when logic is taken to be an applied formalism there is a case for saying that necessity is not entirely a conventional matter. It has been argued that our language—and therefore at least one group of statements which we are obliged to characterize as 'necessary'—is what it is because our minds and our surroundings

[2] There are many who wish to equate the two, but in any case, truth in a general sense obviously cannot be guaranteed without reference to the world.

Both kinds of analyticity are properties of statements only relative to a particular context. No statement has an inherent logical value, that is to say, no statement has the same sort of place in all possible calculi. Frege left the phrase 'laws of logic' unexplained. If a person wishes to maintain that there is only one possible logic—although it may be made to correspond to several calculi—then he will almost certainly be prepared to make analyticity at the very least a sufficient condition of necessity. Of course, in the other sense, most of the statements of a scientific theory are analytic, although this does not mean that they are true in any more general sense than 'derivable'. In neither sense does it seem that 'analytic' and 'necessary' are synonymous, but now analyticity is not even a sufficient condition of necessity.

are what they are. Logical distinctions may be made, it is said, only if there are distinctions there to be made regardless of our making them. It would be out of place to pursue the matter further here, but it is hoped that enough has been said to show that no support is being offered for that brand of 'conventionalism' according to which all necessary statements are so by dint of conventions alone.

Conventionalism, as usually understood, made its greatest impact with the writings of Henri Poincaré at the beginning of the present century. It is important, first, to remember that his concern was with the truths of geometry, and it is as well that one should remember the ambiguity in the meaning of 'geometry'. Since long before Euclid, geometry texts have concentrated on the purely deductive aspects of the subject, and this has tended to obscure the fact that it has been used and often understood as a scientific theory, albeit of remarkable accuracy. We shall see that Poincaré was conscious of this ambiguity, although he occasionally clouds the issue in asserting that geometry is 'not a physical science properly so called'. Secondly, Poincaré was not concerned with the purely deductive aspects of geometry, or indeed with validity within deductive theories generally. He certainly did not adhere to what might be called 'logical conventionalism'. He was chiefly concerned with contrasting the arbitrariness of what would now be understood as 'pure' geometry and the restrictions on our liberty when we make geometrical conventions of practical value. This may be seen from the following brief exposition of his work. Needless to say, we have ignored one or two changes of mind which can be detected in following the development of his ideas. The main conventionalist thesis of relevance to applied geometry he never substantially changed.[3]

Having pointed out the possibility that experience might lead us to reject Euclidean geometry,[4] Poincaré asked himself whether we should conclude that the axioms of geometry are experimental truths. 'But we do not experiment on straight lines and circles', he argued, 'this can only be done on material objects'. Geometry cannot, he thought, be an experimental science: otherwise it would be in need of constant revision. But if the axioms of geometry are not 'experimental facts', what are they? Poincaré's answer was that they are conventions, In choosing them, he admitted, we are *guided* by experience; but our choice is free,

---

[3] See *Science and Hypothesis*, tr. G. B. Halsted (New York, 1929); *The Value of Science*, tr. G. B. Halsted (New York, 1929); and the articles: *Rev. Mét. Morale*, vii (1899) 251 and viii (1900) 73.
[4] *Science and Hypothesis*, Chapter 3.

'and is limited only by the need to avoid all contradiction'. It is tempting to suppose that when Poincaré writes that 'it would be an error . . . to conclude that geometry is, even in part, an experimental science', he is thinking of pure geometry. In fact his entire work is most readily understood if the word 'geometry' is always taken in this sense when it stands alone. When he speaks of experimental laws which are only approximative, 'determining the adoption' of geometrical postulates, he is guilty merely of hyperbole: he means no more than that, when interpreted, some postulates make better sense than others. When he writes that 'the axioms of geometry (I do not speak of arithmetic) are merely disguised definitions', it is clearly an uninterpreted geometrical calculus which he has in mind. Again this is obvious in the arguments offered for his central thesis. He denies meaning to the question 'Is Euclidean geometry true?' One geometry, he says, 'cannot be more true than another: it can only be more *convenient*'. But as if to detract from this revolutionary idea,[5] he assured his readers that he considered Euclidean geometry to be most convenient, and for these reasons:

(i) 'It is simplest, not only in consequence of our mental habits, but in itself—just as a polynomial of the first degree is simpler than one of the second.'

(ii) 'Because it accords sufficiently well with the properties of natural solids . . .'

It was simply that the geometry we need to 'frame our representations' could in principle be different fron the one we do in fact use. To get this point across more clearly he gave an example of a world where experience would 'suggest' a different geometry but which we can equally well describe in the familiar terms of Euclidean geometry.

This brings us to those considerations by which Poincaré was led to his conventionalism. It seems that he was most influenced by the possibility of translating hyperbolic geometry into Euclidean—which suggested that here we are dealing with alternative formulations of one and the same set of fundamental relations. (In retrospect, Poincaré's thesis seems to have been supported on a very slender basis. In his articles written before 1900 he shows a remarkable ignorance of geometries in which other axioms than the parallel postulate were

---

[5] His *commodisme* caught the public imagination, and the word became a fashionable catchword in France at the time. Actually there was nothing absolutely new in the idea, apart from the vigour and cogency with which it was expressed.

systematically dropped. It was not until Hilbert's *Grundlagen der Geometrie* opened his eyes that he saw the full import of his own thesis.) This throws light on what he appears to have meant by 'convention'. Conventions were meant to be such that it was always possible to replace them by different conventions without any consequence except in the empirical content of the geometry. The correspondence-rules of physical geometry were thus at the root of convention in this sense. He realized that by changing them it is possible to turn contingent propositions into definitions, and vice versa. Contrary to a common belief, he did not adhere to the thesis that all necessary truths are true by convention. In so far as geometry was necessary for him, it was not conventional; in so far as it was conventional, it was not necessary. Above all, he did not claim that physical geometry is not empirical in any sense.

Much of the trouble in understanding Poincaré's thesis is that the word 'convention' can reasonably be applied at several points of a scientific theory. Before correspondence-rules are chosen, the highest order hypotheses may be called conventions—for they are at this stage simply formal axioms. There is a freedom of choice in the correspondence-rules themselves—although here, as elsewhere, being possessed of freedom and using it to advantage are quite different matters. An entire scientific theory may also be called a 'convention' when one wishes to reserve judgement on the validity of alternatives to it. At the other extreme, the *ad hoc* assumption, inserted to 'save the phenomena', may be called a 'convention'. One certainly cannot ascribe all these uses of the word to Poincaré, and this opportunity will be taken of clearing up some misconceptions of his point of view.

Although there is an element of convention in all scientific laws, each is made in the light of other laws and, ultimately, of experience. Poincaré never said, as Schlick and others have implied he did, that *all* natural laws are arbitrary conventions, a view which would not be difficult to refute. The essence of a convention is that it can be changed. It does not follow that a statement or set of statements embodying a new convention must needs be empirically acceptable without compensatory changes elsewhere in the theory. To take an example, there is a persuasive argument often levelled against Poincaré according to which the energy Conservation Principle cannot be in any sense conventional. 'The impossibility of perpetual motion is not a matter for us to decide . . .' and so on. The assumption seems to be that Poincaré thought those mortals who make conventions to be infallible. In fact, many of the examples he gives show that this is far from true. Actually, although Poincaré might have disagreed, there may well be good reasons for drawing up a new convention even in regard to energy conservation. If Bondi, for example, found that he could influence the creation rate so as to produce more hydrogen atoms on one half of a vertical wheel than on the other, and hence make the wheel spin, no doubt some physicists would feel an urge to make some such new convention—or to accept Bondi's. Of course it may be possible to make adjustments to the remaining parts of the physics which will allow us to retain a principle of energy conservation even under these circumstances. We recall the introduction of zero-point stress. We are reminded, too, of Eddington's dictum that the Principle of the Conservation of Energy had survived so long because energy is defined as that which is conserved. It will be remembered that Einstein followed the example of classical physics in making the principle an intrinsic part of the foundation of his General Theory.[6] Had the theory been empirically unacceptable it is morally certain that no fault would have been found with the principle of energy conservation—but for historical reasons, and not because that principle is in logical status different from the rest.

It is tempting for a supporter of Poincaré to suppose that all opposition stems from the fear that all truth might be thought subjective if conventionalism were generally accepted. Nothing could have been further from Poincaré's mind, but in some quarters he acquired a reputation for having held that it is impossible to make a statement about the world in geometrical language which has more than subjective import. This is rather like passing from the claim that one is free to graduate a thermometer as one chooses to the claim that one may learn the patient's temperature without reference to the patient. In fact this brings us at last to Poincaré's highly controversial proposition that congruence (spatial and temporal) is of the nature of a definition, a proposition which is, of course, one aspect of the thesis that it is possible to give both Euclidean and non-Euclidean descriptions of the same actual states.[7]

An argument which Eddington directed against the definitional character of congruence was that the idea was trivial, 'the meaning of

[6] See Appendix, Note XVI.
[7] Again this has been misunderstood, having been transformed into the thesis that a resolute Euclidean can always confound those critics who find that Euclidean geometry does not 'fit the facts'. This is a stronger claim than Poincaré's, and we shall consider it in the following section under the heading of the Duhem–Quine thesis.

every word in the language being conventional'. The congruence *relation* he held to be independent of convention: it is only in using the *word* 'congruent', according to Eddington, that the element of convention enters. He resisted the re-introduction of Euclidean geometry into physics, insisting that 'we can graphically represent (or misrepresent) things as we please'.[8] The possibility of using Mercator's projection does not, he said, imply that the earth is really flat. 'It is no disparagement to a square peg to say that it will not fit into a round hole', he wrote, the 'round hole' being Euclidean geometry and the 'square peg' being the 'properties attributed to the material system'. But attributed by whom? Clearly by those adhering to the General Theory of Relativity. The whole question is begged. The contextual element in meaning makes it impossible to speak of the objects of Einstein's Theory as fitting into a permanently Euclidean space. Neither Whitehead nor Milne nor any other principal defender of the merits of Euclidean space could have made such an obvious mistake. Their objects are peculiar to their own theories: they are the round pegs which, it is no disparagement to say, will not fit into a square hole. When Eddington wrote of Milne's results that they were 'surprising only to those who have not realized that if you alter the meaning of words you can make any statement true',[9] he was, in a sense, paying Milne a compliment. For it is simply false that to alter the meanings of a set of scientific statements can of itself ensure their truth. In Milne's case the altered statements seemed to be borne out by experiment, and it was something of an achievement on his part to have made them so.[10]

Many others have written in the same vein as Eddington. Thus we find Bertrand Russell answering Poincaré's arguments at the time of their publication with the claim that space and time have intrinsic metrics. Whitehead, in fact, agreed, offering psychological considerations as support.[11] Elsewhere we find Louis de Broglie[12] being quoted with approval by R. Taton,[13] saying of Poincaré that his 'nominalism

[8] *The Expanding Universe* (Cambridge, 1933), p. 28.
[9] *Sci. Prog.*, xxxiv (1939) 225.
[10] The word 'altered' is prejudicial to Milne's case: there is, of course, no reason why he should be regarded as having formed his theory by somehow distorting Einstein's.
[11] Although Whitehead favoured Euclidean geometry and, as we have seen, did not object to the use of non-Euclidean geometries, he could hardly be called a conventionalist. He objected to the 'natural geometry' of Einstein's theory. ('This, being non-uniform, cannot express "uniform relatedness".') He maintained, as he put it, 'the old division between physics and geometry'.
[12] *Savants et découvertes* (Paris, 1951).
[13] *Reason and Chance in Scientific Discovery*, tr. Pomerans (London, 1957) p. 135.

caused him sometimes to misunderstand the fact that, amongst possible logical theories there are, nevertheless, some which are closer to physical reality or, in any case, better adapted to the physicist's intuition, and therefore more apt to aid his efforts'. The final qualification would almost certainly have been accepted by Poincaré, who was in any case often annoyed when accused of nominalism. Even so, it should be borne in mind that there is a good case for regarding our 'intuitions' as largely a product of intellectual training. To fall back upon an intuitive faculty is rather like venerating a Golden Age—an Age when such authors as Newton and Euclid decided what form our intuition was to take.

All these objections have a great deal in common with Schlick's. From the point of view of epistemology there are superficial differences: Schlick's 'reality', for example, is more or less the set of all true propositions. But each involves the assumption of an underlying structure in space and time, which cannot be found through a mere stipulation. This was discussed in our last chapter. As for the claim of triviality, this is misdirected. Einstein's announcement that the simultaneity of spatially separated events is conventional is an excellent example of the introduction of a non-trivial change in terminology. The innovation was made in the self-same spirit as Poincaré's statement that to ascribe spatial and temporal congruence is a matter of convention. The essence of Russell's reply was that merely because measurement is necessary to discover the equality of two intervals it does not follow that their equality is non-existent without measurement: surely measurement cannot create equality and inequality! To say otherwise might be interpreted as a naïve form of operationalism, and yet the argument involves a misunderstanding. It is not that measurement creates equality, but that a change in our conventions (viz. those according to which we frame our correspondence-rules concerning spatial and temporal congruence) may result in equality becoming inequality, and conversely.

Is this a semantical platitude? It is impossible to give a simple answer to an evaluative question of this sort. There are cases where the answer is obviously Yes. It is certainly something of a platitude to say that John might for practical purposes just as well be called 'Jack', or that all our colour words may be systematically interchanged. When it comes to maintaining that more involved concepts (such as the concept of congruence) may be changed, there is such a large body of theory to be altered that we are inclined to use the word 'platitude' only if we can easily see the changes in which we are involved. This is perhaps an even more difficult problem than that of adapting our ideas to the usages of a completely novel theory. In fact the strongest objections to every single cosmological theory mentioned in Part I were those in regard to changes in the usage of words and concepts ostensibly taken from older theories.

Having indicated some of the misconceptions relating to a conventionalist position, let us attempt to summarize our findings. All will agree that there are conventions, and of many sorts, within all branches of human activity, not least within science. If writers disagree, it is over the question of what is conventional and what is not. We find, therefore, Poincaré arguing that congruence is a matter for definition and Russell denying it. We find Milne (mistakenly) accusing Einstein of not appreciating that there is an element of convention in the notion of a rigid body as one whose rest length is invariant under transport. Recalling Milne's alternative metric descriptions of one and the same model of Kinematic Relativity, we occasionally find a writer asking which of the two is correct. Always it is the *bounds* of convention which are questioned. The fear is twofold. First, it is wrongly believed that the aim of the conventionalist is to put scientific theory above reproach. On the contrary, his attitude militates against apriorism. The second fear is that truth might be wrongly identified with simplicity. The idea arises in this way: having indicated the element of freedom of choice in geometry, Poincaré added, as we saw, some brief suggestions as to how we are to choose between the alternatives which present themselves. 'Convenience', which he thought to be the deciding factor, has somehow been transmuted into 'simplicity'. The following argument has probably carried a good deal of weight: 'The simplicity of the world is comprehensible only because we have chosen laws which are themselves simple. The simplicity of the world is thus of our own creation.' The conventionalist is sometimes then caricatured as going on to say that the world itself is of our own creation. A more subtle caricature makes him affirm that the natural sciences are merely logical constructions, dealing with a world implicitly defined by their hypotheses and rules of procedure. And this is the point at which most philosophers baulk at what they conceive to be conventionalism: it is not so much that they wish to deny the permanent possibility of achieving equal or better predictive and explanatory success in new ways, but rather that they

COSMOLOGY AND SCIENCE

fear for their doctrines of the nature of truth.[14] The scientist, on the other hand, tends to be uneasy with a theory which seems to scorn the permanency of propositions which he has long cherished as fundamental. For him, each case must be argued on its own merits. This chapter will end with what for modern cosmology is perhaps the most controversial case of all. Is the nature of space purely conventional?

It is always rash to maintain the possibility of doing something before it is done—in a scientific context no less than elsewhere. Have we any assurance that it will always be possible to save a given feature of a particular theoretical scheme by suitable amendment elsewhere? Will it always be possible to retain, for example, Euclidean geometry? That this will always be possible is much too strong a claim, but one not entirely without merit, as we hope to show.

2. THE DUHEM-QUINE THESIS

The problem begins with these considerations: on the face of it, the falsity of a theory (T) does follow deductively from a denial of one of its observational consequences (O), whereas the truth of the theory does not follow deductively from the truth of such consequences. This has for long been recognized. Symbolically we can write:

(i) $T \supset O$
O
―――
No conclusions of scientific interest.

and (ii) $T \supset O$
$\sim O$
―――
$\sim T$ (by *modus tollens*).

Duhem pointed out, however, that it is usually an individual hypothesis (h), and not an entire theoretical scheme, which is called in question.[15] Denoting the remainder of T, when h is removed, by H, we can now write:

(iii) $(h \cdot H) \supset O$
O
―――
No conclusion.

and (iv) $(h \cdot H) \supset O$
$\sim O$
―――
$\sim(h \cdot H)$

―――
[14] If mention of these has been avoided, it is because most of the traditional discussion on the subject is aside from the issues which concern this book. Some remarks on the subject are given, however, in the Appendix, Note XVIII.
[15] Pierre Duhem, *The Aim and Structure of Physical Theory* (Princeton, 1954) part ii, ch. 6, especially pp. 183–90. This edition is the translation of the second (1914) edition of a book first published in Paris in 1906.

There is therefore no asymmetry between verification and refutation as far as the isolated hypothesis h is concerned, for one can always throw the blame for refutation on the set of hypotheses H. Quine went further, holding in effect that one can always blame H and always amend it to H' in such a way that the actual observations do follow from the theory. Now Grünbaum has very properly denied that we have any logical guarantee of the existence of an H' of the required kind, that is to say, one which allows us to write

$$h \, . \, H' \supset O',$$

O' being the sentence expressing the actual observations.[16] The availability of a suitable H' must obviously be demonstrated for each individual case. But Grünbaum goes one stage further, arguing that the Duhem–Quine thesis is false and that there are cases in which it is demonstrably not possible to find a suitable H'. Although it is dangerous to claim, in effect, that human ingenuity cannot provide such an alternative—for of necessity our discussion is restricted to existing conceptual schemes—his counter-example is of some interest. But before considering it, let us summarize our position. We can always blame H, rather than h, in the face of contrary evidence. We know that in countless instances it has proved possible to amend H to H'. We know that often it has proved too difficult—one would not dream of saying 'impossible', in most cases—to make such an amendment. We are sure that intractable cases will continue to arise. Thus acceptance of the strong thesis (S) *It is always possible to find a suitable H'* cannot make one jot of difference to scientific procedure. With this important reservation, Grünbaum's argument may be considered.

Grünbaum takes h to be the (applied) geometry and H to be an assertion concerning the freedom from perturbation of the measuring-rod. This assertion, according to him, may be verified by showing two chemically different solids to agree in length everywhere in the region of interest. In the event of counter-observations ($\sim O$) it follows by (iv) that h and H cannot both be acceptable. But H is supposedly above reproach, and therefore it seems that h must be rejected.

The entire argument hinges on the hypothesis H. The chemical difference of the rods is adjudged a matter for rudimentary observation, involving no physical theory. As for 'freedom from deforming influences', it is argued that 'this can be asserted and ascertained independently

---

[16] Article in *Boston Studies in the Philosophy of Science*, ed., Wartofsky, (Dortrecht, 1963).

of (sophisticated) collateral theory'. In short, Grünbaum's procedure is to find a hypothesis H which can be accorded moral certainty. He points out that if it is necessary to make corrections for rod-deformation and these involve the geometry (h), then the schemata (iii) and (iv) will not do. Quite apart from such considerations, however, we can see that Grünbaum's argument is in the wrong spirit. For from the fact that it is conceivable that a hypothesis such as his H may be verified beyond all reasonable doubt it merely follows that it is conceivable that the strong thesis (S) is false. The falsity of S is not an analytic affair; but it is after all, extremely unlikely that Duhem and Quine could have been under the impression that its truth was a logical truth.

Although our main arguments do not rest upon the certainty with which we endow H, yet one or two features of Grünbaum's supposedly irrefutable hypothesis H should be underlined. Why are the rods said to be 'chemically different'? Presumably because if there are perturbing forces to be found, they will act differently on chemically different solids (as would temperature, for example). But what if the forces were indifferent to chemistry (as gravitational forces are supposed to be)? That they are not so, is an assumption which Grünbaum ought to place alongside h and H in schema (iv). Secondly, can one be so certain of H after all? What if a procedure is laid down (within the applied geometry) for comparing the lengths of widely separated objects? Clearly the two rods side by side may agree in length, and yet differ in length from a third, distant, rod with which they agreed on being moved together. Are perturbing forces not at work? Suppose, however, that no such procedure is available, and suppose that someone is resolved to retain Euclidean geometry at all costs. In Grünbaum's example it might seem that such a person is driven to postulate the 'dependence of the rod's length on the independent variables of position or orientation'. Is a remetrization not incompatible with the Euclidean h? Grünbaum thinks so, and objects to the idea that one should tamper with the word 'congruent'. To do so, he writes, is to 'violate the requirement of semantic stability'. But this is to beg the argument against conventionalism. If changes in scientific structure are desirable, why should semantical rules be excepted from change? If the syntactical element in the meaning of a word is to change, as the theory as a whole changes, does it matter that the contribution to its meaning which derives from so-called rules of correspondence should also change? 'Semantic stability' is, of course, to be prized in the short run. On the other hand, to characterize semantical change as evidence of 'instability' would be grossly misleading. Numerous examples of discrete semantical changes were given in Part I—'simultaneous', 'distance', 'force', 'stationary', 'expanding', and so on. Each, in its way, involved a minor triumph. As to the charge of triviality, we have already had something to say on this score in connexion with the Eddington–Milne dispute.

'It will always be possible to find a suitable modified hypothesis or set of hypotheses, H', to account for actual observations'. We have already indicated ways in which this strong thesis is unsatisfactory. It cannot be analytically demonstrated—but did anyone ever suggest that it could ? As for the interpretation of the schemata given above, Grünbaum can surely not be following Duhem's intentions. Where the schemata are useful is in explaining how it is possible to preserve an isolated hypothesis h. To take h, as Grünbaum does, to be the whole of 'applied Euclidean geometry', is to put too great a strain on the simple schemata. Turning to Chapters 8 and 9 of Part I, we find that Milne, Whitehead, Birkhoff, and all those who wished to keep a Euclidean metric had in common a desire to keep no more than the uninterpreted Euclidean calculus. To do justice to their procedure one must fragment h at the very least, into an uninterpreted Euclidean calculus (h_1), an uninterpreted calculus pertaining to any accompanying physics (h_2), and the whole set of correspondence-rules which relate the elements of h_1 and h_2 to the total theory—including its empirical generalizations— (h_3). Faced by observations which conflict with what is predicted, one can now only deny the conjunction of h_1, h_2, h_3, and H. However sure one may feel of H, therefore, h_2 and h_3 still admit of adjustment, if it is wished to retain 'pure' Euclidean geometry (h_1). It may be the case, of course, that we wish to retain h_2 or h_3 or even only a part of either. Thus according to Bergmann and Brunings, it is always possible to set up a coordinate system such that the relativistic equations of motion of any order have Newtonian form.[17] Again, Infeld and Scheidegger have argued that 'it is only a matter of representation whether the relativistic effects are explicitly contained in the equations of motion or in the metric field'.[18] The prevailing opinion is admittedly that the physics should be kept as simple as possible, and that the geometry should be sought accordingly; but this, in its way, is just as extreme as the view of the resolute Euclidean.[19]

[17] *Rev. Mod. Phys.*, xxi (1949).   [18] *Canad. J. Math.*, iii (1951) 195.

[19] Notice that there are snares in all arguments from simplicity—even in regard to claims for the paramount simplicity of a given geometrical calculus. Thus Karl Menger

What conclusions can be drawn from this chapter? Certainly not that it will always be possible, even in principle, to retain a given hypothesis, or set of hypotheses, in any scientifically valuable way. But false statements can often be turned into good maxims. The scientist who, acting as though he believes it possible to hold on to a favoured hypothesis, explores one way after another of doing so, is likely to be in the van of scientific advance. One must disparage the complacency of those who believe that any scientific theory embodies permanent truths. 'The world is always as it seems' (Wittgenstein) is a pernicious principle which, if it were ever consistently followed, would be fatal to natural science. It is the essence of conventionalism that no scientific theory is unambiguously determined by the observations it is to explain. This may seem very much like a platitude which could as well have been given at the outset. But, as Locke taught us, more important than the platitude is the task of clearing the way for its acceptance.

---

argues that hyperbolic (and not Euclidean) geometry is the only one which can be developed from a certain small group of simple assumptions as to 'continuity', 'intersecting' and 'joining'. *Albert Einstein, Philosopher-Scientist,* ed. P. Schilpp (New York, 1949) p. 464. Although he is no doubt right within his chosen terms of reference, it would be rash to claim that the same result follows from all simple sets of primitive terms.

CHAPTER 14

GENERALITY, SIMPLICITY, AND THE COSMOLOGICAL PRINCIPLES

And so far as the nearest fixt stars are from our Sun, so far may we account the fixt stars distant from one another. Yet this is to be understood with some liberty of recconning. For we are not to account all the fixt stars exactly equal to one another, nor placed at distances exactly equal, nor all regions of the heavens equally replenished with them.

NEWTON

'. . . and all places are alike to me.'

KIPLING, The Cat that Walked by Himself

ONE of the regulative principles which perhaps all formal disciplines claim is that which is loosely known as the 'Principle of Simplicity'. The principle is evident in much of Greek astronomy, and explicit in the work of Grosseteste and Duns Scotus: it has, in one form, become widely known as 'Ockham's Razor'.[1] The logical import of Leibniz's Principle of the Identity of Indiscernibles was also clearly that of a principle of economy. We find Newton affirming that 'Nature does nothing in vain'. Nearer our own time, Ockham's Razor has been used (or so some of the Logical Positivists have supposed) to rid philosophy of its metaphysical content. Without going further into the innumerable forms which principles of this sort have taken, it is hoped to trace their influence on recent cosmology, where it is tacitly understood that the 'assumptions' of individual theories should be kept to a minimum. The problem is more involved than in the case of an isolated deductive system, for now one must not lose sight of the exigencies of 'saving the phenomena'. The main difficulty, however, is much the same: it is the difficulty of formulating criteria by which the overall simplicity of a theory may be judged. Is this to be judged by the number of its 'theoretical terms', by the number or complexity of its hypotheses, or by the number of its hypotheses of highest order?

The informal remarks of the various cosmologists would suggest that rarely is the first statement of a new theory granted a minimum of hypothesis. As for the simplicity of the hypotheses themselves, the

---

[1] This was variously phrased by Ockham, but the most famous version which runs *Entia non sunt multiplicanda praeter necessitatem* was given by John Ponce, a seventeenth-century follower of Scotus. See A. C. Crombie, *Augustine to Galileo* (Heinemann, 1957) p. 231.

history of theories of gravitation just before the General Theory of Relativity shows that well-informed investigators will, as often as not, prefer to add new terms to the mathematical equations expressing existing laws, rather than add a complication of principle. Here we see how two distinct aspects of a theory's overall simplicity have to be weighed, one against the other. By what criterion can simplicity in one respect be compared with simplicity in another? Is it possible to measure simplicity in even a single respect? If it is possible to do so, are we justified in saying that the simpler theory or hypothesis is the truer? We shall have to form an opinion on these matters if we are to understand the most frequently quoted argument for the so-called cosmological principles. This chapter will therefore begin with some brief remarks, not by way of answering these questions fully but in order to indicate what appears to be the most reasonable attitude towards them.

1. SIMPLICITY

If one examines the various principles of simplicity historically it soon becomes clear that they can be divided into two categories—those dealing with things and those dealing with signs. It may be that the individual can form some insight into the 'inherent complexity of reality', if it is proper to talk of such a thing; but at least as far as science is concerned, one must begin with words. What might be called the 'semiotic simplicity' of a theory can then be analysed into these parts:

(i) That of the uninterpreted logical and mathematical calculi.

This must itself be a function of the number of the possible sorts of variable, of the number and complexity of the axioms and rules, and of the notational economy. Compactness—as in the tensor notation—sometimes works against the interests of clarity, or so it is widely believed. Quite apart, therefore, from the difficulty of striking the correct balance between the several aspects of this first kind of simplicity, a person wishing to found a metrical study of the subject will have the unenviable task of deciding, as it were, whether to put notational simplicity in the numerator or denominator of his overall simplicity function.

(ii) That of the semantical rules, and other rules of correspondence between the calculus and terms denoting theoretical concepts.

Do we, for example, need many or few such rules? Do they require one-place or many-place predicates?

(iii) That of the ordering of hypotheses into levels, in the resulting physical theory.

It is not at all obvious, of course, that complexity increases simply with the number of levels. The skill with which a concatenation of levels is attained, is also relevant.

(iv) That of computation.

This is clearly bound up with (i), but the considerations placed under that heading are not likely to give the whole story. One differential equation, for example, may be much easier to solve than another, which in the light of (i) is more complex; and degree is obviously more important than order in this case.

This list of respects in which simplicity may be sought does not claim to be exhaustive. Another respect, for example, relates to the proximity of a theory to others which may serve as its model. This connects with what may loosely be called 'presupposition', which is in turn linked with the psychological advantage accruing to any theory.[2] Behind simplicity in all its aspects, however, there appears to be a strong element of practical convenience. But simplicity and convenience are not one and the same; and no right-thinking cosmologist will be prepared to sacrifice universality, ease of confirmation, depth and theoretical unification on the altars of convenience and simplicity.

We can now see that even should it be possible to reach agreement on suitable ways of measuring each kind of simplicity, combining these measures into a function acceptable by all parties is never likely to be achieved. For one thing, not all kinds are compatible. It seems that we must judge and compare one theory with another on an *ad hoc* basis. Nebulous assertions of the need to prefer the simple to the less simple should, at the very least, contain the stipulation *'ceteris paribus'*. Unfortunately, when comparing theory with theory, other things are never equal. Only when it is a question of making piecemeal adjustments to a given theory will the rule be tolerably easy to apply.

Lastly, what of hypotheses of the type 'Nature is simple'? In order to discover their status one must ask by what evidence they could be justified. Could they, perhaps, find support from the fact that there are

---

[2] Science is certainly not distinguished by the absence of questions of personal taste, as it was widely supposed to be. An interesting example concerns Eddington and Lemaître. Eddington preferred the initial expansion to be 'not too unaesthetically abrupt', whereas Lemaître took exactly the opposite view.

simple theories describing Nature? There can be no way of discovering the extent of Nature's simplicity, for there can be no guarantee that another theory will not come along which is in some sense less complex: simplicity is above all, a relative matter. The trouble here is that although many will acknowledge the statement to be true, few think that it behoves them to defend it. In the long run it seems best that we should not assign it a truth-value, but regard it as an injunction: Construct your theory in a way as simple as is compatible with freedom from error!

2. GENERALITY

From considerations of simplicity it is a short step to the subject of generality—indeed, the two notions are often regarded as contraries. The subject as it concerns us is perhaps best introduced by way of example. Consider first Milne's Kinematic Relativity, and the reasons for its neglect long before it was thought to have been refuted by observations. No doubt the theory was largely ignored because it was widely believed that Milne followed a purely deductive path—and Milne encouraged the belief. But perhaps more important than any other reason was the fact that, however simple, the theory was held to be a special case of the General Theory of Relativity, offering less scope for the free adjustment of parameters. It seems that, in the last resort, its 'conceptual basis' counted for very little.[3]

On the other hand, the extreme flexibility of the models of the General Theory of Relativity, to which their adherents so often draw attention, has been held by Milne, Gold, Bondi, and others to be a distinct disadvantage. One of the few arguments Milne offered for his belief that 'the less general is to be preferred to the more general', may be paraphrased thus:[4]

> It is widely held that Kinematic Relativity is a special case of the General Theory of Relativity.
>
> The universe is a particular case.
>
> (Tacit conclusion): Kinematic Relativity is to be preferred to the General Theory.

---

[3] Kinematic Relativity was also thought to be at a disadvantage in so far as it did not place *all* possible observations on an equal footing, as did Einstein's basic theory. When Weyl's Principle is taken into account, however, from the point of view of cosmology the two 'relativities' are not very different; for both now have their privileged fundamental particles.

[4] *K.R.*, p. 226.

It is difficult to decide how important this obvious *non sequitur* was in fashioning Milne's ideas. We have already hinted at Milne's belief that the path he followed did not allow of alternatives. 'There must', he argued, 'be a unique solution to the question "Why?" applied to the universe in respect of each feature.'[5] The reason for the multiplicity of the solutions offered by the General Theory of Relativity, according to Milne, is that it did not begin 'at a sufficiently primitive level'. Like Descartes, he believed it necessary to pursue, starting from first principles, 'a single path towards the understanding of this unique entity, the universe', a path which was correct only if it allowed of no 'bifurcation of possibility'.[6] (It would be interesting to know whether Milne did not regard the variety of possible generating functions ψ as a 'bifurcation of possibility'.) This is, more or less, a repetition of statements to be found in his two earlier books. It is more extreme than Bondi's belief that steady-state cosmology is to be commended on the grounds of its definite, rigid, and vulnerable predictions.[7]

Bondi's argument was quite different from Milne's, for he stressed that his theory should take precedence over the rest because it was in principle the one most easily disproved by observation.[8] (We notice that this argument is not closely bound up with the tenuous aesthetic principles of the last section.) In support, Bondi and Kilmister quote K. R. Popper: 'Once a hypothesis has been proposed and tested, and has proved its mettle, it may not be allowed to drop without "good reason".'[9] Popper allowed as 'good reason' the possibility of replacing the hypothesis 'by another which is better testable'; but it is probable that he did not intend this to be interpreted as here. 'Better testable' may be read as 'more easily refuted', but there are at least two ways in which this may be understood. Popper might be taken to mean that the *kind* of observation required to test the new hypothesis may be easier to perform than that required for the old—in which case it is not at all obvious that his thesis is plausible.[10] Bondi seems to argue, however, that of two theories, the first allowing of any one of a whole range of values for a given quantity, the second indicating a unique value, the

[5] *Modern Cosmology and the Christian Idea of God* (C.P., Oxford, 1952) p. 49.
[6] loc. cit.   [7] *Jubilee R.T.*, p. 152.
[8] Bondi and Kilmister, *B.J.P.S.*, x (1959) 56 (review).
[9] *The Logic of Scientific Discovery* (Hutchinson, 1959) p. 53.
[10] Consider the kind of (independent) observation required to test, for example, the hypothesis of the perturbation of the planets by the matter giving rise to the zodiacal light. It was virtually impossible to check the hypothesis except in terms of the perturbations it was meant to explain, and yet it appears to have been widely entertained. Admittedly one can always claim that those who entertained it had a poor sense of values.

latter is the easier to disprove. This may not be true, however. It is, as a purely contingent matter, easier to refute 'The length of my ruler lies between two and three feet' than 'The length of my ruler is one foot'. But this is because my ruler is of the commonest English variety. Suppose that the word 'ruler' were to be replaced by an anonymous symbol for which any suitable word could be substituted. We cannot even begin to discuss the relative ease of disproving the two propositions. Suppose now that the first sentence were to be replaced by 'The length of my ruler lies between eleven and thirteen inches'. Again as a contingent matter, it would now be the harder to refute. It seems that neither type of proposition can be characterized as permanently the easier or the harder to disprove. It appears that, as with simplicity, not all propositions can have an inherent degree of falsifiability. Moreover, those who take paucity of undetermined parameters to be an infallible guide to simplicity are apparently wrong in at least one of their assumptions—for in both examples the second statement was the rigid one, whilst the first required one more variable for its statement in logical terms.[11]

The examples of the last paragraph achieve their end by restricting the range of the parameter which we suppose to represent the length of the ruler. It does, in fact, seem reasonable to say that where the special prediction is a case of the general, the unique value cannot be the harder to refute. Where observations point to the near-correctness of a specific value which lies, nevertheless, within a range permitted by inequalities belonging to a rival theory,[12] we should still hardly favour the one theory rather than the other unless we believed that the greater virtue attends the greater risk.

(Lest one be left with the impression that the steady-state theories are inflexible and easy to dismiss, we recall the way in which a term allowing for galactic evolution has on several recent occasions been used to remove the very property of inflexibility. Perhaps this was a case of sacrificing virtue to security.)

Gold has argued in a similar vein to Bondi's. Our aim, he believes, is

[11] It has been held that the greater the number of the parameters of a theory, the greater its prior probability. 'Probable' is a notoriously ambiguous word, however, and this claim will be identical with Bondi's if the word is taken to mean something like 'least rigorously testable'. Is the severity with which the theory is to be tested to be judged on practical grounds or 'in principle'? The example above suggests that it is impossible to make any prior judgement of the issue 'in principle'. If 'probability' is given any other meaning, it is not at all obvious that the hypothesis quoted here is irrelevant to Bondi's argument.

[12] We recall instances where the pressure, density and curvature are stipulated to be governed by inequalities of state.

to devise a scheme 'whose economy of hypothesis is so great that it will be judged most improbable for so simple a fitting scheme to exist by chance'.[13] He maintains that the smaller the number of hypotheses, the more likely they are to be valid; and that it is a merit of his own and Bondi's theory to dispense with a number of 'cautious assumptions', replacing them by *one* 'drastic assumption'—that of continual creation. 'Likely' here presumably means 'desirable on the grounds of being severely testable'. Paradoxically, 'likely' will then mean, roughly, 'having a high degree of prior improbability'.

3. Deduction and Extrapolation

Although it seems that to give priority to a theory merely on the grounds of its inflexibility is unwarranted, yet it is not of itself likely to foster the sort of rationalism of which Whitehead, Milne, and Eddington, for example, stand accused. That there exists a very real danger of rationalism in cosmology stems, no doubt, from the small number of links with observation, and from the fact that most of the cosmologists of this era began their formal studies with applied mathematics, carrying over into cosmology many of their former attitudes. They looked for certainty, and could naturally find it only in the formal derivation of their results. When this was seen to be insufficient, some of them sought to introduce an empirical element which was above reproach. Consider, for example, Milne's desire to appeal to no knowledge derived from small-scale experiments, but to use 'only such brute facts, such irreducible facts, as are of the intuitive sort or do not rest on the questionable principle of induction'.[14] Bondi might be thought to have been taking up this idea when distinguishing between the extrapolations of the 'empirical school' and inferences from laws in the form of axioms, as made by 'adherents of the deductive approach'.[15] He clearly thought of the steady-state theory, not to mention the theories of Milne, Eddington, Dirac, and Jordan, as deductive. Now Bondi cannot have thought that these theories were adequately described as following necessarily from a given set of premisses. The system becomes a science only granted an interpretation, as he well realized. Without semantic rules, a theory would be no more than a word-game based on some or other calculus. All physical theories of any value are deductive in form. Rather than contrasting two different kinds of completed theory—'inductive' and 'deductive'—Milne and Bondi are surely rather

---

[13] *Vistas*, ii, p. 1724. [14] *Ap. J.*, xci (1940) 132.
[15] *Cosmology* (2nd ed.), pp. 5–7.

contrasting the orders in which different theories are compounded and, perhaps, subsequently expounded.

Consider Bondi's reference to extrapolation. When a writer expresses a distaste for the processes of extrapolation, he may be thinking of any one of the following sorts.

(i) The assertion of a general statement without having examined all possible instances.

(Milne's worry appears to be that an 'inductive' conclusion formed in this way could at best be only probable—hence his resorting to 'brute facts'. Bondi, on the other hand, clearly had something else in mind.)

(ii) Making a general statement involving 'things' which by their nature may in principle never be observed in any but an oblique sense.

(Here we have regard to the referential use of purely theoretical concepts.)

(iii) Making a general statement whose scope goes beyond what can in practice be observed or measured.

(Here 'beyond' may be in regard to distance, time, or any other property which can be simply ordered. 'In practice' may be read as 'in principle' under certain circumstances—as for example, if we are to extrapolate our laws to cover galaxies beyond the 'horizon' of a model.)

Did Bondi have any of these meanings in mind? It seems more likely that he was distinguishing those cosmological theories which are begun with cosmological problems in mind—as the steady-state theories and, perhaps, Kinematic Relativity—from those which require little or no extension of physical theories already proven on a small scale—as for example relativistic cosmology. Of course, caution dictates that a statement of the range over which a law has been found acceptable is an important part of its credentials. One of the chief concerns of all scientific theories, nevertheless, is to sanction extrapolation (in sense (iii)) and the expression of the distant and unknown in terms of what is familiar.[16] That 'the galaxies may have moved otherwise before exhibiting the present trend', provides all the more reason for our wanting to relate the present trend to a more general one, as all modern theories

---

[16] Interpolation may be included under this same heading. For some reason this is regarded as less dangerous than extrapolation, although one only has to refer to Hubble's Law and to the widely-held belief that the law does not apply within the Local Group, to see that the dangers of interpolation are no less.

of cosmology try to do. Extrapolative, if taken in this sense, is therefore hardly a useful category, for there is no scientific theory of any value which it does not include.

As we have said, it seems probable that Bondi (and to some extent Milne) was less concerned with a logical characterization of two different types of scientific theory than with the way in which cosmological theories have come into being. We recall the gradual way in which the General Theory of Relativity was unfolded, each move being ostensibly concerned with a narrow class of predictions and each being only a short step from a Newtonian analogue. We may compare this tentative means of advance with that of the theories of Whitehead and Birkhoff, and we may contrast them, perhaps, with the sweeping methods of Milne, Bondi, and Gold—and to a lesser extent Hoyle. The distinction is lost, on the other hand, as soon as we take relativistic cosmology as Robertson and Walker presented it. And even if we try to uphold the distinction for its historical value it is not at all obvious that Milne, Bondi, Gold, and Hoyle appealed to 'no knowledge derived from small-scale experiments'. To point to the derivation of a metric from the Cosmological Principle alone is no criterion. To quote Milne on this very issue:[17]

> As our view is that the geometry adopted by an observer is arbitrary, it would be idle to compare geometries; we should get no farther. Likewise it would be idle to compare any propositions stated in terms of coordinates. We must go back to the actual observations.

By the time they are developed so far as to allow of comparison with the statements of the astronomer, both the steady-state cosmology and Kinematic Relativity have made abundant references, not to observation, but to hypotheses of other theories which are known to account tolerably well for observations made on a small scale. (Take, for example, Bondi's assumptions that light follows a null geodesic of a Riemannian space[18] and that free particles follow geodesic paths.)

In summary, before passing to a special case to which these considerations are relevant: simplicity is difficult to define, but it is not always a purely aesthetic requirement. The severe restrictions on generality placed on the several theories discussed in Part I are not aimed at the suppression of unpleasing detail, but at simplifying the kinds of observation which are to be made, as well as at an easing of the mathematics. The last is especially true of the General Theory of Relativity, the mathematics of which its opponents, following Milne, customarily refer

[17] *W.S.*, p. 290.   [18] *Cosmology* (2nd ed.), p. 146.

to as 'cumbrous'.[19] But without some yet higher order principle of selection, there will always be an element of personal taste in deciding between the claims of generality and simplicity. By and large it seems that the former is usually preferred when the case is clear-cut. Indeed, the steady-state theory and the Einstein–de Sitter model of General Relativity have been stigmatized as '*a priori*', taking as they did, respectively, a constant value for h_1 and special values for λ and k.[20] Any account which has the misfortune to be followed by a more general one seems to run the risk of being accused of the same unreasonable restraint. When it is a question of comparing two different models based on the same theory, one can hardly use, in favour of the less general, the argument from rigidity which Milne, Bondi, and Gold proffered for their own theories. But when it comes to rejecting one theory on the grounds that it is less general than a completely alien one, the issues are much harder to settle. It is one thing to say that, for example, Milne's Dimensional Postulate is severely restrictive and that Kinematic Relativity would be better without it, but it is quite another matter to hold that since the Postulate is violated by Einstein's General Theory this must therefore be superior to Milne's Relativity, being the more general of the two.

It is important and relatively easy to compare the explanatory ranges (or 'observational bases') of two theories. One should hesitate, however, to judge theories by their 'order of generality'. One should hesitate, moreover, even to compare the generality of theories which are as disparate, conceptually, as the principal theories of cosmology. Broadly speaking, the more scope there is for the adjustment of a theory in the face of empirical evidence, the more general one will suppose it to be. To make the definition more precise, rules of translation must needs be laid down—as, for example, those between the Friedmann–Lemaître and the neo-Newtonian models, and those between Kinematic Relativity and the General Theory of Relativity. We have, in fact, already indicated that translations are bound to be incomplete if theories are to retain their identity. In the last resort it seems that in making a choice between the many available theories one must proceed on an *ad hoc* basis. A multitude of factors must be taken into account, and to repeat them would be to repeat a substantial part of the first half of this volume.

[19] *W.S.*, p. 316; *K.R.*, p. 226.
[20] See, for example, McVittie, *The Paris Symposium of Radio Astronomy*, paper xcvi.

4. Homogeneity and Isotropy. The Cosmological Principles

Many important steps have been taken in the name of simplicity. Our laws, we are often told, should contain no reference to epoch. 'The simplest view', writes McCrea, 'is that we should suppose there to be nothing to distinguish our position in time from any other'.[21] We have seen many instances where Hubble's Law has been interpreted without regard to this view. Again we are sometimes told that our laws should exclude all reference to the distribution of matter, if they are to be in their simplest form. (Some would have us use the clause 'if they are to be universally valid'.) Once again, there is a strong body of opinion which would, following Mach, put aside the simplification which this proposal offers. But there is one theoretical move which is universally agreed to lead to a high and desirable degree of simplification, and this involves introducing homogeneity and isotropy into the features of the cosmological model.

Before Riemann, the homogeneity of space was not questioned, for it was hardly clear that any meaning could be attached to an assertion of its inhomogeneity.[22] Riemann showed how it was possible to introduce the idea of a non-homogeneous space in a perfectly consistent way. Spaces of constant curvature were, however, the more interesting from a mathematical standpoint, and the more studied. In 1886 F. Schur presented an important theorem of Riemannian geometry:[23]

> If the Riemannian curvature of a space at each point is the same for all orientations then it does not change between one point and another.

In short, an isotropic manifold is also homogeneous as to Riemannian curvature. (The converse is not necessarily true. Gödel's space, for example, it not isotropic, but it is homogeneous.) Schur's theorem is obviously of some relevance to the cosmologist who wishes to make use of Riemannian geometry, for he can observe the world of galaxies in all directions, although from only one place. Of course he is always at liberty to subscribe to the idea of a spherically symmetrical system of galaxies, at the centre of which our own is to be found. As we have seen in other contexts, however, the cosmologist usually prefers to suppose that there is, broadly speaking, nothing unusual about any individual view of things. The assumption is accordingly made that the isotropic

---

[21] *Endeavour*, xvii (1958) 6.
[22] Leibniz's writings might be interpreted as exceptions.
[23] *Math. Ann.*, xxvii (1886) 537, see p. 563.

aspect which the galaxies present to us is typical of all possible aspects. Homogeneity (which for some reason is often said to be 'more fundamental' than isotropy) then follows in accordance with Schur's theorem.[24] It follows from Einstein's field equations that a universe which is homogeneous in regard to curvature allows of homogeneity in regard to pressure and the distribution of matter. In terms of a model universe comprising a continuous perfect fluid, this means that at each instant of time the density and pressure of the fluid are alike in any two equal volumes of space.[25] The idea can be extended to other properties of a model, such as features which correspond, for example, to the numbers, types, masses, absolute magnitudes, and linear sizes of the nebulae. The assumption of uniformity in these respects may be looked upon as more or less independent of uniformity in regard to pressure and density, although there will presumably be theoretical connexions between the two sets of properties, even if the connexions are at the moment undetermined. The model is thus assumed homogeneous in all respects, often in the name of simplicity but also because few observations have been thought to suggest alternatives which it would be reasonable to examine, with the possible exception of those on the apparent clustering of nebulae.

In Chapter 6 we saw the problem of homogeneity and isotropy was attacked on the basis of several different sets of assumptions—notably by Robertson, Weyl, Tolman, Walker, and Milne, before 1935. It was seen, too, how eventually Robertson and Walker made use of the principle which Milne called 'Einstein's cosmological principle'.[26] We recall the Robertson–Walker line-element which is applicable to all major cosmological theories that can be described in Riemannian terms, regardless of whether or not these require the paths of free particles to be geodesic.

Beginning with Einstein and Milne, we may now collect together some of the statements of the so-called cosmological principles. Einstein's 'All places in the universe are equivalent' becomes for Milne 'Not only the laws of nature, but also the events occurring in nature,

---

[24] A. G. Walker rigorously proved that a non-uniform model can present an appearance of spherical symmetry only to a single observer, and that homogeneity therefore follows from local isotropy if ours is not such a special position. (*J.L.M.S.*, xix (1944) 219, 227. Cf. *Obs.*, lxv (1944) 242.)

[25] R. L. Brahmachary considers the properties of a model containing an *imperfect* fluid ($T_1^1 \neq T_2^2$). (*Naturwiss*, xl (1953) 456.) A disturbed-state universe is found to change to one in which density and pressure are functions of the distance from the origin. On the differences between perfect and imperfect fluid see *R.T.C.*, pp. 216–22, 376.

[26] Chapter 8.

the world itself, must appear the same to all observers wherever they be, provided that their space-frames and time-scales are similarly orientated with respect to the events which are the subject of observation'. In Chapter 8 it was shown how Milne interpreted this as a restricted version of a 'principle of covariance'. Bearing in mind some of Milne's remarks, however, we may provisionally distinguish these three respects in which all (or a restricted group of) observers are expected to 'see the universe in the same way':

(i) in regard to the 'laws of nature',
(ii) in regard to the structure they impute to the universe,
(iii) in regard to any specific observation or, perhaps, sequence of observations.

Broadly speaking, (i) and (ii) are concerned with *types* of observation, whereas (iii) is concerned with actual observations. It may be that we should want to coalesce (i) and (ii), although many would certainly wish to retain the distinction between empirical generalizations ('laws' in perhaps the commonest sense) and hypotheses at a higher level. Few people would wish to say, for example, that the chosen expression for the line-element expresses a law: it belongs rather to the second heading. But it is the third heading which is of the most doubtful value. Surely no one can ever have intended to make the hypothesis that, for example, a specific and identified star will at some time or another appear to each observer to have the same properties. Instead one finds substituted something like 'the totality of observations that any observer can make is the same for all possible observers' (presumably also in regard to order); and the fact that the words 'on average' are usually included somewhere in statements of this sort suggests further that 'totality of observations' is to be interpreted not in terms of a conjunction (whether of a finite or an infinite number) of observations, but of a generalization, involving the universal quantifier. It seems, therefore, that only (i) and (ii) are worth retaining. In fact (i), as we left it, is ambiguous, and we shall probably not be far from the truth if we represent 'cosmological principles' as falling under one of these headings:

(i*a*) Principles of covariance (restricted or otherwise).
(i*b*) Laws (empirical generalizations) of aspect. Laws, that is to say, which say something like 'At whatever places (and, perhaps, times) such and such property of matter is determined, the results will tend to be the same'.
(ii) Hypotheses as to structure.

THE COSMOLOGICAL PRINCIPLES

Although hypotheses under the last heading might themselves specify a certain Riemannian metric, more probably it is the hypothesis or set of hypotheses which justify the choice of metric to which the name 'cosmological principle' refers. Broadly speaking these are hypotheses as to metrical isotropy and homogeneity. Through such equations as Einstein's field equations hypotheses of type (ii) might well be used to sanction hypotheses of type (ib): it is perhaps too ambitious to try to distinguish between these two categories in any case, for such properties of matter as its density cannot be measured at distant places except by a chain of inference peculiar to the chosen theory.

With these qualifications, we may briefly characterize (ia), (ib), and (ii) as, respectively, meta-laws, laws, and high-level hypotheses. It was apparently a meta-law which Milne had in mind when he referred to the Cosmological Principle as 'a definition, a principle of exclusion, enumerating the class of systems to be considered and excluding all others' and 'in no sense . . . a "law of nature".'[27] In saying this he antagonized a large number of writers who regarded his Principle as an ordinary law of science, and whose challenges he never satisfactorily answered—as perhaps he might easily have done; for he seems always to have hovered on the verge of a perfectly reasonable hypothetico-deductive account of scientific explanation. Like Eddington he appeared, to the world at large, reluctant to admit the need for recourse to experience in appraising a scientific theory. That the diagnosis was in each case mistaken can, unfortunately, be maintained only on the hypothesis that neither patient was familiar with his own condition. Milne referred to his Cosmological Principle as the 'simplest substitute for the unworkable concept of homogeneity'.[28] We have seen that one sort of homogeneity (in regard to Riemannian curvature) follows from an assumption of isotropy in a Riemannian manifold. On what grounds then does Milne object to the concept of homogeneity?

A system is homogeneous in a physical sense relevant to gravitational cosmology when, in the reckoning of a single observer, the densities at *all* points at a given time are assigned the same value by that observer. All observers will have the same criterion of homogeneity, but, because one such observer will, in general, be in motion relative to another, the events at which the latter judges the density to be the same, *and which he judges to be simultaneous*, will not appear to have this property of simultaneity to the former. 'Equality of density', wrote Milne, 'is not an attribute of a pair of points or particles alone,

[27] *W.S.*, p. 20, note. [28] ibid., p. 124.

but an attribute of a pair of points or particles considered in conjunction with a specified observer'.[29] The argument so far relies upon the Einsteinian denial of absolute simultaneity, but of course Milne realized that coordinate systems may be defined with respect to which two distant events, simultaneous for one, are simultaneous for all. The usual notion of homogeneity will then apply, just as it may, without contradiction, be applied to a static distribution of matter.

Another important observation by Milne on the subject of his Cosmological Principle was that it 'does not prescribe *a priori* whether the total population is infinite or not, whether it is locally homogeneous or not, whether it possesses world-wide homogeneity or not, whether it is included in a finite volume or not'. Lest it should be thought that in application the principle points unambiguously to homogeneity, regardless of the rest of the theory, we are reminded not only of the point made in the last paragraph, but that in Milne's own application of the principle the density-distribution was described as 'increasing outwards in the experience of any observer', although locally homogeneous. Milne's principle, even so, is often referred to as merely an assertion of homogeneity.

In applying his Cosmological Principle, Milne appears to have made a logical mistake which indicates that a view of space somewhat like Newton's is all too easily assimilated. As H. S. M. Coxeter and G. J. Whitrow have pointed out, he was seemingly led astray by an uncritical use of the word 'orientation'.[30] In keeping with his Cosmological Principle, the description of Milne's entire system—all particles of which are in uniform relative motion—from each of three particles A, B, and C, must be identical. The first of these, A, must therefore see a particle to which he assigns the same velocity as C assigns to B. Milne, it was explained, went further and assumed not only that A and C assigned the same *speed* but also the same *orientation* to the two particles. (He inferred that, for this to be possible, A, B, and C must be collinear.) Without identifying some background of reference, it is difficult to see how any meaning can be attached to the idea of orientation. The criticism can be extended to Milne's construction of the substratum of his continuous models and to the associated statistical systems. Rather than rely upon this dubious foundation, Coxeter and Whitrow preferred to replace Milne's Cosmological Principle by an effectively equivalent

---

[29] ibid., p. 66.
[30] *P.R.S.* (A), cci (1950), p. 419.

combination of the sample principle,[31] the principle of spherical symmetry about each fundamental observer,[32] and criteria for the geometry of light paths similar to those of Helmholtz and Lie.[33]

'Cosmological principles' returned to the limelight with the advent of the series of papers on steady-state cosmology in 1948. Bondi and Gold, like Milne, did not distinguish between different aspects of what they conceived to be a single principle. They did not, for example, clearly distinguish between the hypothesis that matter is, on a sufficiently large scale, uniformly distributed, and the regulative principle which is concerned with the 'unrestricted repeatability of all experiments'. This, containing an assertion that the outcome of an experiment is not affected by time and space *per se*, they described as the 'fundamental axiom of physical science'.[34] Bondi, writing in his book *Cosmology* on the subject of their cosmological principle, selects three arguments from the 'great variety of arguments for its validity'. Thus although he still speaks of it as a single principle (for the moment we overlook the fact that he and Gold used the so-called 'perfect' form of it), it is clear from these arguments that in all but name it is more than one. According to the first argument the principle is again said to be presupposed by all physics; for unless 'position in space and time is irrelevant' it will follow that 'the repetition of an experiment becomes quite impossible'.[35] In his insisting that the world shall be 'homogeneous as far as the laws of nature are concerned', Bondi's words are reminiscent of several passages by Whitehead, who insisted upon the homogeneity of space in order to make measurement possible. 'In any theory which contemplates a changing universe', writes Bondi, 'explicit and implicit assumptions must be made about the interactions between distant matter and local physical laws'. But what sort of interaction can obtain between *matter* and *laws*? If to suppose an answer possible is not to make a simple category mistake, this must be a question as to whether any law dealing with matter in motion is allowable if it does not take into account distant matter.

---

[31] Whitrow, *ZS. Ap.*, xii (1936) 47; xiii (1937) 113. The principle involves only scalar quantities, and does not lend itself to Milne's mistake. According to Whitrow, 'the sample principle ensures that local phenomena are everywhere the same' (p. 114. Cf. p. 54 of the first paper.) He went on to show (pp. 115–22) that, in the derivation of the equations of motion of a test particle, the sample principle is as powerful as Milne's Cosmological Principle, but that in constructing supplementary statistical systems it is necessary to augment it with Walker's postulate of spherical symmetry (pp. 122–24).

[32] Walker, *P.L.M.S.*, xlii (1937) 90.

[33] Robertson, *Ap. J.*, lxxxii (1935) 284.

[34] *M.N.*, cviii (1948) 252, section 1.

[35] This and the following quotations are from *Cosmology* (2nd ed.), pp. 11–12.

Milne, for one, disliked the relativistic view that 'space time and material phenomena are interdependent'; and that 'no one of them can be assigned without affecting the character of the others' he very properly described as 'a rule of procedure, not an assertion about some preconceived entity "space".'[36] In a sense, however, he used Mach's idea on one occasion in an analysis of the law of inertia (Newton's first law of motion).[37] (He decided that in his reformulation of the Newtonian system, with a uniform substratum instead of empty space, and so on, the universe would not influence uniform rectilinear motion gravitationally. This would only follow from local inhomogeneities.) Did Bondi think that 'Mach's Principle' was essential to any satisfactory physical theory? As it happens he never had to answer the question, for he and Gold side-stepped it by adopting this 'Perfect Cosmological Principle': 'Apart from local irregularities, the universe presents the same aspect from any place at any time'. Mach's Principle, or any other 'interaction principle' where what interacts is taken into account in the 'aspect', can make no difference to a theory with a constant background of matter.

Of Bondi's three arguments for the cosmological principles, the first, already broached above, is the most interesting. It is not easy, however, to be sure of his precise meaning, and partly for this reason a reconstruction will be given of the sort of argument he seems to have in mind. Taking his other arguments first we start with his claim that the principle that the Earth is not in a central specially favoured position 'has become accepted by all men of science'. It is, he argues, only a small step to the statement that the Earth is in a *typical* position; and with a suitably wide interpretation of the word 'typical', his cosmological principle follows. He presumably had in mind the paper by Walker in which it was shown that since a non-uniform universe can present a spherically symmetrical aspect only to an observer in a central position, homogeneity follows from local isotropy if we assume that the Earth is not in the *special* position.

In this, as well as in his third argument, Bondi treats his cosmological principle as one which concerns only the apparent distribution of matter. The third argument is on the grounds of simplicity—which is assumed to imply uniformity. This is adopted, on the other hand, 'purely as a working hypothesis'—and that Bondi would be prepared to say as much for the 'repeatability principle' seems most unlikely. In any case, the latter is a principle of higher logical order than the

[36] *W.S.*, p. 11.   [37] *P.R.S.* (A), clviii (1937) 324.

hypotheses of the theory, whilst the former is not. If the former is to be sanctioned by the repeatability principle, it is presumably in conjunction with some other principle; and this brings us back to Bondi's first argument. The new premiss is what we called an 'interaction principle', of which the Mach–Einstein principle will be regarded as a special case.

The first argument might then go as follows.

Premiss (i) If an experiment is repeated at different places (or times)* under the same conditions as previously, the same result will be obtained.

Premiss (ii) The distribution of matter is taken into account in the laws which we suppose to govern the results of an experiment—or, as it is usually put, 'which govern local phenomena'.

Conclusion (iii) The distribution of distant matter must at any one time (at all times)* look the same from all places.

(This inference assumes that it is possible to speak of an experiment as being performed at two places at the same time.)

We could, of course, deny the (narrow) conclusion if we were prepared to argue that the result of an experiment not only depends on the distribution of distant matter as 'seen' from any given place, but is such a function of place (in space, time, or both) as exactly to compensate for changes in the distribution of matter. On the other hand, there seems to be no difference between saying this and denying the need for an interaction principle. Also, there is the danger of becoming involved in the Absolute Space issue, which will not arise given a straight denial of an interaction principle.

The first objection to this inference is that if the premisses are supposed inviolable, then it seems impossible that one should be able to adapt one's theory to deal with the non-uniformities of the world. 'Look the same', in (iii), must surely be taken in an exact sense. One might wish to deal with non-uniformities statistically, but this would presumably suggest that (i) be given a statistical interpretation. Such a step would certainly not be to the liking of most physicists. Bondi and Gold were not alone in thinking of (i) (as it is here expressed) as the 'fundamental axiom of physical science'. But how secure is the principle, and how could it possibly be refuted?

* Inclusion of the asterisked phrase leads to the stronger ('perfect') form of cosmological principle.

The repeatability principle is, in a widely accepted sense, a meta-scientific principle. As such it can be found incompatible with others at the same logical level, but it cannot be refuted in the same way as can a simple empirical generalization. Even so, there are some who wish to qualify it in at least one respect. These qualifications will be considered, to see whether they are necessary. In doing so it will be found that the inference is only valid granted a very narrow and unsatisfactory interpretation of (i).

As Herbert Dingle has observed, for a hundred years it has been widely believed that no experiment is exactly repeatable, the entropy of the universe never being twice the same.[38] A first, and minor, observation is that of course no assertion can be made, one way or the other, outside the context of a specified cosmological model. But if a particular model is chosen which either indicates, or is consonant with a theory of thermodynamics which indicates, a progressive change in entropy, this does not mean that the principle (i) must be abandoned. It means rather that if we have any reason for believing that the entropy of the universe is likely to influence the result of an experiment, it should be taken into account in the appropriate law. As with Mach's Principle, we do not say that the entropy of the universe 'influences our laws'.

A difficulty appears to be that if we eliminate entropy from our law (always assuming that we have found how to introduce it) in favour of a theoretically known function of time and other variables, we are left with a time-dependent law, which some would wish to exclude on principle. One reason for excluding reference to time (except differentially) from our laws seems to be this: in 'classical' science no time-involved law is to be found, and therefore the temptation is for those who wish to introduce time into existing laws to speak of the 'constants' of the law as 'varying in time'. The first objection, namely that this is a contradiction in terms, can be countered by a simple change in terminology. A more important objection is one which relies on the stonger version of the repeatability principle (i). By noting the result of a time-dependent experiment we may, it is said, 'find the Absolute time'. And the concept of Absolute time is vaguely held to be indefensible. This sort of argument, however, begs an important question. It assumes that the 'time' which enters into our time-dependent law is somehow 'time-in-itself'. Thus we have an analogy with the supposedly non-sensical concept of place-in-itself, and hence with Absolute space. It

[38] M.N., cxiii (1953) 393.

is doubtful whether anyone would be prepared to object to a place-dependent law if 'place' were interpreted as 'place relative to matter'. Our inference from (i) and (ii) to the weak form of (iii) depended on the assumption that this was a valid concept. In other words we deduced the weak form of (iii) by taking 'place relative to matter' as a single variable rather than as a two-place relation between the matter in the universe and place-in-itself. And similarly the strong form corresponding to the 'perfect' cosmological principle followed an analogous treatment of the time variable. Only in the qualifying paragraph which followed did we consider the second possibility, pointing out that the interaction principle might as well be denied as suppose that (iii) followed from it and (i) by such special pleading as we there considered. The argument does not therefore commit us to the concept of time (or place)-in-itself—in fact quite the contrary. We certainly have said nothing on what can be meant by 'time (or place) relative to matter', but any time determined from an experimental result would correspond with this concept, rather than with that of time-in-itself. We might equally well speak of 'place relative to matter' as determined, say, by the discovery of a systematic variation of the velocity of light from place to place, where places are determined by any reasonable independent criterion. In this sense both time and place could function in physical laws and the repeatability principle could still be upheld.

Another way in which (i) might be brought under attack is by asserting that as a scientific principle it is trivial, being in the nature of a mere definition. It has been suggested that a version of the 'Law of Causality' which coincides, more or less, with this principle is in reality no more than a definition of 'under the same conditions'.[39] By 'conditions' one presumably means those which are assumed relevant to an experimental result but in an unknown way. They are therefore kept constant, as between experiments. 'Conditions' may, of course, be known or unknown, and it seems to be with regard to the latter that the quoted claim is being made. It is said that the only way of deciding whether an experiment has been repeated under the same conditions is to see whether it gives the same results. This is unsatisfactory in more ways than one. At first sight it appears to make nonsense of statistical systems for which no unique experimental result is to be expected. The force of this objection might be questioned, however,

---

[39] This is not unlike the claim made by Philipp Frank in 1907. See ch. 1 of *Modern Science and its Philosophy* (New York, 1955). Amongst those who objected to Frank's views were Einstein and Lenin.

were the 'result' interpreted as 'mean of many results'. The principle (i), on the other hand, if intended as a definition of 'likeness of conditions', is a very poor definition; for where the result depends upon two or more factors there might be compensating changes in these which lead to similar results with changed conditions. If it is assumed that likeness of conditions can be independently judged, then there is no point in interpreting (i) in this way. Although it offers us no assurance that the conditions under which two observations are made are the same, (i) certainly implies that given a change of experimental findings, with all known conditions allowed for or kept constant, then there are unknown 'forces' at work. But this does not make (i) a *definition* of likeness of conditions.

To return to the subject of the inference to the strong or weak version of the Cosmological Principle, using premises (i) and (ii): there is something very unsatisfactory in it when (i) is interpreted as we have interpreted it—and this from the point of view of two distinct groups of theories. Those theories which reject all interaction principles cannot use it, and those which require the strong version of the Cosmological Principle use it only to dispense with one of its premises. (As we have suggested already, interaction with an unchanged background can be presupposed, but not detected.) Let us suppose, however, that the inference, *qua* inference, has been accepted and its conclusion (*a*) accepted as being strictly true only statistically, (*b*) denied, as strictly false. In case (*a*), since Mach's Principle can hardly be given a statistical interpretation, the repeatability principle must. In case (*b*), either Mach's Principle or the repeatability principle has to be styled false. This would seem to be the really important case. Looking over the sorts of reason given for holding to the principles of uniformity like (iii), time and again we find cosmologists saying that if only inhomogeneities could easily be allowed for, they would do so. In other words, they would deny (iii) in a strict sense, at best taking alternative (*a*). It would then be concluded that the only reasonable procedure for any cosmologist is either to reject Mach's Principle or to accept a statistical form of the principle that experiments give the same results under the same conditions, at all times and at all places. The former alternative would appear to involve least in the way of conceptual sacrifice.

But there is a serious flaw in this entire argument. The argument is plausible, admittedly, if (i) is glossed in a not uncommon way—and for this reason we gave it. If '. . . the same result will be obtained' is interpreted, as is often the case, as '. . . the same *numerical* result will be obtained', then the argument seems to hold. But a far more plausible rendering would be '. . . the same *kind* of result will be obtained'. Let us suppose that our law expresses the numerical answer (n) to an experiment as a function (f) of several variables, which include m (place relative to matter) and μ (time relative to matter). Then with an obvious use of suffixes, (i) as interpreted so far requires

$$n_1 = n_2, \tag{1}$$

where
$$n_1 = f(m_1, \mu) \text{ and } n_2 = f(m_2, \mu) \tag{2}$$

(experiment done at two places at the same time),

or where $\quad n_1 = f(m, \mu_1)$ and $n_2 = f(m, \mu_2)$ \hfill (3)

(experiment done on two occasions at the same place).
The only solution of (1) and (2) which is likely to be physically acceptable is

$$m_1 = m_2, \tag{4}$$

which can be taken as the ordinary Cosmological Principle. Adding the similar solution of (1) and (3):

$$\mu_1 = \mu_2, \tag{5}$$

this becomes the 'perfect' cosmological principle. Traditionally this interpretation was sufficient, since neither m nor μ was conceived as entering physical laws in any form. But when m and μ are taken into account it seems far more plausible to require *functional*, rather than numerical, identity of results. We are prepared to admit that the result, say, of an experiment to determine the velocity of light, may vary from place to place and time to time, so long as its functional dependence on place and time is unchanging. And from this consideration alone we can infer nothing whatsoever in the way of a cosmological principle in the material sense. In fact we have simply turned the repeatability principle itself into a cosmological principle in the non-material sense of some of Milne's writings—a principle of covariance, in short.

5. Cosmological Theories and Models

It remains for us to say something of the conflicting attitudes borne towards the status of the cosmological principles (in sense (iii) above) and the unrelenting uniformity which they appear to ascribe to the world we see. Broadly speaking, there appear to be three ways in which writers introduce any one of these principles.

(i) As a good approximation to the truth, and held provisionally.

(ii) As an exact truth, justified only by evidence which favours inferences drawn from theories incorporating it. For example, evidence for the derived (δ,m)-relation would be the sort of indirect evidence required. (This is actually the sort of justification which would be given by those holding to them in sense (i).)

(iii) As an *a priori* principle, requiring no justification apart from appeals to intuition and the like.

In fact the line between (i) and (ii) is hardly worth drawing; for the man who holds to the principle provisionally at least holds to it exactly, provisionally. The important distinction is rather between the first two ways on the one hand, and the third way on the other—the distinction between the belief that science proceeds in a hypothetico-deductive manner and the belief that, granted enough *a priori* principles, the rest will follow unequivocally: that the world is somehow bound to conform.

In the 1930s especially, one of the favourite pastimes of the hard-headed empiricist was to find Milne and Eddington guilty of this sort of *a priorism*. But, as we have already pointed out, neither Milne nor Eddington fits very comfortably into the third category: neither was oblivious of the need for observational support. One source of misunderstanding stemmed from the then orthodox view of science as a study which essentially began with abstraction from observation. Induction was supposed to lead one to a general law and further observations would confirm the law or otherwise. How one proceeded to erect the conceptual 'pyramid' above this stratum of law was usually left unexplained, but the main idea was that the scientist worked systematically upwards in a piecemeal way. A principle of high logical order was thus supposedly perfectly sound if reached in this way. But if the same principle were arrived at by any sort of intuitive leap, even though the principle was to be subsequently 'verified' through its consequences, this principle was stigmatized as a 'mere assumption', an 'arbitrary postulate', or worse.

Our purpose here is not to try to defend these so-called rationalists but to outline certain confusions in regard to the elements out of which cosmological theories are constructed. There seems to be some justification for saying that, as often as not, the mote of rationalism is in the eye of his positivist accuser. The confusion begins when distinctions have to be drawn between the statements of theory, statements in regard to a model, and statements of observation. The confusion is not diminished by ambiguity in such words as 'world', 'universe', 'idealization', and 'model'. For the rest of this chapter we shall concentrate on these ambiguities and on the special role of the models of cosmology.

'If we apply Boltzmann's law of distribution for gas molecules to the stars, by comparing the stellar system with a gas in thermal equilibrium, we find that the Newtonian stellar system cannot exist at all. . . .'[40] In this passage Einstein makes clear the way in which a single formal calculus may be applied to two or more kinds of 'situation'. In this case the 'applications' are not even at the same epistemological level: stars may feature in empirical generalizations, but molecules (in this context) do not. The 'gas in thermal equilibrium' would not, on the other hand, be generally said to provide a *model* for the stellar system in the same way as the well-known picture of a dynamical system of small, perfectly elastic spheres might be said to provide a model for both. Now as examples of cosmological uses of the term 'model', consider the reference by Tolman to 'a study of cosmological models . . . not necessarily agreeing in all particulars with the real universe';[41] and consider McVittie's more explicit statement that 'the class of space-times with metric of a certain form will be said to represent *uniform model universe*'.[42] None of the four senses so far singled out is in obvious agreement with the rest. Without defending one rather than another, therefore, we shall consider some of the ways in which they differ.

A physical theory may be roughly characterized as a small set of hypotheses, 'initial' in a logical sense, leading down to a larger set of empirical generalizations. The inference may involve hypotheses at intermediate 'levels', but rarely will it be possible to stratify a given physical theory in a unique way. Corresponding to the theory is an underlying calculus, with what pass for axioms as corresponding to the initial hypotheses of the theory. Linked essentially to the calculus, but conventionally distinguished from it, is the assumed logic.

The mechanical models of science, real and imagined, have a long history, and from them the very term 'model' was introduced into our methodological vocabulary. But how such models can possibly be of

[40] *P.R.*, p. 78.   [41] *R.T.C.*, pp. 332-33.   [42] *G.R.C.*, p. 139.

any assistance is best understood by considering the derivative usage summarized in this definition: A model for a theory is another theory which can be made to correspond (perhaps only partially) to the underlying calculus of the first, in particular in regard to deductive structure. There is thus a correlation (which may not always be one-to-one, but which should be preferably so) between concepts, sentences, and inferences of a theory and its model. On occasion, this sort of model-theory is referred to as a 'realization' of the theory. In mathematics, where it is seldom an embarrassment if two 'theories' should correspond exactly in set-theoretical structure, a more precise definition than ours can be given. For the moment we are simply content to find close parallels. As examples we may quote those between the quasi-Newtonian theory of Milne and McCrea and earlier relativistic theories; we may refer back to Chapter 3 for the numerous close analogies between various theories of gravitation—in particular Newton's—and theories of hydrodynamics, elasticity, electromagnetism, and so on. We have seen how, without exception, later theories of gravitation were partially modelled upon Newton's.

Mechanical models are at the same theoretical level as the things to which the main theory refers. (For this reason it may seem a trifle odd to refer to them as 'idealizations'.) It does seem that all such models, however simple and easily visualized they may be, are used as such precisely because they carry with them an easily assimilated theory. This is not to say that someone may not suddenly 'see' a resemblance between one kind of objective situation and another, and hence choose to make the one a model for the other. However, as far as modern science is concerned it seems more probable that it is in drawing analogies between theories that the scientist's mechanical model is conceived.

Models, in either of the two senses so far discussed, are in no sense logically necessary adjuncts of the main theory. They may be invaluable from a didactic point of view, especially where only a qualitative account is required. There is, of course, always the danger that their use might lead to a transposition, into the realm of the main theory, of some of those features of the model which are of no relevance to structural similarities. It is, on the other hand, this very aspect of the situation which makes models useful in physics as a whole: they suggest ways in which the main theory may be extended or otherwise modified to advantage. They thus assist in the growth of theories.

We have said that models, in the objective sense, may be looked upon as idealizations. Owing, however, to the imperfect correspondence between the model-theory and the main formalism, there will be important limitations upon any thinking done in terms of the 'ideal'. This sense, moreover, is not the usual sense of the 'idealization' in cosmology, nor are models often provided in the way explained. Rather than provide a theory parallel to the main one, the 'idealized universe' is simply presented as something described directly by the main theory itself. And it is the main theory, or a part of it, which is more often than not referred to as a model. In fact the central theory is not referred to as a model unless it is highly speculative and uncertain or (and this is much more usual) restricted by hypotheses which are felt to be needed only for certain special applications of the theory. Thus one might say that the steady-state theory requires a Riemannian metric of de Sitter form but that this becomes specified as a model when (for example) field equations are laid down which connect the expansion rate with density, pressure, galactic dynamics, and so forth. Conversely, one might say that Einstein's General Theory gives rise to special models when assumptions as to metrical (or physical) isotropy, homogeneity, boundary conditions and so forth are specified. There is obviously a great deal of latitude permissible in such a usage.

Less often the cosmologist speaks of models as idealizations, and it is then tempting to suppose that he is using the term in the more conventional sense. There appear to be very few examples where this is strictly so. One which comes to mind is to be found in a paper by J. L. Synge which deals with mechanical representations (beads on wires, spherical tops and the like) of spaces with positive-definite line-element.[43]

If 'model' is made more or less synonymous with 'idealization', it will most often refer to that hypothetical situation to which the specialized theory (or model in the sense given) refers, without remainder. Tolman combines the two senses, writing of a particular 'distribution of material' together with a 'space-time' as providing a model for 'the actual universe'.[44] Likewise McVittie writes of 'a model universe whose contents have all the uniformity that can be desired',[45] and then goes on to speak of accepting or rejecting a model universe according as it agrees or disagrees with observation. These writers are typical of the majority. At times 'model' may be used as a synonym

---

[43] *Ann. Math.* Princeton, xxxvi (1935) 650.
[44] *R.T.C.*, p. 333.
[45] *Fact and Theory in Cosmology* (Eyre and Spottiswoode, 1961) p. 97.

for 'metric', but there seems to be little justification for this use. Although there are minor ambiguities, we may at least attempt to summarize the main uses of the word 'model' as follows.

(i) Small-scale representation.
(ii) Parallel theory.
(iii) Mechanical model to which (ii) applies.
(iv) Restricted main theory.
(v) Idealization to (iv).

In (ii) and (iii) one is overlooking those aspects of the main theory which are irrelevant, intricate or unfamiliar. In this way a simplification is achieved. In cosmology the simplification is brought about in another way. (The method is not peculiar to cosmology—but only the model-terminology which describes it.) The idealization is there only as intricate as the restricted main theory. Elsewhere it is the intransigencies of an unfamiliar theory which are neglected in laying down a model: it is the irregularities of the world which are being ignored— or so the word 'idealization' seems to imply. The respects in which the idealization permits of being easily visualized are very limited. It is a favourite tenet of those who advocate the liberal use of models in senses (ii) and (iii) that they offer the best, or even the only way, of achieving an understanding of theoretical concepts. (The implication is that each theory must be modelled on another until a sufficiently obvious theory is reached. Clearly a great deal must be lost *en route*.) If this is so, the cosmologist is in an unenviable situation, if we are right in finding few examples of the models (ii) and (iii) in cosmology. In fact, whether or not we are right in this respect, this is a very doubtful account of what is to comprise an understanding of theoretical terms. These take their meaning from their contextual position in the theory— which corresponds to a certain place in the calculus. The model may help to make this position clearer, but it is not logically necessary on pragmatic grounds. There are occasional instances of the use of this sort of model to further the understanding of a cosmological theory. One of the strongest objections to McCrea's notion of zero-point stress was the fact that it was found difficult to visualize a material having such properties as his theory required. The objections abated to some slight extent when he explained how a theory[46] which speaks of matter having positive density and negative pressure may be understood in

---

[46] i.e. one using the field equations of the General Theory of Relativity with the steady-state metric. See Chapter 10, p. 234.

terms of an elastic material which expands in a certain way.[47] McCrea believes that such a way of discussing the problem reduces its novelty and presumably overcomes unreasonable prejudice. In fact, judging by the way in which his ideas were first received, it seems that one of the more telling arguments for devising a mechanical model is that it may be used to counter the objections of those who do not believe a theory to be admissible without one.

We have remarked upon another aspect of the use of models in senses (ii) and (iii): they are capable of assisting in the growth of theories. This fact has been expanded by some recent authors into the thesis that a theory which has been found an accompanying model has not only a greater propensity for growth, but a greater predictive power than would be possessed by the theory alone. If this were true, cosmology would again be at a disadvantage. As a thesis in the psychology of the creative scientist, this may possess a grain of truth. As a thesis to be conclusively demonstrated by reference only to theories and models and their structural adjustments, this can be very easily disposed of. For the whole point of theories of types (ii) and (iii) is that they can be separated from the main theory without any loss to that theory. Suppose, then, that the model leads one to alter the theory, to perceive ways in which new hypotheses and new theoretical concepts may be incorporated so as to lead to new prediction. After its help has been given, the model may be set aside, for the time being at least. There is now no *a priori* means whatever of ruling out the possibility that the theory could not have been amended in exactly the same way by a sufficiently far-sighted individual—who might easily be as much at home with his theory as another individual with a model for it.

We can now see a further reason for some cosmologists' work appearing to be rationalistic. Not only is it much more obvious of some that they proceed by hypothesis and deduction rather than by 'abstraction from the observations', but those who are most clearly aware that they are doing so are precisely those who tend to use the 'idealization' terminology which we have outlined. Milne, for example, often overstated the case. But it is no fault to set up 'an arbitrary postulate about the whole universe' and deduce from it what 'the distribution of

---

[47] The energy-equivalent of the mass leaving a given region is exactly equal to the energy given to that region to perform the expansion within the region. Moreover, stress throughout a region affects the gravitational mass of the contents—to use a hybrid metaphor involving Einsteinian and Newtonian physics. A negative stress will make these contents behave as though having a negative gravitational mass. The repulsion between such masses is pictured as keeping the expansion going.

matter must be in order that it shall be obeyed'. The fault is in believing the material to be an aspect of a factual world. The matter which 'must be' is one aspect of the idealization, and nothing more. No more than a claim to a formal necessity need be made. Thus it is perfectly reasonable to speak of the Cosmological Principles as applying exactly (and necessarily) to the idealization. (When the words 'world' or 'universe' are used instead, they tend to add some confusion.) When it comes to 'applying' the theory, however, in so far as a (lawlike) cosmological principle can be 'directly interpreted'—namely as an item of ordinary language—it is invariably given a statistical interpretation. No cosmologist has ever done otherwise, although some have spoken so carelessly as to make it appear that they might. As statistical laws, the cosmological principles are without meaning when applied to individual objects; and even when applied to the case of many individuals, there is usually the possibility of saving appearances by claiming that the numbers still do not warrant statistical methods. Even so, once a set of procedural rules has been laid down and agreed to, the laws are as susceptible as any to disproof and take their place with the remaining laws of the theory. It is often said that one is driven to them for want of better information, and this is in a sense true. But unless cosmologists can judge of a sort of systematic variation of density, pressure, and so on, with position which is capable of leading to more accurate predictions, the principles accepted by Einstein, Milne, Bondi, and the rest will have to be retained. And even a discrete idealization will have some sorts of symmetry and will require the use of statistics in the interpretation of the theory.

CHAPTER 15

CONCEPTUAL PROBLEMS (i) DISTANCE AND COORDINATES

BEFORE the latter part of the nineteenth century, space was generally supposed to be something in which bodies could be moved and compared, whilst congruence was a fundamental property of geometrical figures.[1] Some mathematicians were never very happy with the so-called 'principle of superposition' (that is, of congruent geometrical figures) and some tried to avoid using it. They felt that the word 'motion' was not to be taken in its material sense—for clearly geometrical objects were not physical ones. But not until the time of Klein and Lie were the geometrical terms 'motion' and 'congruence' explained in a sufficiently abstract way.[2] It was Riemann's geometry which gave Klein the idea that space should be defined, not as a vehicle for motions, but as a set of points with a particular set of properties or 'structure'.[3] Thus geometers might even speak of the possibility of a single set of points having more than one structure, or, what amounts to the same thing, as forming more than one space.

The structure, at the time of the *Erlanger Programm*, was regarded as characterized by means of a transformation group. Affine geometry, for example, would then be taken as the subject dealing with those properties which are invariant under the linear transformation group. The structure may, on the other hand, be regarded as implicit in a set of axioms in terms of which the geometry is specified. Affine geometry, from this point of view, would be the development of those axioms of Euclidean geometry which exclude the idea of congruence. Whether looked at from the first point of view or the second, the structure of any metric space must be largely determined by the definition of distance it uses. What Riemann had achieved was to clarify the concept of distance, and by virtue of the great generality of his conception, to make Klein's programme of the classification of existing metrical geometries

[1] Cf. the fourth proposition of Euclid's *Elements*.
[2] Riemann speaks of the principle of superposition of figures. It is interesting to notice that Pieri later took 'motion', along with 'point' as his only indefinable terms. (*Della geometrica elementare como sistema ipotetico deduttivo*, Turin, 1899.)
[3] Klein, *Vergleichende Betrachtungen über neuere geometrische Forschungen* (Erlangen, 1872). Reprinted in *Math. Ann.*, xliii (1893).

easier. These methods, moreover, were later readily extended to other varieties of geometry.[4]

The greater part of modern cosmology incorporates Riemannian geometry, as we have seen. It should therefore be clear that if modern cosmology incorporates concepts of distance different from those of nineteenth-century geometry, they must have been introduced by a conscious and explicit amendment of older ideas. In the present chapter it is our intention first to examine the central features of the older concepts of distance and then to discover respects in which later ideas fail to do justice to earlier requirements. We shall also consider the question: Can any one of the modern concepts be made fundamental to the rest, so that we may continue to use the word 'distance' in much the same way as it is used in common speech? In the course of the chapter we shall be driven to consider the operationalist doctrine of meaning.

1. Distance, Coordinates, and a Possibly Circular Argument

We have already seen how Riemann took the square of the length between adjacent points (ds^2) to be a homogeneous function of the second degree in the differentials of the coordinates. To appreciate his argument we must list certain features of the concept of distance which were, before the end of the last century at least, common to all geometries, just as they appear to be common to all ordinary uses of the term 'distance' in everyday speech. These can be summarized in the form of the following statements, which will also be useful for reference at a later stage. 'Measure of the distance between the points P and Q' will be represented by '$d(P, Q)$'.

(i) A measure of the distance between any two points exists, in some sense of the word.

(ii) A one-to-one correspondence can be set up between the class of all distance-measures and the positive real numbers.

(iii) $d(P, P) = 0$

(iv) $d(P, Q) > 0$ if P and Q are not the same point.

(v) $d(P, Q) = d(Q, P)$.

(vi) $d(P, Q) + d(Q, R) \geqslant d(P, R)$.

---

[4] There are now known to be Riemannian spaces which, in terms of the *Erlanger Programm*, appear to be without structure. See J. H. C. Whitehead and O. Veblen, *The Foundation of Differential Geometry* (Cambridge, 1932).

(i) DISTANCE AND COORDINATES

The following two postulates may also be mentioned here, for the sake of completeness:

(vii) (*Archimedes' Postulate*) If a and b ($a > b$) are the positive numbers corresponding to two distance-measures, there is always an integer n such that $nb > a$, and nb corresponds to a distance-measure.

(viii) (*Du Bois Raymond's Postulate*)[5] If a is a positive number corresponding to a distance-measure, then for any integer n the number a/n corresponds to a distance-measure.

Now in Riemannian terms, (v) asserts that if the differentials are to change in sign the value of ds will be unchanged, and thus ds^2 must be an even root of a positive function (by (ii) the distance-number must be real) of an even degree in the differentials. By analogy with the Euclidean case, Riemann took it to be the square root of a homogeneous function of the second degree in the differentials. This is not as arbitrary as it at first appears, being the simplest hypothesis compatible with another of his assumptions, namely that if the differentials are all to be increased in a certain ratio then ds must also increase in the same ratio. At least Riemann was not oblivious to the other possibilities.

Shortly after Riemann's death Helmholtz published the results of his own investigations into the form of the distance function, claiming to prove that the line-element is the square root of a homogeneous function of the second degree in the differentials.[6] The evidence which Helmholtz offered for this result hinged upon the common enough view of a geometrical figure as a rigid body capable of being freely moved. In this approach, like Riemann before him, he was more Euclidean than Euclid, who seems—judging from his avoidance of the use of the principle of superposition on all but two occasions—to have disliked the idea.[7] What, after all, is the use of defining metrical properties in terms of a rigidity which can only be attributed to a body if we have independent means of measuring it? Perhaps the question is unfair: geometry for Helmholtz was something which hovered between the empirical and the abstract. Helmholtz deserves mention, however, if only

---

[5] *Allgemeine Functionentheorie* (Tübingen, 1882), p. 46.
[6] 'Über die Thatsachen, die der Geometrie zum Grunde liegen', *Gött. Nachr.* (1868), reprinted in Helmholtz, *Wiss. Abhandlungen*, Bd. ii (Leipzig, 1883), pp. 618–39. A paper on the same subject is reprinted at pp. 610–17.
[7] Helmholtz's prime interest was in the physiological problem of the location of objects in the field of vision.

because his work moved Sophus Lie to inquire into a reasonable geometrical meaning for the terms 'congruence' and 'motion'.[8]

It would probably have been generally agreed at this time (had the proposition ever been put forward) that to speak of geometrical motions was to speak of a one-to-one relationship between, for example, points and points, or lines and lines, such that the metrical properties of the one set of relata are the same as those of the other. Lie, more specifically, introduced the idea of coordinates, and replaced the idea of motion with that of transformations between systems of coordinates. The idea of congruence was then very naturally replaced by that of invariance under the coordinate transformation in question—a subject which we touched upon in chapter 4.

At once a difficulty presented itself to all who gave any thought to the matter. How could Lie introduce and use coordinates before defining congruence? Is the concept of congruence not essential to the definition of a coordinate system?

One means of breaking what seemed to be a vicious circle was suggested by a procedure which von Staudt had developed many years before.[9] He had shown how to define a harmonic range by means of the quadrilateral construction, thereby making it possible for Klein to show how to establish a one-to-one relationship between the points of the line and the system of rational numbers.[10] Mathematicians were slow to realize the significance of von Staudt's work, and although during the early years of non-Euclidean innovation there was a move to recognize the autonomy of more than one geometry, yet the historical development of projective geometry lent weight to the view that there was no distinction to be drawn between metrical geometry and projective geometry. Even harmonic ratios were still being defined in metrical terms long after von Staudt's well-known construction. After nearly a quarter of a century however, von Staudt's new approach to projective geometry was brought to the general notice when first Klein and later Pieri conclusively showed the logical independence of metrical and projective geometry.[11]

Congruence had at last been excluded from projective geometry. It

---

[8] His starting point was Klein's paper of 1872. Cf. Lie, *Theorie der Transformationsgruppen*, Bd. iii (Leipzig, 1893), pp. 437–543.

[9] Von Staudt, *Geometrie der Lage* (Nürnberg, 1847); *Beiträge zur Geometrie der Lage* (Nürnberg, 1856, 1857, 1860).

[10] 'Über die sogenannte Nicht-Euklidische Geometrie', *Math. Ann.*, iv (1871) 573, and vi (1873) 112.

[11] Pieri, *I Principali della Geometria di Posizione* (Turin, 1898); Klein, *Vorlesungen über nicht-Euklidische Geometrie* (Göttingen, 1893).

was also vaguely recognized that, in the correspondence set up between the points of a line and the numbers of arithmetic, the numbers, in the absence of a definition of length, have no metrical significance.[12] Russell, in an early work, seems to have thought differently, for he argues that coordinates used in metrical geometry must themselves be metrical.[13] M. Pasch, on the other hand, had actually shown in 1882 a way in which the coordinates of any point in a plane could be projectively defined, and had only then given, in effect, a non-Cartesian basis upon which the Cayley–Klein metric could be superimposed.[14]

Do the methods introduced by von Staudt, Pasch, and the rest, allow Lie to avoid the accusation of circularity and to define coordinates other than in terms of distance?

Before we become too deeply involved in the search for ways in which circularity may be avoided, it is necessary to point out that an answer will depend on our point of view. The geometer, the analyst, the physicist—each looks at geometry in a different way. To the analyst an n-dimensional Riemannian space (or manifold) may consist of 'points' only in the sense of ordered and unique sets of n independent real numbers. Such 'points' will be ordered by the coordinates, their mutual 'distances' may be defined in the usual way, and so on. But there is no question of identifying points with entities distinct from ordered sets of real numbers. The geometer, on the other hand, may wish to make such an identification—as between, say, the coordinate set and an axiomatically defined projective space. Because this sort of thing is possible, the Riemannian theory in its purely analytical aspects is by custom often given the name 'geometry'. But in this 'geometry' the set of coordinates *is* the point. Although distance, in turn, may be defined as a certain function of the coordinates the question of spatial position never enters: the geometrical metaphor could easily be dropped and the analysis would be essentially the same. The geometer's 'identification' contrasts with the physical identification of a set of coordinates. From this point of view, whether or not there can be found

---

[12] The numbers are no more than *labels* of points. This can easily be seen if we take as an axiom of extension that the open set of points on a line is isomorphic with the set of all complex numbers. A point with a *real* number as its label can easily be given a *complex* number label by simply changing the gauge points. Did the numbers correspond to distances we should be obliged to say that the distance had changed (in this case from real to complex) whilst the system of points remained, by hypothesis, unaltered.

[13] *Foundations of Geometry* (Cambridge, 1897) ch. iii, section B.

[14] *Vorlesungen über neuere Geometrie* (Leipzig, 1882). We shall not here consider a much stronger claim than any of these, which seems to be implied in some of the work of Cayley and Klein. According to this, metrical geometry is purely a derivative of the projective theory.

another theory isomorphic to the coordinate 'geometry' is quite irrelevant. What matters is that anyone wishing to determine the distance of an object from a knowledge of its coordinates must first be prepared to lay down a procedure for assigning coordinates to the object.

It is perfectly well-known that this is not the way in which cosmologists go about determining distances. But conversely it is often thought that in determining distances they are doing no more than assigning coordinates to the objects of their study. And if they are not doing so, where do coordinates come in? If it is possible to find distances without them, why have them at all? Such questions do appear to have epistemological relevance, whereas in analysis and (pure) geometry proper corresponding questions of priority would at the most be logical ones. Historically it seems that three languages—analytical, geometrical, and material—were confused as the result of an admixture of spatial metaphor and appeal to intuition in the first two. Lie might have been justifiably accused of circularity had he supposed his manifold to correspond to a material region to which coordinates were assigned by metrical means. But of course he supposed no such thing. Whether there is any reason for thinking that cosmology has ever been in danger of doing so, will be considered in the following section.

2. Coordinates. Whitehead's Challenge of Circularity

Lest it should be thought that the conceptual difficulties we have indicated are obviously illusory, we may quote from the works of four writers. Each subscribed to a different cosmological theory and each found something unsatisfactory in existing expositions of the way in which Riemannian coordinates were introduced. Whitehead's objections will be set aside until later in the section, but they bear some resemblance to those given by Milne, who said of Einstein's theory that it gave no rule 'for ascertaining the coordinates of an event from observations of it'.[15] We then find Birkhoff objecting that 'even in the simple case of Schwarzschild's solution of the one-body problem, no clear cut physical interpretation of the Schwarzschild coordinates seems to be available'.[16] Lastly we find McVittie attempting a solution of the difficulty, which will now be explained.[17]

McVittie began by showing how an observer, equipped with clock

---

[15] *W.S.*, p. 341.   [16] *Collected Papers*, p. 968.
[17] *Cosmological Theory* (London, 1937), ch. iii.

(i) DISTANCE AND COORDINATES

and rigid measuring-rod, could first express his own changing time-coordinate (t_1) in terms of his clock reading (s_1), using the relation

$$s_1 = \int_0^{t_1} \sqrt{\{g_{44}(t, 0, 0, 0)\}}\, dt. \tag{1}$$

He next explained that if the ends of a rigid scale are regarded as the events (t_0, 0) and (t_0, dx^μ), and therefore the length of the scale is

$$dl = \sqrt{(g_{\mu\nu}\, dx^\mu\, dx^\nu)}, \tag{2}$$

finite values of the coordinates could be obtained were it not for the fact that, the coordinates being independent, it is not possible to integrate the expression in general. As one way of avoiding the difficulty he suggested that distance measurements be taken in a definite order. The coordinates X^1, X^2, for example, of the event (t_0, X^1, X^2, 0) may be defined in terms of measurements l_1, l_2, by the relations

$$l_1 = \int_0^{X^1} \sqrt{\{g_{11}(t_0, x^1, 0, 0)\}}\, dx^1 \tag{3}$$

and

$$l_2 = \int_0^{X^2} \sqrt{\{g_{22}(t_0, X^1, x^2, 0)\}}\, dx^2. \tag{4}$$

'So long as we apply the same convention as to order to the measurement of the coordinates of all events, no real ambiguity can arise.'[18]

From these two special cases McVittie passed to the assignment of coordinates, *all* of which differ from those of the observer. To this end the observer, according to McVittie, can use a second clock 'precisely similar to his own', the law of motion of this being known to him. How it is possible to specify the curves (in space-time) along which the clock is said to travel *before* a coordinate system is set up is not explained; but this is not the greatest of McVittie's problems. In the general case, as before, the observer must know in advance the forms of the metrical coefficients g_{mn} as functions of the coordinates. And this was the sort of apparent circularity in the arguments of General Relativity which had, several years previously, led Whitehead to reject the Einsteinian account.

Whitehead's argument for doing so runs, briefly, as follows: to determine the inhomogeneities of space-time—that is to say the complete behaviour of the functions g_{mn}—Einstein presupposed a knowledge

[18] McVittie, op. cit., p. 37.

of the energy tensor. But the components of the energy tensor contain terms involving those measurements of length and time which are supposedly denied to us without prior knowledge of the inhomogeneities of space-time.[19] In this same connexion, he professed himself to be unable to understand 'what meaning can be assigned to the distance of the sun from Sirius if the very nature of space depends upon casual intervening objects which we know nothing about'.[20] He wanted, as he says, to begin without the need to place reliance upon 'the contingent relations of bodies', which 'we cannot prejudge'. It was not simply that ordinary language made no allowance for the contingent relations of bodies. For Whitehead, space-time had to be homogeneous in order that there should be 'fixed conditions for measurement', that is to say, 'definite rules of congruence applicable under all circumstances'. It seems not unlikely that he was influenced by Riemann's belief that it is both necessary and sufficient for a Riemannian space to have constant curvature 'if measurement is to be possible'. In Whitehead's encyclopaedia article 'Geometry', this Riemannian result is introduced by the remark 'measurement demands superposition and, consequently, is dependent on some magnitude independent of its place in the manufold'.[21] As we saw, Einstein was shortly to indicate how Whitehead's claim for the necessity of this thesis could be undermined.

It is almost impossible to discuss Whitehead's criticism outside the context of his philosophical views in general. Even so, it is not difficult to see that his objections are cogent only if at the outset it is decided that the old division between physics and geometry is to be preserved. In so far as he later fails to prove that there is something intrinsically wrong with Einstein's different convention—in regard to the geometrization of gravitational theory—he begs the question. Briefly, he argues as follows: time clearly cannot be introduced as a relation between permanent objects, and therefore space-time is to be regarded as 'a relatedness between events', with objects no more than 'adjectives of events'.[22] If space-time were a 'relatedness between objects' it would 'share in the contingency of objects and may be expected to acquire a heterogeneity from the contingent character of objects'. Which, of course, is what Einstein thought to be the case. As, according to Whitehead, it has rather 'the character of a systematic uniform relatedness between events which is independent of the contingent adjectives

---

[19] See, for example, *The Principle of Relativity* (Cambridge, 1922), p. 83.
[20] Whitehead, op. cit., p. 59.
[21] *Encyclopaedia Britannica* (11th ed., 1911) pp. 727–28. [22] *P.R.*, p. 58.

(i) DISTANCE AND COORDINATES

of events' we must, according to him, 'reject Einstein's view of a heterogeneity in space-time'.[23] For his belief that 'our experience requires and exhibits a basis of uniformity of spatio-temporal relations', he offers arguments from 'the character of knowledge in general and of the knowledge of nature in particular'.[24] These arguments are not in the least convincing, but it is not necessary to go into them here.[25] For our present purposes it is enough that we should answer his more specific objection to the circularity which he believed he saw in Einstein's procedure.

McVittie was well aware of Whitehead's objections, and his attempt to explain how the vicious circle could be broken is enlightening in so far as it is typical of the relativist's answer to Whitehead. Falling back on a form of the cosmological principle—namely as it relates to a roughly apparent symmetry in the distribution of matter and radiation—McVittie argued that an observer can surmount the 'great weakness' of the theory of relativity 'by assigning to the energy tensor *a priori* values, which appear plausible on general physical grounds, in terms of some abstract coordinate system. . .'[26] The observer then finds the coefficients of the metric from Einstein's gravitational equations and finally, 'by working back to the physical meanings of his abstract coordinates, compares the hypothetical physical situation which he has built up with that which he actually observes'.[27]

Returning later to ideas not unlike these, it will now be suggested, however, that the objection which Whitehead made to the General Theory of Relativity is insubstantial, and that his views on the nature of coordinates are themselves totally unsatisfactory. It is not of prime importance that a full set of coordinates should ever be assigned to a particular event, even though observation of it be important evidence for one's theory. The feeling that it should, at least in principle, be possible to do so generally stems from an over-enthusiastic rendering of an outmoded empiricist account of the function of theoretical terms within science.

Whitehead was certainly no straightforward empiricist, but in this respect he makes the same sort of mistake as one or two later positivists. 'If you are dealing with nature', he tells us, 'your meanings

---

[23] loc. cit.   [24] ibid., preface.
[25] If space-time were not homaloidal, according to Whitehead, induction would be impossible. (op. cit., pp. 14–16, 62–66.) This rests upon a rather bizarre epistemology: to know a factor it is necessary to know its necessary relations with other factors . . . and so on.
[26] McVittie, op. cit., p. 39.   [27] loc. cit., cf. Milne, *W.S.*, section 513.

must directly relate to the immediate facts of observation'.[28] He assures us that it is necessary first to analyse 'the most general characteristics of things observed', and then the more casual contingent occurrences. There can be 'no true physical science which looks first to mathematics for the provision of a conceptual model. . . .' But it is not necessary that each and every theoretical hypothesis should be rendered explicitly in terms of 'the immediate facts of observation'. If it were so, physics would still be on a level with taxonomy. The coordinates used in most cosmological theories are no different from scores of similar terms throughout physics, in lacking a direct interpretation. And yet, as with the rest, this does not mean that they may be dispensed with. The urge to explain the meaning of coordinates in a sequence of practical operations is strong, perhaps, only because in some scientific contexts coordinates are thought to be on the list of so-called 'data of observation'.

It might be argued that the two angular coordinates of a distant galaxy may, to use Whitehead's words, be made to 'directly relate to the immediate facts of observation'. The radial coordinate, on the other hand, certainly cannot be placed in this category. No doubt Whitehead would have liked to identify it with radial distance, and no doubt there are many who would wish to say that distances are in principle directly observable. It is easy enough to claim that it is unnecessary to give a phenomenological explanation of each and every theoretical statement, but this does not mean that Whitehead's objection to Einstein's method has been answered. It is necessary also to show that coordinates are amongst the terms of which it might be unreasonable to ask a direct interpretation. Can the objection be answered by pointing to the historical fact that a large part of relativistic cosmology has already been expressed in a form susceptible to observational test? How can a theory be involved in circularity if it can be used to make specific predictions which are confirmed?

A supporter of Whitehead's point of view might answer in some such vein as this:

> It is not that we hold your method to be incapable of giving rise to prediction, but that it does so in an unsatisfactory way. You argue that it is not necessary to determine the metric tensor explicitly in terms of observable quantities. You explain how, instead, it is possible to make certain restrictive hypotheses of a general sort which

[28] Whitehead, op. cit., p. 39.

(i) DISTANCE AND COORDINATES

are to be tested obliquely through their consequences—(δ,m)-relations and the like. Such consequences will impute a uniformity to nature, but in the wrong respects. They may involve us in the claim that galaxies are distributed isotropically about all points—whereas it is easy to see that they are not. The poverty of the relativistic method is shown in the fact that it cannot proceed from the contingent distribution of matter: rather it must proceed from a series of guesses, each of which may or may not be subsequently verified. This is very much a hit and miss affair.

It is certainly necessary to impute uniformity to nature—but in regard to form rather than content. The best way of avoiding Einstein's vicious circle is by not confusing form (the metric) and content (the actual distribution of energy).

Einstein's case might be represented thus:

We are in the best scientific tradition in so far as we proceed by attempting to verify propositions deduced from a set of hypotheses—the word 'guesses' does not do justice to the careful way in which the hypotheses were chosen.

The line between form and content cannot be drawn in any hard and fast way. Einstein's geometrization of the theory of gravitation meant a departure from the older line of division. But, once again, the change may be justified in terms of the confirmable statements to which his theory leads.

As to principles of symmetry, what are the alternatives? Those who have worked out a cosmology on the basis of Whitehead's theory have used similar principles. All theories use the method of idealization. Cosmological principles are abstractions writ large. Rather than concentrate on the individuals themselves they deal statistically with many individuals, when applied to the world of galaxies.

There is something to be said for both points of view. Here we simply want to point out that Whitehead's objection is not unanswerable and that his request for an interpretation of coordinates within all theories is unreasonable. Each man has a certain liberty of choice in deciding which terms of his own theory are to be directly interpretable and which not.

Although not intended as an answer to Whitehead, it is clear that even within relativistic cosmology coordinates may ultimately be assigned to distant galaxies. The possibility is not challenged, but the *point* of making such an ascription is very obscure. Let us suppose, for a moment, that it is meaningful to talk of one galaxy as being twice as distant as another. Even if the radial coordinate of the first was twice that of the second, by one reckoning, a coordinate transformation could easily be effected which changed the factor from two to three, say. Effecting the same transformation throughout the working for, say, a (δ,m)-formula, there will be no change in the final answer. Only those who have acquired the mistaken belief that coordinates are essentially the same as distances will object to the transformation. Coordinates are the mere scaffolding of theories, and their essential value lies in the opportunity they offer us of looking beyond them. What we said of the radial coordinate could equally have been said of angular coordinates, although they are usually directly interpreted as 'angular distances' with respect to which our definition of material isotropy is drawn up. Why do we not likewise identify the radial coordinate with distance? In the following section we shall see the answer: there are a great many concepts of distance, and cosmologists do not appear to be able to agree as to which shall be made fundamental.

3. CONCEPTS OF DISTANCE. OPERATIONALISM

We begin by reminding ourselves that 'distance' and 'interval' are not to be confused, and that the line-element of a four-dimensional space-time corresponds to the second quantity. In simple three-dimensional Riemannian terms distance is defined, conceptually, in terms of the coordinates and the components of the metric tensor. Mathematically distance is an invariant and, practically, distances will be assessed in terms of some 'standard' which represents this invariant. But when it comes to the determination of intervals, if a rod-standard is to be used it must somehow be supplemented by a timepiece.

All astronomical statements concerned with distances derive a part of their significance from a complex of practical operations. By way of example, we outline such grounds as might be used in deciding the distance of a remote galaxy. These might be:

(i) The galaxy is a member of a cluster, the apparent luminosity of the brightest member of which has been recorded, and

(ii) The intrinsic luminosity of this member is known by extrapolation from physically similar systems in which blue supergiants, novae, globular clusters, &c., of carefully measured apparent luminosity, have been found, and

(iii) The intrinsic luminosity of these objects is known—assuming that they are of the same nature as those found in galaxies containing Cepheid-type variable stars, which are in turn assumed to be physically similar to the corresponding stars of the Galaxy (stars, that is to say, similar in spectrum and period-luminosity function), and

(iv) A period-luminosity function for such stars is known on the basis of the determination of the parallaxes of small groups or clusters of which they are members. The radial velocities and proper motions of the stars of the cluster are assumed known, the first by spectroscopic means and the second by difficult photographic measurements of angle. In the course of the calculation it will be seen whether the stars share a convergent.

These are not the only possible links in the chain of inference. Reference to Cepheid variables could have been omitted, and for the cluster-parallax procedure we could have substituted the visual or photographic determination of trigonometrical parallax, or one of the several variants of the spectroscopic-photometric type of parallax measurement.

(Concerning the reliability of the various stages of the argument astronomers have usually erred on the side of modesty and caution—apart, perhaps, from a short period during which the earliest Cepheid calibration was widely thought to be above reproach. They have recognized that trigonometrical parallaxes beyond 120 pcs are subject to accidental errors comparable with the parallaxes themselves, and in each of the above stages (i), (ii), and (iii), uncertainties of the order of 50 per cent cause little dismay.)

In stages (i)–(iv) the measuring-rod is not mentioned. Although it could be used in the process of calibrating a spectrometer, it is not essential to the measurement of fractional spectral shifts. The dimension of length enters at stage (iv) through the term for the velocity of light, which may thus be regarded as giving us our standard of length. But this, of course, is to play down the theory which relates the observed positions of spectral lines with their wavelengths. It also plays down the fact that the standard rod is used to validate the physical theory in which the astromomer has become involved. But it does point, even so, to the sort of way in which a chronometric definition of distance may be set up.

If reliance is placed on trigonometrical parallaxes, then indeed the measurements are reducible to the measurement of a base-line in terms of a standard rod in a simple way. The base-line for stellar measurements is the diameter of the Earth's orbit round the Sun. The most acceptable of the half dozen ways of measuring this do not, as it happens, involve the measurement of an accessible base-line, except perhaps at an early stage in the subordinate theory (for example, in the measurement of the velocity of radio waves).

This all illustrates one or two aspects of astronomical practice, but it does not do justice to the needs of modern cosmology. First of all, we have ignored the modification which it is necessary to make to the inverse square law, within some theories. Secondly, we have not mentioned the possibility of introducing further concepts of distance—a possibility which has now been appreciated for at least a quarter of a century.

Interest in the possibility of defining 'distance' in new ways appears to have been aroused by Whittaker's paper of 1931,[29] to which we have already referred. His definition was not related to existing astronomical practice in any simple way, although I. M. H. Etherington later showed how it agreed exactly with a definition of distance as estimated from apparent size.[30] Etherington introduced yet another definition, which made distance an absolute scalar invariant (this was not true of Whittaker's definition). Later we find Walker defining the distance of a star as estimated from its luminosity; that is to say, he made the necessary relativistic modifications to what had become the astronomer's most important method of estimating distances.[31] The older concept of distance by parallax was defining by McCrea for the special case of a Robertson–Walker metric[32] and more generally by G. Temple.[33] Another definition is of distance by volume.[34] The theoretical connexions between these definitions will be summarized here. The definitions which we have used are not those for the general form of the metric but only for the Robertson–Walker form:

$$ds^2 = dt^2 - \{R^2/(1 + kr^2/4)^2\}(dr^2 + r^2 d\theta^2 + r^2 \sin^2\theta\, d\phi^2). \qquad (5)$$

[29] *P.R.S.* (A), cxxxiii (1931) 93.
[30] *Phil. Mag.* (7), xv (1933) 761.
[31] *M.N.*, xciv (1933) 159.
[32] *ZS. Ap.*, ix (1935) 290 (equation 10).
[33] *P.R.S.* (A), clxviii (1938) 122 (equation 6.11).
[34] See *G.R.C.*, p. 154.

(i) DISTANCE AND COORDINATES

They may be compared, in effect, as follows :[35]

(i) Distance by apparent size (ξ) and Whittaker's distance function (Δ):
$$\xi = \Delta = r \cdot R(\tau)/(1 + kr^2/4), \tag{6}$$
where τ is defined by the equation
$$\int_0^r \frac{dx}{(1 + kx^2/4)} = \int_\tau^t \frac{dx}{R(x)} \tag{7}$$
and the observation is made at (t, r, θ, ϕ) of a star whose world-line is $r = 0$.

(ii) Etherington's distance function:
$$\Delta_E = \Delta \cdot R(t)/R(\tau). \tag{8}$$
(We notice that the factor $R(t)/R(\tau)$ is the expression for the Doppler shift $(1 + \delta)$.)

(iii) Luminosity-distance:
$$D = \Delta \cdot R^2(t)/R^2(\tau). \tag{9}$$

(iv) Distance by parallax:
$$\Delta_\pi = rR(t)/\{1 - kr^2/4 + rR'(t)\}. \tag{10}$$

(v) Distance by volume:
$$\Xi^3 = 3R^3(t) \int_0^r \frac{x^2 \, dx}{(1 + kx^2/4)^3}. \tag{11}$$

Only ξ, D, and Δ_π were defined with a straightforward operational procedure in mind: this is not readily apparent from the formulae quoted here, but the names reveal the procedures in question which are essentially the traditional astronomical procedures. We notice further that these concepts cannot be introduced into the theory unless they are developed as functions of such quantities as r, t, and $R(t)$. Roughly speaking, these unobservables may then be eliminated between some or all of the equations linking them with observables. Since the distance-term need never occur in such an equation—it certainly denotes nothing

---

[35] For early papers on the definitions of distance, in addition to those cited here, see H. S. Ruse, *P.R.S. Edin.*, lii (1931) 183; liii (1932) 79; McCrea &c., ibid., p. 31; Walker, *Q.J.M.*, iv (1933) 71; Hubble and Tolman, *Ap. J.*, lxxxii (1935) 302; Synge, *Q.J.M.*, vi (1935) 199.

observable—we may well ask why we should ever concern ourselves with it at all. In fact we shall return to this question later in the section.

Before discussing these definitions further it is necessary to settle an important point of principle. We have seen some of the different practical ways in which the astronomer determines distances—and here there is often a belief that these are different ways of determining one and the same sort of distance. We have also given several definitions of distance which are not even theoretically equivalent. It must now be decided whether we are to distinguish between different distance concepts not only in the second case, but also in the first. Are we, for example, to distinguish 'distance-by-photometry' from 'distance-by-triangulation'? If not, are we to go to the other extreme and regard one concept of distance as fundamental to the rest? These problems will be considered in turn.

Measurement cannot be taken as the discovery, by observation, of a certain feature of a world independent of us and of our description of it. No measurement is of any absolute value: its value stems from its being a tacit comparison. In every comparison there are presuppositions, however simple, and every comparison of scientific interest is bound to be theory-involved. However, more extreme arguments have been given for this view of measurement, arguments often closely associated with the doctrines of *operationalism*. According to one version of this, an important part of the meaning of a term is given by the rules relating this term to some event or train of events. In a stronger version, the words 'an important part of' would be omitted. Sentences, moreover, which are not purely formal or extra-scientific, are said to be meaningful when there exists a concrete procedure, 'a set of operations', for determining their truth or falsity. Their structure must, of course, be syntactically sound, but even this is often said to require some sort of operational test. The full implications of the operationalist thesis will not be considered here, but the claim to find a difference of concept wheresoever there is a difference of operational procedure will be much qualified.

It is difficult to place operationalism historically, for although its modern synthesis is due almost entirely to one man, P. W. Bridgman, most of its doctrines have been uttered before in a different guise. In the present century we find that Eddington anticipated some of Bridgman's remarks, affirming that 'a physical quantity is defined by the series of operations and calculations of which it is a result'. He added that 'we do not need to ask the physicist what conception he attaches to "length"; we watch him measuring length and frame our definition according to the operations he performs'.[36] It is hardly likely, even so, that Eddington, with his distinction between the unknowable 'external world' and that human creation the 'physical world', would have wished to be associated with the modern versions of operationalism. How did the strong version ever seem plausible to anyone?

One of Bridgman's earliest arguments, and one which lent a great deal of weight to his thesis, was drawn from Einstein's Special Theory. Einstein had proposed to abandon the naïve concept of absolute and universal simultaneity in favour of an ostensibly artificial definition. It now appeared that in order to judge of the simultaneity of distant events a series of carefully circumscribed operations had to be performed. This example—which a moment's reflection will show to be of a common enough sort in the physical sciences—suggested that no one would claim to understand the concept of simultaneity unless he understood how it was to be judged. But understanding the nature of sequences of operations, from a necessary condition soon became a sufficient condition of understanding meanings. And since operations and meanings were equated, it became natural to find conceptual differences wherever techniques differed. Every scientist knows that when experience discloses a virtually invariable correlation of one concept with another (as, for example, of inertial and gravitational mass) and when the same symbol is accordingly made to correspond to both, it is easy to forget the nature of the original distinction. We should be less liable to forget this if we were to keep before us the maxim 'different operations define different concepts'. But at what price? Should we not be ignoring important criteria for meaningful statement? Are we not tempted to force all statements into the same straitjacket? Bridgman, for example, in an attempt to explain all meaning in terms of operations, included mental operations in his list. In addition, he gave a highly artificial account of the significance of analytical statements: the concepts of mathematics he regarded as operationally defined in a pencil and paper sense.[37] What if we need to use calculation to obtain a result on one occasion, and not on another—are we dealing with the same concept? How do we know where to stop? Is length-by-ruler to be

[36] *M.T.R.*, pp. 3–4.
[37] Cf. his *Logic of Modern Physics* (New York, 1928); *The Nature of Physical Theory* (New York, 1936) and an article in *Psychological Review*, lii (1945). One of the biggest objections to the operationalist philosophy is that it appears to deny meaning even to such apparently harmless dispositional words as 'imperceptible', 'movable', and so on. It is worth remembering that there are some who have argued that *all* universals are dispositional in character!

distinguished from length-by-screw-gauge? Suppose that we outline different procedures giving rise, in the theory as it stands, to the same distance concept, or at least to the quantification of the same symbol. Suppose further that the numerical distances of some readily identified object are found by these procedures to be, respectively, $D_1, D_2, \ldots D_n$. What if these numbers are all different? There is, after all, no *a priori* reason why they should be otherwise. If they were radically different the chances are that we should make the appropriate conceptual distinction and redraft the theory. If the differences were invariably small most of us would be inclined to say that the observations were subject to experimental error and our one symbol stood for a single concept. But (runs the operationalist's argument) even where D_1 and D_2 turn out to be more or less the same and even where, as a matter of convenience, one assimilates the two concepts, in any careful or philosophical account they must always be distinguished.

Referring to the astronomical procedures for determining 'the various kinds of distance' it should be clear that anyone holding strictly to these tenets is committed to using well over a dozen names where only one is currently used. Why, then, do astronomers usually make no such distinction, and simply speak of 'distance'?

So obvious is the answer that it is surprising that it should be so often overlooked. No distinction is made, wherever the whole of the underlying physical theory obviates the need for it. The theory may entail, for example, such a statement as: 'The triangulation-distances of two stars are always in the same ratio as their luminosity-distances'. If this seems to be refuted by experience then other parts of the theory may well be brought to the rescue. In the quoted example allowance may be made for the interstellar absorption of light, refraction by the Earth's atmosphere, and so forth. And if, in the last resort, there are unexplained discrepancies between the results of two kinds of operation then a decision has to be made which is far from trivial. It is not likely to be enough simply to regret the loss of convenience and to proceed to give different names to the two concepts, for the subsidiary physical theories—practical geometry and photometry, in this instance—do not touch, as it were, at only one point. The entire complex of underlying physical theory has been challenged and a revision of the whole may be called for. Of course it is a contingent matter that the same results will be found when taking a measurement by different techniques, even though they are theoretically equivalent. But that is no reason for multiplying the number of concepts unnecessarily. It is just as much a

(i) DISTANCE AND COORDINATES

contingent matter that the same results will be obtained in measuring the length of an object on successive occasions by the *same* technique, but no one would dream of saying that different concepts of length were involved in consequence.

In summary: operationalism does not offer a more than partially satisfactory theory of meaning. In reducing all kinds of meaning to one it implies, for example, that all theoretical terms are on the same level. (Thus we find Bridgman criticizing Einstein's General Theory for failing to give an operational definition of coordinates. As we argued in the last section, this is totally unnecessary.)

Operationalism provides us with a method distinguishing one concept from another which is not precise enough to be of much interest. Likewise it ignores the requirements of theory, within which different concepts may be equivalent. To quote Eddington: 'I define distance as the distance found by parallax observation *or by any other astronomical method accepted as equivalent* to actually stepping out the distance.'[38] There are cases where we do not wish to maintain the equivalence of concepts—as for example with distance-by-volume and distance-by-apparent size—but it is our theory which tells us why this would be unacceptable, just as it is our theory which tells us how they are related.

The operationalist maintains that many more conceptual distinctions are desirable than are customarily made, and that one can only equate different concepts of distance on empirical grounds. The latter part of the assertion is, as we have indicated, true but misleading; for one can equate them on theoretical grounds, where, of course, the theory is an empirical one. As for the case where there are generally supposed to be two practical methods of determining the same quantity—let us say the methods of triangulation and chain-measurement for determining the distance between two places—there seems to be no good reason why a single concept of distance should not be said to take a part of its meaning fron each operational procedure. Only if one makes the mistake of strictly equating operation and meaning is one obliged to draw a pointless distinction.

4. Which Distance-Concept is Fundamental?

Astronomers and others occasionally speak as though there were some real and absolute distance between any pair of objects. It has been pointed out here that there are many different concepts of distance,

[38] Eddington, *The Expanding Universe* (Cambridge, 1933). My italics.

the meaning of which can only be elicited by—amongst other things—laying down the practical means of deciding distances; and that whilst one may be in many ways superior to another, none is in any sense absolute. The root of the trouble is that in physics information concerning lengths has often been fairly easily obtained, and length has consequently been regarded as a very fundamental physical quantity. This has led many physicists to suppose that there must be one *basic* method of length-determination to which all others may be reduced. 'The whole of geometry', wrote Einstein, referring to the operation of moving a measuring-rod parallel to itself a definite number of times, 'may be founded upon this conception of distance'.[39] With Whitehead, we might say that he 'favours one meaning . . . and becomes the servant of that meaning'. Whether, on the other hand, the meaning makes a good master remains to be seen. Again, the theoretical cosmologist for long accepted a situation where astronomers reduced and published their results in an established style, even though this meant introducing a concept of distance which was not best adapted to his theory. He tried, in short, to avoid distorting too severely the forms of existing scientific language. But if there are many ways of defining distance, why should we adhere to those which were historically first? Is there no criterion by which we may judge the value of one concept as opposed to another? It seems that we may note at least five desirable features of a satisfactory concept of distance. These, which are listed below, are clearly not independent, nor are they given in any order of preference: indeed, it is hard to see how we would begin to draw up a table of relative importance.

(i) The concept should have the support of tradition: other things being equal, it should in some respects be familiar.
(ii) It should not be difficult to visualize the ways in which distances, in any new sense of the word, would in practice be determined.
(iii) Their practical determination should be as simple as is compatible with ease elsewhere. . . .
(iv) . . . and, in particular, with theoretical ease. Unfortunately the two are not always compatible. For example, McVittie says of distance-by-volume that there is no 'obvious operational procedure' for finding it; and yet it is not difficult to find an approximate relation between it and spectral displacement.

[39] *The Meaning of Relativity*, 6th ed. (Methuen, 1956), p. 7.

(v) Any concept of distance should as far as possible be consonant with the distance axioms listed in the first part of this chapter. In a sense this is to repeat (i) and (ii), for all will agree that these axioms are in keeping with the requirements of ordinary language. But (i) and (ii) are intolerably vague, whereas here we have a fairly precise means of putting the different distance concepts to the test. This will be done in a later part of this section. Before doing so we shall discuss the concepts of proper-distance and luminosity-distance, which are undoubtedly the most favoured candidates for the title 'fundamental'. This will be followed by a discussion of criticisms offered of Milne's distance-concept, in the light of the criteria listed here.

The foundation of the General Theory of Relativity owed much to the idea that the space-like interval between neighbouring events could be found using a rigid rule. This leads to the definition of an integrated *proper distance*, connecting simultaneous but distant events. It is this last feature which appears to rob the definition of its respectability; for it is a requirement of the Special Theory of Relativity that, of individual observers in relative motion, one might judge two events to be simultaneous whilst another, from the same place, might not. The phrase 'simultaneous but distant events' can be given a meaning in any theory admitting cosmic-time. But this can only be introduced into Einstein's General Theory as the result of some special hypothesis as to the relative motion of observers (as laid down, for example, by Weyl's Principle). Let us for the moment gloss over the thorny problem of cosmic time and consider reasons for retaining the notion of proper distance.

First we point out that the use of rod-measures is certainly not an essential part of the conceptual foundation of the General Theory of Relativity; for we recall that Robertson and Walker on several occasions interpreted the relativistic metric in terms of Milne's light-signal technique.

Why is it that, with Bondi, there are those who believe that proper distance presents a 'mathematically well-defined but physically somewhat nebulous picture'?[40] The integrated proper distance or, as Bondi calls it, the 'absolute distance', between two points of a universe with the line-element (5) is, in our notation,

$$R(t)\int_0^r \frac{dx}{(1+kx^2/4)}. \qquad (12)$$

[40] Bondi, *Cosmology*, 2nd ed., p. 69.

Certainly the idea might conjure up the image of a vast number of yardsticks end to end. 'Physically', we are told, 'this definition is not very valuable'.[41] It seems that our criterion (ii) is at the root of these objections. The operationalist will no doubt be the first to take exception to proper distance. Is the concept not meaningless because no sequence of practical operations corresponds to it even in principle?[42] If the objection is to be circumvented by referring to the ways in which such distances could be calculated from measurements made in other sorts of operation, why, it will be asked, should we bother? For do we not have a perfectly good alternative in the form of luminosity-distance, which is far more readily connected with astronomical practice? Luminosity-distance is defined, very simply, to be such that the apparent intensity of a light source falls off inversely as the square of the distance.[43] It was defined simply in order to preserve astronomical procedure. However, as explained in Part I, there is a great deal to be said about the 'correction' of apparent magnitudes. In point of fact, it is too much to say that the 'reduction procedures' of astronomy remain unaffected, for the corrections which are to be made rest heavily upon the underlying cosmological theory. We saw how this point was overlooked in the 1930s, with the consequence that several important mistakes followed. Even so, this definition of distance has in its favour a fairly straightforward interpretation which is not the case with proper distance. On a small scale, luminosity-distances will agree with proper distances, although they are not theoretically equivalent. (To put the matter another way, the statement that the apparent luminosity of a source varies inversely as the square of the proper distance is incompatible with the General Theory of Relativity.) Of course, one cannot rule out the possibility that there may prove to be a practical and easily visualized means of determining some 'distance' which turns out to be theoretically equivalent to proper distance. It seems not unlikely that the man in the street would then prefer it to luminosity-distance, even though he may have to be initiated into the mysteries of the equivalence (just as at the moment he has to make an effort to understand the equivalence of rod-distance and triangulation-distance).

The protagonist of luminosity-distance might be content without

---

[41] op. cit., p. 107.   [42] For the *use* of a measuring-rod requires time.
[43] The definition previously given in equations (6), (7), and (9) was derived from this definition using a Robertson–Walker metric together with a few widely accepted laws of optics. The derivation would not be universally accepted—as, for example, in Whitehead's theory. But our simple definition could be used in *any* theory.

(i) DISTANCE AND COORDINATES

any analysis of the concept in terms of 'basic-experience', 'intuitive appeal' or 'ordinary speech'. This is not an unreasonable attitude as long as it is consistently maintained, as it is clearly enough except in expositions of a general nature. There the inclusion of this clause is rarely found: 'by "distance" you and I might well mean different things'. Here, once again, is the problem of the theoretical term, with the very important difference that the term is now a familiar word of everyday occurrence.

It seems that behind the question 'What concept are we to take as fundamental?' there are two sorts of feeling. The first is that the laws of cosmology should be made easily understandable, so that we should have no trouble in understanding, say, Hubble's Law. The second feeling is operationalist inspired: whatever practical operations we find ourselves able to perform should be linked directly with the concepts of our theory. The first feeling connects with criteria (i), (ii), and (iv) at least. It is very tempting to argue that it is reasonable enough to criticize a theory on the grounds that it is impossible to explain it in everyday terms. But this does not do justice to the present case. Even if we go to the extreme and say that 'distance' is to be interpreted as a theoretical concept like, say, 'electronic mass', we still do not rule out the possibility of understanding statements involving the term. We may use, if it is felt to be necessary, a model—as explained in the last chapter.

As to the second feeling, Bondi, for example, seems to think that since existing theories do not contain the concept of luminosity-distance as 'a primitive element . . . this must be counted against them all'.[44] This is to ignore all criteria for the value of a theoretical concept other than the practical criterion (iv), and seems to be altogether undesirable.

One alternative to using a word inconsistently is to avoid it altogether. Although it might well be possible to draw up a cosmology devoid of any distance concept, it would hardly be likely to resemble existing schemes. As things stand, however, it is not clear that any of the definitions of distance is clearly theoretically superior to the rest, in the sense of providing us with a 'middle term' leading to the most concise and easily derived overall account. This being so it was natural to turn, consciously or otherwise, to the requirement of (astronomical) familiarity. But Milne chose a different course.

[44] Bondi, op. cit., pp. 127-28.

As we saw, Milne defined the distance between one object and another in terms of the velocity of light, the time of emission of a signal from one to the other, and the time of receipt of a returned signal. We saw that the justification for this unusual definition was his belief that he must relate all distant events to occurrences in the immediate consciousness of the observer, and that temporal experience (for example, of the apparent simultaneity of events) being the most *immediate* mode of experience should consequently be made fundamental. He drew attention also to the logical priority of time measurement over distance measurement. To explain distance measurement, he said, it is necessary to explain the meaning of the phrase 'two points at the same time', and this requires a prior explanation of time. (He was thinking of rod-measurement.)

Milne rejected entirely the reduction of his own conception of distance to one requiring 'a so-called rigid measuring-rod'; for he observed that we cannot say of this 'what is meant by its maintaining the *same* length when transported or when pointed in different directions'.[45] There is little doubt that here he does less than justice to both Poincaré and Einstein, who were both clearly of the opinion that the invariance of the length of a 'rigid scale' is a matter of convention.

One objection raised against Milne's view was that it is impossible to understand the meaning of 'velocity of light' without first understanding what is meant by 'distance'. Once again the issue devolves on the question of understanding the primitive elements of our theory. Is it meant to be a matter of psychology that we cannot immediately comprehend the concept of velocity, whereas we can immediately comprehend the concept of distance? In any case, the argument is forestalled by pointing out that Milne's constant positive number c is referred to as representing a velocity only after 'distance' has been defined.

Milne's definition of 'distance' came in for heavier criticism on the grounds that, apart from using light as its instrument, it bore no resemblance to astronomical practice.[46] Even astronomers have lives far shorter than the time intervals involved.

Born's argument is often quoted with approval, even by those who profess to admire Milne's analysis. As a matter of exegesis, it was said, there can be found no theoretical connexion between Milne's definition of distance and astronomical procedure. He appeared simply to accept uncritically the distances supplied uncritically by contemporary

[45] K.R., p. 6.   [46] M. Born, *Experiment and Theory in Physics* (Cambridge, 1943).

(i) DISTANCE AND COORDINATES

observers. This is unfortunate because there is, for his theory, a definite link with the astronomical practice of deriving distances from luminosity measurements. It illustrates well a point already made, namely that it is an unfortunate tradition which makes the 'conceptual basis' of a theory appear to be its most important link with observation.

Suppose Milne to be comparing the apparent luminosities L_α and L_β (energy received per unit area per unit time) of two sources of photons of the same kind and emitted at the same rate, these being received at rates of N_α and N_β, per unit time. Then, the energy of these photons being an invariant of the kinetic theory,[47]

$$(L_\alpha/L_\beta) = (N_\alpha/N_\beta)\,(r_\beta^2/r_\alpha^2), \tag{13}$$

where r_α and r_β are the distances of the nebulae from the observer at the time of emission. Thus the ratio (L_α/L_β), which is essentially what the astronomer uses as one criterion of distance, gives the conventional inverse square of the ratio of the distances multiplied by the factor (N_α/N_β). It can also be written

$$(L_\alpha/L_\beta) = \{(1 + \delta_\beta)/(1 + \delta_\alpha)\}\,(r_\beta^2/r_\alpha^2), \tag{14}$$

with the usual notation for the spectral shift.

The fact that this formula suggests a possible definition of luminosity-distance[48] is of less importance than the fact that it readily provides a procedure for calculating Milne's ordinary distance (r) from observations on the received light regardless of the procedure outlined in his 'conceptual foundation'. The 'correction factor'—as the astronomer would be inclined to call $(1 + \delta_\beta)/(1 + \delta_\alpha)$—is different from that required by the generally accepted treatment. To compare the two accounts, that is, of Kinematic and General Relativity, we may write down an equation from the latter theory:

$$(L_\alpha/L_\beta) = \{(1 + \delta_\beta)/(1 + \delta_\alpha)\}^2 \cdot (\bar{r}_\beta^2/\bar{r}_\alpha^2), \tag{15}$$

where, referring back to the coordinate r used in the earlier part of this chapter,

$$\bar{r} = r/(1 + kr^2/4). \tag{16}$$

This last result must not be too closely compared with that given above, for Milne's terms r_α and r_β were interpreted as distances, whereas $\bar{r}_\alpha$ and $\bar{r}_\beta$ are simply coordinates with no such significance.[49]

---
[47] See *K.R.*, ch. viii.
[48] This would merely have to be proportional to $r/\sqrt{(1 + \delta)}$.
[49] In a universe with constant curvature it can be seen from equations (6), (9), and (16) that they are proportional to luminosity-distances.

The history of the 'correction factors', the one the square of the other, is very confusing largely owing to a misconception on Hubble's part. Hubble thought that his observations favoured the former expression and wrongly supposed that the General Theory of Relativity could be made compatible with it. Milne seized upon Hubble's findings as support for his own theory, and the passage in which he does so provides further evidence that his thought was not, as Born would have us believe, remote from current astronomical practice. Milne's theory is not remote from astronomy, nor is it remote from the 'immediate consciousness of the observer'. But theory and 'immediate consciousness' are far from being linked at the level of Milne's 'conceptual foundation'; except perhaps in principle (a useful phrase for saving epistemological appearances!).

As we have already seen, Milne reduced the observational foundation upon which he wished his thesis to stand to three or four quantitative laws. With Tolman he claimed the distinction of being the first to lay down explicitly the 'directly testable consequences' of a cosmological theory. There is hardly a term occurring in a cosmological treatise which refers to anything directly apprehended. There is, however, a class of terms which, for the sake of convenience, might be regarded as in some sense primitive, and the values ascribed to them are loosely termed the 'observational data'. Spectral shifts, number-counts, apparent magnitudes and clock-readings would fall under this head: in laboratory physics 'distance' might also have been included.

Once a tradition is established whereby a term is accepted as primitive a general feeling is likely to prevail that it must always be so. Thus it is with 'distance'. One can draw something of a moral, perhaps, from geometry. There it has become gradually less and less imperative to make the concept of distance an essential part of the foundation of individual systems. Scientific theories change but much of their vocabulary persists. In cosmology it seems that 'distance' can be introduced easily enough as a function of observational procedures; but what is the point of doing so? To take a specific example, if apparent magnitude can be expressed as a function of the Doppler shift without the mediation of a term for luminosity-distance, what is the point of introducing luminosity-distance as a function of apparent magnitude? Why do we not dispense entirely with the concept and all like it?

The all too frequent characterization of science as merely the correlation of experience, is, if interpreted as a demand for the total

(i) DISTANCE AND COORDINATES

elimination of what it is usual to term 'theoretical concepts', very incomplete. It is true that a programme of excision may to some extent be possible in the sciences as they are found at any one time, but it would be all too easy to carry out the programme and to overlook the need for facility of calculation, for brevity of exposition and, perhaps most important of all, for perspicuity. The conciseness of expression allowed for by the liberal use of nominal definition means that the various concepts of distance might be found a part to play, regardless of their resemblance to each other or to past concepts, and regardless of their theoretical equivalence to such concepts or to everyday notions. The sceptic may argue that the part played by the distance-concept was a purely historical one. It may have led us to our (δ, m)-relations and the like, but we can—he may prove—now see our way more clearly without it. He may acknowledge the need for conciseness of expression and he may require the liberal use of coordinates to that end, and yet may be able to offer a perfectly convincing demonstration that (existing) distance-concepts do not make for such conciseness. How is he to be answered? Without any doubt the majority would fall back on our need for laws which we can easily comprehend. It is not for nothing that Hubble's Law is still the starting point of most popular accounts of the achievements of cosmology.

It seems reasonable, therefore, to ask whether the nebular distances to which we so often loosely refer are radically misnamed. As already explained, there is a test available to which the unfamiliar concept of distance may be submitted, and which is able to provide some sort of assurance that the common word 'distance' is not being abused. This test involves discovering whether or not, under the various interpretations of 'distance', the following statements are true:

(i) The measure of the distance between any two points exists.

(ii) A one-to-one correspondence can be set up between the class of all distance-measures and the positive real numbers.

(iii) $d(P, P) = 0$.

(iv) $d(P, Q) > 0$ if P and Q are not the same point.

(v) $d(P, Q) = d(Q, P)$.

(vi) $d(P, Q) + d(Q, R) \geqslant d(P, R)$.

(vii) Archimedes' Postulate.

(viii) Du Bois Reymond's Postulate.

Of these conditions, (iii) and (iv) are satisfied by all definitions of distance in current cosmology. Likewise (i), (ii), (vii), and (viii) are generally acceptable, although they may require amendment should the world be supposed finite in extent, or in the aggregate number of those things made to correspond in the theory to the points of the above statements. Some might also wish to modify them on epistemological grounds, bearing in mind the 'horizon' properties of many models. ('It is meaningless to speak of assigning a distance to an object which is in principle unobservable'—and so on.) It is the remaining conditions (v) and (vi) which are the most interesting. It should be clear that they fall in with most of our ideas, but it is not at all clear that they are satisfied by the definitions of distance previously discussed. Condition (v) is relatively straightforward. We make the assumption that $r_{PQ} = r_{QP}$ (and hence $\bar{r}_{PQ} = \bar{r}_{QP}$) at all cosmic times. It is then possible to say that a necessary condition of (v) holding for Δ_E, Ξ, Δ_π and l is that either $R(t)$ is constant or the distances are evaluated for the same cosmic time—that is the time of reception of light by the one from the other. In the case of ξ, Δ, and D it is necessary also that the time of travel of light is the same in both directions. In short, there is little for the man in the street to find fault with, as far as the reflexivity of these concepts is concerned.

Condition (vi) is much more difficult to deal with. As it turns out to be practically unsatisfiable when the sign of equality is taken, we feel justified in ignoring the case of inequality. Now that time enters the problem there is a difficulty of interpretation. For all but the concept of proper-distance we shall interpret (vi) in the following way: we suppose three objects to be A_1, A_2, and A_3 on the same light path. The first sends a light signal at time t_1 which passes A_2 at t_2 and reaches A_3 at t_3. As before, we assume that $r_{PQ} = r_{QP}$ at all times. The coordinate separation of A_1 and A_2 is r_α (or $\bar{r}_\alpha$) and of A_2 and A_3 it is r_β (or $\bar{r}_\beta$), and these are constant. With an obvious notation, the three distances in (vi) are in a typical case (luminosity-distance) given by

$$\left. \begin{aligned} D_{12} &= (R_2^2/R_1)\bar{r}_\alpha, \\ D_{23} &= (R_3^2/R_2)\bar{r}_\beta, \\ D_{13} &= (R_3^2/R_1)(\bar{r}_\alpha + \bar{r}_\beta). \end{aligned} \right\} \quad (17)$$

[50]

It is easily seen that a necessary condition that

$$D_{12} + D_{23} = D_{13} \quad (18)$$

---

[50] [I would now here accept a remark by the late G. C. McVittie to the effect that the combination of luminosity distances calculated by observers located at different places and times is an illegitimate use of the concept.]

(i) DISTANCE AND COORDINATES

is that $R_1 = R_2 = R_3$—that is, $R(t)$ is in general constant. This is an altogether unpalatable supposition for modern cosmology, and we cannot therefore suppose that the concept of luminosity-distance—at least within the very general terms of reference laid down by Robertson and Walker—is an acceptable substitute for the notion of distance commonly held. Exactly the same sort of argument can be given for ξ, Δ, Δ_E, Ξ, and Δ_π.

The way in which we chose to interpret (vi) was not the only way open to us. Consider, first, the notion of proper-distance. This is, if we allow it at all, a theoretical construct. In the expression for it, namely $R(t) \int_0^r \{1/(1 + kx^2/4)\}\,dx$, the time for which the function $R(t)$ is to be calculated is the time at which the proper-distance is required. It has nothing to do with a time of reception or emission of light which is essential to the performance of some distance-determining operation—as are the functions $R(t)$ in ξ, D, and Δ_π, for example. Integrated proper-distance is 'instantaneous-distance', and in this respect, if no more, it corresponds with our everyday ideas. It is not clear whether those who have used them intended Ξ, Δ, and Δ_E to have the property of holding in the instant, as it were. Are we, for example, justified in interpreting (vi), taken for Δ, as

$$\bar{r}_{PQ} \cdot R(t) + \bar{r}_{QR} \cdot R(t) \geqslant \bar{r}_{PR} \cdot R(t), \tag{19}$$

where P, Q, and R are no longer necessarily on the same light path and t can be any time we choose? This was clearly intended of only one of the concepts we listed, namely that of integrated proper-distance. Concentrating on this case, we thus have to ask whether it is plausible to assert

$$R(t) \cdot I(0, r_{PQ}) + R(t) \cdot I(0, r_{QR}) \geqslant R(t) \cdot I(0, r_{PR}), \tag{20}$$

where

$$I(0, r) = \int_0^r \{1/(1 + kx^2/4)\}\,dx. \tag{21}$$

The inequality (20) will hold if and only if

$$f(r_{PQ}) + f(r_{QR}) \geqslant f(r_{PR}), \tag{22}$$

where $f(r) = 2\arctan(r/2)$, r, or $2\operatorname{arc\,tanh}(r/2)$, according as $k = 1$, 0, or -1. It is not difficult to show that, suitably restricting the range of r-values, (22) is equivalent to

$$r_{PQ} + r_{QR} \geqslant r_{PR} - k(r_{PQ} \cdot r_{QR} \cdot r_{PR})/4, \tag{23}$$

where the last term on the right is omitted if $k = 0$. It is perfectly easy to find counter-examples to (23) within the specified ranges, except when $k = 0$, and thus we conclude that (vi) is not satisfied by integrated proper-distance except possibly in the simple Euclidean case—at least with this interpretation.

If P, Q, and R are on the same curve $\theta = $ const., $t = $ const., then (vi), with the sign of equality, can be interpreted

$$R(t) \cdot I(0, r_{PQ}) + R(t) \cdot I(r_{PQ}, r_{QR}) = R(t) \cdot I(0, r_{PR}), \qquad (24)$$

which is an identity regardless of the means of assigning coordinates to Q and R. This, it seems, is as near as we shall come to finding a theoretical concept of distance which conforms to our everyday ideas. The much scorned notion of integrated proper-distance has this one saving grace. What is more, we have not found it necessary to speak of operations being performed at widely separated points 'at the same cosmic time'.

CHAPTER 16

CONCEPTUAL PROBLEMS (ii) ABSOLUTE AND RELATIVE

He that is giddy thinks the world goes round
The Taming of the Shrew, V, ii, 20

RIGHTLY or wrongly a stage has been reached in cosmology where 'absolute', as used of space and time, is a polemical term. The aim of this chapter is to explain how this state of affairs has come about, and to suggest that cosmologists put an end to the all too simple dichotomy of Absolute and Relational philosophies of space and time.

Roughly speaking, the situation is this: Leibniz, following older writers, held that space is a set of relations amongst material objects. Newton, following in particular the Cambridge Platonist Henry More, believed that space had an existence independent of matter, being in this sense absolute. Similar views were held with regard to the concept of time. Newton was able to counter Leibniz's philosophical arguments with arguments from dynamics, that based on the rotating pail experiment being the most famous. Although many philosophers may have inclined towards Leibniz's views, and although Newton's arguments were to be undermined by, for example, Berkeley and Mach, it was not until the Einstein-Poincaré theory of relativity became widely accepted that a so-called 'relational' view of space and time came into its own. The restricted and general theories of relativity are now so widely accepted, in their foundations if not in their details, that it has become a maxim of the scientific strategist to acknowledge 'relational' space and time, and to profess horror of anything 'absolute'. All historical distinctions between the terms are conveniently overlooked. Then, again, it is all too frequently forgotten that 'absolute' and 'universal' have different connotations, and so the idea of cosmic (or universal) time is often ostensibly rejected, or accepted grudgingly, as the result of this quasi-philosophical prejudice. Seldom is it asked whether 'relative' and 'absolute' are mutually exclusive predicates—not, after all, an irrelevant question. We shall begin, therefore, by attempting to draw what appear to be the most important historical distinctions between the terms 'absolute', 'universal', and 'relative'. We shall then show the absurdity of attempting to fit modern writers into the seventeenth-century categories.

1. Newton and Leibniz on 'Space'

The historical problem has unfortunately been obscured by one or two writers who, in the first quarter of the century, saw in the Special Theory of Relativity a vindication of the Leibnizian arguments, and who saw Newton's absolute space in the aether of nineteenth-century physics. On the other side, it could be said that Newton's writings have been too generously interpreted. It is often held that Newton's absolute space is a certain set of inertial frames none of which can be distinguished from the rest. On this view there would be, therefore, no absolute uniform velocities, but only absolute accelerations (and rotations). Newton thought differently:

> It is from their essence or nature that they [that is things] are places; and that the primary places of things should be movable, is absurd. These are therefore the absolute places; and translations out of these places are the only absolute motions.
> But because the parts of space cannot be seen . . . instead of absolute places and motions, we use relative ones . . .[1]

He goes on to say that there may be 'no body really at rest' although 'the cases by which true and relative motions are distinguished, one from another, are the forces impressed upon bodies to generate motion'. There follow the well-known accounts of the experiments involving, first, the revolving pail of water and, secondly, the revolving system of connected globes.[2] He claimed thereby to have identified absolute motions (accelerations and rotations) in the sense that it was thought unnecessary to refer to any other object in order to attach a meaning to the statement that a body rotates. Absolute *space* in the sense of the totality of positions which an object may have, was not 'identified'. How could it be identified as were 'absolute motions' when, on Newton's admission, 'the parts of space cannot be seen'? In Book III, Proposition xii of the *Principia*, Newton argues that 'the common centre of gravity of the Earth, the Sun, and all the planets, is to be esteemed the centre of the world'. It is doubtful whether this should be interpreted (as it usually is) as an identification of that set of inertial frames which we now tend to equate with what Newton denoted by 'absolute space'. It seems rather that he was simply looking for a place which could be labelled 'centre of the solar system'.

When a scientific writer chooses to speak of something which cannot be perceived as being 'real', it is usually more profitable to ask what

---
[1] *Principia*, Scholium following Definition viii.   [2] loc. cit.

part it plays in the scientific system under review than to dismiss it at once as 'metaphysical'. It is well known that Mach, believing in the need to eliminate all metaphysical concepts fron physics, and placing absolute space under this heading, proposed that inertia be regarded as dependent upon the large-scale distribution of matter. It is interesting to conjecture upon the likelihood of his not having taken this view, had Newton not written the informal and unnecessary preface to the *Principia*, from which we have already quoted. Newton and his followers undoubtedly wrote much which was philosophically dubious. But it is not at all obvious that to use Newtonian dynamics, or to accept a part of Newton's method, is to become necessarily involved in Newton's philosophical doctrines.

Undoubtedly Euler's argument in favour of Newton's view of space was historically of the first importance: it was important because it persuaded Kant to abandon his earlier Leibnizian position in favour of the Newtonian. Euler argued that the first law of Newton's highly successful dynamics was incompatible with the former view, since the law could be given content even if there were only one existing inertial object. The motion of such an object could not, *ex hypothesi*, be specified in terms of relations. Kant's philosophical shadow fell across the whole of the nineteenth century, and it is therefore of some significance that he added yet another ('geometrical') demonstration of the need for absolute space. A blunt positivistic principle was ultimately used to beat aside these subtleties. Kant's argument rested on the possibility of 'incongruent counterparts'. Only the similarity, and not the difference between right and left hands can be 'accounted for' on the simplest relational view, according to Kant. (It is easily seen that the nature of the problem—of 'moving' the one hand into the other—changes if four dimensions are allowed.) With an absolute background, Kant supposed that each point of the object had its place, and that the aggregate of places must be different as between the two hands. Newton's idea of space as a background or *locus quo* is still clearly present.

Consider now the positivist position. All will agree (Newton not least) that the logic of the word 'moves' is such that 'P moves' is essentially incomplete. Explicitly or implicitly it must be made clear, when such a statement is made, how it could be completed so as to read 'P changes its position relative to Q'. To assert that 'all motion is relative' is to assert a simple logical rule. But it has been held in effect by very many writers—including Berkeley (*De Motu*, 1721) and

Mach—that it is to assert that 'P changes its position relative to Q' must be verifiable. It was further implied that it must be somehow directly verifiable; that is to say, P and Q must be readily experienced. There are very many statements of science which are not verifiable, except perhaps through their consequences. (The last might be an unacceptable concession to positivism, for a true conclusion can be drawn from false premisses.) May this statement not be one of them? If we substitute 'absolute space' for 'Q', may we not regard the statement as one containing a theoretical concept? Historically this way of speaking was not used,[3] for Newton and those after him always thought of the deformation of the water surface in his pail experiment as suggesting absolute (accelerated) motions with respect to absolute space, the existence of which was thereby supposed demonstrated more or less directly. When Berkeley and Mach pointed out that the acceleration might just as well be considered as relative to the fixed stars, they might equally have regarded themselves as discovering what Newton's Absolute space really was. They had found something which could be directly apprehended, and which would give our statement involving P and Q an empirical appearance. But the curious fact remains that even in its new guise the statement is unverifiable; for how can we know the effect on the pail experiment of removing or altering the fixed stars? The answer, given after Mach's death, was: erect a plausible theory of a changing universe using Mach's principle.[4] As suggested in past chapters, however, most of the arguments for some such interaction principle as that in the Mach–Einstein form do not appear to be in the least conclusive. Even judged in the light of Ockham's Razor (upon which Mach rested his argument) we can see that all Mach did was to eliminate a 'metaphysical' concept concerning a stable entity in favour of a physical concept covering a changing and unstable one. One does not have to agree that this represented any gain in simplicity if one ignores his unduly broad definition of 'metaphysics'. It is a very unfortunate fact that many cosmologists have tacitly accepted it, and have been

[3] It is worth noticing that Berkeley admitted what he called 'mathematical hypotheses' and distinguished between those (force, attraction, gravity, &c.) which work well in the process of computing the positions and motions of objects, and those like absolute space and absolute motion for which not even this much can be said. For 'A note on Berkeley as precursor of Mach and Einstein' by K. R. Popper, see *B.J.P.S.*, iv (1953) 26.

[4] Simple experimental tests have been proposed on the assumption that the oblate form of the Galaxy would lead to an anisotropy of inertia in the laboratory. See, for example, G. Cocconi and E. E. Salpeter, *Nuovo Cim.*, x (1958) 646; *Phys. Rev.*, iv (1960) 176 (letters; for replies see op. cit., p. 342): R. W. P. Drever, *Phil. Mag.*, vi (1961) 683. R. H. Dicke argues that the experiments were misconceived (*Phys. Rev.*, vii (1961) 359).

(ii) ABSOLUTE AND RELATIVE

led unconsciously to the view that the Einsteinian form of his principle must govern any satisfactory system of mechanics.[5]

Above all, any exegesis of Newtonian ideas will be plainly inadequate which ignores the way in which space was regarded as something *in which* objects were situated. Space was the aggregate of all places, occupied and unoccupied. Clarke, for example, who is best remembered as Newton's supporter in the controversy with Leibniz, believed that it made sense to speak of 'moving the universe from one place to another', whilst he thought it a merit of this view of space that it allowed the possibility of a vacuum—a place, that is to say, unoccupied by matter.

Why did Newton, who clearly recognized our logical need to relate the motion of one thing to another, hesitate to take movement as being relative to physical objects? The reason is given in the Scholium following the eighth definition: 'True motion is neither generated not altered, but by some force impressed upon the body moved; but relative motion may be generated or altered without any force being impressed upon the body.' My pen moves relatively to the Moon, and therefore the Moon to my pen; but the Moon experiences no force in consequence of this relative motion. It is not, in Newton's terms, 'true motion'. A Machian would reply that Newton had simply failed to find the significant sort of 'relative motion'. The fact remains that what Newton did find was a system of dynamics capable of occupying mathematicians for at least three centuries: Leibniz's relational view of space was a relative failure, despite its logical appeal. His doctrine of space as an 'order of coexistences' or an 'order of situations' and time as an 'order of successions' of phenomena was never of itself connected with any useful dynamical theory.

Before discussing the relational view in general it is interesting to observe the inconsistencies in Leibniz's writings. Most of them originate with his lapsing into a Newtonian way of speaking. Leibniz spoke of the existence of 'no real space out of the material universe',[6] but he also spoke of the 'parts of space'[7] and of space as that 'which results from places taken together'.[8] Why are we, like Leibniz, so liable to slide into these ways of speaking? The source of most of the difficulty is that the term 'space' is capable of being used in so many different

---

[5] It is worth adding that although Mach did not die until 1916, he opposed Einstein's Special Theory of Relativity to the end of his life. As Popper has pointed out (op. cit., footnote 8), in the preface to the 1912 German edition of his *Mechanik* he alluded to this theory only by way of complimenting Hugo Dingler, Einstein's opponent. Neither Einstein's name, nor that of the theory, was mentioned.

[6] Leibniz–Clarke *Correspondence*, Leibniz's fifth letter, section 29.

[7] ibid., section 42.   [8] ibid., section 47.

ways. It finds a use as a virtual synonym for 'volume' or 'displacement', 'distance' or 'interval', 'place', 'vacant place', 'totality of geometrical points or coordinate values', and 'metric' (that is, 'a method of deriving distances from coordinate values'), to name a few. But all uses are concerned to some degree with the *location* of objects. As we have already indicated, there are few who will not agree with Aristotle and Leibniz—and Newton—that one must specify the location of an object by reference to other things. But having specified the location of an object in this way, it is natural to speak of 'replacing the object by another', that is, of 'putting an object in a place formerly occupied by another'. Space looked upon as the set of all places then becomes, as it were, a set of variables each of which may or may not attach to an object at any one time.

To return to Leibniz's central doctrine: 'in order', he wrote, 'to have an idea of place, and consequently of space, it is sufficient to consider . . . relations [of things amongst themselves] and the rules of their changes, without needing to fancy any absolute reality out of the things whose situation we consider.'[9] Space 'denotes, in terms of possibility, an order of things which exist at the same time, considered as existing together'.[10] It is, in short, an *'order of coexistences'*[11] and 'being neither a substance, nor an accident, it must be a mere ideal thing'.

Now a common modern interpretation of the relational concept of space makes space something 'determined solely by the fundamental particles'. Some will go so far as to identify space in some way with the material of the universe itself. This may well be a plausible rendering in its own right, but it does not do justice to Leibniz's meaning. Consider, for example, the statements:

(i) 'There is nothing outside the material universe.'
(ii) 'There is nothing outside space.'

For Newton the first would probably have been judged false, it being probable that empty space stretched beyond the limits of the universe. The negation of the second would have been regarded as self-contradictory. On the latter-day relational view, both (i) and (ii) would presumably merge, and their negations be self-contradictory, the universe being all-comprehending. But for Leibniz, although this might

---

[9] ibid., Leibniz's fifth letter, section 47.
[10] ibid., Leibniz's third letter, section 4. Cf. Ockham's characterization of place in its formal aspect, as an *ordo ad universum*.
[11] loc. cit. We notice that to speak (outside political circles) of coexistence, means committing oneself to defining some sort of world-wide simultaneity.

have been said for (i) (if by 'outside' we mean 'not comprised in') it would certainly not have been said of (ii). If by 'space' we mean an 'order of coexistences', and no more, then (ii) appears to be meaningless. It seems that we must distinguish at least two kinds of relational view (later we shall make further distinctions). For Leibniz and most of his followers, space is not something which inheres in objects and which must consequently move with them. If such a belief is widely held at the present time, it is probably encouraged by Weyl's Principle and the notion of co-moving coordinates. Admittedly there have been philosophers who have spoken of space as a 'real entity'—but they were not thinking of a material entity. They were, in fact, usually spiritually closer to Newton than to Leibniz.

2. Absolute and Universal Time

Before turning to modern trends in the development of these arguments concerning the nature of space, it will be necessary to discuss briefly the concepts of absolute and universal time. 'Absolute, true and mathematical time, of itself, and from its own nature, flows equably without relation to anything external.' This well-known statement by Newton has often been attacked, in particular on the grounds that 'flows equably' cannot be understood except in terms of the time-concept itself. We saw that Newton appreciated the logical point that motion can only be specified relatively to something, but here he appears to forget it. How else could he speak of time flowing 'without relation to anything external'? The strongest reason for Newton's giving the above definition was almost certainly that he wished to achieve a symmetry of exposition as between time and (absolute) space. Other factors in influencing his thought (not least the teachings of Isaac Barrow)[12] are discussed at length by Whitrow.[13] There is evidence that Newton was confused by the unlikelihood of there ever being an 'equable motion' in nature, by which time could be accurately measured. But what criteria could he possibly have given for finding such an equable motion? All he could have done would be to compare one periodic phenomenon with another. This in a sense is what he did in practice. Having introduced 'mathematical time' into his dynamics, he used the resulting system to instruct in the relation between intervals of time so-defined and the periodic times of simple physical systems. When

[12] Cf. the remark of Barrow, whose lectures Newton attended, 'whether things move or are still, whether we sleep or wake, time pursues the even tenour of its way'. *Lectiones Geometricae*, trans. E. Stone (1735) p. 4.
[13] *The Natural Philosophy of Time* (Nelson, 1961), pp. 130 ff.

Newton is accused (justifiably it seems) of reifying time, he is doing much the same thing as the present-day cosmologist who set up an idealization corresponding to the terms of his theory and then speaks of it as though it were real.

Leibniz, too, appears to have worked by analogy from his doctrine of space. Time, for Leibniz, was 'not something distinct from temporal things'. 'Instants', he said, 'considered without the things, are nothing at all'; and '. . . they consist only in the successive order of things'.[14] The argument used proceeds by a *reductio ad absurdum*. Why should God have created the world at one time rather than another? (This would be possible for one who accepted Newton's supposition that time is something external to things.) According to Leibniz, the Principle of Sufficient Reason shows that God could never have created the world under those circumstances. That the world exists is supposed to show the absurdity of the premiss concerning absoute time.[15] (A similar argument was used against absolute space.)

The strongest objection to the views on time and space as developed by Leibniz was that they did not appear to lend themselves to quantification.[16] His ideas might without much difficulty have been developed into a theory of the temporal ordering of events, but they gave no hint at all of ways to effect a metrical comparison of different durations of time. We have seen how Robb, Milne, and others tackled this problem in ways of which Leibniz would probably have approved. It would be mistaken, however, to suppose that Newton's formal definition of 'time' contributed to the great success of his dynamical theory. This is almost certainly indifferent to philosophical considerations of the sort brought out in the Leibniz–Clarke correspondence; and, as we shall suggest in the last section, it seems that as much could be said of recent physical theory.[17]

There is one distinction, however, which it is important to draw, namely, that between universal and non-universal time. It is unfortunate that modern writers often confuse 'universal' and 'absolute', as applied to time. In fact it is clear that both Leibniz and Newton

---

[14] Leibniz–Clarke *Correspondence*, Leibniz's third letter, section 6.

[15] See p. 390, *infra*.

[16] Accused of this, Leibniz feebly replied: 'Order also has its quantity; there is in it, that which goes before and that which follows; there is distance or interval. Relative things have their quantity, as well as absolute ones. For instance, ratios or proportions in mathematics . . .' *Correspondence*, Leibniz's fifth letter, section 54.

[17] Newton's ideas on time are today often unfortunately rendered as: Absolute time is independent of events (rather than of 'things' or 'happenings'). This now has a self-contradictory appearance, where 'event' is understood as 'thing at a certain time'.

(ii) ABSOLUTE AND RELATIVE

would have subscribed to a doctrine of universal time, for no alternative was clearly appreciated before Einstein dismissed the concept of absolute simultaneity. Leibniz's doctrine of pre-established harmony suggests that time imposed the same pattern of preception on all minds. Consider, for example; 'Minds are . . . images of the Divinity Himself . . . and are capable of knowing the system of the universe' (section 83 of the *Monadology*), and also the well-known metaphor of the divine clock-maker. On the other hand, although Leibniz does not appear to have explicitly considered the possibility of time-perspectives, he did maintain that each soul has always, from its beginning, a point of view peculiar to its own position in the universe. Kant (in his early relational views a follower of Leibniz) is sometimes said to have found inconsistencies in the idea of a universal time. Whilst he did indeed claim that the idea of time is not appropriate to the universe itself, he assumes throughout his work that time is universal in the sense that it is 'world-wide'.[18] It was precisely because this was the attitude of so many, that when Einstein held that arbitrarily-moving observers were incapable of agreeing upon a common time his claim was regarded with such dismay. In 1905 Einstein said that there can be no universal agreement as to the simultaneity of spatially separated events. Twelve years later he was to reintroduce (as a feature of his cosmological theory) the notion of a universal time. This was done in his first paper on relativistic cosmology, and clearly it was an outcome of the highly restrictive nature of his 'cosmological principle'. Yet as soon as local inhomogeneity could be allowed for, it was clear that this universal time property would disappear. Even so, it might be reintroduced statistically on a cosmic scale—hence the current phrase 'cosmic time'.[19]

3. Cosmic Time and Modern Cosmology

Eddington was quick to notice Einstein's apparent change of mind, and in a typical flight of fancy suggested that 'a being coextensive with the world might well have a special separation of space and time natural to him'.[20] The introduction of such a being was a concession to those who looked for an absolute distinction between past, present, and future. We might add that only some such being could override the

---

[18] The series of causes, necessitating the sequence of effects, was said 'to render valid, both universally and for all time—and in consequence objectively—the empirical cognition of the relations of time'. (*The Critique of Pure Reason*, Transcendental Analytic, II, ch. ii, second 'Analogy of experience', last sentence.)

[19] This seems to have been first used by Milne.

[20] *Space, Time and Gravitation* (Cambridge, 1920) p. 163.

'horizon' difficulty. Since no two particle-observers in the model are centred upon the same field of events, 'universal' would not seem to be the best description of the time by which they divide up those events.

If Einstein's restricted relativity is accepted, a common time can be locally constructed, but only as long as the frames of reference are not accelerated relatively to the mean distribution of matter in the neighbourhood. As far as relativistic cosmology is concerned, one can easily judge whether a model utilizing a given metric may be said to introduce the idea of cosmic time. Using co-moving coordinates and thus setting $(dr/ds) = (d\theta/ds) = (d\phi/ds) = 0$, we are left with an expression for (dt/ds). If this is unity (or $1/c$, according to the units chosen), coordinate time will agree with the proper time as reckoned at each fundamental particle of the model; and both may then be called 'cosmic time'. Thus the Minkowski metric and the metric of Einstein's (original) model allow of this interpretation, as does that of Robertson and Walker. The original form of the de Sitter metric does not yield this result, nor does either of the two well-known forms of the Schwarzschild metric with cosmological term. If the de Sitter model be taken in its Lemaître–Robertson form, however, it does allow of a cosmic time, and there is no doubt that this fact contributed to the relative popularity of the later version.[21]

Milne, at an early date, took exception to the coordinate time of the models of the General Theory of Relativity.[22] The circumstance that the nebulae (as represented by fundamental particles) were assigned fixed coordinates he found unreasonable; for 'at rest with respect to the system of nebulae' could not, he said, be given any 'objective meaning'.[23] (Mach's comments on this statement would have made interesting reading.) 'Absolute rest' and 'absolute velocity' had, he regretted to say, reappeared. Robertson's reply to this criticism was not very illuminating: he held that this was simply a means of *defining* rest and velocity 'relative to the mean motion of matter occupying that limited portion of the space-time universe in the neighbourhood of the body whose state is in question'.[24] He then carried the argument into the

---

[21] 'This is the physically significant form', wrote Weyl in 1927—simply meaning, no doubt, that the new coordinates were more convenient to work with than the old. Later commentaries which have said much the same thing have done so for a different reason: the mutual separation of any two points of the later version (i.e. with fixed spatial coordinates) increases exponentially as the (cosmic) time. Likewise, the system of galaxies is thought to expand. Whether it does so in the same way is entirely overlooked, however, in talking of the 'greater physical significance' of the later models. This common phrase thus seems inappropriate here.

[22] *ZS. Ap.*, vi (1933) 1.   [23] op. cit., p. 16.   [24] *ZS. Ap.*, vii (1933) 153.

enemy camp, pointing out that Milne's own theory employed a cosmic time in the form of the proper times (τ) of the events—namely the times between 'the singular event' and each subsequent event.[25] 'The curved 3-spaces $\tau =$ constant may then', he said, 'be described in terms of 3 spatial coordinates $x^\alpha(\alpha = 1, 2, 3)$, and these together with τ constitute a most natural curvilinear reference frame for all events P lying within the light cone whose vertex is the unique event O'. Using this cosmic time τ, Robertson purported to equate Milne's model to a relativistic model with a metric of the sort he and Walker derived for relativistic cosmology.

Milne's answer was that he objected to the use of the 'cosmic time' τ as an *observed time*: as a coordinate time it was harmless enough.[26] (He could have profitably extended this insight to the case of co-moving coordinates, to which he had objected in the first instance.) It will be recalled that Milne's particle-observers can agree on a common time only under certain conditions. Despite Robertson's claim it remains true to say that Milne's completed theory had no cosmic time—in the sense that the time assigned to a distant event by an observer associated with a fundamental particle is not in general the same as the proper time of the event. Bondi speaks of the existence of 'a universal or cosmic time' for all members of an equivalence, but this is not the usual convention.[27] In Milne's relatively stationary equivalence there is certainly an absolute simultaneity, the various members of the equivalence assigning the same epoch to any event.[28] With a uniform motion equivalence, however, there is no cosmic time. In any case, Milne later abandoned even the idea that there was a unique natural time associated with each fundamental observer, as soon as he appreciated the implications of the logarithmic time-transformation.[29]

To specify a class of privileged observers is the rule rather than the exception. Following Weyl, the exponents of relativistic cosmology did so explicitly. Einstein, in his restricted Theory of Relativity had done so, as had Milne. An argument against accepting as a cosmic time the aggregate of local times associated with a class of privileged observers has been put forward by K. Gödel.[30] To define any such cosmic time depends, he said, on the determination of the mean motion of matter in each region of the universe. He considered that it was

---

[25] op. cit., p. 157. This use of τ is not to be confused with Milne's later use of the symbol.
[26] ibid., p. 180.   [27] *Cosmology*, 2nd ed., p. 128.
[28] *K.R.*, pp. 48–50.   [29] See pp. 169–70 *supra*.
[30] *Albert Einstein, Philosopher-Scientist*, ed. P. A. Schilpp (Evanston, 1949).

'doubtful whether there exists a precise definition which has so great merits that there would be sufficient reason to consider exactly the time thus obtained as the true one'.[31] But this state of affairs is nothing new, for here is the central problem in assigning coordinates. The problem in defining 'cosmic time' (Gödel refers to it as 'absolute time') is said to be 'one of obtaining a precise definition' by introducing an 'arbitrary element', as, for example, 'the size of the regions or the weight function to be used in the computation of the mean motion of matter'. Contrary to Gödel's belief, the real difficulty does not arise in this way. There is certainly a practical difficulty to be encountered, but this comes at the end of the problem rather than at the beginning. When a theory has been built up to incorporate a cosmic time coordinate, and conclusions have been drawn which it is hoped may be astronomically confirmed, there comes the astronomical problem of deciding upon a suitable region—say that determined by a cluster of galaxies—within which a mean velocity of matter can be ascertained. There are other ways of dealing with the concept of time than by beginning 'time is . . .' or 'time is not . . .'. Part of the trouble is an ambiguity in the word 'definition'. We are not interested in establishing intergalactic agreement on time-measurement. Here, indeed, arises the problem of the significance of coordinates all over again.

As a consequence of the apparent success of theories in which cosmic time is accepted, Jeans concluded that there is no reason to abandon the intuitive idea of an objective and universal time.[32] Gödel, by virtue of his having found cosmological solutions which do not involve cosmic time, has emphasized the fact that we are not obliged to agree with Jeans.[33] It should not be thought, however, that he found any inconsistency in the idea of a cosmic time. His procedure was simply to reject the condition that the one-parameter system of 3-spaces should be everywhere orthogonal to the world-lines of matter,[34] and in doing so to reject the opportunity—but not to deny the possibility—of making use of a cosmic time coordinate.[35] Gödel's line-element was as follows:

$$ds^2 = a^2\{dx_0^2 - dx_1^2 + (1/2)\exp(2x_1)\,dx_2^2 - dx_3^2 \\ + 2\exp(x_1)\,dx_0 dx_2\}, \quad (1)$$

[31] op. cit., p. 560, note.
[32] 'Man and the Universe', Sir Halley Stewart Lecture (1935), pp. 22–23; see also *Physics and Philosophy* (Cambridge, 1942) pp. 63–68.
[33] *Rev. Mod. Phys.*, xxi (1949) 447.
[34] Cf. Robertson's conditions and Einstein's and Weyl's before that.
[35] Gödel, op. cit., cf. also *M.T.R.*, p. 16.

where x_0 is a time coordinate. He proved Einstein's field equations to be satisfied if
$$(1/a^2) = 8\pi\gamma\rho \tag{2}$$
and
$$\lambda = -(R/2) = -(1/2a^2) = -4\pi\gamma\rho. \tag{3}$$

The four-dimensional space-time which this defines has many strange properties: it is stationary and spatially homogeneous, and it has rotational symmetry. (There exists, that is to say, a one-parameter group of transformations which carries any point of the space-time into itself.) A 'positive direction of time' can be consistently introduced, making it possible to decide for any two neighbouring points on any world-line (of matter or light) which of the two is the earlier; yet it is not possible to assign a time coordinate t to each point (of space-time) in such a way that t always increases if one moves in a 'positive time-like direction'. This follows from the most surprising property of the model, namely that it contains closed time-like lines connecting any two points (P and Q, P preceding Q) of a world-line of matter (itself always open) on which Q precedes P. It is therefore in principle possible to travel into the past and the future, and to influence either.[36] This property must be judged an absurdity by anyone committed to the ordinary modes of speech: the logical difficulties associated with a closed personal time are well known. In fact not even cosmologists were tempted to change their habits of speech, for the model yields no spectral displacement, whilst it possesses the feature that 'matter everywhere rotates relative to the compass of inertia with the angular velocity $2\sqrt{(\pi\gamma\rho)}$'.[37] We shall return again to the notion of a 'compass of inertia'.

4. The Exaggerated Philosophical Involvement of Theories of Natural Cosmology

We earlier saw how different aspects of Newton's views on absolute space were emphasized by different writers. For Newton, his space was something in which objects were situated and—with a very free interpretation of his ideas—the set of all inertial frames. Kant concentrated on the former and even today some would wish to argue that the idea is of psychological assistance. Applied mathematicians have obviously

---

[36] Gödel made light of this property, for the velocities required are greater than $(1/\sqrt{2})c$ and this, he maintained, was unlikely ever to become feasible. See Schilpp, op. cit., p. 561.

[37] See *Rev. Mod. Phys.*, xxi (1949) 447. Gödel's notation has been changed here.

tended to dwell on the second aspect of his controversial pronouncements on the subject. But the fragmentation of the concept of absolute space has become even more complete during the present century. As explained at the beginning of the chapter, this would not matter in the least were it not the case that discovering any sort of 'absolute' in a theory is often regarded as tantamount to disproving it.

To begin with the Mach–Einstein principle, the 'principle of the relativity of inertia'. This does not in any earlier sense implement the relational view of *space*. It does not pronounce on the nature of space in any obvious sense, although for a theory accepting it the Newton–Euler account is made redundant. But the fact that Einstein laid down a connexion between matter and what *he* called 'space', does not mean either that such a step is essential to any satisfactory theory, or that the meaning of the word 'space' so determined is inherently more correct than any other.

With the Special Theory of Relativity, we are again not dealing with a theory of the nature of space, but of motions of particles relative to each other and to the set of inertial frames. The triply-infinite set of inertial frames connected by the Lorentz transformations is often referred to as a new 'Absolute', but even though thought of as a permanent setting for events, it can hardly be looked upon as a place in which events come about.

With the General Theory of Relativity, Newton's absolute space becomes not only redundant (on the basis of Mach's ideas) but unacceptable; and this because 'there are to be no privileged spaces' (Einstein). At least one of Einstein's reasons for holding this view was inspired by Mach's positivism: as explained in Chapter 4, arguing that privileged spaces would imply 'factitious causes', and that factitious causes cannot be responsible for influencing 'observable facts' (cf. Newton's arguments against relational space!), he concluded that there can be no privileged spaces. But what does he mean by 'spaces' here? Certainly not what Newton or Leibniz meant. A reasonable synonym might be 'coordinate systems'; and on the subject of coordinates Einstein was very clear. First of all, it is impossible for even the most ardent Leibnizian to gloss Leibniz's (spatial) 'relations' as simply differences of spatial coordinates in Einstein's theory; for the fact that arbitrary transformations (even involving coordinate-time) are there permissible would make nonsense of the idea. Of course this likewise rules out the equation of Newton's 'space' with Einstein's. 'The introduction of a system of reference serves no other purpose than to facilitate the description of the totality of . . . coincidences [of material points which in making measurements we are verifying]'.[38] Newton's space was a good deal more important to him than the 'systems of reference' between which Einstein wished to choose freely, and then forget. Did Einstein's 'space-time' become a new 'Absolute'? The point has frequently been made of late. Einstein certainly spoke often of assigning coordinates to the 'space-time continuum', and this can no doubt be thought of as an Absolute, namely that which comprises the aggregate of past, present, and future events. Once again, however, it bears no comparison with the immaterial absolutes which had gone before it.

As the General Theory of Relativity developed, it was said to have been found to harbour other sorts of Absolute, to the great glee of those who opposed it. Gamow was probably the first to hint that a 'rotating universe' may be represented by the group of anisotropic solutions of the field equations.[39] Gödel's solution has already been mentioned, the 'compass of inertia' with respect to which the rotation occurred being called 'absolute' by many. (It will be remembered that this solution is such that, if a transformation is taken which locally reduces the metric to that of the Special Theory of Relativity, then distant matter possesses rotation, and Mach's Principle is not satisfied.) Gödel's solution was generally regarded as revealing blemishes on Einstein's theory, although it is not clear why such an easily identifiable absolute as this should have been regarded as one of them. More serious, or so it was believed, was the existence of the closed time-like lines and C. Y. Fan's discovery that the angular momentum of the model would not be conserved in an expansion.[40] Why the fact that it permits the identification of a frame of reference more *convenient* than the rest— even one which rotates relatively to the fixed stars—should be an embarrassment to relativity theory is not clear: in the early days of Einstein's theory its supporters never tired of pointing out that the existence of inertial frames did not in the least affect the relativity principle. A moment's consideration will show that using any theory which requires two nearby particles to have a relative velocity smaller than a certain finite value it is possible to set up a frame of reference which does not rotate relatively to galaxies at very great distances; for as the distance increases, the effect of a finite velocity on the observed

[38] *P.R.*, p. 117. [39] Letter to *Nat.*, clviii (1946) 549.
[40] *Phys. Rev.*, lxxvii (1950) 140. O. Heckmann and E. Schücking derived a Newtonian analogue of Gödel's model, exhibiting both expansion and rotation. (See, for example, *Jubilee R.T.*, p. 114.) L. Ozsváth and Schücking have found a relativistic model similar to Gödel's but without the closed time-like lines (cf. *Nat.*, cxciii (1962) 1168).

(relative) proper motion of two objects diminishes. Relative to this frame, we can have no *a priori* assurance whatsoever that the nearer distributions of matter will not rotate. Whatever the blemishes on Einstein's theory disclosed by Gödel, the possibility of a model exhibiting a rotation with respect to the compass of inertia was not of them.

What of the charges brought against Milne to the effect that he had, tacitly or otherwise, subscribed to a version of Newton's absolute space? Milne's opinions on the matter were voiced in a language which at times gave them a Newtonian air, but it does seem that this was often misleading. If a cosmic time is not available within a theory, it is hard to see how an absolute space could possibly be introduced. We have seen that in Milne's theory such was not generally available. What, then, was Milne's '*substratum*'? In his *Kinematic Relativity* he explains, after the manner of his earlier book, the kinematics of this, 'the idealized system of mutually separating particles' which, having no preferred members, move in such a way as to make the Lorentz formulae apply. It was explained that the 'swarm' filled the whole of the interior of an expanding sphere which, in keeping with the properties of the Lorentz transformation, could be regarded as having any member of the swarm at its centre. The boundary of the substratum is completely inaccessible to any particle within it, however great its velocity. When Milne refers to the substratum as having 'all the properties of infinite space',[41] he has this feature in mind and he is not thinking of the substratum as something in which material objects are situated. Indeed, he goes on to refer to the substratum as 'a system of frames of reference in motion'. As such, his particle-observers must be regarded as prior to it, rather than conversely. Milne, in effect, used the term 'substratum' as synonymous with 'homogeneous equivalence'—'homogeneous', that is to say, in Milne's very general sense and not in the instantaneous present of an observer.[42] (We saw in Chapter 8 that the two senses coincide when there is an 'absolute simultaneity' as, for example, when using a logarithmic time regraduation to pass from a 'uniform motion substratum' to a 'relatively stationary substratum'.)[43]

Milne objected to the idea of absolute space; he frequently spoke of the freedom of an observer to choose his space arbitrarily and he also spoke of some 'spaces' as 'private'—none of which Newton would have recognized. 'In assigning structure to space', he wrote, 'in restoring structure to structurelessness, mathematical physicists had in effect

[41] *K.R.*, p. 8. Cf. *ZS. Ap.*, vi (1933) 77.
[42] *K.R.*, pp. 51–56.   [43] See p. 169 *supra*.

(ii) ABSOLUTE AND RELATIVE

reintroduced an ether'.[44] Here he objected to an alternative Einsteinian view of space, and in particular to Jeans's picture of galaxies as 'mere straws floating in the stream of space'. On the other hand, he can scarcely be classed with Leibniz, for he frequently speaks of 'embedding events in Euclidean space', and of fundamental particles 'moving against a background of space'. Is this merely a slip of the pen? One of the more explicit discussions of the subject is given in answer to the questions: 'What is "outside" the system?' and 'Do not the receding nebulae require external space in which to expand?'[45] The answer he gives is that the questions are meaningless. The background to his model is 'everywhere dense' and the existence of a horizon to his model means that there is no 'window into outer space'. We can, he says, by no conceivable method observe any object in this space. Thus, by the tenets of 'logical' positivism, in which Milne had been caught up, he was bound to deny the significance of any hypothesis as to the existence of outer space. In fact here he seems to be arguing that statements about space take their significance from statements about observable objects which are *situated in* the space: the space is not itself directly observed. Why, then, should it matter one way or another whether we postulate the 'existence' of space outside the system? It does not matter, according to Milne, 'whether the space exists *a priori* or whether the space is created by the system as it expands—immaterial because either view has no verifiable consequences and our whole philosophy is that an unverifiable proposition [concerning the world of nature] is meaningless.' As though this might seem too severe, however, he pointed out that objects beyond the boundary could be 'mathematically described'. So too could infinite space. But neither is of interest to the physicist, in Milne's view, for physics 'demands verifiable answers to its questions'.[46]

Milne has been quoted here at some length in order to show how absurd it would be to class with Newton all who adhere to the old division between geometry and physics. No doubt there are features in common, but it seems to be a purely contingent affair that many of those who have retained this division in their cosmological writings have tended to think of space as a subsistent background to matter— and certainly Whitehead was as notable an exception as Milne. It seems to be possible, in fact, to distil seven or eight different points of view with regard to space from the sorts of discussion referred to in this chapter. What is of more immediate interest, however, is that in certain

[44] *W.S.*, p. 2.   [45] ibid., p. 131.   [46] *W.S.*, p. 132.

cases these different attitudes, despite the fact that they may be thought 'philosophically' incompatible amongst themselves, are separately compatible with the same cosmological theory.

As mentioned already, a common rendering of 'the relational view of space' makes space something determined solely by the fundamental particles. A more extreme view is that according to which space is to be identified with the material of the universe. This is an almost Cartesian attitude, which makes space a set of relata rather than relations, and which is often associated with the notion of 'embedding'. It is more extreme than the first relational view of space only if we are considering the material correlates of the fundamental particles of a model rather than the latter. Writers are not in the habit of drawing any distinction here.

From the Leibnizian stream of thought, three quite closely related attitudes to space can be extracted, all differing from the two already mentioned. The first, which could be summarized in the sentence 'Space is a class of relations between coexistent objects', is far from precise. (Instead of 'a class of relations' we could substitute 'the set of all relations', but this would do less than justice to Leibniz's intentions; for he was clearly concerned only with the ordering relation.)

No doubt 'relation' is usually thought of as more or less synonymous with 'dyadic propositional function'. If so, the proposition follows from each of the remaining neo-Leibnizian propositions to which we have referred:[47]

> Space is the class of such relations between coexistent objects as are summarized in any assignment of coordinates, and
>
> Space is the class of such relations between coexistent objects as can be expressed in terms of distances.

Each of these is to be found implicit in much present-day cosmological writing, and the last is a common rendering of Leibniz's position, given by those philosophers (for example, C. D. Broad) who wish to interpret generously his vague remark that 'even relations have their quantity'.

Since distances and coordinates are often not sufficiently well distinguished, one finds writers who appear to identify the two corresponding propositions. If it is assumed that distances are functionally related to coordinates, and that any topological requirements are clearly understood, then this may presumably be done.

---

[47] If relations of greater complexity are envisaged, this is clearly not possible.

(ii) ABSOLUTE AND RELATIVE

Finally, we shall suppose that in place of the proposition that space is absolute, most people would accept the following:

> Space is that in which material objects are situated and through which they move. It is a background for objects of which it is independent. Any measure of the distances between objects within it may be regarded as a measure of the distances between its corresponding parts.

Anyone subscribing to this need not consider space to be infinitely extended, although it was natural for those who lived before the time of Riemann to think so.

We shall be content with these six or seven divisions within the 'philosophy' of space. They are far from exhaustive: for example, Kant finds no place here, and we have not mentioned the belief that space is an attribute or adjective of substance or matter. The categories between which we have distinguished do, however, appear to cover the requirements of most modern writers. For the sake of clarity they will be collected together here. They are, in the order given above, categories within which philosophers and others may be placed who believe that space is one of the following:

> determined solely by the fundamental particles;
> to be identified with the material universe;
> a class of relations between coexistent objects;
> a class of ordering relations between coexistent objects;
> a class of coordinate-relations between coexistent objects;
> a class of distance-relations between coexistent objects;
> absolute, in the senses outlined above.

These will be referred to as the weak material, the strong material, the weak relational, the simple Leibnizian, the coordinate-relational, the distance-relational and the absolute views of space.

These views are not quite so exclusive as might first appear. For example, one might subscribe to the last as philosophically enlightening, and yet wish to determine one's metric (one would not, of course, refer to it as the 'metric of space') from considerations suggested by the first, fifth, or sixth views. Again, one might believe either the simple Leibnizian or weak relational view to be philosophically helpful, and yet at the same time wish to maintain within cosmology the connexion between matter and metric suggested by one or other of the material views, with 'metric' for 'space'. As they stand, however, the three types of proposition will be generally interpreted as mutually inconsistent. How close, then, are the ties between them and the methods of modern cosmology?

Whitrow is one of the few to have dealt with this question explicitly. He appears to wish to make the 'expanding-space technique' the 'natural concomitant of the relational concept of space'.[48] Going further, he associates what he calls the 'kinematic technique' with the idea of absolute space. 'Thus in the one case', he writes, 'there is motion *of* space and in the other motion *in* space'. In sum, he wishes to find a 'vital philosophical difference' between the two theoretical techniques, despite the close connexion which is to be found between the underlying mathematics. He hopes to exhibit the differences in the course of a (remarkably simple) derivation of the Robertson–Walker metric.[49] Whitrow assumes that each observer can choose a scale of length measurement which will allow him to regard the 'spatial' part of this as static. By this he simply means that the mutual separations of fundamental particles shall not change. He does not appear to have any proposition in mind such as that which we called the 'distance-relational'. Assuming rather that 'the material universe in its large-scale feature can be identified with world-space', he appears to accept the strong-material view. But since his attention is not directed to the physical properties of the material particles of the model, but to the possibility of making certain kinds of observations from a restricted class of them (such as observations of the distances of other particles of the class), perhaps the weak material view comes closer to his own. In fact this is a suitable place to point out that a seventh category is needed if we are to represent the opinion embodied in Einstein's General Theory. There we find a *connexion* between space (in the sense of metric) and matter effected through the field equations, but there is no question of identifying the two.

Is Whitrow's interpretation essential to the results he obtains? The first of these is that 'the relational concept of the universe, according to which there is no independent spatial background against which systematic changes in the geometrical structure of the universe occur, implies the existence of cosmic time'.[50] Since the concept of cosmic time, in some sense or another, is a necessary condition for the concept

---

[48] *The Natural Philosophy of Time*, p. 245.
[49] ibid., pp. 248–51. [50] ibid., p. 251.

of absolute space, Whitrow's analysis cannot thus far be philosophically very exclusive. He has, on the other hand, shown that a belief in absolute space is not a necessary condition for the use of cosmic time.

The severest test of the philosophical neutrality of his argument comes when the metric

$$ds^2 = dt^2 - (1/c^2) \cdot R^2(t) \, d\sigma^2 \tag{4}$$

is finally obtained.[51] 'The fundamental particles . . . are now regarded as embedded in a space of variable scale-factor $R(t)$'. This would surely be without meaning for one holding to the distance-relational view. For to say that *all* distances expand in the same ratio—the length of the measuring-rod being included—is to say that no distance changes, distance being essentially relative to some standard. But there seems to be no reason why a person adhering to the belief that space is absolute should not regard (4) as a rule for calculating the movements of objects and the intervals between events. His background space will remain indifferent as ever to the contingencies of material objects. In a sense this is merely to repeat that it is not necessary to use the words 'space' and 'metric' interchangeably.[52] But now it should be possible to detect some important differences between the seven or more attitudes to space. First, the apparent theoretical involvement of the absolute view is so slight as to make it extremely doubtful whether adherence to it could ever lead one to modify one's theory. This is not true of the distance-relational view, and it is even less true of the usual forms of the material interpretations, as Whitrow has shown. It is this theoretical neutrality of the *central* doctrine of absolute space which makes it so uninviting, except as embodying one small aspect of common speech. In this sense it may have a philosophical value, but only when something more is added to it do we find it scientifically challenging. This 'something more' was the sort of thing Newton added. He made absolute space more than a background for motions: he made it a theoretical concept which featured essentially in the steps leading to specific empirical predictions. We must recognize, therefore, yet another attitude towards the nature of space, stronger than that embodied in the statement given on p. 367. And since Newton's way of using 'absolute space' as a theoretical term is obviously not the only way, one must be prepared to encounter other extensions of these beliefs concerning the nature of space.

[51] Whitrow, op. cit., p. 250.
[52] The person who espouses an operationalist metaphysics will probably disagree.

In conclusion: three broad groups of attitudes as to the nature of space have been distinguished, and at least nine more specific divisions were recognized. Even so, it was necessary to leave them expressed in propositions insufficiently precise to serve as useful theoretical hypotheses, although it seems likely that at least two of them—namely those nearest to views held by Leibniz and Newton themselves—are in a form totally unsuited to this end. It is partly due to their scientific neutrality, and is not merely an accident of history, that these two are the propositions which most philosophical discussions on the nature of space take as their starting-point. Such discussions tend to be of four kinds. The first, historically speaking, began with the question: What is space really like? Before discussions on Reality gave way to discussions of the linguistic habits of the man in the street, the question was taken over by the philosopher-psychologist. Following this period, which covered most of the last century and some of this, and which it is impossible to deal with here, the philosopher began instead to ask after what manner the word 'space' is used in the different sectors of our language; neither did he tend to ignore the natural sciences, which provided him with his most rewarding examples. For historical reasons, some of which have been indicated, he was reluctant to dispense with the simple absolute/relative dichotomy. But since the rationalist concentrated on a simple logical truth, and the Newtonian on the unprecendented success of his dynamics, the conflict was always so indecisive as to leave a high degree of philosophical bewilderment in its wake.

In our list of propositions we began, simply because we were representing traditional attitudes: 'Space is . . .'. The sort of explicit and therefore exclusive definition which results is likely to lead to its own brand of unclarity in the minds of those who wish to concentrate on the various ordinary uses of the term. Here, as elsewhere, legislation which is contrary to custom is bound to inspire reaction. The time has come when the cosmologist must avoid using the word 'space', *qua* theoretical term, in more senses than one if he is to subscribe to a given 'philosophy' of space. And if he does so, he should distinguish between that part of his philosophy which is essential to the understanding of his theory and that which is not. In this way the terms 'absolute' and 'relative' would surely cease to be regarded as necessarily mutually exclusive; and linguistic philosophy would be left with a problem the more clearly defined for having been deprived of trappings acquired from long-discarded physical theories.

CHAPTER 17

CONCEPTUAL PROBLEMS (iii) INFINITY AND THE ACTUAL

THE idea of an actual infinite is not far to seek in cosmology. It occurred, for example, in Charlier's work on the infinite hierarchic form; in McCrea's discussion of the 'arbitrarily great' universe; in the development of the cyclic model by Friedmann and others; in the postulation of spatial and temporal infinities by Bondi, Gold, and Hoyle; and it is involved in acceptance of the hyperbolic model for which McVittie has long argued. Now although for several centuries is has been commonly held that the universe is infinite, yet there has always been an undertone of dissatisfaction with the logical credentials of the notion of an actual infinity. To begin with Aristotle: it is well known that he, faced with a widespread belief in some sort of actual infinite, disputed it whenever the occasion arose, and offered in its place his conception of 'potential infinity'. The notion of a realized infinite was unpopular during and long after the Middle Ages for yet another reason: infinitude was deemed an attribute possessed only of God.[1] But here again logical arguments were used to support the thesis to which Aristotle had given his blessing. Kepler, furthermore, was to reject the idea on what he conceived to be empirical grounds, although at times he appears to want to go so far as to say that it is without meaning. A great many similar protests could be quoted: not least of these would be from the writings of Kant, who refuses to speak of either an infinite or a finite world on the grounds that 'an absolute limit is impossible in experience'. (His qualification will be referred to again later.) Admittedly the Newtonian scheme of mechanics was housed comfortably enough in infinite Euclidean space, even though occasional spectres appeared in the shapes of infinite divisibility (first of matter and then of energy), the infinite potentials of a Newtonian universe, and indefinitely large velocities in dynamics (before the Special Theory of Relativity). At last, just over a century ago, Riemann's distinction between 'infinite' and 'unbounded' provided a rational alternative to Euclidean space. But by merely presenting an alternative, the infinity of the

---

[1] Thus a denial of this tenet was numbered amongst the errors for which Giordano Bruno died. One must remember Aristotle's pronouncement: 'It is clear that there is neither place nor void nor time beyond the heavens'. (*De Caelo*, 279a 12.)

actual world was not shown to be conceptually unacceptable.[2] By what arguments might the idea of an actual infinite be shown logically untenable?

1. The Concept of a Potential Infinity, and its Inadequacy

'There are, King Gelon, some who think that the number of the sand is infinite in multitude . . . but . . . of the numbers named by me and given in the work I sent to Zeuxippus, some exceed not only the number of the mass of sand equal in magnitude to the Earth filled up in the way described, but also that of a mass equal in magnitude to the universe.'[3] Archimedes believed he could fulfil this claim because both Earth and universe were thought to be of finite volume,[4] and the real problem for him was one of writing down sufficiently large finite numbers. He clearly thought the concept of an actual infinity to be meaningful. Lucretius, however, later argued that this involved a contradiction in terms. He maintained that 'nothing can have an extreme point unless there be something beyond to bound it', and that 'since we must admit that there is nothing outside the universe, it can have no extreme point, and therefore lacks an end and a measure'.[5] No better quotation than this can be found to serve as a historical introduction to the all-important distinction, drawn properly only by Riemann and those who followed him, between the unfinite and the unbounded.

[2] Even so, a great many philosophers of our century, having tired of the seemingly endless discussion on the infinite, have been glad to clutch at this Riemannian straw. We thus find a professional philosopher 'relieved to feel that our space may still be amenable to the law of the limit'.

[3] 'The Sand Reckoner' in *The Works of Archimedes*, ed. T. L. Heath (Cambridge 1912) p. 221. The work sent to Zeuxippus (the *Principles*) is lost.

[4] Aristarchus is Archimedes' authority. He could have quoted Herakleides, who taught that the Cosmos was infinite.

[5] *De Rerum Natura*, i, 959–64. Lucretius follows with an illustration: If space were finite and a man were to hurl a javelin from its furthest coasts, the javelin would either continue—in which case 'it does not proceed from the extremity'—or be checked by something beyond—when the same conclusion is reached. This resembles Locke's argument for the proposition that 'space is not a body' (*Essay Concerning Human Understanding*, Bk. II, ch. xiii, sect. 21). The Cartesian definition of matter as *res extensa*, with its concomitant doctrine of the impossibility of 'space void of body', a doctrine which Locke was attacking, seems to have been misunderstood by him. Could a man at the 'extremity of corporeal beings', he asked, 'not stretch his hand beyond his body?' If he could do so 'then he would put his arm where there was before space without body'. The Cartesian would presumably have denied the last statement, but Locke thought that, if consistent, he would be obliged to hold that 'body is infinite'. And this, as he said at the outset, he thought no one would affirm. Until the end of the last century the majority would have agreed with Locke, although the Cartesians did say that matter was without limits only in a sense which should be obvious from their definition of it.

(iii) INFINITY AND THE ACTUAL

Although our concern is with cosmology rather than mathematics, there is no point in our dwelling on the numerous astronomical discussions of the finitude or otherwise of the universe which accept uncritically a concept which may be logically unsound. At the outset, therefore, two important traditions in the theory of the mathematical infinite will be indicated, and ways in which they have been enlarged upon in recent years. In the rest of the chapter we shall point out what seems to be the only satisfactory way of interpreting infinity in regard to the numbers of objects within the universe of discourse of any cosmological theory.

We may begin with Aristotle's argument that is is absurd to hold that a line contains an infinite number of points, even though one can always indicate *any* finite number. Aristotle's solution was that the number of points on a line is *potentially* infinite. Eubulides of Miletus disliked the idea, and argued against it with a well-known paradox which it is not necessary to consider in detail here.[6] The essence of the argument is that the number of points on a line is either infinite or finite; and if finite in number they may be crossed off, one by one, until none remains—which appears absurd. It seems that Eubulides, and all who have since argued in the same way, have missed the whole point of falling back on the concept of potential infinity. Those who do so appear to want to do no more than keep the class under review *open* in the sense that no matter how many members may be found to fall within the class, it is to be logically possible that there should be others. By 'logically possible' it is simply meant that the meanings of the sequences of predicates which determine the class do not of themselves rule out the existence of further members. (For the moment we ignore the different sorts of existence with which the potential infinity is being associated.) In this sense, therefore, an open class may have a finite number of members, and Eubulides' dichotomy finite/infinite is unacceptable when one is concerned with the potentially infinite.

Whether or not this is a reasonable interpretation of Aristotle, the idea of a potential infinity has survived, in one form or another, to the present day. Thus we find A. C. Ewing maintaining that 'space is potentially but not actually infinite', and that the 'assertion of infinity

---

[6] See Diogenes Laertius, *De Vitis*, vii, 82; Cicero, *Academia*, ii, 49. Cf. Aristotle, *De Soph. elenchis*, 179b 34. There was a related controversy which was waged most eloquently in the fourteenth century: Are lines, areas and volumes made up of points or of finite and indivisible magnitudes? Nicolas of Autrecourt took the second view, but the majority sided with Aristotle. (See J. R. Weinberg, *Nicolaus of Autrecourt* (Princeton, 1948).)

does not amount to saying that there is some actual thing, space, extending indefinitely, but only that we, or matter, might go further and further'.[7] Again we find Lemaître saying much the same thing: 'everything in act is finite, potential being is the only legitimate field of application of the infinite or transcendental numbers of the mathematicians'.[8] Later in the present chapter we hope to show that Lemaître was wrong.

On the mathematical side it is possible to detect the idea of a potential infinity in the work of those mathematicians who, from the time of the investigations of infinite series in the seventeenth century, have read 'infinite' as 'indefinitely increasing'. It can also be seen in the intuitionistic mathematics of L. E. J. Brouwer and his followers, for whom the infinite is 'becoming', 'constructive' and 'potential'. A revulsion against the idea that it may correspond to something actual is in fact to be found in mathematicians as renowned as Gauss and Weyl. In a letter to Schumacher in 1831,[9] Gauss protested at the 'use of the infinite as something consummated, as this is never permitted in mathematics'. 'The infinite', he continued, 'is but a figure of speech; an abridged form for the statement that limits exist which certain ratios may approach as closely as we desire, while other manifolds may be permitted to grow beyond all bounds.'[10] As recently as 1946 Weyl wrote that 'the sequence of numbers which grows beyond any stage already reached by passing to the next number, is a manifold of possibilities open towards infinity; it remains forever in the status of creation, but is not a closed realm of things existing in themselves'.[11] Many similar utterances can be found in the works even of mathematicians who are not conscious of having fallen under Brouwer's spell.

It is not suggested that the beliefs of Aristotle, Gauss, Brouwer, Weyl and the rest can all be comfortably fitted into the one category; but our analogy between open classes and potential infinities seems to bring out what is most important in all of them, and at the same time

---

[7] A. C. Ewing, *The Fundamental Questions of Philosophy* (Routledge and Kegan Paul, 1951), p. 151. The language is Aristotlelian but the doctrine is not, for Aristotle believed an infinite universe to be not even a potentiality. The alternatives were not explored in the Middle Ages until the end of the thirteenth century. See A. C. Crombie, *Augustine to Galileo*, pp. 236, 271. Sentiments much the same as Ewing's were expressed by Bishop Barnes. Cf. p. 134 *supra*.

[8] *The Primeval Atom* (Van Nostrand, 1950), p. 27. See also pp. 24–26. Lemaître will not admit the equivalence of whole and part, which the Cantorean concept of infinity requires.

[9] For the original see Gauss, *Werke*, viii (1900), p. 216.

[10] T. Dantzig's translation from *Number, The Language of Science* (1940), p. 211.

[11] *American Mathematical Monthly*, liii (1946) 2.

shows the futility of regarding 'finite' and 'potentially infinite' as mutually exclusive. On the other hand, if mathematicians had claimed to find only potential infinities in their subject—and the natural sciences would then have presumably done no more—there would have been little room for controversy. It was because mathematicians customarily spoke of infinity as essentially completed (that is, 'actual' in a sense peculiar to mathematics) that controversy arose. It was for the same reason that Bolzano, Weierstrass, Dedekind, and Cantor sought to frame a suitable definition for the concept, free from the traditional 'Paradoxes of the Infinite'—to take the title of Bolzano's well-known book.[12] Cantor's great achievement in the form of the mathematical theory of transfinite cardinals will shortly be outlined, for it is obvious that if it is at all possible to introduce an actual infinity into cosmology we must turn to some such concept as his.[13] It is instructive, even so, to consider some of the illogicalities in the notions of infinity which were current when Bolzano was writing and which he pilloried with some success.[14]

Cauchy and others defined the infinite in terms of a variable quantity whose value increases without limit, beyond any predetermined quantity no matter how large. The limit of the unlimited growth is then said to be the infinite quantity—and this, according to Bolzano, is a contradiction in terms. In fact this definition appears to be an unhappy mixture of potential and actual.

Spinoza's definition, which many had followed, made infinite anything 'incapable of further increase'. Whatever its intrinsic merit, this definition certainly cannot do justice to mathematics wherein quantities may multiply and be added to infinite quantities. What, then, of 'that which has no limit' as a substitute for 'infinite'? Bolzano was well ahead of his time in clearly distinguishing 'finite' and 'bounded'. As he pointed out, there are many instances in mathematics of bounded infinities (for example, 'the space between the arms of an angle') and of unbounded finite quantities (for example, the circumference of a circle which is of finite length but without 'bounding point'). The first example is weak, but the idea is there.

---

[12] *Die Paradoxien des Unendlichen* (Leipzig, 1851) 2nd ed., 1889. Bolzano was a Czech theologian and amateur mathematician whose modest mathematical fame was posthumous. His political beliefs were no less radical than his mathematics, and for the former he was removed from his position in the church.
[13] If not always on the side of the angels, Bolzano, Cantor, and their followers had the support of St. Augustine. Cf. *De Civitate Dei*, xii, 18.
[14] Bolzano, op. cit., sections 11–14, &c.

'The commonest formulation of the infinitely great', wrote Bolzano, 'is "that which is greater than any *assignable* quantity".' He then considered the meanings for 'assignable': it may simply mean 'possible', that is, 'capable of attaining actuality at some time or other'. (The introduction of the idea of time, and the tenor of the argument generally, show that Bolzano, like most of his contemporaries, mixed mathematical and physical elements freely.) In this case, he argued, 'linguistic custom' is not satisfied, for we apply the notions infinite and finite 'not only to objects enjoying actuality (as above all to God) but to others where we cannot so much as speak of existence'. Under the last heading he put 'fundamental propositions and truths' and 'ideas-in-themselves' (*Vorstellungen an sich*): 'we assume both finite and infinite sets of these'. The Aristotelian reply would no doubt be that it is not difficult to assume the truth of something which on closer examination turns out to be self-contradictory. In fact Bolzano next asked whether 'assignable' might not be taken to mean 'non-self-contradictory', but his objections to this are not very clear.[15] Lastly he asked whether 'assignable' could mean 'applying to all and only those things which can *in some way or other be given for us*, that is, become the objects of *our experience*'. This he rejects because he believed that usage requires 'finite' and 'infinite' to be predicates relating to certain intrinsic properties of things—regardless of our 'power of knowledge' and 'faculties of sensation'.

Bolzano was determined to attempt to formalize the notion of an actual infinite, which he believed to be required both by ordinary and mathematical usage. Before he could do so he had to answer those who claimed to find evidence for the impossibility of 'the objective existence' of infinite sets. 'An infinite set can never be collected together in thought', runs one of the oldest arguments of all. But this is not true, according to Bolzano, who pointed out that he could think of the set of all inhabitants of Prague or Pekin without forming a separate representation of each separate inhabitant. We shall return to this point in due course. His other arguments are in much the same vein and all appear to hinge on the irreducibility of actual to potential: 'nothing can derive its capacity for existing from its *capacity for being thought of*'.

Bolzano paved the way for an adequate formalization of the concept of a (mathematically) actual infinite, but it is to Cantor's theory that one must turn for the realization of this. The theory of transfinite

---

[15] Into this very definition, he wrote, we insinuate the non-existence of the infinite, 'since a quantity supposed greater than any non-self-contradictory quantity would have to be greater than itself, which is of course absurd'.

cardinals is almost entirely of his invention, although Adam of Balsham, Galileo, Leibniz, Bolzano, and Dedekind came very near to the pattern of some of his argument. Galileo realized that we *compare* before we *evaluate*. In the words of his character Salviati who has just drawn attention to the one-one correspondence between numbers and their squares: 'neither is the number of squares less than [the totality of] all numbers, nor the latter greater than the former; and finally, the attributes "equal", "greater" and "less" are not applicable to infinite, but only to finite quantities.'[16] This line of thought was anticipated in a remarkable way by Adam of Balsham in 1132, who had, clearly enough, grasped the idea of a set equinumerous with another, and yet containing it as a proper subset (to use modern phraseology).[17] Where Adam and Galileo saw danger, Leibniz saw contradiction. Since every number can be doubled, the even numbers are, as he saw, as numerous as the entire class of natural numbers. The whole, it seemed, was no greater than the part.[18] The older passages were ignored, whereas Leibniz's argument met with general approval. The efforts of Bolzano, who chose to treat this property of infinite numbers as involving no contradiction, met with little more recognition than Galileo's. Bolzano believed that the principle that the whole is greater than the part must be qualified, and Dedekind later showed how this might be done.[19] Bolzano, however, was concerned with series, that is to say *ordered* classes, and it was left to Cantor to give the first satisfactory account of classes regardless of order. As is well known, two mathematically definable sets may have the same cardinal number but different ordinals, the *arrangement* of the terms being all-important. Arrangement seems to be irrelevant when speaking of a class of like objects (as, for example, galaxies). In any case, the ordinal infinite is generally explained in terms of Cantor's theory of transfinite cardinal numbers.

2. CANTOR'S TRANSFINITE CARDINALS

Opposition to the idea of an actual infinite on the part of mathematicians was so strong in the last quarter of the nineteenth century

[16] *Two New Sciences*, cf. Thomas Salusbury's translation in *Mathematical Collections and Translations*, ii (London, 1665), p. 27.
[17] See *Twelfth Century Logic*. (Twelfth century logic, texts and studies.) L. Minio-Paluello (Rome, 1956). The passages in question are discussed by Ivo Thomas in *Journal of Symbolic Logic*, xxiii (1958) 133.
[18] *Phil. Werke*, ed. C. J. Gerhardt (1875), i, p. 338.
[19] Bolzano, op. cit., sect. 9: 'I propose the name *infinite multitude* for one so constituted that every single finite multitude represents only a part.' Dedekind's *Was sind und was sollen die Zahlen?* first appeared in 1888.

that Cantor's work was effectively suppressed until long after he had ceased writing mathematics.[20] Whether or not the opposition was justified, it is to Cantor's work that we must turn if we are to sanction the introduction of completed infinities into cosmology.

Two classes or sets which can be brought into a one-to-one correspondence Cantor termed 'equivalent'. The cardinal number of a given set is then defined as an object common to all and only those sets which are equivalent to the given set.[21] This definition, in a modified form, has since been made familiar by Frege and Russell. Amongst the *finite* cardinal numbers are those of the various kinds of finite well-ordered series (regardless of the order of their members). Cantor went further and spoke of the cardinal numbers of those discrete series possessing a first element but no last, one of the simplest examples of such a 'progression' being the sequence of natural numbers. Taking that series as a standard, an infinite set is said to be *denumerable* if it can be placed in one-to-one correspondence with the natural numbers. An infinite set is shown to be denumerable by an indication of the way in which its members can be listed, without repetitions, the first corresponding to 0, the second to 1, and so on. It can be shown that the integers, the rational numbers, the ordered n-tuples of members of sets which are themselves denumerable (n a positive integer), and the algebraic numbers are all denumerable. As any two examples of such sets can be shown equivalent, it follows that they all have the same cardinal number (symbolized as aleph-zero: $\aleph_0$), the simplest of the infinite set of transfinite cardinal numbers which Cantor subsequently defined.[22]

In 1874 Cantor showed that there are many infinite sets which are not denumerable; these include the set of the real numbers, the real transcendental numbers, and the set of the sets of natural numbers. The cardinal assigned to these sets is $2^{\aleph_0}$. In Cartesian geometry, the real numbers are identified with the points of the (real) Euclidean line. This set of points having long been called the (linear) 'continuum', the

[20] Cantor's work was repeatedly rejected by editors of mathematical journals and (except in England) was not widely known to mathematicians as a whole before his death in 1914.

[21] Many of these ideas came from Bolzano. The best account by Cantor is in *Math. Ann.*, xlvi (1895) 481. (An English translation by P. E. B. Jourdain is to be found in *Contributions to the Founding of the Theory of Transfinite Numbers* (1915).) The English translations of Cantor's terms vary: equivalent classes are often said, for example, to have the same power. The word '*Cardinalzahl*' was first used by Cantor in 1887. He had used '*Mächtigkeit*' with the same sense in his writings of the previous decade. See E. V. Huntington, *The Continuum* (Harvard, 1917), p. 75, note.

[22] A well-known theorem tells us that all infinite sets have a denumerably infinite subset. Consequently no infinite cardinal can be less than aleph-zero.

(iii) INFINITY AND THE ACTUAL

cardinal $2^{\aleph_0}$ was called the 'power of the continuum'. One can extend the result to other continua by showing the equivalence of their corresponding coordinate systems to the real numbers. The set of n-tuples of real numbers (or the points of (real) Euclidean n-dimensional space) falls under this heading, as does the set of real continuous functions of a real variable.

These few results are sufficient to permit a discussion of the way in which we might legitimately speak of actual infinities in cosmology. We must of necessity ignore certain contradictions which, as it turned out, Cantor's work contained: these do not appear to affect the elements of the theory outlined here.[23]

3. Infinities in Cosmology

We might begin by considering the common denial of the contention that whole and part may have the same number of terms. 'That the whole is greater than the part is one of our most immediately evident axioms', writes one philosopher. It would be futile and perverse to try to alter the familiar uses of the words 'whole' and 'part'; but when it is a question of elaborating the cosmologist's conception of the infinite, it is clear that, no less than in mathematics, one must decide between discarding the idea altogether and discarding some of the usages of everyday speech. Cosmologists have taken the second alternative so often before that it might be thought unlikely that one further occasion would be noticed. Yet Lemaître, who is one of the very few writers to have broached the subject, rejected the idea of an infinite aggregate of stars. His conclusion seems to rely upon the confusion of the *act* of counting an infinite set of stars with *rules* for counting them. Given a finite class, reference may be made to both whole and part by a straightforward enumeration of their members: the classes, that is to say, may be defined extensionally. It is equally clear that in most cases they may be defined by intension: an example might be 'The class of elliptical galaxies with apparent magnitude less than $20^{m\cdot}3$'. The members of the class might just as well be named and listed. In the case of the 'infinite whole' there is no possibility of a definition by extension. But that is no reason why an intensional definition should not suffice—as, for example, in specifying the class of elliptical galaxies

---

[23] Poincaré held that the paradoxes of the theory of sets were due to Cantor's attempt to treat infinite sets as completed wholes. See W. and M. Kneale, *The Development of Logic* (Clarendon Press, 1962) pp. 655–56, 672 ff.

regardless of apparent magnitude.[24] (There are those, of course, who hold this class to be finite.)

The three features with regard to which the universe is most frequently held to be infinite are duration in time, the greatest distance between two objects, and the number of objects of a given kind. Which of Cantor's cardinals is appropriate to each? An answer will largely depend on one's attitude towards the nature of space and time. To our three features we could have added a fourth, namely spatial 'content'. We might be satisfied to assert that the number of points of space has cardinal $2^{\aleph_0}$, avoiding reference to distance and thinking only of coordinates. But this step, being devoid of physical content, would add nothing to Cantor's original discussion of the power of the continuum. There are two things which make distance and time more interesting than coordinates:

(i) To make an object 'infinitely distant', in time or space, by a mere transformation of scale will cut across customary definitions of these quantities.

(ii) Times and distances may be supposed to have a particulate nature. It is well known that new interest was aroused in the possibility of a shortest length by the discovery that the electron and proton have diameters of the order of 10^{-13} cm. Other 'fundamental' lengths can be constructed out of the fundamental constants of physics.[25] A shortest time-unit ('chronon') would then be the time for light to traverse the shortest distance ('hodon').

The significance of (i) is simply that no-one is going to be impressed by talk of 'actual infinities' in cosmology unless the quantities we are dealing with have a stake in reality. Whether distances and times qualify for this privilege is a moot point; but, allowing the idea, it should be clear that if our model predicts infinite distances, infinite past times or infinite future times, we shall say that the appropriate cardinal is $\aleph_0$ with a particular space or time and $2^{\aleph_0}$ with a continuum. Once again, however, our epistemology will affect our answer. If we admit a horizon at a finite distance and if we reject all forms of inference to distances beyond it, we shall still hold to $2^{\aleph_0}$ with the continuum

---

[24] Cf. Locke, op. cit., Bk. II, ch. xvii, sect. 13. Locke thought that the absurdity of the claim to have a 'positive idea of infinite' was clear as soon as the impossibility of adding to it was realized. Here it is Locke's idea of addition which needs amending.

[25] For example, $\sqrt{(\gamma h/c^3)} \simeq 10^{-32}$ cm.

(iii) INFINITY AND THE ACTUAL

hypothesis; but in the other case we shall hold the number of possible distances to be finite.[26]

Bearing in mind both the remoteness of the astronomer's concepts of distance from those of everyday, and the element of convention in our choice of a geometry,[27] it is not surprising if one is less disturbed by the prospect of an actually infinite space than by the idea of an infinite number of objects contained in it.[28] With the assumption of uniformity in the distribution of matter, the hyperbolic and Euclidean spaces of the general relativistic, quasi-Newtonian, kinematic, Whiteheadian, and steady-state models do seem to imply the existence of an infinite number of objects. (Once again, the implications of a horizon are being ignored.) What is the nature of this infinity?

Reasons will later be given for thinking that many cosmological theories do in fact involve denumerable infinities. First, however, it is pointed out that with the sorts of geometry envisaged in cosmology, any infinite cardinal of a class of objects, each being of finite size, cannot be of higher order than $\aleph_0$. This follows since it is possible to define a simple procedure whereby one may (conceptually) count spatial cells which are of such size that they could at most accommodate a definite finite number of objects of the sort in which we are interested.[29]

The number may, of course, be finite, and in a sense we are dealing with a potential infinity after all. What matters is that the potentiality of the infinity should not be regarded as ruling out the possibility of an actual infinity, in the Cantorian sense. If it were a question of counting the number of words in a book, one would not hesitate to say that the answer was a contingent matter: 'The answer turns out to be such and such'. How—it may be asked—can we possibly say that the number of atoms in the universe may 'turn out to be infinite'? Is the infinity not bound to remain a potential one by virtue of the fact that we can never get to the end of a counting process? A first answer might be that the counting was done conceptually, and that it was conceptually shown that there might be an existent infinity of material objects. The

---

[26] Cf. the quotation from Robertson's 1928 paper at p. 135 *supra*.

[27] In our last paragraph a Riemannian manifold was assumed. Different results might have been obtained with topologically non-separable spaces.

[28] In the past we should perhaps expect to find the two problems linked, as in Halley's 'objection . . . rather of a Metaphysical than Physical nature; . . . that the number of Fixt Stars is not only indefinite, but actually more than any finite Number; . . . which seems absurd *in terminis*, all Number being composed of Units, and no two Points or Centers being at a distance more than finite.' (*Phil. Trans.*, xxxi (1720) 23.) Halley seems to think of this as a paradox, involving a contradiction. He is reluctant to discard his premisses, even so!

[29] One such procedure is outlined, by way of example, in the Appendix, Note XIX.

382 CONCEPTUAL PROBLEMS

neo-Aristotelian would perhaps then argue that the phrase 'there might be' throws the matter into the province of empirical decision procedures: the idea embodies in this phrase, it might be said, is taken directly from the thesis which it is used to oppose. Not until 'there might be' is legitimately replaced by 'there are' has the Cantorian cosmologist made out a case.

The reply to this challenge is essentially very simple. There are grounds for thinking the infinity to be actual if it can be shown on the grounds of a specific cosmological theory that every nth cell in the counting sequence (n finite) is occupied by a part of at least one object. Cosmological theories do have rules of this sort, in the form of laws concerning the homogeneity of the distributions of matter to be found in the world. In order to bring a method of defining the spatial cells into line with such laws we should have to introduce the concept of distance into the definition, but this would not be difficult. The only obstacle of any importance is encountered in the introduction of statistical methods. What are we to mean, for example, by the assertion 'The average density of matter is one galaxy for every n cells'? It would be meaningless to speak of averaging over a universe of infinite volume and mass. But if we obtain a value for the mean density of the galaxies ($1/n$ galaxies per cell-unit of volume, say) how is it to be included in our counting procedure? Can we simply say that every nth cell is to be put into correspondence with one of the natural numbers? Statistical laws as usually interpreted do not sanction this sort of precise step. Should we rather say that in a sufficiently large number (N) of cells there will be (N/n) galaxies, and hence modify our conceptual procedure by counting N at a time? But if this is a truly statistical distribution, are there not grounds for saying that however large N may be there will always be deviations from the theoretically predicted mean, so long as N is finite—and by the nature of what we are trying to prove, we can only assume that N is finite. There seems to be a way round this difficulty, however. Statistical hypotheses can be laid down which suggest expectations of a definite upper limit (say D) to the deviation from the mean after a definite number of cells are counted.[30] (Expectations of one sort or another are what concern us in physics generally. One cannot reasonably object to statistical hypotheses on the grounds that they are not purely mathematical.) Suppose that we treat the problem as one involving a mean distribution of $((1/n) - D)$

[30] For a mathematical model of phenomena showing 'statistical stability', see Harald Cramér, *Mathematical Methods of Statistics* (Princeton, 1946) ch. xiv.

(iii) INFINITY AND THE ACTUAL

galaxies per cell-unit of volume. Since we can show the cells to be denumerable as before, we can say that the total number of galaxies cannot be less than $((1/n) - D) \times \aleph_0$. Under such circumstances, therefore, it seems that there is no *a priori* reason for refusing to allow a cosmologist to assign Cantor's least transfinite cardinal to the total number of galaxies in a Riemannian space which is unbounded and not of finite volume. The infinite set of galaxies must not, of course, be spoken of in the same way as finite sets. Whereas one can speak both intensionally and extensionally of a finite set, now only the former is legitimate. It might be said that one can never *refer* to the infinite totality. The traditional distinction between sense and reference appears to have been too sharply drawn, although this is not the place to argue the matter. In any case, it is always possible, given what may appear to be a straightforward reference to an actual infinity, to expand it in terms of rules.

We are reminded of the intuitionists. The members of this group, as we said earlier, insist that intuitively understandable construction is the only valid way to a proof of mathematical existence. Some of them would reject all concepts that cannot be constructed by a finite process; some, with uncommonly clear intuition, would allow a compromise, rejecting only those which are not reducible to finite processes by means of a finite number of rules. This seems to be the only reasonable step. Above all, when it comes to asking whether statements containing the phrase 'infinite number' are empirically determinate, one should not give the answer 'No', merely on the grounds that one cannot examine each of an infinite number of objects. One may 'verify' the existence of an infinite number of objects, intensionally understood, in the sense that one may test the validity of rules of correspondence (object-to-place) over a limited range of measurements. And this sort of procedure is nothing new to cosmology.

CHAPTER 18

CONCEPTUAL PROBLEMS (iv) CREATION AND THE AGE OF THE UNIVERSE

In Part I we encountered several cosmological models which contained no singularity at a finite time in the past which would have seemed to indicate the non-existence of all matter at that time. These models, having an 'infinite time scale', fell, broadly speaking, into four groups: asymptotic, cyclical, static, and steady-state. The last group of models avoids the creation problem in its familiar form, only to reintroduce it in a new guise. The static models we can presumably ignore, as being without relevance to the existing pattern of galactic movement. The asymptotic models fall squarely under the infinite time-scale heading. They approach an Einstein-state asymptotically from the point of view of one looking backwards in time, although they are nevertheless often derived from consideration of the perturbation of a supposedly pre-existent Einstein-state. This initial disturbance is not, therefore, the sort of event which cosmologists have in mind when they speak of 'Creation', even though some have turned to it as a plausible substitute. The cyclical models have their own problems, as Einstein, Tolman, and others realized.[1] May it not be impossible to justify the idealization on which the analysis is founded when we come to deal with the lower limit of the contracting phase? To assert this impossibility is one way of parrying the many difficult questions which the theory would otherwise have to face. One might then avoid, for example, such a question as 'What happens to the energy of the system when its volume is zero?' But there are other difficulties: Is it not logically absurd to speak of the recurrence of identical states of the universe? Is not 'identical' too exacting a word? If time is to be defined (as some would have us believe) only by reference to the state of the universe, must not exactly similar states occur, by definition, at the same moment of time?

To begin with the last question: the identity of which we spoke was an identity of idealizations. The world is to be the same on more occasions than one in certain respects only—namely those with which cosmology deals. It would lead to absurdity to suppose that these are the respects in which time is to be gauged: two states which differ in

[1] Cf. Chapter 6.

(iv) CREATION AND THE AGE OF THE UNIVERSE

other respects, but not in these, would have to be said to occur at the same time. The sort of recurrent universe described by Friedmann must, needless to say, be distinguished from that involving a statistical recurrence. (Lucretius made use of this theme, and even then it was not new.)[2] The most obvious difference is that a periodic time for the older sort of recurring universe cannot be defined, whereas relativistic (and Newtonian) cosmology predicts a precise periodicity. The idea of a statistically recurrent universe is unlikely to carry much weight, by reason of increasing entropy. Once again there will be logical difficulties for those holding to certain views of the nature of time. If there is to be a recurrence in only a restricted number of respects—and it is hard to see how the idea could be made precise if this assumption were not made—then time must not be decided solely by the state of the universe in these respects, unless the survival of evidence of previous states be ruled out. In fact, as seen already, the idea is untenable that there is a unique entity, 'the universe', which passes successively through a series of states, each associated with a given time. (It is doubtful whether anyone would wish to rephrase the statistical argument in terms of a universe considered as an ordered aggregate of *events*.) Yet another objection might proceed from Ockham's Razor and another might be positivistic: if there is a supra-sensible Being who, as it were, plays the same game twice (with, perhaps, a time-piece for distinguishing between the two occasions) then the thesis can never be verified by us; whilst if there is no such Being it can be verified by nobody. Clearly there are problems other than scientific ones associated with the recurrent universe idea. It is worth adding that for centuries the idea has had its appeal largely because the recurrence was *not* presupposed in all respects: perhaps in the next world we might make fewer mistakes than in this . . . and so on. Despite the objections raised here, one can say this of those periodic models which contract to a minimum (finite) radius and then expand: so long as we restrict the respects in which history is to repeat itself, and so long as we do not identify time with the state of the universe in these respects, the models may be consistently described as being unending in time, of being without beginning in time, and of requiring no process of continual creation.

What of the non-periodic models which predict a singularity at a finite time in the past? It is well known that their demise has given rise to talk of creation and, consequently, of the age of the universe. We shall discuss these two questions in this last chapter, introducing

[2] See Appendix, Note XX.

at the same time that 'solution' of the problem of creation offered by exponents of the steady-state theories.

1. The Age of the Universe

There is no doubt that the announcement of Hubble's Law provided a great stimulus to discussion of the possibility of a rational treatment of the problems of creation and the age of the universe. We recall how natural it seemed to be to pass from the (approximate) relation

$$D = (1/h_1) \cdot \delta \qquad (1)$$

to a simple linear 'velocity-distance relation'. It was equally natural for those familiar only with the empirical results to presume that the universe must have originated at a point, $(1/h_1)$ years ago. The general trend of the function $R(t)$, as permitted by the field equations of General Relativity and by Newtonian cosmology,[3] shows that this interpretation is not justified: the length of the subtangent will not in general be the same as the time from the point-singularity to the present. The latter may in principle be inferred, granted sufficient information about the present trend of $R(t)$ and its general form. In some cases, as for example in the Einstein–de Sitter model, the connexion is simple; but generally it is not. But whether mathematically simple or otherwise, the problem of interpreting the occurrence of a singularity in the function $R(t)$ remains. First, however, some further remarks will be made on the interpretation of the phrase 'age of the universe'.

Where we have no prior information about the form of $R(t)$ (such as might be obtained, for example, from highly restrictive hypotheses) it should be obvious that a knowledge of h_1 and, for that matter, h_2, cannot supply the form of the function. To make the dogmatic identification of $(h_1)^{-1}$ and the time from $R(t_0) = 0$ to the present is, of course, equivalent to assuming that $R(t)$ is of the form $a(t - t_0)$, and hence that $h_2 = 0$. In fact the indications of the best astronomical data at present available are that, on the basis of relativistic cosmology, this is not so. The case is unsatisfactory in other respects, for it can be shown that the requirement of positive or zero density and pressure here implies that $\lambda = 0$ and that space is hyperbolic.[4] Whether or not it is reasonable to accept these values for λ and the sign of the curvature is an open question, but it would be very unreasonable to accept them on the strength of an argument such as this.

If it is assumed, instead, that $1/h_1$ is simply proportional to the time

[3] Cf. Fig. 1 p. 131 *supra*.   [4] *G.R.C.*, p. 183.

of the expansion, there follow similar restrictions on λ and the sign of the curvature, and once more the argument seems to follow quite arbitrary lines.[5]

Another misleading assumption as to the nature of 'Hubble's constant' is to be seen in its very name. If h_1 is to be constant, then integration of $h_1(=R'/R)$ gives $R = R_0 \exp(h_1 t)$, and $(1/h_1)$ is now the time required for R to increase by a factor ε.[6] The naïve interpretation of $(1/h_1)$ is now replaced by another which not only places restrictions upon the values of λ and the sign of the curvature, but also makes it necessary that $h_2 = h_1^2$ — for $(d/dt)(1/h_1) = (1 - (h_2/h_1^2))$. This, the solution of the steady-state theories, is once again regarded as unsatisfactory by most relativistic cosmologists, because it reduces the number of parameters, thereby making the model easier to refute. In any case, this is not relevant to the present discussion, for the function corresponding to $R(t)$ in this model is at no finite time zero.

(The absence of year-to-year variation in the spectral shifts of distant nebulae led Milne to postulate the constancy of the Hubble factor. Although he recognized the hypothesis as involving a 'gross extrapolation', there is no doubt that the simplicity of the resulting model greatly influenced his work, with its emphasis on uniform motions.)[7]

It is plainly wrong to suppose that the reciprocal of the Hubble parameter is in general likely to provide a measure of anything which could be termed 'the age of the universe'. This is still more evident when we bear in mind that in the cases of the quasi-periodic model and the perturbed Einstein model there are previous histories to consider. It might be answered that the Einstein state and the previous cycles are of smaller intrinsic interest than the present phase of expansion and the evolution of the present constituent parts of the universe, but this answer does not justify the use of the phrase 'age of the universe'.

It should be clear that one cannot properly speak of 'the age of the universe' and ascribe to it a unique value, unless it is proper to speak of a universe and unless it is possible to formulate a significant procedure by the use of which all observers, however distant, might in principle agree upon the epoch to be assigned to any event. Such observers would, in other words, be using universal time. In Chapter 16 we saw how, in relativistic cosmology, absolute simultaneity was reinstated in a very restricted sense, and then only for a limited class of observers moving in a certain way. This qualification has to be

---

[5] See McVittie, *A.J.*, lviii (1953) 129.
[6] Cf. the non-static form of the de Sitter universe.   [7] Cf. *W.S.*, pp. 75–76.

added to any theory which accepts the Special Theory of Relativity locally. It will not apply to Newtonian cosmology in its usual form, but the latter is unsatisfactory for the very reason that it cannot account for particle behaviour at high velocities. Granted, however, that it is possible to introduce cosmic time in a logically consistent way, at the very least it is necessary to qualify the phrase 'age of the universe' with the words 'as judged by one of a privileged group of observers': only for the members of the privileged group is cosmic time defined. Only they can reasonably speak of the successive existences of universe-states of an agreed sequence. It is not to be denied that observers other than these may experience the passage of time, but it would be held that they cannot combine their experience in a coherent theory without relating their movements to the fundamental observers.

There is nothing really odd about saying that the age of a thing can be agreed upon only by a restricted class of observers, but it will be generally believed very strange to maintain in connexion with the supposed entity 'the universe', that 'exists now' is for most observers to be meaningless except as a sophisticated theoretical term. With these theories in mind one seems bound to speak in this way. Are we to refuse to allow these considerations to dictate to us how we should speak? It might be answered that a scientific theory, which is presumably carefully constructed and confirmed, can instruct us in ways of avoiding inconsistency in our ordinary language. But it must be admitted that in our ordinary dealings *esse est percipi* with a minimal addition in the way of theoretical construction.

We have so far assumed that 'age of the universe' was to be understood in terms of the function $R(t)$ or, in theories which do not incorporate the function, as the time which has elapsed since all matter was at its closest separation. Instead, we may speak—as in Chapter 11—in terms of an upper bound to the possible ages of constituent objects. Admitting cosmic time, one could then characterize the universe as a class of objects. There need be no misgivings about predicating 'age' of the class in the same way as we should predicate it of the members of the class, so long as we realize that in doing so we are simply speaking obliquely of an upper limit to the potential ages of objects. If '$A(x)$' is to be read as 'age of x' and '$O(x)$' is to mean 'x is an object', then

(i) 'The age of the universe is T'

would on this account be rendered as

(ii) '$(x)(t)[\{O(x) \cdot A(x) = t\} \supset (T > t)]$'.

(iv) CREATION AND THE AGE OF THE UNIVERSE

This paraphrase will permit us to speak of a finite age for the universe on the assumption of the Eddington–Lemaître model, that is, that which begins with the Einstein state, so long as this is supposed to comprise undifferentiated matter without anything qualifying as an object and so long as the expansion begins with some finite velocity—that is, not asymptotically. The definition could easily be extended to deal with the assertion of an infinite age for the universe. It has what might be thought in some quarters an advantage over a definition in terms of $R(t)$: of the evidence which is used for stating an age for the universe, that which applies to astronomically familiar objects is widely thought to be unequivocal—that is, not theory-involved. Quite apart from the fact that this is an illusion, the propositions (i) and (ii) cannot be equivalent, for even though (ii) must be taken as derivable from (i), the converse does not follow for this reason: the value of T in (ii) could be replaced by any figure in excess of the value obtained from an argument from the form of $R(t)$, and the proposition would remain true, whereas no one would wish to assert (i) with this new value of T.

Any limitation on the past duration of physical objects or on that of the expanding phase raises the question of initial conditions. To assume a genuine singularity in $R(t)$ is to become immediately embroiled in the question of creation. Apart from avoiding this issue by falling back upon a model with infinite time-scale (steady-state, cyclical, Eddington–Lemaître, and so on) it is always possible to relegate the problem to metaphysics or theology. Lemaître's 'primeval atom' was something of a compromise: unfamiliar laws were to be devised for the highly compact mass, the single quantum of energy; but there was no question of discussing the length of time for which it subsisted before disintegrating, nor was there any question of Lemaître's presuming to lay down a physics of creation. There have been those, on the other hand, who have believed it possible to make meaningful and scientific assertions about the origin of matter, and in the next section we shall consider whether their beliefs were justified.

2. The First Event

To introduce the problem we shall discuss one of the most interesting passages in Kant's *Critique of Pure Reason*, but first we may put this in its historical setting by referring to arguments proposed by Leibniz in his correspondence with Clarke. Leibniz believed that the universe was created at a finite time, but he was clear as to the alternative:

> For since God does nothing without reason and no reason can be given why he did not create the world sooner, it will follow [to one who believes in an absolute time] either that He created nothing at all, or that He created the world before any assignable time, that is, that the world is eternal.[8]

On theological grounds this was repugnant to him—the eternity of Aristotle's universe had given cause for widespread criticism in the Christian world.

Kant's general position in regard to the Leibniz–Clarke controversy was at first—at least as early as 1755—that of Leibniz. Less than ten years later he abandoned this position in favour of the Newtonian account, but he soon found himself unable to subscribe to the view that time and space were real entities. The most important outcome of his dissatisfaction with both sides of the controversy was that section of his *Critique of Pure Reason*[9] which deals with the so-called 'Antinomies of Pure Reason'.[10] It is the first of these with which we are primarily concerned.[11]

The thesis of the antinomy runs as follows: *The world has a beginning in time and is also limited in regard to space.* The proof of the first part is a *reductio ad absurdum*. It is first assumed that the world has no beginning in time and concluded that an eternity must have elapsed before every moment of time, and that therefore an infinite series of successive conditions or states of things in the world must have happened. But 'the infinity of a series', we are told, 'consists in the fact that it can never be completed by successive synthesis'; and an infinite past series of occurrences is therefore said to be impossible. Thus 'a beginning of the world is a necessary condition of its existence'. The (similar) argument for space does not concern us here.

The antithesis of the antinomy is: *The world has no beginning and no limits in space, but is—in relation both to time and space—infinite.*

---

[8] Leibniz–Clarke *Correspondence*, Leibniz's fourth paper, section 15.
[9] First edition, 1781 (A); second edition, 1787 (B).
[10] (A), p. 426; (B), pp. 454 ff.
[11] To understand them there is no need to become involved in Kant's doctrine of 'transcendental idealism' of which they are virtually independent. T. D. Weldon makes this point: 'It is immediately clear that Kant considers the theses to be the *a priori* contentions of rationalist cosmology, while the antitheses represent the empiricist attack on it, and also that the truth of the theses rather than that of the antitheses is desirable both on practical and speculative grounds. For the theses, in so far as they are true, involve the existence both of God and a moral capacity in man.' (*Kant's Critique of Pure Reason* (Oxford, 1958) p. 204.) It will shortly be clear that the final remarks are not true of the first Antinomy.

(iv) CREATION AND THE AGE OF THE UNIVERSE

Again Kant proceeds by a *reductio ad absurdum*. If it is granted that the world had a beginning, then there must have been a time in which the world did not exist—that is to say, a 'void time'. For a beginning, according to Kant, is 'an existence which is preceded by a time in which the thing does not exist'. In a typically Leibnizian vein he argues that nothing can originate in a void time, 'because no part of any such time contains a distinctive condition of being, in preference to that of non-being'. This, he adds, is the case 'whether the supposed thing originate of itself or by means of some other cause'. In consequence, 'many series of things may have a beginning in the world, but the world itself cannot have a beginning, and is therefore infinite in relation to past time'.

If we are to avoid the antinomy it is sufficient to show the flaws in either thesis or antithesis, but if large parts of present-day cosmology are not to be rejected, both arguments must be shown at fault. This is not difficult. To begin with the antithesis: Kant's 'Observations on the First Antinomy' show that he was himself aware of a way of escaping the conclusion. To do so, one must deny 'the existence of an absolute time before the beginning of the world, or an absolute space extending beyond the actual world'. Space, for Kant, was merely 'the form of external intuition', and not a real object which can be 'externally intuited'.[12] So far, so good: but Kant cannot, it seems, sustain the Leibnizian position—we recall that Leibniz often found it difficult—and like a true Newtonian he goes on to say that 'it is nevertheless indisputable that we must assume these two non-entities, void space without and void time before the world, if we assume the existence of cosmical limits relatively to space or time'. This proposition, far from being indisputable, is false. Even before the advent of a geometry of elliptical space, a thoroughgoing supporter of Leibniz could argue that it is *meaningless* to speak of totally void spaces and times. One cannot have relations between objects when there are no objects. One is reminded of the well-known passage in St. Augustine's *De Civitate Dei* (Bk. xi, ch. 6) according to which 'the world was made with time and not in time, for that which is made in time is before some time and after some time'.[13] This does not mean that the nature of possible spatial and temporal limits to the universe is made any clearer, but Kant's argument for the antithesis does appear to break down.

The argument for the thesis rests heavily on Kant's view of the

---
[12] Kant is here emphasizing that space contributes what is *permanent* in perception.
[13] This is not unlike a passage in Plato's *Timaeus*.

nature of infinity. He denies that 'infinite quantity' means 'quantity such that none greater can exist', preferring the concept of potential infinity. The infinite is that which exceeds anything which can be given or completed by successive synthesis.[14] One is first tempted to ask more of 'successive synthesis'. Is it a time-consuming addition of unit to unit?—and so on. But it is possible to see, without answering such questions, that Kant's argument is tautologous: with the definition of 'infinite' which he accepts, it would be a contradiction in terms for him to assert that a (potential) infinity could be made to correspond to an actual thing. The thesis is thus analytically true. But its contingency and therefore the plausibility of its denial can be ensured, as we shall once more explain, given a suitable definition of 'infinite'.

That this is possible has recently been denied by Whitrow in two or three passages which will be quoted here as going straight to the heart of the matter:

> Kant's argument . . . essentially concerns successive acts occurring *in time*. It says nothing about the possibility of an infinite series of acts in the future, but asserts the impossibility of an infinite series of acts having occurred already. An elapsed infinity of acts is a self-contradictory concept. This conclusion, I believe, must be accepted.[15]

And later in the same work (in connexion with the Achilles paradox):

> Even if we remain fully convinced of the correctness of Cantor's analysis of the continuum . . . we must not assume that, in actual fact, that is, in *time*, any infinite sequence of operations can be performed.[16]

Whitrow is here wishing to distinguish between the infinite set of positions which Achilles must pass (in one version of the paradox) and the acts of passing them—acts which he likens to acts of *counting* them in time. He is content to regard the former as a totality, but the latter only as 'an indefinitely growing, dynamic, or uncompleted infinite'.[17] He points to a difference between Zeno's paradox and Kant's antinomy: in the former, 'the infinite series of successive acts is a purely conceptual one resulting from our method of analysis', whilst in the latter case, 'the infinite series is presumed *ex hypothesi* to have actually occurred'.[18]

In the last chapter we denied that the notion of an elapsed infinity of acts was self-contradictory. It is so only if the possibility of applying the notion of an actual infinity to acts or events is to be denied. There is no reason to deny that this is impossible. It does not involve

[14] Cf. 'Observations on the First Antinomy' (thesis): (A) pp. 430–34; (B) pp. 458–62.
[15] Whitrow, *The Natural Philosophy of Time*, p. 32.
[16] Whitrow, op. cit., p. 148.   [17] loc. cit.   [18] op. cit., p. 152, note.

(iv) CREATION AND THE AGE OF THE UNIVERSE

contradiction, and in fact it has already been explained how such a notion may be applied—namely by laying down a rule for enumeration. If such a rule can be laid down, and the class which interests us be therefore shown denumerable and—on theoretical grounds—unlimited, then there is no reason why we should not talk of a 'completed infinite'. Russell held that Kant ignored the possibility of a series having no first term—as, for example, the series of negative integers ending with minus one. Whitrow believes that this is beside the point, for 'the only way in which we can actually produce such a series *in time* is by counting backwards, that is by *beginning* with minus one'.[19] But the 'producing' which *we* do is conceptual. We are not, in using descriptive rather than referential language, obliged to produce a first member of the infinite sequence from somewhere in the past: our correspondence-rule of unrestricted application can just as well proceed from the present into the past.

In saying that the rule is of unrestricted (that is, of indefinite) application, we may be accused of using the notion of potential infinity—which indeed we are. It is necessary if the very concept of a rule is to be understood, and using it involves us in no contradiction. The function of Cantor's transfinite analysis was quite different. It showed us how to condense our descriptive language and provide this with conceptual objects which it is not inappropriate to describe as 'actual', granted that the enumerated objects are accepted as actual. One would not dream of claiming that the infinite world and, *a fortiori*, infinite time are actual in the same sense as that in which galaxies and lifetimes are actual. When we deny that there is a contradiction in the notion of an elapsed infinity of acts, we are saying no more than that Cantor has shown how, up to a point, it is possible to handle symbols for infinite classes as though they stood for finite classes. We are certainly not to be understood as claiming that the 'actual infinities' of cosmology are directly accessible to the senses.

Before leaving the subject of Kant's antinomies, it must be added that he himself believed the arguments to be fallacious. To understand his reasons it is necessary to become involved in his transcendental idealism: 'phenomena are nothing, apart from our representations'. In the Seventh Section of his comments on the antinomy he presents it again, now in this form:

> If that which is conditioned is given, the whole series of its conditions is also given; but sensuous objects are given as conditioned, consequently . . . etc.

[19] op. cit., p. 32, note 1.

But he now points out that only in regard to things-in-themselves can we speak of the necessity of conditions leading back in infinite regress from 'given' and 'conditioned' things. Phenomena, or 'mere representations', are not 'given' unless I 'attain to a cognition of them'. In their case, although 'a regress to the conditions of the conditioned . . . is enjoined', it is not 'given'. The fallacy in the syllogism is now clear for Kant:

> . . . in the major [premiss of the syllogism] all the members of the series are given as things-in-themselves—without any limitations or conditions of time—while in the minor [premiss] they are possible only in and through a successive regress, which cannot exist, except it be actually carried into execution in the world of phenomena.

Anyone who believes, however, that phenomena may be identified with Kant's things-in-themselves will not on this score find the antinomies fallacious. Moreover, in claiming, in effect, that the syllogism exhibits a category-mistake, Kant's view is not ours. Although there is a similarity—as for example, when we distinguish between referential (cf. his 'given') and descriptive (cf. his 'possible') uses of the word 'infinite' —for us the former use is possible only through the employment of descriptive language whereas for Kant it has nothing whatsoever to do with language.

In the Ninth Section on the antinomy of pure reason, Kant discusses the problem more specifically, and again his general philosophical position obtrudes. He first presents as a 'regulative principle of reason' the proposition that 'in our empirical regress no experience of an absolute limit, and consequently no experience of a condition which is itself absolutely unconditioned, is discoverable'. The grounds for this proposition are thoroughly disconcerting in so far as they beg several important questions. Its truth is said to rest upon the consideration that experience of an absolute limit 'must represent to us phenomena as limited by nothing or the mere void, on which our continued regression of perceptions must abut—which is impossible'.[20] This in turn rests upon the denial of any such thing as an uncaused event, a denial to be found in Kant's axiom that 'every condition . . . must itself be empirically conditioned'. We have already explained how the notion of a limit may be introduced into the concepts of space and time. Kant's philosophy, however, obliged him to look to perception for evidence of absolute limits. But he believed that no experience of

---

[20] Kant does not, of course, mean that we cannot in practice experience the absolute limit.

(iv) CREATION AND THE AGE OF THE UNIVERSE

either a void time or a void space was possible, and therefore finally came down on the side of the antithesis of the antinomy (with suitable philosophical modifications).

We notice that Kant's whole case rests on the illogicality of the concept of a First Event—we discount the confessed sophistry of the first antinomy. He does not seriously concern himself with any argument involving the implausibility of a finite *duration* for the universe. Since a finite period of time can always be transformed into an infinite period by effecting a simple regraduation of the timekeeping process, there might be something to be said for attempting to rephrase any claim to a finite age for the universe in terms of the occurrence of a First Event. We shall now, therefore, consider whether this is a reasonable shift of emphasis.

Let us suppose that we are to admit the concept of a first event. Does the implied uniqueness of such an event require that all who determine its epoch through some cosmological procedure must arrive at the same result? It would obviously be absurd to suppose so, unless all were in principle able to use a cosmic time. That this is not available in a given model is, even so, no argument for the illogical nature of the concept. If, on the other hand, it is so much as possible to affirm that different observers label a certain event differently, it must be possible for them to identify that event independently. It seems sufficient to consider the concept form the point of view of a single observer.

As is to be expected, modern writers have tended to fall back on older arguments, for the simple reason that the ingredients of the problem are conceived to be so few: time, existence, and non-existence. Consider, for example, Michael Scriven's argument that no verifiable claim can be made either that the universe has a finite age or that it has not.[21] His argument hinges upon the 'logical fact' that we are allowed meaningful reference to any past time, which may or may not correspond to a physical state. An empty state is said to be unverifiable for 'it cannot give evidence of its date'. He maintains that, unless the concept of 'time without things' can be given a meaning, one can give none to 'time of the first things'; and hence his argument follows.

The 'logical fact' is admittedly one which might be supposed to be embodied in ordinary speech, but many have thought, with Leibniz,

---

[21] *B.J.P.S.*, v (1954) 181. Cf. Kant: '... the world, as not existing ... independently of the regressive series of my representations, exists likewise neither as a whole which is infinite nor as a whole which is finite in itself.'

that on this score ordinary speech was 'unphilosophical'. To one who equates meaning and verifiability, it will indeed seem that the impossibility of an empty state's giving evidence of its time is an argument against allowing meaningful reference to 'time without things'. (For different reasons, Leibniz would have agreed with the conclusion whereas Newton would not.) But one can only pass from this to the conclusion that 'time of the first things' is without meaning if one supposes that 'the First Event precedes all others' and 'nothing existed before the First Event' are equivalent; for only the second can be meaningless if time previous to the First Event is not defined—and that it must be so defined is simply untrue.[22]

E. H. Hutten appears to have fallen into the same error in a recent article:[23]

> The zero-point of time is arbitrary, unlike the zero-point of temperature: we can always ask, What happened *before*? Moreover the zero of time represents a singularity that, unlike that of temperature, is *not asymptotic*: it must have been reached, otherwise the universe would not have begun to exist. Singularities can be tolerated within a physical theory, provided that they are *asymptotic*.

It should be clear that in so far as the zero-point of time is arbitrary, it has nothing whatever to do with the representation of any singularity of interest to cosmology. In so far as it corresponds to a singularity it is not arbitrary. That the universe exists is a necessary condition only of the fact that the *non*-arbitrary time origin has been reached. Our arbitrary time conventions can have nothing to do with material existence. We have already qualified the assertion that it is always open to us to ask 'What happened before?' As for the implication that only asymptotically approached singularities can be tolerated in physics, this is a dogma which, even if it were borne out in all contexts but ours (which it is not), would not of itself prove that a singularity in the function $R(t)$ was illogical.

Assuming a First Event, can we say that it is 'of the kind we know now'? This question is introduced here because of late it has become popular to provide it with the answer 'No'. The temporal origin of the universe, according to Whitrow, is not an event in the same way as

---

[22] To suppose otherwise might be called the Lucretian fallacy. In connexion with space, Lucretius argued that 'nothing can have an extreme point, unless there be something beyond to bound it, so that there is seen to be a spot further than which the nature of our sense cannot follow it'. (*De Rerum Natura*, i, 960–63.) Cf. p. 372 *supra*.

[23] *The Monist*, xlvii (1962) 110.

(iv) CREATION AND THE AGE OF THE UNIVERSE

subsequent happenings.[24] Again, according to R. Harré, 'no first event could have been of the kind we know now'.[25] Both are clearly right to insist that the First Event must be without antecedents (overlooking the fact that, on one view of the nature of time, it may be meaningless to say as much). What is not clear is how this event in consequence differs *otherwise* from all others. Events are obviously not being thought of as connected by causal links of the kind against which Hume preached so effectively. Whitrow went on to say that the role of the First Event was 'analogous to that of a number which is the limit of a sequence, although not itself a member of the sequence'. Harré, likewise, is determined to set this event apart from the rest: 'If it is an event having no antecedents, then we *cannot know which event it is*, for we can have no laws of nature which we can now discover which would enable us to retrodict to an event which has no antecedents.'

Richard Schlegel objected in a similar vein to the sort of metrical discontinuity which some models require, on the grounds that a unique-event creation is 'counter to scientific experience', with its 'continuity of process'.[26] But continuity of process is, if anything, a feature not of the world, but of the world as described by certain physical theories which as a historical matter have been widely accepted. This does not mean that discontinuities cannot be consistently described by any scientific theory.

Although we are not convinced that it is necessary to distinguish the First Event from the rest in any but the obvious respect (lack of antecedents), yet there is what at first sight appears to be a flaw inherent in the idea, as will be explained. Consider any of the expanding models which, according to the Friedmann–Lemaître equations, expands monotonically from $R(t) = 0$. With a continuous time-variable the difficulty arises that when $R(t) = 0$, nothing yet exists, and the corresponding epoch cannot be that of the First Event. Yet the epoch for which $R(t)$ first becomes positive cannot be named, assuming a continuous variable.

There are at least two possible solutions of this difficulty. The first would be to introduce the concept of atomic time. The limit of the corresponding mathematical series would then belong to the series, corresponding to the chronon of the First Event. A second solution would be to deny the applicability of the field equations to the case of the unlimited degree of material concentration in the neighbourhood of the

---

[24] *B.J.P.S.*, v (1954) 215.   [25] ibid., xiii (1962) 110.
[26] ibid., v (1954) 226.

singularity. This solution would be accepted by any cosmologist who believed the models in question to be otherwise acceptable—we recall in particular Lemaître's hypothesis of the primaeval atom. But then it seems that all we have shown is that what is commonly taken as indicating the occurrence in the past of a First Event, namely the singularity in the function R, indicates nothing of the sort on any reasonable set of hypotheses as to the validity of our fundamental theories in its neighbourhood. We have not shown that the notion of a First Event is either acceptable or otherwise.

Hume's thoughts on a related subject are still relevant here. For Hume, the idea of causation must be derived from some relation among objects, and essential to this idea were the relations of contiguity and succession in time.[27] He realized, however, that an object may be contiguous and prior to another without being considered as its cause.[28] When causation has somehow been established, it draws its importance, according to Hume, from its allowing the mind to go beyond what is immediately present to the senses: it allows, in short, forms of inference amongst which must be numbered those known as 'scientific'. But why, Hume asks, is it necessary that everything whose existence has a beginning should also have a cause?[29] (If it has none, then, he might have asked, is this a ground for saying that the thing is not of the kind we know now?) His answer was that this is neither intuitively nor demonstrably certain—for there is nothing contradictory or absurd in separating the idea of a cause from that of a beginning of existence. To say that every effect must have a cause, because it is implied in the very idea of effect, is to evade the question—a point which is even today often overlooked. For it must be supplemented, as Hume saw, by the hypothesis that 'every existent is an effect'. Some would wish to introduce such a hypothesis, or its modern equivalent, and then it would follow that the First Event is not of the kind we know now in two respects. On the other hand, this all seems quite unnecessary. Events may be treated as items ordered into a structure by theory. Within this structure the First Event occupies a special position. But there seems to be no point in speaking of it as different in *kind* from all other events.

[27] Hume, *Treatise of Human Nature*, ed. L. A. Selby-Bigge (C. P., Oxford, 1888) pp. 73–82, 157, 172.
[28] From the time of the Greeks until about a century ago, causes were generally spoken of as *objects*: more recently the idea of causation has been linked with those of energy, action and events.
[29] op. cit., pp. 78 ff, 157, 172.

3. Creation

In speaking of the First Event we are only a short step from thoughts of the Creation, and yet—apart from an occasional unguarded utterance —writers on cosmology have shown caution in changing these thoughts into words. They are conscious, perhaps, of the verbal gymnastics in which previous ages have indulged. Generally speaking, by 'act of creation' we mean an act whereby some or other material is 'transformed', 'form' being the only new thing.[30] What is the material which is transformed in the creation of the universe? The difficulty of accepting the idea of a creation *ex nihilo* was the source of much medieval casuistry. One finds that, as a frequent compromise, a view was adopted close to Plato's as expressed in the *Timaeus*. There, creation consists in the creator's endowing primal matter with form and organization. In answer to the medieval question 'Who created God?' God was said to be *causa sui*, owing His Being to no other. So, later, in the words of Descartes, God is said to be the efficient cause of His own existence; in Spinoza's words, God's nature cannot be conceived as not involving existence; and so on.

A brief reference is made here to this sort of reasoning because many a cosmologist seems to have something rather like it at the back of his mind. To speak of 'creation *ex nihilo*' is to use a form of words which acknowledge that the subject of transmutation must be mentioned and in which, since no subject is evident, an empty phrase is substituted to stand grammatical proxy. Today this seems pointless rather than harmful: it is innocuous, however, only if it purports to answer no question of a causal nature. Aquinas was well aware of this: '*produced from nothing* is in the sense of *not produced out of anything*, as when we say, he speaks of nothing, meaning that he does not speak of anything.'[31] Many critics of the steady-state theory could well take Aquinas's lesson to heart.

'Produced from nothing' is not part of a causal proposition: can it be that it is out of place to ask for a cause of the First Event? Certainly the normal sense ascribed to the notion of cause is one of production or generation. In this sense it seems paradoxical to ask for the cause of the First Event; for the notion of temporal priority of cause over

---

[30] But cf. Aquinas, *Summa Theologica*, 1a, xlv, 2, 5. 'In mutation there is a constant subject before and after. Creation is therefore not mutation, except by an extension of the meaning. . . . Creation works without raw material requiring to be prepared by an instrumental cause.'

[31] op. cit., 1a, xlv, 3. See Appendix, Note XXI.

effect is usual. This is not so, however, if we follow in the tradition begun by Hume, and interpret 'cause' in terms of *law* rather than in terms of *production*. At the same time the principle that 'any event must be caused' loses what self evidence it might ever have had.

It seems that there is a certain awe to be experienced in 'dealing scientifically with the question of creation'. Bondi, for example, believes that in the steady-state theory 'the problem of the origin of the universe, that is, the problem of creation, is brought within the scope of physical inquiry and is examined in detail instead of, as in other theories, being handed over to metaphysics'.[32] We saw how he and Gold could discuss the *rate* at which matter appears for the first time and its *distribution* at the time of its appearance. They could discuss the *kind* of matter (for example its chemical or electrical nature) and its initial *motion*.[33] On the other hand, a recurrent theme of so many discussions of this topic is that of the impossibility of explaining creation scientifically, some rather outmoded views on causation being usually called in as evidence. Creation, the Beginning, the First Event—not all have the same meaning, but all involve a discontinuity which has often been said to constitute a *limit to that which may be known*. Scientific explanations of them, in the old causal sense, are out of the question. But cosmology might be said to have explained the beginning of the universe once it has decided that the occurrence of such a state of affairs in the past follows from a theory which finds support in current observations. There is no doubt, however, that acceptable as this account may be, it would not satisfy the great majority for whom, other aspects of causation apart, no occurrence would be said to be explicable in terms of *later* occurrences. There is also presumptive evidence for thinking that those who accepted the ideas of continual creation found as one of their strongest arguments one which rested upon a somewhat restrictive view of the nature of scientific explanation. Hoyle, for example, complains that 'it is against the spirit of scientific inquiry to regard observable effects as arising from "causes unknown to science",' and that this is what a unique creation implies.[34] Ostensibly in opposition to Hoyle, others have held that to multiply the *occasions* of creation (as did Hoyle) does not remove his objections. The attitude which this counter-objection reveals is no more warranted than Hoyle's. Both the belief that science deals with causes and the requirement

[32] *Cosmology*, 2nd ed., p. 140.
[33] Bondi and Gold, *M.N.*, cviii (1948) 252. See section 4 (pp. 266–68) for an uninhibited treatment of these matters.
[34] ibid., p. 372.

(iv) CREATION AND THE AGE OF THE UNIVERSE

that any event must be preceded by a cause, are misleading. In the steady-state models it is, of course, possible to relate the sum of all events prior to the occurrence of the new entities to events which follow their appearance and in which they are involved. But it is not clear why, in explaining one thing in terms of others, temporal priorities should be considered as an essential part of the explanation.

It was perhaps inevitable that the singularities found by cosmologists should seem to mirror a traditional picture.[35] Friedmann used the creation-terminology in his first contribution to cosmology. Lemaître was cautious and spoke of the 'initial state', but the creation idea was soon to be very frequently and quite casually applied to the appropriate discontinuity. The notion of a transmuted material was always absent: if anything this was thought to be creation *ex nihilo*, although later there came McCrea's indication of a source of the created material. The steady-state theories, capable of introducing creation in the sense of a transmutation, are seemingly less radical than those which are unable to use the creation metaphor. (Admittedly, neo-scholastics are unlikely to accept zero-point stress as the *materia prima* of the world.) Whether dealing with substance, matter or energy, men have always looked for the intellectual anchorage which conservation principles can offer. Any other advantage of McCrea's interpretation is methodological: it acknowledges that it is easier to think of a certain law as a conservation law than as one which (systematically) violates conservation principles.

The question of the origin of the stress energy has never been discussed, of course, any more than were the origins of prime matter. These are simply not legitimate mysteries, it seems: one has to stop somewhere. . . . But how are we to choose what shall correspond today to the prime matter of the middle ages? Where are we to stop? The question of what shall count as fundamental to the 'created' world is even more important for the cosmologist who does not wish to fall back on the transmutation metaphor. Since 'universe' cannot mean more than 'universe of discourse', is he not bound to admit that there may well have been events, with which his theory is not concerned, before the epoch of 'creation'? (On a relational view, time would have to be defined in terms of the type of event with which the theory does *not* deal.) This unassuming way of talking is unlikely to appeal to cosmologists, who presumably exclude the possibility of a body without

---

[35] It is a matter for conjecture whether this would have happened had the Augustine-Aquinas attitude to creation prevailed. According to this, the miracle of creation was a continuing one: the world would not stand for a single instant were God to withdraw His support. (Augustine, *De genesi ad litteram*, iv, 2; Aquinas, *Contra gentes*, iii, 65.)

size, mass, or energy, having any other property at all. Yet the cosmologist's conception of mass is not at all as simple as our intuitive ideas of matter. In fact, since mass is so seldom conserved in cosmological models, whether or not it is a *sine qua non* of all other properties of matter, the concept of mass is not capable of assuming the traditional role of that of substance; of something, that is, which retains its identity despite changes in its form, and which in its totality can be equated with the universe. The question of what shall count as our prime matter is still without an answer. Perhaps this is because it strikes most cosmologists as uninteresting, metaphysical or even misleading; or perhaps it is because they can see a way of avoiding the question entirely.

Negative cosmic stress has been attacked as illogical, since it cannot be conceived[36]—presumably as the result of some psychological disability. But the cosmologist is on the horns of a dilemma: the alternative creation-theories have been attacked precisely because they offered nothing by way of a *materia prima*, conceivable or inconceivable. Mario Bunge, for example, would have us believe that 'either creation hypothesis smuggles magic into cosmology, thus turning it into science-fiction'.[37] He gives three reasons. The first two need not concern us here.[38] The third is so fundamental to any discussion of the problem of creation that we shall take it as the starting-point of our final section.

4. Cause and Creation

> The hypothesis conflicts with the whole 'spirit' of modern science which abhors creation *ex nihilo* (magic) and accepts, on the other hand Lucretius' genetic principle, according to which nothing comes out of nothing or goes into nothing. In assuming that the emergence of matter, though lawful, is determined by nothing (indeterminate), the steady-state theory endorses radical indeterminism—or, to put it bluntly, it endorses magic. Logicians may not be impressed by an ontological argument such as this . . . but they ought to be persuaded by the fact that the creation hypothesis is *ad hoc* in the worst sense of the word.[39]

---

[36] For other objections, see Appendix, Note XXII.
[37] *The Monist*, xlvii (1962) 126.
[38] The first involves a misreading of the paper by Bonnor and McVittie (*M.N.*, cxxii (1961) 381), which does not—as Bunge believes—show that Hoyle's theory is untestable. The second is that the creation law cannot be tested 'either in the laboratory or by fairly direct astronomical observation'. (p. 118.) This argument is true but trivial. Why the test should be 'fairly direct' is not clear.
[39] Bunge, op. cit., pp. 130–31.

(iv) CREATION AND THE AGE OF THE UNIVERSE

The first sentence might be compared with the two following (by the late Viscount Samuel and Herbert Dingle respectively):[40]

> For the idea of 'creation out of nothing' is not science at all: it is mere magic.

and

> Those who profess to believe in creation out of nothing could not, of course, use this argument against you, but genuine scientists would be free to do so.[41]

We are clearly meant to believe that creation out of nothing cannot be an element of genuine science. But has anyone ever claimed that it could? Dingle was perfectly well aware of the distinction between the senses of 'created' according to which it may be used either to indicate nothing more then the simple fact of appearing, or to make the positive assertion that something may come to exist out of nothing.[42] Likewise Samuel had earlier quoted Hoyle as saying, not that material comes from nothing, but that 'it does not come from anywhere'. We recall Aquinas's words on the subject, quoted earlier in the chapter. We have no wish to deny that cosmologists do occasionally use the less happy alternative, but in doing so they are certainly not erecting it as a profound metaphysical principle. The essence of magic is, no doubt, violation of the 'genetic principle'—as Bunge implies. But that view of the nature of science which makes the 'genetic principle' appear sacrosanct is so unsatisfactory as to take the force out of Bunge's argument, as may easily be seen.

It is Bunge's contention that the notions of *productivity* and *lawfulness* are essential to all types of determinacy—he made the point at length in his well-known book *Causality*.[43] Before asking whether this is reasonable or not, we should point out that since he explicitly rejects two established meanings of 'determination' on the grounds that they lack the sense of 'productivity', what he is trying to effect is nothing less than a piece of semantic legislation.[44] So far as science is to be called deterministic, Bunge's account of the word is to be rejected. Reasons for thinking the idea unsound will be given shortly. So far as the traditional ideas at the basis of causal determinism are concerned, Bunge gives a fair account of the situation: but causal determinism is neither an essential nor a desirable feature of modern science. To advocate the

---

[40] *A Three-fold Cord*—a discussion between the two authors (Allen and Unwin, 1961) pp. 120, 132.
[41] The nature of the argument in question is of no importance here.
[42] op. cit., pp. 127–28.   [43] Harvard University Press, 1959.
[44] op. cit., pp. 8–12.

emergence of matter where there was none before is not 'radical indeterminism', unless by this one simply means that the notion of production is missing. Bunge himself acknowledges that 'laws are conceivable which could "govern" the appearance of things, or the emergence of properties, out of nothing',[45] but again he insists that 'what is logically possible need not be causally possible'.[46]

Bunge's positive arguments for retaining the notion of production within the province of scientific explanation are scattered and elusive. He very properly attacks the claim that causation is the *sole* category of determination; for he wishes to preserve the concepts of statistical, teleological and dialectical determinacy. He even goes so far as to admit that causal determinism is a special elementary and rough version of general determinism, expressly to exclude 'irrational and untestable notions (such as creation out of nothing)'.[47] The principle that the latter should be excluded he refers to as 'the ancient principle', suggesting that its antiquity is yet another argument in its favour. Later in the work, arguing for the ontological, rather than logical nature of the causal problem ('for it is supposed to refer to a trait of reality') he offers this statement as proof: 'the laws of nature, whether causal or not, are by no means logically necessary'.[48] But the question Bunge should be asking is: Are any laws of nature 'causal' which cannot be paraphrased so as to be interpreted otherwise? The distinction ontological/logical begs the question: it is quite the wrong distinction. The laws of nature do not have to be logically necessary to be interpreted as other than causal. Reason and rationalism do not inevitably go together, nor is the rejection of causalism necessarily the same as the acceptance of indeterminism.

There are several reasons for not recognizing the causal nexus in science. (This is not to say that the category of cause was without historical importance to the natural sciences, nor that a discussion of it is without interest to linguistic analysis.) The first reason is that in situations where it is natural to make one physically significant variable a many-place function of others, the cause-effect terminology becomes very artificial. A related objection is that even such traditional candidates as Newton's law can hardly be referred to as 'causal', involving as they do the *inter*action of particles. The second reason is that—as Bunge is at pains to argue—the causal principle is ontological in scope. That is to say, it implies that to a given effect there corresponds a

[45] ibid., p. 24. [46] ibid., p. 240.
[47] ibid., p. 24. [48] Bunge, op. cit., p. 239.

(iv) CREATION AND THE AGE OF THE UNIVERSE

unique, final, and unalterable cause. It is part of the very essence of the sciences—as we have argued in earlier chapters—that we should never suppose a scientific theory to attain any sort of finality. Our attitude is thus totally incompatible with the causal principle in its usual interpretation. The usual concept of cause is therefore rejected as an acceptable scientific category.

Where the concept of creation *ex nihilo* is held to be irrational it is usually on the grounds that the First Event can have no conceivable physical cause, causes being antecedent to their effects. It is strange, therefore, that Bunge should draw the same conclusion, and yet deny that the antecedence of causes is essential to the causality idea. Causality, on his formula, is compatible with the instantaneous linkage of cause and effect. Furthermore, Bunge will admit that spatial contiguity is not essential to the notion of cause. It is hard to see what reasons he can have, therefore, for rejecting an account of continual creation such as was provided by McCrea, even by the light of his own principles. But speculating in this way on a possible reconstruction of his case does not seem likely to be very profitable.

In conclusion, we may defend the concepts of age, First Event, and creation as they are encountered in all current cosmological contexts; but in all cases with important reservations. There is a time parameter, perhaps only for members of a restricted class, which may be all too loosely interpreted as indicating the 'age of the universe'. Unfortunately, any full elaboration of the phrase must needs introduce so much circumlocution and so many qualified metaphors, all to no purpose unless it be exegetic, that it is probably best avoided. Admittedly, cosmology would then lose one of its chief attractions for many people. There is no inherent contradiction, however, in the notions either of a finite or an infinite value of the parameter itself. Several theories may lead one to suppose the occurrence of a First Event. There are two or three senses in which this may be understood: it may suggest an event prior to which there was, in absolute time and space, nothing to which physics would wish to ascribe existence; it may be understood to imply that an assertion of prior events would be meaningless; or it may suggest the occurrence, as it were, of a cataclysm in a hitherto dormant existence. Each is associated with theoretical problems, but none is clearly illogical.

When it comes to creation there are many more facets to the problem, and we must plead guilty to having ignored those which are outside the province of natural cosmology. Creation requires a Creator and a *materia prima*. Paradoxically, it is only when the nature of neither of these is physically obvious that we use the term 'creation' in cosmology: there we never find the creation metaphor, simply because it involves a transformation of one sort of material into another; and if both could be theoretically described, we should at once tend to drop all reference to creation. It is absurd, therefore, to suppose that cosmology has anything to offer theology in any form other than pantheism or, perhaps, immanent theism. A physical theory either can stand without invoking the supra-physical, or it cannot. In the latter case it is not acceptable as science, and in the former it can offer no criticism of deistic theology. It would have been to the advantage of both theology and natural cosmology had the latter entirely avoided the term 'creation'. In its place there are the perfectly adequate expressions 'First Event', which will cover most cases, and, for the steady-state theories, 'spontaneous occurrence'. Both sorts of event may be described without inconsistency, without the language of cause and effect, and without recourse to the supernatural.

CONCLUSION

It was not our intention to use this history to illustrate a single theme, but one which it is hoped stands out clearly from the rest is this: it is an illusion that the natural sciences advance by a series of steadily improving approaches to the truth. The illusion is fostered by the realization that in formulating a new theory it is virtually impossible to begin afresh, shaking off all the conceptual shackles of previous theories. The dynamics of continuous media and the idea of a field of force; Poisson's equation and Einstein's earliest formulation of the gravitational field equations; the General Theory of Relativity and Hoyle's theory of continual creation: these and a score of other examples show how—in cosmology at least—theories are conceived in the midst of others, each capable, to a greater or lesser degree, of pulling its successors into its own shape. The force-dynamics of Whitehead, Birkhoff, and Milne, which follow a classical tradition tempered by Minkowski's treatment of the Special Theory of Relativity; the General Theory of Relativity, the form of which was largely determined by Riemannian geometry and the calculus of tensors; current concepts of distance, and the attempts to accommodate earlier methods of determining distances: in these and in so many other ways, we have seen the influence, conscious and unconscious, of one phase of thought upon another. But although metempsychosis of this sort is far from rare, we have found no evidence of immortality in the natural sciences. None has an absolute and permanent value. The individual theory of cosmology is neither true nor false: like any other scientific theory, it is merely an instrument of what passes for understanding.

APPENDIX

Note I (p. 22). *The clustering of galaxies*

The surveys of the 1930s suggested that the galaxies were randomly distributed with a few small clusters providing the exceptions rather than the rule. As methods of observation improved, the galaxies assigned to the Local Group increased in number and the 'local metagalaxy' took its place with the other known clusters. By the late 1930s it had become an open question as to whether, with upward of twenty known members (according to some), the Local Group was no more than a detail of a much larger system. E. Holmberg (*Lund. Ann.*, vi (1937)) and A. Reiz (*Lund. Ann.*, ix (1941)) gave preliminary evidence for this view, and in 1951 Mrs V. Cooper Rubin claimed to have found evidence for a differential rotation of 'the inner metagalaxy' (*A.J.*, lvi (1951) 47). The whole issue is extremely involved: a knowledge of the *extent* of a system of this kind has so far proved elusive in the absence of knowledge as to the lower limit of galactic size and luminosity and of the proportion of galaxies of various luminosities (that is, the galactic luminosity function). A knowledge of its movements is founded, almost entirely, upon the very sparse results of observations of radial velocities, published by Humason (*Ap. J.*, lxxxiii (1936) 10 = *Mt. Wilson Cont.*, no. 531. Cf. *Ap. J.*, lxxiv (1931) 35 = *Mt. Wilson Cont.*, no. 426) and Humason, Mayall and Sandage (*A.J.*, lxi (1956) 97). One of the most careful analyses of the evidence so far given is that by de Vaucouleurs (*A.J.*, lviii (1953) 30; also *Sci. Am.*, cxci (1954) 30) His conclusion, supported by early radio-astronomical studies, was that the Galaxy is one of some tens of thousands of similar objects arranged in a flattened but otherwise irregular group. The Local Group he classified as a discrete sub-system near its edge, other sub-systems comprising the nebular complex in Leo, the Ursa-Major cloud and the Virgo cluster with its extension in Centaurus. He denied that there is any strong rotational symmetry in the Local Group. The flattening ascribed to the 'Supergalaxy' as a whole he interpreted as suggesting a general rotation, but it is generally agreed that no analysis of this feature which ignores magneto-hydrodynamics will be at all satisfactory.

If this trend in astronomical thought had continued it might have been accompanied by a renewal of interest in an hierarchical model of the grouping of galaxies. Many astronomers seem to be sceptical of the clustering of systems like the Local Group and the Virgo cluster. The findings of C. D. Shane and C. A. Wirtanen (*A.J.*, lix (1954) 285) indicated that most galaxies are indeed found in clusters, generally of fewer than a thousand members each, but that these appear to be randomly distributed. (Jerzy Neyman and E. L. Scott, *Ap. J.*, cxvi (1952) 144. Cf. *P.N.A.S.*, xl (1954) 873 and (with Shane) *Ap. J.*, cxvii (1953) 92.) McVittie has since given the relativistic version of Neyman and Scott's account of the distribution of the galaxies in space. (*A.J.*, lx (1955) 105.) These results were generally admitted at the time of their announcement, but doubt has been cast even upon them by the recent findings of Miss P. R. R. Leslie (*M.N.*, cxxii (1961) 28). In a statistical examination of a number of surveys of radio sources she found that amongst associations of sources with an angular separation of between $3'·5$ and $200'$ arc no more than about one tenth of these occurred in clusters.

APPENDIX

For a very extensive bibliography on the subject of galactic clustering see G. de Vaucouleur's article in *Vistas*, ii, pp. 1604–6, where 137 items are listed. To this may be added a short paper by the Bulgarian astrophysicist, Nicola St. Kalitzin, in which a relativistic hierarchical model is described. (*M.N.*, cxxii (1961) 41.)

Note II (p. 69). *Schwarzschild space-time and the two-body problem in the General Theory of Relativity*

Schwarzschild states the problem thus: 'To discover a line-element with coefficients satisfying the requirements of the field equations [following Einstein he writes these as

$$\sum_{\alpha}(\partial/\partial x_\alpha)\Gamma^\alpha_{\mu\nu} + \sum_{\alpha\beta}\Gamma^\alpha_{\mu\beta}\Gamma^\beta_{\nu\alpha},$$

for points outside the central sphere], the determinantal equation [$g = -1$], and these four requirements.' The requirements to which he is referring are

(i) All components g_{mn} are to be independent of time (the condition for a static world).
(ii) $g_{\mu 4} = g_{4\mu} = 0$.
(iii) The solution is to be spatially symmetrical about the origin of coordinates, in the sense that an identical solution is obtained when the spatial coordinates are subjected to an orthogonal transformation (rotation).
(iv) The components of g_{mn} are to vanish 'at infinity', with the exception of g_{44} ($= 1$), g_{11}, g_{22}, and g_{33} (each being equal to -1).

Schwarzschild's 'exterior solution'

$$ds^2 = (1 - 2m/r)dt^2 - (1/c^2)\{dr^2/(1 - 2m/r) + r^2 d\theta^2 + r^2 \sin^2\theta d\phi^2\}$$

(where $m = (\gamma M_0/c^2) = 1\cdot 48$ km for our Sun) is, of course, very well known. There is clearly a singularity at the points $r = 2m$, and the 'physical significance' of this singularity has been the subject of heated discussion on several occasions. E. Rabe, for example, supposed he had found the 'physical meaning' of the singularity in showing that it corresponds to the point of the gravitational field at which a freely falling point-particle would acquire the velocity of light were it to fall freely from the outer limits of the gravitational field. (*Astr. Nachr.*, cclxxv (1947) 251.) The singularity is often regarded as defining an *impenetrable* region, for as r tends to $2m$ from above, the element of radial 'distance' ($d\bar{r}$) given by the equation

$$d\bar{r}^2 = dr^2/(1 - 2m/r)$$

tends to infinity. This way of speaking is obviously quite arbitrary, in so far as the chosen definition of length depends upon a central mass which, in the world at large, is not unique. It is also hard to see what impenetrability can have to do with metric properties as such (but see Appendix, Note XVI for more recent views on this question of interpretation).

Mie and Hilbert both thought that it was possible to avoid this vexed question entirely. The constant of integration to which Schwarzschild gave the value $2m$ was made zero by Hilbert on no better grounds than that the expression for the gravitational potential is thereby simpler than it would otherwise be. (*Gött. Nachr.*, (1917) 53, read December, 1916). Mie had an alternative way of solving the 'exterior problem' in terms of what he called a 'rational coordinate system'; but his method of excising the singularity is no less arbitrary than Hilbert's (*Ann.*

Phys. Lpz., lxii (1920) 46). The singularity cannot, in fact, be completely removed so long as the static condition is retained. As Serini found, there can be no non-singular solutions of the field equations for empty spaces which are static and have the form $g_{mn} = 1 - \text{const}/r$ at infinity (*Rend. Lincei* (5), xxvii (1918) 235. Cf. Einstein and Pauli, *Ann. Math. Princeton*, xliv (1943) 131). Lemaître later showed how to remove the singularity by introducing a transformation leading to a *non-static* form of the line-element, namely

$$ds^2 = dt'^2 - (1/c^2)\{(2m/r)dr'^2 + r^2 d\theta^2 + r^2 \sin^2\theta d\phi^2\}$$

where $r = (9m/2)^{1/2}(r' - ct')^{2/3}$ (Lemaître, *Ann. Soc. Sci. Brux.* (A), liii (1933) 51 and in particular p. 82. Cf. J. L. Synge, *P.R.I.A.*, liii (1950) 83; *Nat.*, clxiv (1949) 148).

More recent papers on the removal of the singularity are:

D. Finkelstein, *Phys. Rev.*, cx (1958) 965; C. Fronsdal, ibid., cxvi (1959) 778; M. D. Kruskal, ibid., cxix (1960) 1743; R. W. Fuller and J. Wheeler, ibid., cxviii (1962) 919.

Finkelstein's paper has been shown by Miss E. A. Hilton (unpublished Ph.D. thesis, London, 1963) to have an unusual and presumably undesirable property. The Finkelstein space-time purports to be an alternative to Schwarzschild's, and yet whereas the latter manifests an event-horizon at $r = 2m$ (see Appendix, Note XVI) the former does not. It was shown, however, that if we consider the time-reversal of the Finkelstein metric, there will be an event-horizon at $r = 2m$ in the resulting space-time (Finkelstein's transformation leaving r unchanged).

The differential equations of the path followed by a free particle in a Schwarzschild space-time (namely those of the geodesics) are easily found, but their integration requires, it seems, the use of elliptic functions or worse. This problem attracted the attention of a great number of mathematicians over the course of the next ten or fifteen years, but has remained a somewhat isolated one (de Sitter, *Proc. Acad. Amst.*, xix (1916) 367; Levi-Civita, *Rend. Lincei*, xxvi-xxviii (1917–19) twelve papers; A. R. Forsyth, *P.R.S.* (A), xcvii (1920) 145; F. Morley, *Am. J. Math.*, xliii (1921) 29; C. de Jans, *Mém. Acad. Brux.*, vii (1923).) Several other special solutions of Einstein's field equations were given, following Schwarzschild's example. Most were of purely academic interest but two of especial importance were the Schwarzschild 'interior solution' (*Berlin Sitz.* (1916) 424). This might be called the first paper in relativistic hydrodynamics. Owing to the assumptions of symmetry and static state, the hydrodynamical side of the problem was much simplified. The first really comprehensive work on this subject was given by Synge (in *P.L.M.S.* (2) xliii (1937)), and the solution of the corresponding problems when the free particle has both mass and charge. (Two early papers from a numerous list are: H. Reissner, *Ann. Phys. Lpz.*, 1 (1916) 106 and H. Weyl ibid., liv (1917) 117). The study of the relative motion of even only two bodies of comparable mass soon revealed the forbidding complexity of the General Theory of Relativity. The main source of the difficulty is that, unlike the classical field theories, this theory is *non-linear*. Both share another difficulty, namely that on the world-line of a point particle the gravitational field (or metric tensor) is singular. In the classical theory the usual way out of the conceptual tangle is to insert in the equations of motion of a particle a 'background' field from which the particle's own field is subtracted. In the General Theory of Relativity the principle of superimposing or dividing fields in this simple way is ruled out by the non-linearity of the equations. Schwarzschild's treatment of the two-body problem for the case where one of the masses is insignificant by comparison with the other avoids this difficulty, but it cannot be avoided when the masses are of comparable size.

Until 1937 no method, it was generally believed, had yet been discovered to which this problem would yield. Einstein and Grommer had made a little headway when in 1927 they pointed out that since the field equations are incompatible with a *static* field the motion of a particle might be somehow implicit in the field equations (*Berlin Sitz.* (1927)). In 1937 Einstein, Infeld, and Hoffmann, using only the field equations in empty space, investigated solutions which, along some world-lines, have simple singularities representing simple mass particles (*Ann. Math. Princeton*, (2), xxxix (1938) 65). They showed that such solutions can exist only if the singular world-lines satisfy certain differential equations, the equations of motion. (Using an approximation method which applies only to slowly varying fields.) In this way the equations of motion for the many-body problem were derived from the field equations. By linearly superimposing solutions with *different* particles as 'sources', a new solution of the field equations is obtained with *unchanged* particle motions. If the motion of the sources were determined by *linear* field equations they could not, therefore, be said to 'interact' at all. The equations of motion are made to appear as a necessary consequence of the non-linearity of the field equations.

It is not, in fact, true that the problem could not be satisfactorily formulated in terms devised before the advent of this last memoir, for Eddington and G. L. Clark, independently of Einstein, Infeld, and Hoffmann, obtained the relativistic approximation to the two-body problem by the older methods, using the geodesic equations and the energy-momentum tensor (*P.R.S.* (A), clxvi (1938) 465. Cf. Clark, ibid., clxxvii (1941) 227 for a different approach). As Eddington remarked elsewhere, 'the problem of two bodies in the General Theory of Relativity is as difficult mathematically as that of three bodies on ordinary dynamics'.

The literature of the subject is by now very extensive—it was so even in 1937. The following works deserve mention:

(i) W. de Sitter, *M.N.*, lxxvi (1916–17) 699; lxxvii (1917) 155 (errata p. 481); lxxviii (1918) 3.

These give an early and full account of the astronomical consequences of Einstein's theory. The second paper treats in great detail the resultant field of Sun and Earth and the consequent orbit of the Moon.

(ii) V. A. Fock, *J. Phys. U.S.S.R.*, i (1939) 81.

(iii) A. Papapetrou, *Proc. Phys. Soc.*, lxiv (1951) 57.

These two authors use similar approaches to the derivation of equations of motion from the field equations alone. Their method is quite different from that used by Einstein, Infeld, and Hoffmann, and is usually said to be considerably simpler.

Note III (p. 85). *Eddington and the quantum interpretation of the Einstein world*

Arguing for the existence of the cosmological constant, Eddington fell back upon the Einstein model—presumably on account of its simplicity and the fact that it allowed the possibility of having its constituent particles ($2N$ in number) in an eigenstate (this last condition being necessary for the use of the wave equation of quantum mechanics). (*P.R.S.* (A), cxxxiii (1931) 605.) Assuming the usual relations between the mass and radius of this Einstein world, and assuming that the total mass of the electrons (equal in number to the protons) could be ignored in the expression for the mass of the whole ($M_0 = N \cdot m_p$) he argued for the introduction of a new term $((i\sqrt{N})/R)$ in the orthodox wave equation for the electron by the following 'mere analogy':

> If the energy of a specified electron due to a singularity at distance r is i/r the energy of N indistinguishable electrons due to a singularity at distance R (the centre of the spherical universe) will be $((i\sqrt{N})/R)$. (op. cit., p. 606. This 'mere analogy' is amplified in section 6.)

Lastly, identifying the term $((i\sqrt{N})/R)$ with the term 'normally attributed to the proper mass of the electron' (imc^2/e^2) he inferred that

$$\lambda = (2\gamma m_p/\pi)^2 \cdot (mc/e^2)^4 = 9{\cdot}79 \times 10^{-55},$$

with $R = 1{\cdot}01 \times 10^7 \text{ cm} = 328 \times 10^6 \text{ pcs},$

using current laboratory data. It is difficult to know what importance Eddington placed on this sort of half-intuitive argument. He was careful to notice that 'the present radius of our expanding universe is greater than the radius R of the Einstein world; but, inasmuch as it does not correspond to an eigenstate, it is irrelevant to the theory of the wave equation'. This argument led him to a value for the Hubble recession-parameter (465 km sec^{-1} mpc^{-1}) close to Hubble's figure of 528. Both figures were subsequently revised. From this time until his death in 1944, Eddington elaborated his argument, using the concept of interchange energy and the exclusion principle. All forms of physical interaction are, according to him, to be traced ultimately to the indistinguishable nature of the elementary particles. These ideas are too involved for further discussion here. They have so far had little influence on cosmology as such.

Note IV (p. 86). *Weyl's cosmological terms and matter*

We have briefly explained in Chapter 16 how it was Weyl's aim to demonstrate the way in which the 'metrical structure of the world' has its origin in both gravitational and electro-magnetic fields. His account of the structure was given not only in terms of a Riemannian quadratic form ($g_{mn}dx^m dx^n$) but also in terms of a linear differential form ($\phi_m dx^m$). (*Berlin Sitz.* (1918) 465 and *Ann. Phys.* Lpz., lix (1919) 101. Cf. also *S.T.M.*) Weyl developed this theory from a Hamiltonian principle as his 'initial physical law', which he expressed as follows:

> 'The change in the *Action* $\int \mathbf{W} \, dx$ for every infinitely small variation of the metrical structure of the world that vanishes outside a finite region is zero.' (*S.T.M.*, p. 286.)

The action was to be an invariant and hence $\mathbf{W}$ a scalar density ('derived from the metrical structure'). Mie, Hilbert, and Einstein had all adopted the same view, but Weyl's conception of *invariance* was broader than theirs. Weyl demanded invariance with regard to the process of recalibration (in which ϕ_m and g_{mn} are replaced by $\phi_m - (1/\lambda)(\partial\lambda/\partial x^m)$ and λg_{mn}), over and above the ordinary invariance with respect to transformations of coordinates.

As Weyl maintained—after providing a thorough investigation of the mathematical restrictions upon $\mathbf{W}$—the assumption which is simplest 'for purposes of calculation' is that

$$\mathbf{W} = -(F^2/4)\sqrt{g} + \alpha l.$$

(*S.T.M.*, p. 295. Cf. W. Pauli, *Phys. ZS.*, xx (1919) 457.) Here α is a pure number, l denotes Maxwell's action of the electromagnetic field, whilst F is the scalar of curvature in general metrical space (formed analogously to the 'R' of Riemannian space). Working from this expression Weyl showed that a static and *uniform* distribution of neutral matter over a spherical space is a possible equilibrium state. He emphasized that the cosmological term which Einstein had added to his field equations at a comparatively late stage was a part of his own theory from the very beginning ('early' and 'late' are not meant in a historical sense). In point of fact, if an electromagnetic field is present, Einstein's cosmological term, according to Weyl, must be supplemented by a similar term dependent on the electromagnetic potentials. This additional term had an important consequence for Weyl's concept of matter, for it meant that a material particle could be conceived without the need for a 'mass horizon'.

Weyl had previously explained away matter as no more than a 'real singularity of the field'; there was, he said, a sphere of a certain radius which (centred on the electron) enclosed zero mass. (*S.T.M.*, p. 261.) Actually the mass enclosed by a sphere of radius r was given as $(m_0 - e_0^2/(2c^2r))$ where m_0 and e_0 were the mass and charge producing, respectively, the gravitational and electrical field of the electron. The mass of the electron, according to his main theory, 'is to permeate the whole of the field with a density that diminishes continuously'. This feature he reckoned unsatisfactory when taken in conjunction with the idea that the external field was entirely free of electrical charge. The singularity, using the *two* cosmological terms, was not removed, but simply exchanged for a *point* singularity with which, for some reason, Weyl was much happier. (See *S.T.M.*, pp. 298–300.)

Note V (p. 90). *Taub and the Mach–Einstein Principle*

More specifically, Taub showed that, for the spatially homogeneous Riemannian space-times which admit the three-parameter group of the Euclidean translations and have no singularity of R_{nst}^m along the time axis, the equations $R_{nst}^m = 0$ follow from the equations $R_{mn} = 0$. For all other spatially homogeneous space-times the g_{mn} have ' "spatial singularities" in the sense that in a special coordinate system they are not bounded for all values of the spatial coordinates (including infinite ones)'. Examples are given of space-times admitting transitive three-parameter groups for which the equations $R_{nst}^m = 0$ do *not* follow from $R_{mn} = 0$, and yet for which the g_{mn} are finite for all finite values of the time. Taub remarked (cf. Eddington, *M.T.R.*, p. 165) that the Mach–Einstein principle could only be jeopardized when a decision was made as to what sort of singularity the General Theory of Relativity is to interpret as corresponding to the presence of matter.

Note VI (p. 128). *Bonnor and perturbations of the relativistic model*

Bonnor's method (*ZS. Ap.*, xxxv (1954) 10) was to imagine the Friedmann model to be subjected to a distortion which both preserved spherical symmetry and isotropy and satisfied the boundary conditions. He found that if the matter of the model were to satisfy the same equation of state at all times, then not only would it be stable to spherically symmetrical perturbations, but it would permit no such disturbance. He found that if the equation of state changed at a given time there would be no discontinuities in the pressure and its time derivatives, and a different Friedmann model would result. Given no equation of state from a given moment, inhomogeneities could develop, in which case the stability of the model would depend on the nature of the deformation. Bonnor indicated, moreover, a way in which a contracting Friedmann model could pass from contraction to expansion without passing through a singular state. (ibid., p. 20.)

Bonnor used similar methods in a later paper on the instability of the Einstein universe (*M.N.*, cxv (1955) 310). He began by questioning a point of procedure which Einstein himself mooted in his earliest presentation of this model. In short, he assumed its homogeneity to be associated with the *averaged* inhomogeneities (of both pressure and density) of the world. 'But for this interpretation to be permissible,' wrote Bonnor, 'one must be sure that inhomogeneities could in fact exist in the Einstein universe.' (ibid., p. 314.) As it turns out, several static and inhomogeneous solutions have for long been known. It is occasionally assumed that the existence of condensations is a sufficient condition for an unstable model, but that there are static solutions (except when $p = 0$) featuring condensations disproves this. (Cf. references under Note XIV.) Raychaudhuri (*Bull. Calc. Math. Soc.*, xliv (1952) 31) a few years before, had given exact static solutions of the field equation corresponding to an Einstein model with spherical pockets of radiation, the density of which may vary. Bonnor, following his method, showed (rather more generally) how to construct models in which static, spherically symmetric regions of arbitrary density are embedded in Einsteinian space. (op. cit., p. 315.) He showed how a disturbance would, in these models, entail discontinuities in the pressure or its time derivatives. He also showed that if the matter present satisfies an equation of state, 'no physically plausible disturbance' of the initial state is possible, and the model is stable: if there is no equation of state small disturbances will, except in special cases, lead to instability. Bonnor ended by claiming that the new boundary conditions appear to work effectively in the problems he discussed. 'All the functions', he added, 'which they leave arbitrary have a reasonable physical interpretation and the data required to determine the solutions are very much what one would expect on classical grounds. This has certainly not been true of sets of boundary conditions used in some of the previous work on this subject.' (op. cit., p. 322.)

In his earlier paper (*ZS. Ap.*, xxxv (1954) 10) Bonnor confined his attention to the state of the model immediately after the disturbance. Raychaudhuri later found it possible to integrate the field equations completely under the same circumstances (*ZS. Ap.*, xxxvii (1955) 103). Once again the boundary conditions followed were those given by O'Brien and Synge (Dublin Institute, 1952).

Note VII (p. 147). *Recalibration of the Cepheid P-L relation*

H. Mineur had earlier (*Ann. astrophys.*, vii (1944) 60) expressed dissatisfaction with the calibration of the period-luminosity relation, in the course of applying some new determinations of radial velocity by A. Joy. (See *Ap. J.*, lxxxix (1939) 356.) He did not, however, appreciate the reason for the discrepancies obtained. (See *C.R.*, ccxxxv (1952) 1607.) Baade's conclusion was soon confirmed by others, for example A. D. Thackeray and J. Wesselink of Pretoria (*Obs.*, lxxv (1955) 33). A very thorough revision of the absolute magnitudes of a group of classical Cepheids within the Galaxy was made by A. Blaauw and H. R. Morgan in 1954 (*B.A.N.*, xii (1954) 95) on the basis of their statistical parallaxes, using new information on the correction to be made for absorption. Despite the thoroughness of this investigation, statistical parallaxes are treated with more reserve than was once the case, for errors of measurement are appreciable fractions of the small measured proper motions. 'Pulsation parallaxes', originally proposed by Baade in 1928, might well find a use in the future. It is worth remembering that the distances of the globular clusters were found using RR Lyrae variables, and the adjustments to their curve were slight. Intragalactic distances were therefore little affected. Interstellar absorption is the difficult problem here.

Note VIII (p. 176). *Kinematic Relativity: its axiomatic formulation*

Another who investigated a generalized form of Kinematic Relativity by introducing an arbitrary Riemannian space-time was McVittie. (*M.N.*, xcv (1935) 270; *ZS. Ap.*, x (1935) 382 and *Cosmological Theory* (London, 1937) ch. v.) The way in which McVittie's method differs from Walker's is explained by Walker in *P.L.M.S.*, xlvi (1940) p. 125. A three-cornered controversy on the subject is to be found in *Obs.*, lxiv (1941) 11 (discussion between McVittie, Milne and Walker). McVittie also gave an axiomatic treatment (*P.R.S. Edin.* (A), lxi (1942) 210) to which Whitrow objected strongly (ibid., p. 298). Walker attempted an axiomatization of the theory some years later (*P.R.S. Edin.* (A), lxii (1948) 319). Whitrow in 1937 listed a series of 'axioms' for the theory, showing how Milne's account could be simplified (*ZS. Ap.*, xiii (1937) 113). For some reason Kinematic Relativity seems to lend itself to a simple axiomatic formulation.

Note IX (p. 164). *Models with a discrete material content*

We have seen that Milne appeared to prefer the idea of a model with discrete point-concentrations of matter to the hydrodynamical alternative. Relativistic cosmology had concentrated on this alternative. But Milne believed there to be insuperable difficulties involved in generalizing the discrete model from one to three dimensions. (The former is found at pp. 56–7, *W.S.* The system is fully determined, given the relative motion of two of its members.) That Milne was mistaken was eventually to be shown by Coxeter and Whitrow in 1950 (*P.R.S.* (A), cci (1950) 417).

Relatively little progress has been made in the study of theories of cosmology of any type which incorporate hypotheses of any kind of non-homogeneity, and, *a fortiori*, those whose models comprise a system of discrete particles. As instances of such theories we have one proposed by McCrea in 1931 (*Proc. Ed. Math. Soc.*, ii (1931) 158), Milne's of 1933 (*Q.J.M.*, v (1934) 30; *W.S.*, p. 28, note), that of Coxeter and Whitrow, referred to above, and one put forward by G. C. Omer a year or so earlier (*Ap. J.*, cix (1949) 164. Results intimated in *Phys. Rev.*, lxxii (1947) 744; lxxiv (1948) 1564).

McCrea's model has as 'spatial section' the surface

$$f(x, y, z, w) \equiv x^{2n} + y^{2n} + z^{2n} + w^{2n} - a^{2n} = 0$$

where n is a positive integer (compare Einstein's universe where n is unity). As n increases, this surface approximates more and more closely to the (single) four-dimensional cube at whose sixteen vertices—or so the gravitational field equations suggest—'the matter necessary to give rise to this curvature is . . . largely concentrated.' (McCrea, op. cit., p. 158. In other words, the curvature is small except at these sixteen points.) It emerges that in this model of sixteen 'nebulae', any observer 'would see the six nearest in orthogonal directions and eight others symmetrically placed. The remaining one, that diametrically opposite in the four-dimensional representation, will be seen in all directions, providing a continuous (though not uniform) background in the sky.' (op. cit., p. 163.)

Omer's model was assumed a spherically symmetrical distribution of matter about a central observer. However 'philosophically objectionable' the introduction of a central observer, Omer held that 'with the present observational data there is no other logical position in which to locate him'. Omer probably meant that he had been led to consider the model on the grounds of its mathematical simplicity and that as the nebulae appear to be more or less symmetrical about our own Galaxy, we might as well identify this with the unique point of the model. Such a view is notoriously difficult to discredit, but it is not likely to find many supporters.

Whitrow and Coxeter (op. cit.) refer to an earlier result obtained by Whitrow: each equivalence can be transformed into another, by an appropriate regraduation of the time-scale. (See *Q.J.M.*, vi (1935) 249 at p. 256.) Following upon this result, Milne gave his well-known logarithmic transformation, relating a uniformly expanding equivalence to a static one. (*P.R.S.* (A), clviii (1937) 324; clix (1937) 171. Cf. *K.R.*, pp. 36–7. For this, and some of the following results, see Chapter 8.) Had the 'particle-observers' been relatively stationary, Milne's problem of distributing them in three-dimensions in accordance with the Cosmological Principle would merely have been that of finding a homogeneous 'honeycomb' in the space considered—namely Euclidean space. They were not originally relatively stationary—but now the logarithmic transformation of time had provided a means of regarding them so. With Milne's convention for distances, the uniformly expanding ('simple kinematic') model is transformed into a stationary model with hyperbolic space of constant curvature. (See *K.R.*, pp. 46–50. In 1939 Whitrow (*Q.J.M.*, x (1939) 313) showed that Milne's work had in part been anticipated by V. Varicak (*Phys. ZS.*, xi (1910) 577) who, without formulating a transformation of time scale, had nevertheless shown the Lorentz formulae to be consonant with a hyperbolic velocity-space.) Milne's problem had thus been reduced to one of finding a geometrical construction for a homogeneous 'honeycomb' in hyperbolic space. Whitrow, who seems to have been the first to see Milne's problem in this light, pointed out that its geometrical counterpart had been solved in 1883 by V. Schlegel (*Nova Acta Leopoldina*, xliv (1883) 343–459. See p. 444, where four different solutions to the problem are given.) Coxeter and Whitrow now gave a more complete discussion of the problem in both hyperbolic and spherical space. As for hyperbolic space, the ratios of the distances between vertices to the radius of curvature of space suggested (with rough empirical estimates of the latter and of the separations of galaxies and clusters) that the vertices of the lattice could only be identified with super-clusters—for which there was no satisfactory evidence. As for spherical space, it was shown that the greatest possible number of points in a non-trivial homogeneous distribution is exactly 14 400. The mesh, to which this case corresponds, although the 'finest' possible for spherical space, was, for obvious reasons, thought 'too coarse to be of direct cosmological significance'.

Note X (p. 190). *Fernández and Birkhoff's cosmology*

In this note, Fernández explains that he has developed a cosmology on the lines laid down by Birkhoff. Assuming that 'the metagalaxy is perfectly symmetrical'—that the mass-density, that is to say, at all galaxies is the same function of time (or, as Fernández says, 'the same function of the age of the universe') —he finds a universe 'with finite volume and infinite mass'. (By 'metagalaxy' he is presumably referring to the entire system of galaxies.) Both the distribution function for the velocities of the galaxies and the number-magnitude relation were also said to have been obtained. It seems that the only later reference to any possible application of Birkhoff's theory on a galactic scale was made by Alba Andrade. He solved Birkhoff's equations of motion to a first order approximation for small particle velocities. (*Revista Mexicana Fisica*, i (1952) 38. (Fernández writes on Birkhoff's theory on p. 11).) He found that a particle in a circular orbit would be expected to have a small acceleration in a direction perpendicular to the orbit's plane.

Note XI (p. 208). *Projective Relativity*

The work of Jordan and his collaborators is collected in *Schwerkraft und Weltall, Grundlagen der theoretischen Cosmologie* (Braunschweig, 1952, 2nd ed, 1955). This work does not connect the simple mathematics of the genesis of matter with the remaining 'Projective Relativity' (but see pp. 174, 185). At p. 198 Jordan explains Wegener's continental drift hypothesis in terms of the secular decrease in the constant of gravitation! On Jordan's theory of the creation of matter with new conditions imposed on the original 'projective' theory, see *Jubilee R.T.*, p. 165 and G. Ludwig and K. Just, *ZS. Phys.*, cxliii (1955) 472.

Note XII (p. 217). *Zero-point stress with a hyperbolic model. Gravitinos*

Using Whittaker's and McCrea's idea of gravitational mass ($\sigma = \rho + 3p/c^2$; σ is an invariant) and restricting his attention to what he termed a 'gravitationally steady-state' in which σ is held to a constant value, McVittie found that $R(t)$ satisfies the equation (*P.R.S.* (A), ccxi (1952) 295)

$$R''/R = \lambda/3 = (8\pi\gamma\sigma/6).$$

Selecting for further consideration the hyperbolic solution

$$R = A \cosh(\eta t) \text{ with } \eta^2 = \lambda/3 - (8\pi\gamma\sigma/6)$$

(this is of interest in that the corresponding model exhibits both expanding and contracting phases) it is found that

$$(d\rho/dt) = -(3/c^2)(dp/dt) = (6\eta/8\pi\gamma)(\eta^2 - kc^2/A^2)\tanh(\eta t)\operatorname{sech}^2(\eta t).$$

The conversion of stress-energy to mass is thus evident during the expanding phase ($t > 0$). Further restrictions were placed upon the solution in order to obtain a close parallel with the steady-state 'creation process', and in a later paper McVittie listed other forms of the function $R(t)$ for which, granted a suitable sign of the curvature, this parallel can be found. (*A.J.*, lviii (1953) 129. See p. 133.) These contributions make the source of this surprising conversion perfectly clear. With a transformation of both spatial and temporal coordinates, the new coordinate-density, being equal to the time-like component of the transformed energy tensor, will in general involve both ρ and p. This is not so in Newtonian hydrodynamics, The original stress, as McVittie pointed out, will thus contribute to the new density.

The concept of 'zero-point stress' has been developed yet one stage further by F. A. E. Pirani who has ingeniously suggested that it may be explained in terms of the contribution of 'gravitinos' to the energy-momentum tensor (*P.R.S.* (A), ccxxviii (1955) 455). These are entities of zero rest mass and negative energy which accompany the continual creation of matter so as to guarantee the conservation of 4-momentum at each creation event. (These, as Pirani indicates, resemble Synge's 'attractive impulses'.) It is shown that it may be possible to identify the gravitinos with neutrinos of negative energy. Although they are obviously capable of annihilating matter, an argument is presented which shows that the probability of their doing so is slight in an expanding universe.

Note XIII (p. 225). *The dissolution of galactic clusters*

In 1934 Bok showed that 'tidal' forces of galactic rotation must also accelerate the dissolution of certain types of galactic cluster (*Harvard Circular*, no. 384 (1934) and *Obs.*, lix (1936) 76). Shortly afterwards came criticism of those arguments which Jeans had based upon the statistics of binary orbits (Ambarzumian, *Nat.*, cxxxvii (1936) 537; Kuiper, *P.A.S.P.*, xlvii (1935) 201). Amongst the constructive theories produced over the next ten years, which were concerned with the stability of galactic clusters and binary systems are: Mineur, *C.R.*, ccviii (1939) 631; Chandrasekhar, *Science*, xcix (1944) 133 and *Ap. J.*, xcviii (1943) 54; xcix (1944) 54 (cf. *Principles of Stellar Dynamics* (1942) ch. v); Spitzer, *M.N.*, c (1940) 396; Kuiper, *Ap. J.*, xcv (1942) 212.

Note XIV (pp. 225 and 256). *Condensations in an expanding universe*

McVittie, *M.N.*, xci (1931) 274, xcii (1932) 500, xciii (1933) 325; Lemaître, *C.R.*, cxcvi (1933) 903, 1085, *P.N.A.S.*, xx (1934) 12, 366; Sen, *Bull. Calc. Math. Soc.*, xxix (1937) 185; Moghe, ibid., xxxi (1939) 19.

These works of the pre-war decade are mostly concerned with the special features of *single* condensations in an expanding universe. The subject is of the greatest astrophysical importance, for in principle it ought to be possible (i) to outline some mechanism for the formation of local condensations which does not lead to a general instability in the accepted cosmological model (see the reference to Bonnor in Note VI for one such mechanism):

(ii) to explain the spiral nebular form (see Narlikar and Moghe, op. cit.);

(iii) to explain the movement of stars within the nebulae, to calculate their velocity of escape and so on. Sen and Moghe, op. cit., suggest that the latter are much smaller than the classical values;

(iv) to account for the formation of stars in the history of the galaxy;

(v) if it is claimed that radically different conditions obtain in the 'initial stages of the expanding universe', to explain these physical conditions;

(vi) to account for the relative abundance of the various chemical elements'.

These topics have all been very extensively discussed, especially since the war, and occasional reference is made to them in the main text.

Note XV (p. 226). *The clustering of nebulae—accidental?*

Lemaître afterwards persisted in following up a suggestion put forward by de Sitter in 1933. ('The Astronomical Aspect of the Theory of Relativity', *University of California Publications in Mathematics* (1933) ii, p. 166). This was to the effect that the existing clusters are 'not real clusters bound together by the gravitational attraction of their members, but just accidental and temporary irregularities in the homogeneous distribution of the galaxies over space'. (Cf. *Rev. Mod. Phys.*, xxi (1949) 357.) To this Lemaître added the hypothesis that although they would tend to disperse in 10^8 or 10^9 years, yet nebulae of these clusters are replaced by others from the field of nebulae outside. (*Bull. Acad. Brux.*, xxxiv (1948) 551; xxxvii (1951) 291.) However well integrated with his theory as a whole, it should be clear that no conclusions can be drawn from this hypothesis as regards a 'lower limit to the age of the universe'. When one considers the possible number of variants of theories, not only of dissolution but of the formation of nebular clusters, the uncertainty in any quoted value was very great. This was especially true since the effect of interstellar matter was until recently almost invariably ignored.

Note XVI (pp. 209 and 274). *Horizons*

The time-horizon typical of the de Sitter model, which achieved a certain notoriety at the time of its first announcement, was not again discussed at length until the steady-state models of Bondi, Gold, and Hoyle (with the de Sitter metric) revived the issues involved. Whitrow (*Obs.*, lxxiii (1953) 205) began by indicating the existence of a paradox whereby an event occurring at the 'horizon' of one fundamental observer in the steady-state theory's homogeneous expanding model, will not occur at the horizon of another such observer. Bondi, Gold, Whitrow, and Hoyle all contributed to a discussion of the paradox (*Obs.*, lxxiv (1954) 36, 172, 235; cf. *Nat.*, clxxv (1955) 69, 382, 808) but little headway would have been possible without Pirani's distinction between ambiguities in the phrase 'observable region'. (*Obs.*, lxxiv (1954) 172.) For Whitrow this meant 'events which are or have been visible at a given epoch (to an observer)'; for Hoyle it meant 'events which are or will be visible to the observer'. S. Gupta (*Current Science*, xxiv (1955) 263) discussed the paradox, using yet another definition. Not surprisingly, the accepted concepts of time and distance were of first importance in all this. W. Rindler gave a very careful analysis of the horizon idea, not confining his attention to the de Sitter metric (*M.N.*, cxvi (1956) 662 and *Obs.*, lxxvii (1957) 12) but taking instead all models based on the Robertson–Walker metric. The most important aspect of Rindler's account was its distinction between two very different concepts of horizon:

(i) *Event-horizon*. For a given fundamental observer A, this is a hypersurface in space-time dividing events into those that have been, are, or will be observable by A and those that are never observable by A. (The de Sitter universe has such a horizon.)

(ii) *Particle-horizon*. This, for a given fundamental observer A and cosmic time t_0, is a surface in the instantaneous space $t = t_0$, dividing all fundamental particles into those already observable by A at time t_0 and those not already observable at this time. (Dirac's model and that proposed by Einstein and de Sitter jointly are examples of models incorporating a particle-horizon.)

Lemaître's model possesses both types of horizon. The above distinction can be roughly characterized as follows: with an event-horizon the rate of expansion is such that some photons moving towards A never reach A; with a particle-horizon the initial rate of expansion of the model is greater than the speed of the photons.

It was said that the necessary and sufficient conditions for the existence of, respectively, event- and particle-horizons are that $\int^{\infty} (c/R) dt$ and $\int_0 (c/R) dt$ should converge (the lower limit of the latter integral being replaced by $-\infty$ in the case of models defined for negatively unbounded values of t). The proper distances of event- and particle-horizons are, respectively, $R \int_t^{\infty} (c/R) \, dt$ and $R \int_0^t (c/R) \, dt$ (as before, the lower limit in the last integral may need to be replaced by $-\infty$).

In an unpublished thesis (op. cit., *Note* II) Miss Hilton, having examined more carefully the necessity and sufficiency of Rindler's conditions, has found it necessary to qualify the latter for the case of a closed model if his definitions of the horizon concepts are to be retained.

If Rindler's observer A does not move with a fundamental particle then it can still be shown that if the model possesses both kinds of horizon there are still events absolutely unobservable by A.

For a partial treatment of these problems by the methods of information theory, see an article by A. W. K. Metzner and P. Morrison (*M.N.*, cxix (1959) 657).

Miss Hilton has elaborated upon many of the points discussed here (op. cit.). For example, she finds that the singular surface $r = 2m$ in the Schwarzschild space-time is an event-horizon, and moreover one that is invariant in the sense of being an event-horizon for *every* observer in the region $r > 2m$. Contrast this invariance with horizons in Robertson–Walker models where each observer has a horizon peculiar to himself. For the consistency of the Robertson–Walker models it is necessary (on account of the postulated homogeneity) that the class of unobservable events should be non-empty. Miss Hilton points out that this is not necessarily so in the Schwarzschild case. [In 1966 I heard from her (Elizabeth [Hilton] Harte) that she had revised her earlier sufficiency conditions for the existence of a particle horizon.]

Note XVII (pp. 235 and 243). *The K-term*

A shift in the spectrum of light from the nebulae means a new energy distribution for the received spectrum. The total energy registered from a distant nebula —corresponding to which is an apparent magnitude (all magnitudes given below are photographic)—can be found granted only a great deal of information as to the area, the temperature, the spectral distribution function of the source, and the so-called extinction function for the material intervening between it and the observer. It can be shown that

$$\log_{10} D = 0{\cdot}2(m - K - M) + 1$$

where K is a function of δ. (See McVittie, *General Relativity and Cosmology* (Chapman and Hall, 1956) pp. 154–8. McVittie's definition of the K-term agrees with that found in Hubble and Tolman's paper (*Ap. J.*, lxxxii (1935) 302.) The details of McVittie's method of calculating the correction are criticized in Davidson's paper (*M.N.*, cxix (1959) p. 60). De Sitter, assuming all sources to be black bodies at the same temperature, calculated that for photographic magnitudes, K was given, correct to the second order in δ, by

$$1{\cdot}81\delta + 1{\cdot}54\delta^2 \quad (6\ 000°\text{K})$$

or
$$2{\cdot}96\delta + 1{\cdot}34\delta^2 \quad (5\ 000°\text{K}).$$

Some material of Hubble's, unpublished at the time of his death, (see McVittie, op. cit., p. 157), suggests the (nearly linear) empirical relation

$$K = 3{\cdot}55\delta.$$

At the time of Hubble's controversial paper of 1936 (*Ap. J.*, lxxxiv (1936) 517) he assumed that the nebular spectra were similar to that of the Sun, and were black body spectra at 6 000°K. (Contemporaries thought this to be on the high side. Cf. G. L. Greenstein, *Ap. J.*, lxxxviii (1938) 605, who arrived at a figure of 5250.) Sandage (Humason, Mayall and Sandage, *A.J.*, lxi (1956) 97), using

Stebbins' and Whitford's extinction function, calculated the coefficients of the δ-expansion of K. The results were:

$$K = 4\cdot 38\delta + 0\cdot 34\delta^2 \text{ for } E\text{-type galaxies}$$
$$K = 5\cdot 02\delta - 0\cdot 34\delta^2 \text{ for } Sb\text{-type galaxies}$$
$$K = 2\cdot 70\delta + 2\cdot 00\delta^2 \text{ for } Sc\text{-type galaxies.}$$

The only conclusion so far reached seems to be that the coefficient of δ is not less than $2\cdot 7$.

Note XVIII (p. 285). *Philosophical theories of truth*

Most of us appear to want to maintain that truth implies correspondence with the facts, but as philosophy discovered long ago, the idea of a correspondence between statement and fact is exceedingly difficult to elucidate. Wittgenstein, in the *Tractatus*, nevertheless argued for the thesis that a proposition is a picture or projection of the fact which it is meant to describe, having the same structure as the fact itself. Some years later we find Tarski advocating a correspondence theory of truth, but now offering some such definition of 'correspondence with the facts' as is implicit in this (meta-) statement:

> The statement 'Grass is green' corresponds to the facts if, and only if, grass is green.

This way of proceeding is very odd for two reasons. First, it seems that one who followed Tarski in his philosophy and Milne in his cosmology would be obliged to say—for example—that there is a fact corresponding to the flatness of space just as there is a fact corresponding to the greenness of grass. Second: introducing the notion of fact adds nothing either to our science or to our ordinary discourse. Like that of truth, it is a metaphysical notion, and it is of some interest to ask to what extent the two are independent of one another.

To answer this question, one will naturally begin with the principal rival of the correspondence theory—the coherence theory, according to which a statement is true only if it is compatible with a corpus of knowledge or belief. This is clearly only acceptable as part of a theory of truth; for how can we arrive at a corpus of knowledge without a prior notion of truth? And if we are simply to refer a statement to a corpus of *belief*, then it may turn out to be true in one country or epoch and false in another. There are philosophers who hold that it is reasonable to speak in this way, namely the pragmatists. Truth is defined by them in terms of utility, workability, practical consequences, or the like. Truth tends to be discussed by the pragmatists in terms of the sources and outcomes of beliefs. Thus some would hold that a statement cannot be true if no one ever believed it. Others might concede that truth can somehow be a function of *hypothetical* utility; and so on.

It would be misleading to describe the views represented in this book as 'pragmatist' or 'instrumentalist'. Nowhere have we so much as attempted to give a metaphysical theory of truth. Nowhere is it asserted that the words 'true' and 'false' can always be translated into pragmatic terms. Admittedly we have preferred to speak of certain acts of choice between hypotheses as being made on the grounds of convenience (for want of a better word) where others would have spoken of preferring the true to the false; but this is not to say that we are giving a translation of their words. Rather we are taking one of the universally acknowledged characteristics of truth—namely its enduring quality—and pointing out the inconsistency in ascribing durability to ideas which are plainly surrounded by workable alternatives. A retort might be that of course one can never be quite sure that the truth has been correctly discerned, but that simply 'there is something there' to aim at. It might never be known (it will be said) when the objective is reached; but by dint of barking our shins against obstacles along our dimly-lit path, we can certainly learn when we have taken a wrong turning.

These are the views of the truth-despairing, truth-loving sceptic, and they are harmless enough so long as they are not taken as implying that science is thwarted in some grand ambition. The very notion of a goal involves the notion of our being capable of knowing when the goal has been attained. If we despair of ever finding general criteria of truth except, perhaps, for certain basic statements, then how can we hold truth to be our goal? Perhaps—or so some philosophers have hoped—we may restrict our ambition, and concentrate on a few self-evident truths, in order to break out of the sceptic's nightmare. But this is no way to answer the majority of scientific questions—it would be useless, for example, with such a question as 'Is space truly flat?' A goal which is in principle unattainable in regard to the bulk of what we call knowledge is no goal at all.

What do we offer in its place? Science cannot be the search for truth, since so-called crucial experiments can never finally establish a theory. They can refute it, however, and if science is not the search for truth, it is at least the retreat from falsehood.

Note XIX (p. 381). *The number of galaxies is not more than denumerably infinite*

We begin with three hypotheses:
(i) Either it is possible to lay down a definition of what is meant by 'object', in such a way that objects of less than a certain size are excluded; or it is possible to predict a minimum size on theoretical grounds (as in many astronomical contexts).
(ii) It is possible to divide up the objects in such a way that the surfaces of the resulting objects are entirely convex and contain no discontinuities.
(iii) It is possible to decide (as in (i)) what is the smallest radius of curvature on any part of a resulting bounding surface. (This will be called r.)

(These hypotheses are reasonable if we are interested in, say, ellipsoidal galaxies. Any object with an anticlastic surface will present difficulties, but there is no reason why in such cases we should not consider the fundamental physical particles as our smallest units.)

The volume of a sphere of radius r cannot contain more than one entire object, but it may contain parts of many objects. Even so, the number of our objects which can intersect a sphere of radius r must be less than 12. (This is the closest-packing problem.) The cube inscribed in such a sphere has side $a = 2r/\sqrt{3}$, and fewer than eight of our objects can intersect any one such cube at any one time.

Consider now any division of the three-dimensional Riemannian 'continuum' (it may be allowed to be essentially discrete) by three families of surfaces, spaced in such a way that the largest cell can be completely contained within a cube of side a. (It is assumed that our objects are small enough—for example of galactic size or less—to allow us to speak significantly of 'cubes'.) Then each cell must contain parts of fewer than eight objects. Each cell may be coordinated with an ordered triple of members of the set of *rational* numbers. For if one of the three is a real but irrational number, this can always be exchanged for a rational number corresponding to a point within the same cell, for any finite a. To do so we simply have to terminate the irrational's decimal at a place suitably far along. But the ordered triples of a denumerable set are themselves denumerable, and as the set from which we are taking our ordered triples is an infinite (denumerable) subset of the rational numbers, we conclude that the total number of cells is $\aleph_0$. The number of objects cannot be greater, therefore, than $(7\aleph_0 =)\aleph_0$.

Note XX (p. 385). *The recurrent universe*

According to Aristotle, Heracleitus believed in a universal cycle (*De Caelo*, III, i, 18, &c.). Virgil had similar ideas, showing a sound logical caution in denying memory to those who revisited 'the vault above' (*Aeneid*, vi, 749–51). Lucretius, echoing an Epicurean doctrine, held that the world was continually being destroyed and re-created, the only continuity being supplied by the 'atoms'. (*De Rerum Natura*, tr. C. Bailey, Clarendon Press, Oxford, 1910.) Our world, we are told, is formed not by design but by a chance concourse of atoms. These, 'by trying movements and unions of every kind, at last they fall into such dispositions as those whereby our world is created and holds together.' (Book i, 1029–31.) The world which is 'preserved from harm through many a mighty cycle of years' is not alone—there are other worlds than ours, produced from time to time by chance meetings amongst the infinite number of atoms in infinite space (1048–77). For Lucretius, in short, 'the world' signified a very restricted system, and it would be wrong to press the analogy with the modern 'cyclical universe' too far.

Note XXI (p. 399). *Lucretius and creation ex nihilo*

Lucretius has an interesting argument on this score: 'for if things came to being from nothing, *every kind might be born from all things*, nought would need a seed. First men might arise from the sea . . .' (Book i, 160.) Expressed otherwise: If men are created from nothing then they might equally well be created from nothing and the sea. Lucretius took the conclusion to be false, implying the falsity of the premiss. If the word 'men' be replaced by the words 'hydrogen atoms', we have an argument which might be used against current doctrines of continual creation. The supporters of these doctrines would not, however, deny the corresponding conclusion.

Note XXII (p. 402). *McCrea's treatment of the steady-state solution*

W. B. Bonnor has recently held that McCrea's method of applying Einstein's field equations (without λg_{mn}) to the steady-state metric—which, as we saw in Part I, requires the material of the model to have certain strange physical properties—leads to an indeterminacy in the motion of the material. (*M.N.*, cxxi (1960) 475.) This is the case (as previously noticed by Davidson, *M.N.*, cxix (1959) 309) when a de Sitter type metric with central mass m is taken. Bonnor interpreted this as indicating that matter with McCrea's equation of state $\rho + p = 0$ is 'indifferent to a gravitational field'. In particular, he argued that there is no justification for supposing matter in either case to move on a geodesic. With a Newtonian interpretation he found a similar indeterminacy. His conclusion was then that those who wish to work with the steady-state theory must use a dynamics specifically designed for it, and must not fall back upon relativistic or Newtonian dynamics.

In a later paper, Bonnor and McVittie argued further that, given the metric and density of material present, its motion is not uniquely determined by Hoyle's field equations. (*M.N.*, cxxii (1961) 381.) Again, it was suggested that the motions need not be given by geodesics. More recently, McCrea and R. L. Agacy have suggested that the criticism is unwarranted. (*M.N.*, cxxiii (1962) 383.) Taking a more general expression than Hoyle's for the creation tensor, which in both versions is made derivable from a scalar field ψ, they find that ψ can have one of two forms. They give, further, a coordinate transformation which—in a restricted region of space-time—transforms each of the corresponding solutions for the metric into the other. Hoyle's is a special case of one of these solutions. It is shown that the world line of any particle of the model is a geodesic. They write: 'It is meaningless to say that the motion given by [equations giving the components of the velocity four-vector of the (pressure-free) material of the model as functions of position and time] is or is not represented by a geodesic, for this motion describes only the velocity field in the fluid. The problem is first to infer . . . the world-line of a particle of the fluid and then to ask if this is a geodesic.' (op. cit., p. 389.)

SELECT BIBLIOGRAPHY

There is no single extensive bibliography dealing with the cosmological writings of this century. Most of the important work before 1940 is listed in

(1) H. P. Robertson, 'Relativistic Cosmology', *Rev. Mod. Phys.*, v (1933) 62–90.

(2) O. Heckmann, *Theorien der Kosmologie* (Springer, Berlin, 1942).

It is unfortunately almost impossible to obtain a copy of (2).

An excellent bibliography of early writings on relativity is

(3) M. Lecat (with Mme Lecat-Pierlot), *Bibliographie de la rélativité* (Bruxelles, 1924).

The most comprehensive work on non-Euclidean geometry is

(4) D. M. Y. Somerville, *Bibliography of Non-Euclidean Geometry* (Univ. of St. Andrew's: London, 1911).

There is no single collection of reprints of important papers on modern cosmology, but for a useful compendium of extracts of writings from Greek times to the present century see

(5) M. K. Munitz, *Theories of the Universe* (Chicago, 1957).

The first important work on the Special Theory of Relativity and the General Theory of Relativity is comprised in

(6) H. A. Lorentz, A. Einstein, H. Minkowski, and H. Weyl, *The Principle of Relativity*, with notes by A. Sommerfeld; trans. by W. Perrett and G. B. Jeffery (Methuen, 1923, reprinted by Dover Publications, Inc.).

Useful collections of recent papers, each with a single theme, are

(7) *Jubilee of Relativity Theory*, Helvetica Physica Acta, Suppl. IV (Basle, 1956). Contains papers by Von Laue, Alexandrov, Tits, van Dantzig, Møller, Trumpler, Heckmann and Schücking, McCrea, Papapetrou, Corinaldesi, Robertson, Klein, Hoyle, Bondi, Jordan, Ludvig and Just, Rosen, Lichnerowicz, Tonnelat, Pirani, Infeld, Fock, &c.

(8) *Institut International de Physique Solvay: Onzième Conseil de Physique: La Structure et l'Evolution de l'Univers* (1958). Contains papers by Lemaître, Klein, Hoyle, Gold, Wheeler, Schücking and Heckmann, Oort, Lovell, van de Hulst, Morgan, Ambartsumian, &c.

There is no general history of the subject. The following are perhaps the best introductory works, and of these reference (9) is in part historical. Reference (12) is mainly philosophical.

(9) G. J. Whitrow, *The Structure of the Universe* (Hutchinson, 1950). The second edition (1959) is entitled *The Structure and Evolution of the Universe*.

(10) P. Couderc, *The Expansion of the Universe* (London, 1952).

(11) G. C. McVittie, *Fact and Theory in Cosmology* (Eyre and Spottiswoode 1961).

(12) M. K. Munitz, *Space, Time and Creation* (Chicago, 1957).

More specialized works are

- (13) G. J. Whitrow, *The Natural Philosophy of Time* (Nelson, 1961).
- (14) M. Jammer, *Concepts of Space* (Harvard: Oxford, 1954).

Over the last thirty years a large number of articles have been written surveying the subject as a whole. These would be an excellent historical source did each author not tend to interpret all other theories in terms of his own. Amongst the best of these are

- (15) A. S. Eddington, 'The Expansion of the Universe', *M.N.*, xci (1931) 412–16.
- (16) A. S. Eddington, 'The Cosmological Controversy', *Sci. Prog.*, xxxiv (1939) 225.
- (17) H. Bondi, 'Review of Cosmology' (Council Report on the progress of astronomy), *M.N.*, cviii (1948) 104.
- (18) G. Lemaître, 'Cosmological Applications of Relativity', *Rev. Mod. Phys.*, xxi (1949) 357.
- (19) W. H. McCrea, 'Cosmology', *Rep. Prog. Phys.*, xvi (1953) 321.
- (20) H. P. Robertson, 'Cosmological Theory', *Jubilee R.T.*, 128. (Paper of 1955.)
- (21) W. H. McCrea, 'Cosmology', *Endeavour*, xvii (1958) 5.
- (22) G. C. McVittie, 'Cosmology and the Interpretation of Astronomical data', *Proceedings of the Symposium 'Les théories rélativistes de la gravitation'* (1962).

An interesting discussion in which Sir James Jeans, Milne, de Sitter, Eddington, Millikan, Bishop Barnes, General Smuts, Lemaître, and Sir Oliver Lodge took part can be found in

- (23) 'Discussion on the Evolution of the Universe', *Report of the British Association* (1931).

A first-hand account of much of the early observational work is contained in

- (24) E. Hubble, *The Realm of the Nebulae* (Oxford U.P., 1936).
- (25) E. Hubble, *The Observational Approach to Cosmology* (Oxford U.P., 1937).

These have been criticized on theoretical grounds.

The best critical discussions of the latest data of observation are references (11), (22), and

- (26) G. C. McVittie, 'Distance and Time in Cosmology: the Observational Data', *Handbuch der Physik*, liii (Springer, Berlin, 1959).

The principal texts devoted especially to cosmology are reference (2) and

- (27) R. C. Tolman, *Relativity, Thermodynamics and Cosmology* (Oxford, C.P., 1934).
- (28) E. A. Milne, *Relativity, Gravitation and World Structure* (Oxford, C.P., 1935).
- (29) G. C. McVittie, *Cosmological Theory* (Methuen, 1937).
- (30) E. A. Milne, *Kinematic Relativity* (Oxford, C.P., 1948).
- (31) H. Bondi, *Cosmology*, 2nd ed. (Cambridge U.P., 1960).

SELECT BIBLIOGRAPHY

(32) G. C. McVittie, *General Relativity and Cosmology* (Chapman and Hall, 1956).

(33) E. Schrödinger, *Expanding Universes* (Cambridge U.P., 1956).

Relativistic cosmology is covered by all the above references except (28) and (30). Milne's works may be supplemented by references (2) and (31). These are the most comprehensive works on the subject as a whole.

An excellent account of the history of theories of gravitation before 1926 is to be found in

(34) E. Whittaker, *History of the Theories of Aether and Electricity*, vol. i, 'The Classical Theories' (Nelson, 1910, revised 1951); vol. ii 'The Modern Theories' (Nelson, 1953).

Volume ii also contains by far the best history of Einstein's Special and General Theories up to that time.

A recent survey of relativity is

(35) H. Bondi, 'Relativity', *Rep. Prog. Phys.*, xxii (1959) 97.

Valuable texts dealing with the General Theory of Relativity are reference (27) and

(36) H. Weyl, *Space-Time-Matter*, trans. H. L. Brose (4th ed., 1922, reprinted by Dover Publications, Inc.).

(37) A. S. Eddington, *The Mathematical Theory of Relativity*, 2nd ed., (Cambridge U.P., 1924).

(38) E. Schrödinger, *Space-Time Structure* (Cambridge U.P., 1950).

(39) C. Møller, *The Theory of Relativity* (Oxford, C.P., 1952).

(40) J. L. Synge, *Relativity: the General Theory* (Amsterdam, 1960).

Eddington's later works are more easily understood with the help of

(41) N. B. Slater, *The Development and Meaning of Eddington's 'Fundamental Theory': Including a compilation from Eddington's unpublished manuscripts* (Cambridge U.P., 1957).

INDEX

Abbe, C., 6
Abbreviations, ix–xvi
Abraham, M., 31, 54 f.
Absolute differential calculus, see Tensors, calculus of
Absolute space and time, chapter 16 passim
Abundance of elements, relative, 227 f., 256–8
Action at a distance, chapter 3 passim, chapter 9 passim
Adam of Balsham, 377
Aether, gravitational, 30 ff., 41 f.
Agacy, R. L., 424
Age of the Earth, 226
Age of universe, 224 ff., 386–9
Airy, G. B., 31 f.
Aleph-zero ($\aleph_0$), 378 ff.
Alexander, S., 191
Allen, C. W., 234
Alpher, R. A., 227 f., 256–8
Ambarzumian, V. A., 418
Analytic, a priori, and synthetic, 276–8
Anti-matter, 37, 220 ff.
Antinomies of Pure Reason (Kant), 390 ff.
Aquinas, St. Thomas, 399, 401, 403
Archimedes, 372
Archimedes' Postulate, 321, 345 ff.
Argeländer, F. W. A., 9
Aristarchus, 372
Aristotle, 59, 354, 370, 373 f., 390, 423
Armelini, G., 232
Arnot, F. L., 231
Aronhold, S., 62
d'Assa-Montdardier, Compte, 76
Atkinson, R. d'E., 230
Augustine, St., 375, 391, 401

Baade, W., 147 f., 414
Background light of sky, 261 f., see also de Chéseaux-Olbers paradox
Bailey, C., 423
Baltzer, R., 59
Band, W., 155
Barajas, A., 187, 189
Barbarin, M. P., 73
Barkas, W., 230
Barnes, Bishop, 134, 271, 374
Barrow, I., 355
Bateman, H., 56, 63 f.
Baum, W. A., 252
Beer, A., xvi
Beltrami, E., 60 f., 81, 108

Bence Jones, 31
Bentley, R., 25, 27
Berenda, C. W., 188
Bergmann, A., 288
Berkeley, Bishop, 42, 349, 351 f.
Bernoulli, D., 25
Bernoulli, John, 25
Bertrand, G., 48
Bertrand, J., 46–48
Bessel, F. W., 5
Bethe, H., 227
Bettie, H. A., 237
Biesbroek, G. van, 68
Bigay, J. H., 246
Birkhoff, G. D., 174, 186–91, 196, 288, 298, 324, 407, 416 f.
Birkhoff's cosmology, 416 f. (Fernández)
Bjerknes, C. A., 32–34, 36, 41
Blaauw, A., 146, 414
Boisson, H., 53
Bok, B. J., 225, 418
Boltzmann's law, 313
Bolyai, J., 59, 60 f., 74, 76
Bolzano, B., 375–8
Bond, G. P., 8
Bondi, H., 18, 19, 122, 128, 152, 182–4, 198, 205, 208 f., 211 f., 214, 218, 219, 232, 244, 249, 251, 260, 274, 281, 293–9, 305–7, 318, 339, 341, 359, 371, 400, 418, 426 f.
Bonnor, W. B., 128, 133, 259 f., 402, 413 f., 423
Bonola, R., 59
Boole, G., 62
Borel, E., 20
Born, M., 56, 205, 342, 344
Boscovich, R. G., 32, 40 f., 43
Boscovich and action at a distance, 40, 43
Brahmachary, R. L., 301
Brewster, Sir David, 27
Bridgman, P. W., 334 f., 337
Broad, C. D., 191, 366
Broglie, L. de, 282
Brose, H. L., xvi
Brouwer, L. E. J., 374
Brown, E. W., 173
Browne, T., 151
Bruggencate, P. ten, 230
Brunings, J. H. M., 288
Bruno, G., 371
Buc, H. E., 230
Bunge, M., 402–5
Bunsen, R., 8

INDEX

Burbidge, E. M. and G. R., 228, 258
Burbidge, G. R., 220
Buscombe, W., 147

Calinon, A., 113
Cameron, A. G. W., 258
Cantor, G., 375–80, 383, 392 f.
Cantor's transfinite cardinals, 377–9
Carnelley, 37
Carslaw, H. S., 59
Cassini, J. and J. D., 45
Cauchy, A. L., 26 f., 375
Causality, 42, 398–406
  laws of, 309
Cavendish, H., 45 f.
Cayley, A., 62, 77, 108, 323
Cayley-Klein metric, 107, 323
Celoria, G., 9
Cepheid measurements, 11, 13 f., 144, 146 f., 238, 331, 414 f.
Chalmers, J. A. and B., 232
Chandrasekhar, S., 225, 258 f., 418
Charlier, C. V. L., 20–22, 371
Chazy, J., 113
Christoffel, E. B., 60 f.
Cicero, 373
Cipoletti, D., 40
Clairaut, A. C., 45
Clark, G. L., 193, 411
Clarke, S., 25, 353, 356, 389
Classification of relativistic models, 129–35
Clausius, R., 39
Clebsch, A., 62
Clemence, G. M., 68
Clerke, A., 6
Clifford, W. K., 16, 37, 60, 72 f., 77
Clock-graduation, Milne's theory of, 153 ff.
Closed universe, 72 ff.
Cocconi, G., 352
Code, A. D., 244 f.
Compass of inertia, 361, 363
Compton, A. H., 199
Comte, A., 7
Congruence, 319–22
Conservation of mass and energy, 182, chapter 10 *passim*
Continuous entry, hypothesis of, 199 f.
Conventionalism, xxiv, 150, 196, chapter 13 *passim*
Coordinates, conceptual problems in assignment of, 151 ff., chapter 15 *passim*
Cosmic rays, 199
Cosmic time, *see* Time, universal and cosmic
Cosmological principles (including the so-called 'Perfect Cosmological Principle'), 156–8, 171, 179, 208, 211 ff., 249, 301–18

Cosmological term(s), 83–87, 174, 177, 212, 214, 216, 386 f., 412 f.
Cosmological theories—*see, for example*, Birkhoff's cosmology, Jordan's cosmology, Kinematic Relativity, Newtonian cosmology, Relativistic cosmology, Relativistic models, Rayner–Whitehead cosmology, Steady-state cosmology
Costa, A., 20
Couderc, P., 425
Coxeter, H. M., 164, 304, 415 f.
Cramér, H., 382
Crombie, A. C., 290, 374
Coulomb's law, 30
Covariance, principle of, 52 f., 56–58, 104, 302
Creation, 217 ff., 399–406
  continual, 198–208, 400, 405, 417 f.
Creation field (Hoyle's), 214 f., 423 f.
Curtis, H. D., 12
Cyclical models, *see* Periodic models

Dantzig, T., 374
Darboux, C. A., 74
Darkness of sky, *see* de Chéseaux-Olbers paradox
Darwin, G. H., 20
Davidson, W., 237, 243, 245, 248–50, 252, 423
Davy, Sir Humphry, 34
de Cheseaux, J.-P. de L., 18–20, 198, 260
de Cheseaux-Olbers paradox, 18, 198, 260 f.
de Sitter, W., 72, 78–81, 86–93, 97, 99, 103, 105–8, 110–15, 118–20, 122, 124, 126, 129, 130, 134–6, 143, 151, 216, 223 f., 410 f., 419 f.
de Sitter effect, 92–104
de Sitter line-element (and transformations of it), 111–13, 212, 214, 221, 423
de Sitter's model, 87–110, 212, 221, 358
  geometry of, 106–9
Decombe, L., 48
Dedekind, R., 375, 377
Descartes, R., 294, 399
Dicke, R. H., 352
Dingle, H., 127, 308, 403
Dingler, H., 353
Diogenes Laertius, 373
Dirac, P. A. M., 134, 201–5, 207 f., 220, 296, 418
Dirac's model universe, 134, 202–5
Discrete models, 415 f.
Disproof of theories, 294 ff., 422
Dissolution of galactic clusters, 418
Distance by apparent size, 332 ff.
Distance by parallax, 94, 330 ff.
Distance by volume, 332 ff., 338

INDEX

Distance concepts of, 94 f., chapter 15 *passim*
Distance, postulates for, 320 f., 345–8
Donati, G. B., 8
Doolittle, C. L., 48
Doppler effect, transverse, 138
'Doppler formula', 136–41, 153 f., 171, 195, 212
  see also Nebulae, spectra of
Drever, R. W. P., 352
Dreyer, J. L. E., 3
Du Bois Raymond's Postulate, 321, 345 ff.
Dualism of matter and field, 42 f.
Duhem, P., 285, 287 f.
Duncan, J. C., 13
Dynamics, Milne on notion of, 170

Easton, C., 9
Economy, Principle of, *see* Simplicity
Eddington, Sir Arthur S., xvi, 10, 65, 85 f., 88, 95–97, 99 f., 109 f., 120, 122–7, 135 f., 142, 149, 151, 191, 201, 205, 224, 232, 235, 246, 281 f., 292, 296, 303, 309, 312, 334 f., 337, 357, 411–13, 426 f.
Eddington and the Lemaître model, 122 ff., 130
Ehrenfest, P., 80 f.
Eigenson, M. S., 233
Einstein, A., xvi, 20–22, 24, 31, 42, 48, 51–58, 61–73, 80–92, 104–9, 111–17, 119, 122, 130, 132–5, 150, 155–7, 187, 189, 191 f., 210, 212, 214, 246, 281, 283 f., 301, 313, 318, 325–7, 329, 335, 338, 342, 353, 357–60, 362 f., 384, 407, 409–13, 419, 423, 425
Einstein model, 81–83, 130, 358, 411 f.
Einstein tensor, 84
Einstein–de Sitter model, 131, 134, 177, 205, 246, 299
Einstein–Mach Principle, *see* Mach's Principle
Einstein's field equations, 65–67, 70 f., 82 f., 105, 212
Einstein's Theory of gravitation, basis of, chapter 4, 362 f.
Elastic-solid Theory (of aether), 26 f.
Engel, F., 59, 76
Entropy, 211, 385
Eötvös, R. von, 52
Equilibrium theory, *see* Abundance of elements
Equipartition of stellar energy, 226
Equivalent observers (and equivalent nebulae—in Milne's sense), 154 f., 160, 162, 364
  cf. Principle of Equivalence

Erlanger Programm, 319 f.
Etherington, I. M. H., 332 f.
Eubulides, 373
Euclid, 58 f., 283, 319, 321
Euler, L., 351
Euler's equations of motion, 182
Ewing, A. C., 373 f.
Event-horizons, *see* Horizons
Expansion of universe, 21, 37, 117 ff., 122, 211, chapter 11 *passim*
  its cause, 122 f., 127
Extrapolation, 296 ff.

Fabry, C., 53
Fact, concepts of, 268 ff.
Fan, C. Y., 363
Faraday, M., 27 f., 43
Farr, C. C., 39
Fatio, N., 38
Fermat's Principle, 176
Fermi, E., 228
Fernández, G., 189, 416 f.
Fessenkoff, B., 20
Field theories of gravitation, chapter 3 *passim*
Finkelstein, D., 410
Finzi, A., 51
First Event, notion of, 125, 224, chapter 18 *passim*
Freundlich, E. Finlay, 21, 67 f., 230
Fitzgerald, G. F., 36, 41
Fletcher, A., 246
Florides, P. S., 254
Fock, V. A., 116, 411
Föppl, A., 37
Forsyth, A. R., 410
Fournier d'Albe, E. E., 20
Fowler, W. A., 228, 248, 258
Frank, P., 309
Frankland, W. B., 80
Fraunhofer, J., 8, 11
Freedericksz, V., 117
Frege, G., 277
Fresnel, A. J., 26
Fricke, W., 236, 243
Friedmann, A., 109–12, 114, 116–23, 129 f., 132, 135 f., 145, 370, 385, 401
Friedmann's model universes, 113 ff.
Fronsdal, C., 410
Fubini, G. G., 121
Fuller, R. W., 410

Galaxies, *see* Nebulae
Galileo, 377
Gamow, G., 228, 256–9, 363
Gans, R., 40
Gauss, K. F., 47, 59, 61 f., 74 f., 77, 80, 374
Generality, 293–6, 298 ff.
Generating functions, 165 ff.

INDEX

Geodesic principle, 65
Geometry, non-Euclidean, 58–61
  projective, 321 ff.
Gerhardt, C. J., 377
Gerling, C. L., 75
Gheury de Bray, G. E. J., 231
Ghost images, 79 f., 107 f.
Gilbert, C., 205
Giovanelli, R. G., 127 f.
Gödel, K., 359–61, 363 f.
Gödel's line-element, 360 f., 363 (Newtonian analogue), 364
Gold, T., 198, 208 f., 211 f., 215, 218, 219, 220, 232, 244, 249, 251, 259, 273, 293, 295, 298 f., 305–7, 371, 400, 418
Goldhaber, A., 221
Graef, C., 187, 189
Grant, R., 6
Graphs (classification of relativistic models), 131
Grassmann, H. G., 61
Gravitation, Abraham on, 31, 54–56
  absorption of, 21
  Birkhoff's theory, 186–90
  change in time of, 38
  Einstein's early ideas, 52–65
  electrical theory of, 39 f.
  electrodynamic analogies, 46 ff.
  Le Sage and Thomson on, 38–41
  Maxwell on, 30 f.
  Milne on, 158 ff., 170 ff.
  Newton's theory of, chapters 2 and 3 passim
  Whitehead's theory of, 190 ff.
Gravitational constant, 163, 200 f.
  variation of, 173
Gravitational force, expressions for, 17–18, 21, 43–51
Gravitinos, 417 f.
Green, G., 17, 27
Greenfield, M. A., 237
Greenstein, G. L., 420
Grommer, J., 411
Grossteste, R., 290
Grossmann, M., 56, 63
Grünbaum, A., 286–8
Gunn, R., 232
Gupta, S., 418
Guthrie, F., 32

Haar, D. ter, 258
Haas, A., 202, 205, 207
Hall, A., 17
Halley, Edmond, 18, 44, 381
Halm, J., 226
Halsted, S., 59, 278
Hamilton, Sir William, 27
Hanbury-Brown, R., 250
Harkins, W. D., 227, 256

Harré, H. R., 396
Hasenöhrl, F., 52
Hastie, W., 3
Hazard, C., 234, 248
Heath, Sir Thomas, 59, 372
Heckmann, O., 120, 129, 134 f., 180, 183, 243, 363, 425
Hegel, G. W. F., 273
Heisenberg, W., 205
Helmholtz, H. von, 28, 35, 81, 305, 321
Heracleitus, 423
Herakleides, 372
Herman, R. C., 227 f., 256–8
Herschel, Sir John, 5 f., 9
Herschel, Sir William, 3–7, 9
Herz, H., 28
Hicks, W. M., 33–36
Hierarchical universe, 19–22, 182, 408
  see also Clustering of nebulae
Hilbert, D., 60, 66, 280, 409, 412
Hill, M. J., 36
Hilton, E. A., 410, 419
Hinton, C. H., 81
Hoffmann, B., 411
Holmberg, E., 408
Holzmuller, G., 46
Homogeneity, 175, 300 ff., 326 f., 361, 364, 414
Horizons, 118, 194, 346, 357 f., 380, 410, 419 f.
Hoskin, M. A., 5
Hoyle, F., 198, 211–22, 228, 232, 248, 250, 252, 258 f., 273, 371, 400, 402 f., 407, 418, 423 f.
Hubble, E., 11, 14, 21, 109 f., 118–20, 140, 142, 144–6, 148, 153, 159 f., 172, 224, 229 f., 233–7, 246 f., 252, 333 f., 412, 420, 426
Hubble's constant, see Recession factor
Hubble's Law, see Velocity-distance relation
Huggins, Sir William, 8 f.
Humason, M. L., 145 f., 148, 224 f., 408, 420
Hume, D., 42, 155, 397 f., 400
Huntington, E. V., 378
Hutten, E. H., 396
Hydrodynamical analogues of gravitation, 32–38, 47

Idealization, notion of, 89, 312–18, 329, 384
Identical clocks, notion of, 154
Identity of indiscernibles, 290
Inertia, problem of, see Mach's Principle
Inertial mass and gravitational mass, 52 ff., 83, 229 (light)
Infeld, L., 42, 175, 288, 411
Infinities in cosmology, **379–83**, 422 f.

Infinity, concepts of, 71 f., chapter 17 *passim*, 392–4
Interpolation, 297
Invariance and covariance, concepts of, 61 f., 217 f.
Island Universe, theory of, 3–15
Isotropy, 88 f., 300 ff., 329, 413 f.
Itimaru, K., 201
Ives, H. E., 188
Iwatsuki, T., 201

Jammer, M., 149, 426
Jans, C. de, 410
Jeans, Sir James, 11–13, 200, 224–8, 233, 258 f., 360, 418
Jeffery, G. B., 425
Jewell, L. F., 53
Jordan, P., 205–8, 206, 417
Jordan's cosmology, 205–8
Joule, Sir James P., 34
Jourdain, P. E. B., 378
Joy, A., 414
Just, K., 417

K-term (in Wirtz's sense), 142 ff.
(in later sense), 420 f.
Kalitzin, N. St., 409
Kant, I., 3, 19, 42, 59, 351, 357, 361, 367, 371, 389–95
Kapteyn, J. C., 10
Kelvin, Lord, *see* Thomson, W.
Kennedy, R. J., 230
Kepler, J., 30, 371
Kermack, W. O., 173 f.
Kilmister, C. W., 294
Kinematic Relativity, chapter 8 *passim*, 251, 284, 293 f., 343 f.
comparison with General Relativity, 173–6
axiomatic formulation of, 415
'Kinematic technique', 368 f.
Kirchhoff, G. R., 8, 28, 32 f., 36
Klein, F., 59, 62, 73, 107 f., 112, 286, 319, 322 f.
Kneale, M. and W., 379
Knott, C. G., 27
Kobold, H., 9
Kohlrausch, R., 28
Korn, A., 34
Kronecker, L., 81
Kruskal, M. D., 410
Kuiper, G. P., 418
Kustaanheimo, P., 189

Lagrange, J. L., 62
Lambert, J., 3, 19, 74 f.
Lanczos, K., 111–14
Laplace, P. S. Marquis de, 17, 44–7, 49, 68
Larmor, J., 43

Lassell, W., 7
'Laws of nature', 158, 302–11, 403 f.
Layzer, D., 180–2, 190
Leahy, A. H., 33 f., 36
Leavitt, H. S., 13
Lecat, M., 425
Lecornu, L., 48
Leibniz, G., 25, 38, 290, 300, 349 f., 353–7, 362, 365 f., 370, 377, 389–91, 395 f.
Lemaître, G., 86, 94, 109 f., 112 f., 117–20, 122–4, 126 f., 129 f., 135 f., 145, 180, 212, 225, 227, 232, 255, 292, 374, 379, 389, 401, 410, 418 f., 426
Lemaître's model universes, 117 ff.
Lenard, P., 68
Lenin, 309
Lense, J., 48
Leontovski, M., 167
Le Sage, G. L., 25, 32, 35, 38 f.
Le Sage's corpuscles, 38 ff.
Leslie, P. R. R., 408
LeVerrier, U. J. J., 47 f., 62, 69
Levi-Civita, T., 58, 65, 89, 105, 111, 193, 410
Levy, M., 47 f.
Lichnerowicz, A., 128
Lie, S., 305, 319, 322–4
Liebmann, H., 60
Lifschitz, R. O., 128 f.
Lilley, A. E., 239 f.
Lindblad, B., 173, 225
Lindemann, A. F. and F. A., 68
Linear equivalences, 165 ff.
Linsley, J., 260
Lipschitz, R., 60 f., 65
Lobachevsky, N. I., 59–61, 74–8
Local Group (of galaxies), 21, 148, 297, 408
Locke, J., 42, 289, 372–80
Lorentz, H. A., 39, 49 f., 52, 105, 425
Lorentz invariance and gravitation, 49–51, 191 f.
Lubbock, C. A., 5
Lucretius, 372, 385, 396, 402, 423
Ludlam, W., 59
Ludwig, G., 417
Luminosity distance, 332 f., 340 f., 346 f.
Lundmark, K., 10–12, 21, 143–5
Lyra, G., 241

Maanen, A. van, 10–13
McClain, E. F., 239 f.
McCrea, W. H., 22, 126 f., 173 f., 177, 179, 182 f., 215–17, 221 f., 238, 250 f., 254 f., 300, 314, 316 f., 332 f., 370, 401, 405, 415, 417, 423 f., 426
MacCullagh, J., 27

INDEX

Mach, E., 57, 83, 89, 91, 155, 300, 349, 351-3, 358
Mach's Principle, 57, 71 f., 83 f., 87-92, 306-11, 351-3, 362, 413
MacMillan, W. D., 198 f.
MacVigar, J. G., 40
McVittie, G. C., xvi, 122, 126 f., 214, 217, 223, 225, 235-7, 243-5, 247, 252, 255, 260, 299, 313, 315, 324 f., 327, 338, 371, 387, 402, 408, 415, 417 f., 420, 423, 425-7
Magellanic Clouds, 5 f.
Malmquist, K. G., 246
Martineau, H., 8
Mass of the universe, concepts of, 82-85, 123-5, 135, 200-2
Material content (Einstein and de Sitter worlds), 104-6, 118 f.
Maupertuis, P. L. M. de, 45
Maxwell, J. C., 28-32, 35, 41, 412
Mayall, N. U., 148, 408, 420
Mechanism for gravitation, Chapter 3 *passim*
Menger, K., 288
Mercury, anomalous perihelion advance of, 48, 50, 68 f., 187 f.
Merrill, P. W., 208
Messier, C., 4
Metagalaxy, 20, 408
Metrical field, concept of, 63 ff., 89 f.
Metzner, A. W. K., 260, 420
Michell, J., 4
Microphysics, links with cosmology, 201 ff.
Mie, G., 55 ff., 409, 412
Milford, S. N., 238, 241
Milky Way, 3-15
Millikan, R. A., 199
Milne, E. A., xvi, 104, 122, 149-85, 196, 203 f., 213, 224, 232, 237, 243, 250 f., 260, 282, 284, 288, 293 f., 296-9, 301-6, 312, 314, 317, 318, 324, 327, 339, 341-4, 356-9, 364 f., 387, 407, 415 f., 421, 426
Milner, S. R., 150, 163
Mimura, Y., 201
Mineur, H., 414, 418
Minio-Paluello, L., 377
Minkowski, H., 49 f., 52, 56, 63, 407, 425
Minkowski space-time, 63 f., 358
Minkowski, R., 239
Mitchell, S. A., 68
Mitra, K. K., 251
Model, concepts of, 311-18
Moghe, D. N., 122, 225, 256, 418
Møller, C., 427
Monck, W. H. S., 9
Moon, secular acceleration of, 45
More, H., 349
Morgan, H. R., 146, 414

Morley, F., 410
Morrison, P., 420
Mossotti, O. F., 39
Munitz, M. K., 272 f., 425

Narlikar, J. V., 221, 250
Narlikar, V. V., 86, 122, 225, 418
Navier, C. L. M. H., 26
Nebulae, angular sizes of, 242, 252-4
clustering of, 21 f., 247, 250, 252, 408 f., 418 f.
see also Dissolution of galactic clusters
collisions of, 248 f.
congestion of, 254
counts of, 161, 171, 234 f., 237, 245-55, 344
distances of, 11-13, 143 ff., 330 f.
evolution of, 170-3, 243 f., 250, 418
formation of, 126-9, 219 f., 252, 255 ff., 258-60, 418
luminosity function of, 246
magnitudes of, 242-50
parallaxes of, 10-12
see also Nebulae, distances of
spectra of, 8, 10-11, 97, 99, 199 f., 142 ff., 229-54, 331 ff.
progressive reddening hypothesis, 229-41
spiral, 7, 12, 20, 99, 172 f., 200, 418
status of, 3-15
velocity-distance relationship
see Velocity-distance relation
Nernst, W., 39
Neumann, C., 16, 20, 179 f.
Neumann, F., 27
Newcomb, S., 48, 73, 78-80, 108, 261
Newton, Sir Isaac, 17, 25, 27, 32, 41, 45 f., 68, 283, 290, 314, 349-56, 361-5, 369 f., 396
Newtonian cosmology, attempts at, 22 f., 176-85
Newtonian theory of gravitation, chapters 2 and 3 *passim*, 314
see also Poisson's equation
Neyman, J., 246 f., 408
Nicolas of Autrecourt, 373
Non-Euclidean space development of idea of, 58-61, 72-81
Non-static line-element, chapter 6 *passim*
Non-static universe, hypothesis of, 19, 22, 115 ff.
Nordström, G., 187
Notation, xvii-xix
Numerical coincidences, 201-3, 205-8

O'Brien, S., 128, 414
Observational tests of Einstein's theory of gravitation, 67-69
Observational tests of various models, chapter 11 *passim*

INDEX

Ockham, W., 290
Ockham's razor, 290, 352, 385
Olbers, H. W. M., 18–20, 198, 260
Olbers' paradox, see de Chéseaux-Olbers paradox
Omer, G. C., 122, 415 f.
Oort, J. H., 147, 225
Open models, 134 f.
Operationalism, 152, 320, 334–7
Öpik, E., 21
Oscillatory models see Periodic models
Ozsváth, L., 363

Page, L., 166
Pahlen, E. von der, 230
Papapetrou, A., 411
Parsons, W., see Rosse
Particle-horizons, see Horizons
Pasch, M., 323
Pauli, W., 410, 412
Pearson, K., 36 f.
Peirce, C. S., 77
Periodic models, 131–4, 384 f., 423
Perret, W., 425
Perturbations of relativistic models, 65–9, 413 ff.
Picart, A., 38
Pieri, M., 319, 322
Pilowski, K., 233
Pindar, 276
Pirani, F. A. E., 417 f.
Planck, M., 52, 210
Plato, 106, 276, 391, 399
Poincaré, H., 49 f., 52, 77, 278–84, 342, 379
Poisson's equation, 17, 29, 66, 71, 83, 182
Pomerans, A. J., 282
Ponce, J., 290
Popper, Sir Karl, 294, 352 f.
Potential infinity, 373–5, 381, 393
Power of the continuum, 378 f.
Preston, S. J., 39
Prévost, P., 38
Proclus, 59
Proctor, R. A., 6, 44
Primeval atom, 126, 130, 389
Principle of Covariance, see Covariance
Principle of Equivalence, 52–56
Projective Relativity, 417
Proper distance, 98, 332 ff., 339–48 passim

Quadrilateral construction, 322
Quasars, 248 f.
Quine, W. v. O., 287

Rabe, E., 409
Radio-astronomical criteria, 239, 248–50, 408
Radius of curvature of space, limits to, 76–80
Randall, D. G., 175
Rankine, W. J. M., 34
Raychaudhuri, A., 128, 414
Rayner, C. B., 191 f., 194 f.
Rayner-Whitehead cosmology, 194–7
Reality of expansion, 229 ff.
Recession factor ('Hubble's constant'), 202–4, 224 f., 299, 386 f., 412
Redeker, 38
Red-shift, gravitational, 53, 55, 187 f.
Red-shifts, see also Nebulae, spectra of, de Sitter effect, K-term, etc.
Reid, J. B., 166
Reissner, H., 410
Reiz, A., 408
Relative space and time, notions of, chapter 16 passim
Relativistic cosmology, chapters 5, 6, and 11 passim
Relativistic models, classification of, 129–35
Relativity, General Theory of, see Einstein's Theory of Gravitation
Rendell, J. R., 3
Repeatability of experiments, 306–11
Revision of Newton's theory, reasons for, chapters 2 and 3 passim, 179
Revision of recession factor, 145–8
Reynolds, J. H., 172
Ricci, C. G., 58, 60, 62, 65
Riemann, B., 16, 27, 47, 60–62, 73, 75, 80, 108, 300, 319–21, 367, 371 f.
Riemannian geometry, 60–62
and relativity, 56, 58–67 passim
and Kinematic Relativity, 173–6
Rigid body, notion of, 151, 342
Rindler, W., 419 f.
Ritchey, G. W., 12
Robb, A. A., 155 f., 356
Roberts, I., 7
Robertson, H. P., 88, 94, 103, 108, 112–15, 117 f., 120–2, 129, 136, 145, 157, 173–6, 193, 211 f., 224, 243, 301, 305, 339, 347, 358 f., 360, 381, 425, 426
Robertson's line-element, 121 f.
Robertson-Walker line-element, 175 f., 211, 301, 332
Rosse (W. Parsons), Third Earl of, 7,
Row, C. K. V., 190
Rubin, V. Cooper, 408
Ruse, H. S., 333

INDEX

Russell, B., 59, 151, 282–4, 323, 378, 393
Rutherfurd, L. M., 8
Ryle, M., 249, 250, 253

Saccheri, G., 74
Salpeter, E. E., 352
Salusbury, T., 377
Sambursky, S., 232
Samuel, Viscount H., 403
Sandage, A. R., 148, 408, 420
Schatzman, E., 237
Schechter, A., 117
Scheidegger, A. E., 288
Scheiner, J., 8
Schilpp, P. A., 86, 155, 289, 359, 361
Schlick, M., 280, 283
Schild, A., 175, 193
Schlegel, R., 397
Schlegel, V., 416
Schlüter, A. J., 86
Schmidt, M., 248
Schönfeld, E., 9
Schrödinger, E., 230, 241, 427
Schücking, E., 22, 180, 363
Schumacher, H. C., 374
Schur, F., 81, 300
Schuster, Sir William, 37, 221
Schwarzschild, K., 69 f., 73, 78–80, 105, 108, 112 f., 409–11
Schwarzschild line-element, 358, 409–11, 420
    modification of, 113
Schweikart, F. W., 75
Sciama, D. W., 219, 244, 250, 252, 259
Scott, E. L., 246 f., 408
Scotus, Duns, 290
Scriven, M., 395
Seares, F. H., 226
Secchi, A., 8
Seeliger, H. von, 9, 16, 20, 48, 50, 179 f.
Selby-Bigge, L. A., 398
Selety, F., 20 f.
Sen, D. K., 241
Sen, N. R., 128, 225, 256, 418
Serini, R., 410
Shane, C. D., 12, 246 f., 408
Shapley, H., 10–14, 80, 143, 146 f., 225, 246
Shelton, H. S., 230
Sibata, T., 201
Signal functions, 165 ff.
Silberstein, L., 20, 80, 97, 99, 101–4, 109, 118, 199
Silberstein-Weyl controversy, 100–4
Silvester, J. J., 62
Simplicity, 290–3, 295 f., 298
Slater, N. B., 427
Slipher, V. M., 11, 99, 142–5

Smart, W. S., 228, 257
Smith, S., 51, 225
Soldner, J., 68
Somerville, D. M. Y,. 425
Sommerfeld, A., 425
Space, Newton and Leibniz on, 350 ff.
Spinoza, B., 375, 399
Spitzer, L., 258, 418
Stability of Einstein world
    see Perturbations of relativistic models
Stäckel, P., 59
Star counts, 4 f., 10
    see also Nebulae, counts of
Star-gauges (Herschel's), 4
Static world, assumption of, 72, 88, 110
Stationary and static, distinction between, 111–13, 115 f., 121, 221
Staudt, C. G. C. von, 322 f.
Steady-state cosmology, Chapter 10, *passim*, 245, 249–54, 259–61, 295
    Bondi-Gold Theory, 208–12
    Hoyle's theory, 212 ff., 402
    McCrea's treatment, 423 f.
Stebbins, J., 243, 245, 421
Stebbins-Whitford correction, 243–5
Stellar evolution, 226 f.
Stellar parallax, 5–7
Stewart, J. Q., 202, 230
Strauss, E. G., 122
Strömberg, G., 22, 142 f.
Struve, F. G. W., 5 f., 18
Struve, O., 18
Sub-atomic particles, 84 (Einstein), 413 (Weyl)
Substratum (Milne's), 168 f., 364
Supernovae, 13
Suzuki, S., 227, 257
Swedenborg, E., 3, 35
Synge, J. L., 128, 136, 190–3, 195, 225, 256, 315, 333, 410, 414, 417, 427

Tait, P. G., 26, 28, 32, 35, 41, 210, 231
Takeno, H., 201
Takéuchi, T., 133, 231
Tamarkine, 113
Tannery, P., 37
Tansley, I., 3
Tarski, A., 421
Taton, R., 282
Taub, A. H., 90 f., 403
Taurinus, F. A., 75
Teller, E., 228, 256
Temple, G. J., 191, 332
Tensors, calculus of, 58 ff.
Thackeray, A. D., 414
Thomas, I., 377
Thomson, J. J., 52

INDEX

Thomson, W. (Lord Kelvin), 27 f., 30, 32–36, 38–41, 210, 231
Time coordinate, interpretation of, 111 f.
Time, distinction between t- and τ-scales, 167–70
  universal and cosmic, 355 ff., 387 f.
Time-scale difficulty, 223–9
Tisserand, F., 46–48
Tolman, R. C., xvi, 102, 104 f., 111, 114, 121–4, 126, 128, 132 f., 136, 175, 199 f., 227, 233–5, 242, 246, 252, 257, 301, 313, 315, 333, 344, 384, 420, 426
Trefftz, E., 113
Truman, A. H., 142
Trumpler, R. J., 11
Truth, philosophical theories of, 421 f.
Tuberg, M., 225
Turbulence, see Nebulae, formation of

Unification within physics, 30, 37, 41, 55 f., 64
Uniformity of distribution of stars, 19
  see also Homogeneity
Uniqueness of universe, 271–89 *passim*
Universal time, see Time, universal and cosmic
Universe, concept of, 181, chapter 12 *passim*, 354

Val, P. du, 107 f.
Vallarta, M. S., 187, 189
Varicak, V., 416
Vaucouleurs, G. de, 244, 408 f.
Veblen, O., 320
Velocity-distance relation (and associated formulae), 12, 22, 117 ff., 120, 136–41, 143 ff., 159 f., 180, 223 ff., 230 f., 235 f., 239 f., 242–5, 344 f., 386 ff.
Velocity of light, 155 f., 169, 342
  secular variation of, 230–2
Velocity of propagation of gravitation, 44–49
Very, F. W., 10
Vescan, T., 201
Vienna Circle, xxv
Virgil, 423
Voltaire, 45
Vortices, and the theory of gravitation, 25, 34–37
Vrkljan, V. S., 231
Vyssotsky, A. N., 226

Wacker, H., 50
Walker, A. G., 122, 157 f., 163, 173–6, 211, 228, 301, 305, 332 f., 339, 347, 358 f., 415
Wallis, J., 59

Ward, M., 133
Wartofsky, M. L., 286
Waterton, J., 44
Weaver, H. F., 11
Weber, H., 40
Weber, W., 28, 39, 46 f.
Weichert, E., 48
Weierstrass, K. F. T., 375
Weinberg, J. R., 373
Weizsäcker, C. F. von, 227 f., 258
Weldon, T. D., 390
Wesselink, J., 414
Weyl, H., xvi, 56, 83, 85 f., 100–3, 111, 113, 118, 174 f., 188 f., 301, 358–60, 374, 410, 412 f., 425
Weyl's Principle, 92–104, 137–40, 211, 213 f., 355
Wheeler, J., 410
Whitehead, A. N., 59, 150, 186, 190–6, 232, 282, 288, 296, 298, 305, 324–9, 338, 365, 407
Whitehead's theory of gravitation, 190–4
  see also Rayner-Whitehead cosmology
Whitehead, J. H. C., 320
Whitford, A. E., 243–5, 421
Whitrow, G. J., 83, 156, 160, 164–6, 168, 175, 232–4, 237, 239–41, 260 f., 304 f., 355, 368 f., 392 f., 396 f., 415 f., 418, 425 f.
Whittaker, Sir Edmund T., xvi, 52, 97–99, 103, 107 f., 199, 332 f., 417, 427
Whittaker's distance function, 98, 322 ff., 346 f.
Wilkinson, D. H., 232
Williams, E. T. R., 226
Wilson, A. G., 146
Wilson, O. C., 238 f.
Wirtanen, C. A., 247, 408
Wirtz, C., 142–5
Wittgenstein, L., 289, 421
Wold, P. I., 230 f.
World-map and world-picture, distinction between, 170
Wright, T., 3, 19
Wyatt, S. P., 260

Yallop, B. D., 260 f.
Young, T., 26

Zaycoff, R., 133, 200
Zero-point stress, 216 f., 222, 316, 401 f., 417 f.
Zeuxippus, 372
Zöllner, F., 39
Zwicky, F., 13, 51, 225, 229 f., 233

A CATALOG OF SELECTED
DOVER BOOKS
IN SCIENCE AND MATHEMATICS

A CATALOG OF SELECTED
DOVER BOOKS
IN SCIENCE AND MATHEMATICS

QUALITATIVE THEORY OF DIFFERENTIAL EQUATIONS, V.V. Nemytskii and V.V. Stepanov. Classic graduate-level text by two prominent Soviet mathematicians covers classical differential equations as well as topological dynamics and erqodic theory. Bibliographies. 523pp. 5⅜ × 8½. 65954-2 Pa. $10.95

MATRICES AND LINEAR ALGEBRA, Hans Schneider and George Phillip Barker. Basic textbook covers theory of matrices and its applications to systems of linear equations and related topics such as determinants, eigenvalues and differential equations. Numerous exercises. 432pp. 5⅜ × 8½. 66014-1 Pa. $8.95

QUANTUM THEORY, David Bohm. This advanced undergraduate-level text presents the quantum theory in terms of qualitative and imaginative concepts, followed by specific applications worked out in mathematical detail. Preface. Index. 655pp. 5⅜ × 8½. 65969-0 Pa. $10.95

ATOMIC PHYSICS (8th edition), Max Born. Nobel laureate's lucid treatment of kinetic theory of gases, elementary particles, nuclear atom, wave-corpuscles, atomic structure and spectral lines, much more. Over 40 appendices, bibliography. 495pp. 5⅜ × 8½. 65984-4 Pa. $11.95

ELECTRONIC STRUCTURE AND THE PROPERTIES OF SOLIDS: The Physics of the Chemical Bond, Walter A. Harrison. Innovative text offers basic understanding of the electronic structure of covalent and ionic solids, simple metals, transition metals and their compounds. Problems. 1980 edition. 582pp. 6⅛ × 9¼. 66021-4 Pa. $14.95

BOUNDARY VALUE PROBLEMS OF HEAT CONDUCTION, M. Necati Özisik. Systematic, comprehensive treatment of modern mathematical methods of solving problems in heat conduction and diffusion. Numerous examples and problems. Selected references. Appendices. 505pp. 5⅜ × 8½. 65990-9 Pa. $11.95

A SHORT HISTORY OF CHEMISTRY (3rd edition), J.R. Partington. Classic exposition explores origins of chemistry, alchemy, early medical chemistry, nature of atmosphere, theory of valency, laws and structure of atomic theory, much more. 428pp. 5⅜ × 8½. (Available in U.S. only) 65977-1 Pa. $10.95

A HISTORY OF ASTRONOMY, A. Pannekoek. Well-balanced, carefully reasoned study covers such topics as Ptolemaic theory, work of Copernicus, Kepler, Newton, Eddington's work on stars, much more. Illustrated. References. 521pp. 5⅜ × 8½. 65994-1 Pa. $11.95

PRINCIPLES OF METEOROLOGICAL ANALYSIS, Walter J. Saucier. Highly respected, abundantly illustrated classic reviews atmospheric variables, hydrostatics, static stability, various analyses (scalar, cross-section, isobaric, isentropic, more). For intermediate meteorology students. 454pp. 6⅛ × 9¼. 65979-8 Pa. $12.95

CATALOG OF DOVER BOOKS

RELATIVITY, THERMODYNAMICS AND COSMOLOGY, Richard C. Tolman. Landmark study extends thermodynamics to special, general relativity; also applications of relativistic mechanics, thermodynamics to cosmological models. 501pp. 5⅜ × 8½. 65383-8 Pa. $11.95

APPLIED ANALYSIS, Cornelius Lanczos. Classic work on analysis and design of finite processes for approximating solution of analytical problems. Algebraic equations, matrices, harmonic analysis, quadrature methods, much more. 559pp. 5⅜ × 8½. 65656-X Pa. $11.95

SPECIAL RELATIVITY FOR PHYSICISTS, G. Stephenson and C.W. Kilmister. Concise elegant account for nonspecialists. Lorentz transformation, optical and dynamical applications, more. Bibliography. 108pp. 5⅜ × 8½. 65519-9 Pa. $3.95

INTRODUCTION TO ANALYSIS, Maxwell Rosenlicht. Unusually clear, accessible coverage of set theory, real number system, metric spaces, continuous functions, Riemann integration, multiple integrals, more. Wide range of problems. Undergraduate level. Bibliography. 254pp. 5⅜ × 8½. 65038-3 Pa. $7.00

INTRODUCTION TO QUANTUM MECHANICS With Applications to Chemistry, Linus Pauling & E. Bright Wilson, Jr. Classic undergraduate text by Nobel Prize winner applies quantum mechanics to chemical and physical problems. Numerous tables and figures enhance the text. Chapter bibliographies. Appendices. Index. 468pp. 5⅜ × 8½. 64871-0 Pa. $9.95

ASYMPTOTIC EXPANSIONS OF INTEGRALS, Norman Bleistein & Richard A. Handelsman. Best introduction to important field with applications in a variety of scientific disciplines. New preface. Problems. Diagrams. Tables. Bibliography. Index. 448pp. 5⅜ × 8½. 65082-0 Pa. $10.95

MATHEMATICS APPLIED TO CONTINUUM MECHANICS, Lee A. Segel. Analyzes models of fluid flow and solid deformation. For upper-level math, science and engineering students. 608pp. 5⅜ × 8½. 65369-2 Pa. $12.95

ELEMENTS OF REAL ANALYSIS, David A. Sprecher. Classic text covers fundamental concepts, real number system, point sets, functions of a real variable, Fourier series, much more. Over 500 exercises. 352pp. 5⅜ × 8½. 65385-4 Pa. $8.95

PHYSICAL PRINCIPLES OF THE QUANTUM THEORY, Werner Heisenberg. Nobel Laureate discusses quantum theory, uncertainty, wave mechanics, work of Dirac, Schroedinger, Compton, Wilson, Einstein, etc. 184pp. 5⅜ × 8½.
60113-7 Pa. $4.95

INTRODUCTORY REAL ANALYSIS, A.N. Kolmogorov, S.V. Fomin. Translated by Richard A. Silverman. Self-contained, evenly paced introduction to real and functional analysis. Some 350 problems. 403pp. 5⅜ × 8½. 61226-0 Pa. $7.95

PROBLEMS AND SOLUTIONS IN QUANTUM CHEMISTRY AND PHYSICS, Charles S. Johnson, Jr. and Lee G. Pedersen. Unusually varied problems, detailed solutions in coverage of quantum mechanics, wave mechanics, angular momentum, molecular spectroscopy, scattering theory, more. 280 problems plus 139 supplementary exercises. 430pp. 6½ × 9¼. 65236-X Pa. $10.95

CATALOG OF DOVER BOOKS

ASYMPTOTIC METHODS IN ANALYSIS, N.G. de Bruijn. An inexpensive, comprehensive guide to asymptotic methods—the pioneering work that teaches by explaining worked examples in detail. Index. 224pp. 5⅜ × 8½. 64221-6 Pa. $5.95

OPTICAL RESONANCE AND TWO-LEVEL ATOMS, L. Allen and J.H. Eberly. Clear, comprehensive introduction to basic principles behind all quantum optical resonance phenomena. 53 illustrations. Preface. Index. 256pp. 5⅜ × 8½.
65533-4 Pa. $6.95

COMPLEX VARIABLES, Francis J. Flanigan. Unusual approach, delaying complex algebra till harmonic functions have been analyzed from real variable viewpoint. Includes problems with answers. 364pp. 5⅜ × 8½. 61388-7 Pa. $7.95

ATOMIC SPECTRA AND ATOMIC STRUCTURE, Gerhard Herzberg. One of best introductions; especially for specialist in other fields. Treatment is physical rather than mathematical. 80 illustrations. 257pp. 5⅜ × 8½. 60115-3 Pa. $4.95

APPLIED COMPLEX VARIABLES, John W. Dettman. Step-by-step coverage of fundamentals of analytic function theory—plus lucid exposition of 5 important applications: Potential Theory; Ordinary Differential Equations; Fourier Transforms; Laplace Transforms; Asymptotic Expansions. 66 figures. Exercises at chapter ends. 512pp. 5⅜ × 8½. 64670-X Pa. $10.95

ULTRASONIC ABSORPTION: An Introduction to the Theory of Sound Absorption and Dispersion in Gases, Liquids and Solids, A.B. Bhatia. Standard reference in the field provides a clear, systematically organized introductory review of fundamental concepts for advanced graduate students, research workers. Numerous diagrams. Bibliography. 440pp. 5⅜ × 8½. 64917-2 Pa. $8.95

UNBOUNDED LINEAR OPERATORS: Theory and Applications, Seymour Goldberg. Classic presents systematic treatment of the theory of unbounded linear operators in normed linear spaces with applications to differential equations. Bibliography. 199pp. 5⅜ × 8½. 64830-3 Pa. $7.00

LIGHT SCATTERING BY SMALL PARTICLES, H.C. van de Hulst. Comprehensive treatment including full range of useful approximation methods for researchers in chemistry, meteorology and astronomy. 44 illustrations. 470pp. 5⅜ × 8½. 64228-3 Pa. $9.95

CONFORMAL MAPPING ON RIEMANN SURFACES, Harvey Cohn. Lucid, insightful book presents ideal coverage of subject. 334 exercises make book perfect for self-study. 55 figures. 352pp. 5⅜ × 8¼. 64025-6 Pa. $8.95

OPTICKS, Sir Isaac Newton. Newton's own experiments with spectroscopy, colors, lenses, reflection, refraction, etc., in language the layman can follow. Foreword by Albert Einstein. 532pp. 5⅜ × 8½. 60205-2 Pa. $8.95

GENERALIZED INTEGRAL TRANSFORMATIONS, A.H. Zemanian. Graduate-level study of recent generalizations of the Laplace, Mellin, Hankel, K. Weierstrass, convolution and other simple transformations. Bibliography. 320pp. 5⅜ × 8½. 65375-7 Pa. $7.95

CATALOG OF DOVER BOOKS

THE ELECTROMAGNETIC FIELD, Albert Shadowitz. Comprehensive undergraduate text covers basics of electric and magnetic fields, builds up to electromagnetic theory. Also related topics, including relativity. Over 900 problems. 768pp. 5⅜ × 8¼. 65660-8 Pa. $15.95

FOURIER SERIES, Georgi P. Tolstov. Translated by Richard A. Silverman. A valuable addition to the literature on the subject, moving clearly from subject to subject and theorem to theorem. 107 problems, answers. 336pp. 5⅜ × 8½. 63317-9 Pa. $7.95

THEORY OF ELECTROMAGNETIC WAVE PROPAGATION, Charles Herach Papas. Graduate-level study discusses the Maxwell field equations, radiation from wire antennas, the Doppler effect and more. xiii + 244pp. 5⅜ × 8½. 65678-0 Pa. $6.95

DISTRIBUTION THEORY AND TRANSFORM ANALYSIS: An Introduction to Generalized Functions, with Applications, A.H. Zemanian. Provides basics of distribution theory, describes generalized Fourier and Laplace transformations. Numerous problems. 384pp. 5⅜ × 8½. 65479-6 Pa. $8.95

THE PHYSICS OF WAVES, William C. Elmore and Mark A. Heald. Unique overview of classical wave theory. Acoustics, optics, electromagnetic radiation, more. Ideal as classroom text or for self-study. Problems. 477pp. 5⅜ × 8½. 64926-1 Pa. $10.95

CALCULUS OF VARIATIONS WITH APPLICATIONS, George M. Ewing. Applications-oriented introduction to variational theory develops insight and promotes understanding of specialized books, research papers. Suitable for advanced undergraduate/graduate students as primary, supplementary text. 352pp. 5⅜ × 8½. 64856-7 Pa. $8.50

A TREATISE ON ELECTRICITY AND MAGNETISM, James Clerk Maxwell. Important foundation work of modern physics. Brings to final form Maxwell's theory of electromagnetism and rigorously derives his general equations of field theory. 1,084pp. 5⅜ × 8½. 60636-8, 60637-6 Pa., Two-vol. set $19.00

AN INTRODUCTION TO THE CALCULUS OF VARIATIONS, Charles Fox. Graduate-level text covers variations of an integral, isoperimetrical problems, least action, special relativity, approximations, more. References. 279pp. 5⅜ × 8½. 65499-0 Pa. $6.95

HYDRODYNAMIC AND HYDROMAGNETIC STABILITY, S. Chandrasekhar. Lucid examination of the Rayleigh-Benard problem; clear coverage of the theory of instabilities causing convection. 704pp. 5⅜ × 8¼. 64071-X Pa. $12.95

CALCULUS OF VARIATIONS, Robert Weinstock. Basic introduction covering isoperimetric problems, theory of elasticity, quantum mechanics, electrostatics, etc. Exercises throughout. 326pp. 5⅜ × 8½. 63069-2 Pa. $7.95

DYNAMICS OF FLUIDS IN POROUS MEDIA, Jacob Bear. For advanced students of ground water hydrology, soil mechanics and physics, drainage and irrigation engineering and more. 335 illustrations. Exercises, with answers. 784pp. 6⅛ × 9¼. 65675-6 Pa. $19.95

CATALOG OF DOVER BOOKS

NUMERICAL METHODS FOR SCIENTISTS AND ENGINEERS, Richard Hamming. Classic text stresses frequency approach in coverage of algorithms, polynomial approximation, Fourier approximation, exponential approximation, other topics. Revised and enlarged 2nd edition. 721pp. 5⅜ × 8½.
65241-6 Pa. $14.95

THEORETICAL SOLID STATE PHYSICS, Vol. I: Perfect Lattices in Equilibrium; Vol. II: Non-Equilibrium and Disorder, William Jones and Norman H. March. Monumental reference work covers fundamental theory of equilibrium properties of perfect crystalline solids, non-equilibrium properties, defects and disordered systems. Appendices. Problems. Preface. Diagrams. Index. Bibliography. Total of 1,301pp. 5⅜ × 8½. Two volumes.
Vol. I 65015-4 Pa. $12.95
Vol. II 65016-2 Pa. $12.95

OPTIMIZATION THEORY WITH APPLICATIONS, Donald A. Pierre. Broad-spectrum approach to important topic. Classical theory of minima and maxima, calculus of variations, simplex technique and linear programming, more. Many problems, examples. 640pp. 5⅜ × 8½.
65205-X Pa. $12.95

THE MODERN THEORY OF SOLIDS, Frederick Seitz. First inexpensive edition of classic work on theory of ionic crystals, free-electron theory of metals and semiconductors, molecular binding, much more. 736pp. 5⅜ × 8½.
65482-6 Pa. $14.95

ESSAYS ON THE THEORY OF NUMBERS, Richard Dedekind. Two classic essays by great German mathematician: on the theory of irrational numbers; and on transfinite numbers and properties of natural numbers. 115pp. 5⅜ × 8½.
21010-3 Pa. $4.95

THE FUNCTIONS OF MATHEMATICAL PHYSICS, Harry Hochstadt. Comprehensive treatment of orthogonal polynomials, hypergeometric functions, Hill's equation, much more. Bibliography. Index. 322pp. 5⅜ × 8½.   65214-9 Pa. $8.95

NUMBER THEORY AND ITS HISTORY, Oystein Ore. Unusually clear, accessible introduction covers counting, properties of numbers, prime numbers, much more. Bibliography. 380pp. 5⅜ × 8½.   65620-9 Pa. $8.95

THE VARIATIONAL PRINCIPLES OF MECHANICS, Cornelius Lanczos. Graduate level coverage of calculus of variations, equations of motion, relativistic mechanics, more. First inexpensive paperbound edition of classic treatise. Index. Bibliography. 418pp. 5⅜ × 8½.   65067-7 Pa. $10.95

MATHEMATICAL TABLES AND FORMULAS, Robert D. Carmichael and Edwin R. Smith. Logarithms, sines, tangents, trig functions, powers, roots, reciprocals, exponential and hyperbolic functions, formulas and theorems. 269pp. 5⅜ × 8½.
60111-0 Pa. $5.95

THEORETICAL PHYSICS, Georg Joos, with Ira M. Freeman. Classic overview covers essential math, mechanics, electromagnetic theory, thermodynamics, quantum mechanics, nuclear physics, other topics. First paperback edition. xxiii + 885pp. 5⅜ × 8½.   65227-0 Pa. $17.95

CATALOG OF DOVER BOOKS

HANDBOOK OF MATHEMATICAL FUNCTIONS WITH FORMULAS, GRAPHS, AND MATHEMATICAL TABLES, edited by Milton Abramowitz and Irene A. Stegun. Vast compendium: 29 sets of tables, some to as high as 20 places. 1,046pp. 8 × 10½. 61272-4 Pa. $21.95

MATHEMATICAL METHODS IN PHYSICS AND ENGINEERING, John W. Dettman. Algebraically based approach to vectors, mapping, diffraction, other topics in applied math. Also generalized functions, analytic function theory, more. Exercises. 448pp. 5⅜ × 8¼. 65649-7 Pa. $8.95

A SURVEY OF NUMERICAL MATHEMATICS, David M. Young and Robert Todd Gregory. Broad self-contained coverage of computer-oriented numerical algorithms for solving various types of mathematical problems in linear algebra, ordinary and partial, differential equations, much more. Exercises. Total of 1,248pp. 5⅜ × 8½. Two volumes. Vol. I 65691-8 Pa. $13.95
Vol. II 65692-6 Pa. $13.95

TENSOR ANALYSIS FOR PHYSICISTS, J.A. Schouten. Concise exposition of the mathematical basis of tensor analysis, integrated with well-chosen physical examples of the theory. Exercises. Index. Bibliography. 289pp. 5⅜ × 8½. 65582-2 Pa. $7.95

INTRODUCTION TO NUMERICAL ANALYSIS (2nd Edition), F.B. Hildebrand. Classic, fundamental treatment covers computation, approximation, interpolation, numerical differentiation and integration, other topics. 150 new problems. 669pp. 5⅜ × 8½. 65363-3 Pa. $13.95

INVESTIGATIONS ON THE THEORY OF THE BROWNIAN MOVEMENT, Albert Einstein. Five papers (1905-8) investigating dynamics of Brownian motion and evolving elementary theory. Notes by R. Fürth. 122pp. 5⅜ × 8½. 60304-0 Pa. $3.95

NUMERICAL METHODS FOR SCIENTISTS AND ENGINEERS, Richard Hamming. Classic text stresses frequency approach in coverage of algorithms, polynomial approximation, Fourier approximation, exponential approximation, other topics. Revised and enlarged 2nd edition. 721pp. 5⅜ × 8½. 65241-6 Pa. $14.95

AN INTRODUCTION TO STATISTICAL THERMODYNAMICS, Terrell L. Hill. Excellent basic text offers wide-ranging coverage of quantum statistical mechanics, systems of interacting molecules, quantum statistics, more. 523pp. 5⅜ × 8½. 65242-4 Pa. $10.95

ELEMENTARY DIFFERENTIAL EQUATIONS, William Ted Martin and Eric Reissner. Exceptionally, clear comprehensive introduction at undergraduate level. Nature and origin of differential equations, differential equations of first, second and higher orders. Picard's Theorem, much more. Problems with solutions. 331pp. 5⅜ × 8½. 65024-3 Pa. $8.95

STATISTICAL PHYSICS, Gregory H. Wannier. Classic text combines thermodynamics, statistical mechanics and kinetic theory in one unified presentation of thermal physics. Problems with solutions. Bibliography. 532pp. 5⅜ × 8½.
65401-X Pa. $10.95

CATALOG OF DOVER BOOKS

ORDINARY DIFFERENTIAL EQUATIONS, Morris Tenenbaum and Harry Pollard. Exhaustive survey of ordinary differential equations for undergraduates in mathematics, engineering, science. Thorough analysis of theorems. Diagrams. Bibliography. Index. 818pp. 5⅜ × 8½. 64940-7 Pa. $15.95

STATISTICAL MECHANICS: Principles and Applications, Terrell L. Hill. Standard text covers fundamentals of statistical mechanics, applications to fluctuation theory, imperfect gases, distribution functions, more. 448pp. 5⅜ × 8½. 65390-0 Pa. $9.95

ORDINARY DIFFERENTIAL EQUATIONS AND STABILITY THEORY: An Introduction, David A. Sánchez. Brief, modern treatment. Linear equation, stability theory for autonomous and nonautonomous systems, etc. 164pp. 5⅜ × 8¼. 63828-6 Pa. $4.95

THIRTY YEARS THAT SHOOK PHYSICS: The Story of Quantum Theory, George Gamow. Lucid, accessible introduction to influential theory of energy and matter. Careful explanations of Dirac's anti-particles, Bohr's model of the atom, much more. 12 plates. Numerous drawings. 240pp. 5⅜ × 8½. 24895-X Pa. $5.95

ORDINARY DIFFERENTIAL EQUATIONS, I.G. Petrovski. Covers basic concepts, some differential equations and such aspects of the general theory as Euler lines, Arzel's theorem, Peano's existence theorem, Osgood's uniqueness theorem, more. 45 figures. Problems. Bibliography. Index. xi + 232pp. 5⅜ × 8½. 64683-1 Pa. $6.00

GREAT EXPERIMENTS IN PHYSICS: Firsthand Accounts from Galileo to Einstein, edited by Morris H. Shamos. 25 crucial discoveries: Newton's laws of motion, Chadwick's study of the neutron, Hertz on electromagnetic waves, more. Original accounts clearly annotated. 370pp. 5⅜ × 8½. 25346-5 Pa. $8.95

INTRODUCTION TO PARTIAL DIFFERENTIAL EQUATIONS WITH APPLICATIONS, E.C. Zachmanoglou and Dale W. Thoe. Essentials of partial differential equations applied to common problems in engineering and the physical sciences. Problems and answers. 416pp. 5⅜ × 8½. 65251-3 Pa. $9.95

BURNHAM'S CELESTIAL HANDBOOK, Robert Burnham, Jr. Thorough guide to the stars beyond our solar system. Exhaustive treatment. Alphabetical by constellation: Andromeda to Cetus in Vol. 1; Chamaeleon to Orion in Vol. 2; and Pavo to Vulpecula in Vol. 3. Hundreds of illustrations. Index in Vol. 3. 2,000pp. 6⅛ × 9¼. 23567-X, 23568-8, 23673-0 Pa., Three-vol. set $38.85

ASYMPTOTIC EXPANSIONS FOR ORDINARY DIFFERENTIAL EQUATIONS, Wolfgang Wasow. Outstanding text covers asymptotic power series, Jordan's canonical form, turning point problems, singular perturbations, much more. Problems. 384pp. 5⅜ × 8½. 65456-7 Pa. $8.95

AMATEUR ASTRONOMER'S HANDBOOK, J.B. Sidgwick. Timeless, comprehensive coverage of telescopes, mirrors, lenses, mountings, telescope drives, micrometers, spectroscopes, more. 189 illustrations. 576pp. 5⅜ × 8¼. 24034-7 Pa. $8.95

CATALOG OF DOVER BOOKS

SPECIAL FUNCTIONS, N.N. Lebedev. Translated by Richard Silverman. Famous Russian work treating more important special functions, with applications to specific problems of physics and engineering. 38 figures. 308pp. 5⅜ × 8½.
60624-4 Pa. $6.95

OBSERVATIONAL ASTRONOMY FOR AMATEURS, J.B. Sidgwick. Mine of useful data for observation of sun, moon, planets, asteroids, aurorae, meteors, comets, variables, binaries, etc. 39 illustrations 384pp. 5⅜ × 8¼. (Available in U.S. only)
24033-9 Pa. $5.95

INTEGRAL EQUATIONS, F.G. Tricomi. Authoritative, well-written treatment of extremely useful mathematical tool with wide applications. Volterra Equations, Fredholm Equations, much more. Advanced undergraduate to graduate level. Exercises. Bibliography. 238pp. 5⅜ × 8½.
64828-1 Pa. $6.95

CELESTIAL OBJECTS FOR COMMON TELESCOPES, T.W. Webb. Inestimable aid for locating and identifying nearly 4,000 celestial objects. 77 illustrations. 645pp. 5⅜ × 8½.
20917-2, 20918-0 Pa., Two-vol. set $12.00

MODERN NONLINEAR EQUATIONS, Thomas L. Saaty. Emphasizes practical solution of problems; covers seven types of equations. ". . . a welcome contribution to the existing literature. . . ."—*Math Reviews.* 490pp. 5⅜ × 8½. 64232-1 Pa. $9.95

FUNDAMENTALS OF ASTRODYNAMICS, Roger Bate et al. Modern approach developed by U.S. Air Force Academy. Designed as a first course. Problems, exercises. Numerous illustrations. 455pp. 5⅜ × 8½.
60061-0 Pa. $8.95

INTRODUCTION TO LINEAR ALGEBRA AND DIFFERENTIAL EQUATIONS, John W. Dettman. Excellent text covers complex numbers, determinants, orthonormal bases, Laplace transforms, much more. Exercises with solutions. Undergraduate level. 416pp. 5⅜ × 8½.
65191-6 Pa. $8.95

INCOMPRESSIBLE AERODYNAMICS, edited by Bryan Thwaites. Covers theoretical and experimental treatment of the uniform flow of air and viscous fluids past two-dimensional aerofoils and three-dimensional wings; many other topics. 654pp. 5⅜ × 8½.
65465-6 Pa. $14.95

INTRODUCTION TO DIFFERENCE EQUATIONS, Samuel Goldberg. Exceptionally clear exposition of important discipline with applications to sociology, psychology, economics. Many illustrative examples; over 250 problems. 260pp. 5⅜ × 8½.
65084-7 Pa. $6.95

LAMINAR BOUNDARY LAYERS, edited by L. Rosenhead. Engineering classic covers steady boundary layers in two- and three-dimensional flow, unsteady boundary layers, stability, observational techniques, much more. 708pp. 5⅜ × 8½.
65646-2 Pa. $15.95

LECTURES ON CLASSICAL DIFFERENTIAL GEOMETRY, Second Edition, Dirk J. Struik. Excellent brief introduction covers curves, theory of surfaces, fundamental equations, geometry on a surface, conformal mapping, other topics. Problems. 240pp. 5⅜ × 8½.
65609-8 Pa. $6.95

CATALOG OF DOVER BOOKS

ROTARY-WING AERODYNAMICS, W.Z. Stepniewski. Clear, concise text covers aerodynamic phenomena of the rotor and offers guidelines for helicopter performance evaluation. Originally prepared for NASA. 537 figures. 640pp. 6⅛ × 9¼.
64647-5 Pa. $14.95

DIFFERENTIAL GEOMETRY, Heinrich W. Guggenheimer. Local differential geometry as an application of advanced calculus and linear algebra. Curvature, transformation groups, surfaces, more. Exercises. 62 figures. 378pp. 5⅜ × 8½.
63433-7 Pa. $7.95

INTRODUCTION TO SPACE DYNAMICS, William Tyrrell Thomson. Comprehensive, classic introduction to space-flight engineering for advanced undergraduate and graduate students. Includes vector algebra, kinematics, transformation of coordinates. Bibliography. Index. 352pp. 5⅜ × 8½. 65113-4 Pa. $8.00

A SURVEY OF MINIMAL SURFACES, Robert Osserman. Up-to-date, in-depth discussion of the field for advanced students. Corrected and enlarged edition covers new developments. Includes numerous problems. 192pp. 5⅜ × 8½.
64998-9 Pa. $8.00

ANALYTICAL MECHANICS OF GEARS, Earle Buckingham. Indispensable reference for modern gear manufacture covers conjugate gear-tooth action, gear-tooth profiles of various gears, many other topics. 263 figures. 102 tables. 546pp. 5⅜ × 8½. 65712-4 Pa. $11.95

SET THEORY AND LOGIC, Robert R. Stoll. Lucid introduction to unified theory of mathematical concepts. Set theory and logic seen as tools for conceptual understanding of real number system. 496pp. 5⅜ × 8¼. 63829-4 Pa. $8.95

A HISTORY OF MECHANICS, René Dugas. Monumental study of mechanical principles from antiquity to quantum mechanics. Contributions of ancient Greeks, Galileo, Leonardo, Kepler, Lagrange, many others. 671pp. 5⅜ × 8½.
65632-2 Pa. $14.95

FAMOUS PROBLEMS OF GEOMETRY AND HOW TO SOLVE THEM, Benjamin Bold. Squaring the circle, trisecting the angle, duplicating the cube: learn their history, why they are impossible to solve, then solve them yourself. 128pp. 5⅜ × 8½. 24297-8 Pa. $3.95

MECHANICAL VIBRATIONS, J.P. Den Hartog. Classic textbook offers lucid explanations and illustrative models, applying theories of vibrations to a variety of practical industrial engineering problems. Numerous figures. 233 problems, solutions. Appendix. Index. Preface. 436pp. 5⅜ × 8½. 64785-4 Pa. $8.95

CURVATURE AND HOMOLOGY, Samuel I. Goldberg. Thorough treatment of specialized branch of differential geometry. Covers Riemannian manifolds, topology of differentiable manifolds, compact Lie groups, other topics. Exercises. 315pp. 5⅜ × 8½. 64314-X Pa. $6.95

HISTORY OF STRENGTH OF MATERIALS, Stephen P. Timoshenko. Excellent historical survey of the strength of materials with many references to the theories of elasticity and structure. 245 figures. 452pp. 5⅜ × 8½. 61187-6 Pa. $9.95

CHALLENGING MATHEMATICAL PROBLEMS WITH ELEMENTARY SOLUTIONS, A.M. Yaglom and I.M. Yaglom. Over 170 challenging problems on probability theory, combinatorial analysis, points and lines, topology, convex polygons, many other topics. Solutions. Total of 445pp. 5⅜ × 8½. Two-vol. set.
Vol. I 65536-9 Pa. $5.95
Vol. II 65537-7 Pa. $5.95

FIFTY CHALLENGING PROBLEMS IN PROBABILITY WITH SOLUTIONS, Frederick Mosteller. Remarkable puzzlers, graded in difficulty, illustrate elementary and advanced aspects of probability. Detailed solutions. 88pp. 5⅜ × 8½.
65355-2 Pa. $3.95

EXPERIMENTS IN TOPOLOGY, Stephen Barr. Classic, lively explanation of one of the byways of mathematics. Klein bottles, Moebius strips, projective planes, map coloring, problem of the Koenigsberg bridges, much more, described with clarity and wit. 43 figures. 210pp. 5⅜ × 8½. 25933-1 Pa. $4.95

RELATIVITY IN ILLUSTRATIONS, Jacob T. Schwartz. Clear non-technical treatment makes relativity more accessible than ever before. Over 60 drawings illustrate concepts more clearly than text alone. Only high school geometry needed. Bibliography. 128pp. 6⅛ × 9¼. 25965-X Pa. $5.95

AN INTRODUCTION TO ORDINARY DIFFERENTIAL EQUATIONS, Earl A. Coddington. A thorough and systematic first course in elementary differential equations for undergraduates in mathematics and science, with many exercises and problems (with answers). Index. 304pp. 5⅜ × 8¼. 65942-9 Pa. $7.95

FOURIER SERIES AND ORTHOGONAL FUNCTIONS, Harry F. Davis. An incisive text combining theory and practical example to introduce Fourier series, orthogonal functions and applications of the Fourier method to boundary-value problems. 570 exercises. Answers and notes. 416pp. 5⅜ × 8½. 65973-9 Pa. $8.95

THE THOERY OF BRANCHING PROCESSES, Theodore E. Harris. First systematic, comprehensive treatment of branching (i.e. multiplicative) processes and their applications. Galton-Watson model, Markov branching processes, electron-photon cascade, many other topics. Rigorous proofs. Bibliography. 240pp. 5⅜ × 8½. 65952-6 Pa. $6.95

AN INTRODUCTION TO ALGEBRAIC STRUCTURES, Joseph Landin. Superb self-contained text covers "abstract algebra": sets and numbers, theory of groups, theory of rings, much more. Numerous well-chosen examples, exercises. 247pp. 5⅜ × 8½. 65940-2 Pa. $6.95

GAMES AND DECISIONS: Introduction and Critical Survey, R. Duncan Luce and Howard Raiffa. Superb non-technical introduction to game theory, primarily applied to social sciences. Utility theory, zero-sum games, n-person games, decision-making, much more. Bibliography. 509pp. 5⅜ × 8½. 65943-7 Pa. $10.95

Prices subject to change without notice.

Available at your book dealer or write for free Mathematics and Science Catalog to Dept. GI, Dover Publications, Inc., 31 East 2nd St., Mineola, N.Y. 11501. Dover publishes more than 175 books each year on science, elementary and advanced mathematics, biology, music, art, literary history, social sciences and other areas.